微服务项目实战派

从Spring Boot到Spring Cloud

姜桥◎著

電子工業出版社
Publishing House of Electronics Industry
北京•BEIJING

内 容 简 介

随着微服务概念的兴起，如何快速实施微服务成了一个重要话题。在 Java 服务端领域，Spring Cloud 微服务体系事实上已经成为整个 Java 后端架构的标配。如果你打算从事 Java 服务端开发，或者你所在的公司正打算转型微服务，那么 Spring Cloud 是无论如何都绕不开的存在。

本书以项目实战的形式来介绍 Spring Cloud 微服务体系。书中所有实例都来自作者多年工作实践，内容覆盖构建 Spring Cloud 微服务所需的绝大部分内容——包括微服务工程搭建、微服务网关、熔断限流、分布式任务调度、自动化 CI/CD 构建、Kubernetes 容器化部署、微服务监控系统、分布式链路追踪等。

本书适合希望快速提升项目经验的 Java 初学者、正在转型微服务架构的传统项目研发人员、希望提升 Spring Cloud 微服务全栈技术经验的高级开发人员、对特定系统设计感兴趣的产品经理及研发人员。无论读者是否接触过微服务开发，只要具备一定的 Java 开发基础，都能通过本书的学习快速掌握实际场景中的微服务开发技巧，并快速提升项目实战经验。

图书在版编目（CIP）数据

微服务项目实战派：从 Spring Boot 到 Spring Cloud/姜桥著. —北京：电子工业出版社，2022.1
ISBN 978-7-121-42267-6

Ⅰ. ①微… Ⅱ. ①姜… Ⅲ. ①互联网络－网络服务器 Ⅳ. ①TP368.5

中国版本图书馆 CIP 数据核字（2021）第 220883 号

责任编辑：吴宏伟
印　　刷：三河市良远印务有限公司
装　　订：三河市良远印务有限公司
出版发行：电子工业出版社
　　　　　北京市海淀区万寿路 173 信箱　　邮编：100036
开　　本：787×980　1/16　印张：34.5　字数：828 千字
版　　次：2022 年 1 月第 1 版
印　　次：2022 年 1 月第 1 次印刷
定　　价：139.00 元

凡所购买电子工业出版社图书有缺损问题，请向购买书店调换。若书店售缺，请与本社发行部联系，联系及邮购电话：（010）88254888，88258888。

质量投诉请发邮件至 zlts@phei.com.cn，盗版侵权举报请发邮件至 dbqq@phei.com.cn。

本书咨询联系方式：010-51260888-819，faq@phei.com.cn。

前言

写作背景

Spring Cloud 微服务体系在国内真正落地是在 2017 年左右，那时我所在的公司——摩拜单车风头正盛，后端系统也刚完成了从早期单体应用到 Spring Cloud 微服务架构的转型。在这次大规模的微服务架构实践中，我深刻感受到微服务架构给高速发展中的摩拜单车，在后端架构、系统负载、开发方式、组织架构等方面所带来的好处。

于是，我便有了将这些实践经验通过文章输出的想法，而当时我的同事兼好朋友、现在已经成为公众号知名技术博主的“程序员小灰”在得知这个想法后，便鼓励我开通微信公众号“无敌码农”。此后，我便陆续在公众号上输出了一些关于 Spring Cloud 微服务的实践文章，而这些内容也得到了出版界老师的关注，这便是这本书写作的由来。

写作过程并非一帆风顺。因为工作繁忙，所以本书的写作从 2019 年 4 月整整持续了两年多的时间。技术更新是非常快的，这样的写作时长存在着技术滞后的风险，不过幸好目前 Spring Cloud 微服务体系依然是主流，虽然新一代微服务架构 Service Mesh 也在迅猛发展，但短期内并不会完全取代 Spring Cloud。

虽然写作过程进展不快，但这也正好有了更多的思考时间，因此书稿内容也都处于持续优化中，根据技术的变化进行同步。所以，本书在技术上不仅不滞后，反而具有一定的前瞻性。

本书特色

本书以项目实例的形式来展示 Spring Cloud 微服务生命周期所涉及的技术，具有以下特点。

（1）精选业界普遍采用的技术方案进行详细介绍。

虽然 Spring Cloud 微服务的技术生态非常丰富，但在实际应用中，并不是所有技术都是必需的。所以，本书并没有像其他某些 Spring Cloud 图书那样囫囵式地将各个技术组件都介绍一遍，而是以目前互联网业界普遍采用的技术方案进行详细介绍。

例如，关于注册中心的选择，大部分互联网公司并不会直接采用 Eureka，而是会选择性能更高、支持微服务数量规模更大的方案，如 Consul 或 Nacos 等。而对于像微服务的分布式链路追踪方案，一般也不会选择 Spring Cloud Sleuth，而是选择更流行的 SkyWalking 方案等。

（2）覆盖 Spring Cloud 微服务体系的绝大部分内容。

本书覆盖构建 Spring Cloud 微服务的绝大部分内容——包括微服务工程搭建、微服务网关、熔断限流、分布式任务调度、自动化 CI/CD 构建、Kubernetes 容器化部署、微服务监控系统、分布式链路追踪等。

（3）循序渐进，实战性强。

本书对于微服务技术并不是枯燥地叙述，而是根据每个项目实例的特点，“从原理到实践”手把手地教学，并且每章的实例都可以独立学习。书中实例都提供了完整的源码，且精确到具体的包路径。

（4）实例具有较强的实用参考性。

本书所有实例都是作者从多年的工作实践中整理出来的真实项目，用户系统、SSO 授权系统、车辆电子围栏系统、电子钱包系统、支付系统、A/B 测试系统等，都是目前互联网业务系统中真实存在的。本书给出了这些系统的详细设计方案，以及具体代码实现。

（5）代码规范，注重编程思想的提炼。

本书实例中，注重编程规范及软件分层架构。通过学习本书，读者不仅能够快速掌握 Spring Cloud 微服务开发技术，还能感受到良好的编程思想，从而在潜移默化中培养良好的编程习惯，提升编程水平。

（6）技术前瞻，紧跟发展潮流。

本书所涉及的技术具有一定的前瞻性，特别是最后两章所涉及的 Kubernetes 容器编排、监控系统及分布式链路追踪等内容，都是当前流行及今后会流行的技术，也是下一代微服务架构 Service Mesh 所依赖的平台基础。

由于笔者能力有限，错漏之处在所难免，欢迎读者批评、指正。

您也可以通过以下方式直接联系作者：

关注微信公众号“无敌码农”

邮箱：1468325120@qq.com

致谢

感谢博文视点的吴宏伟老师，以及出版社各位工作人员为本书的出版所做出的努力。

感谢我的父母、妻子及可爱的女儿。由于撰写本书，我牺牲了很多陪伴家人的时间，在此感谢他们对我的理解和支持。

最后，感谢微信公众号“无敌码农”的各位读者，你们持续、正向的反馈给了我很多写作灵感。

涉及的技术及版本

本书所采用的技术及相关版本较新，请读者将相关开发环境设置成与下方所列的配置，或不低于本书所列的配置。

- JDK 1.8。
- Apache Maven 3.6.1。
- Spring Boot 2.1.5.RELEASE。
- Spring Cloud Greenwich.SR1。
- Docker 19.03.5。
- Consul 1.9.1。
- MySQL 5.7。
- Redis 3.2。
- PostgresSQL 10.0。
- PostGIS 2.4。
- Spring Cloud Config Server 2.1.1. RELEASE。
- Spring Cloud Hystrix Dashboard 1.4.7.RELEASE。
- Vue 2.9.6。
- Npm 6.13.4。
- MyBatis Plus 3.3.0。
- Apache Zookeeper 3.7.0-bin。
- Elasticjob-Lite 3.0.0-RC1。
- Elasticjob-Lite-UI 3.0.0-RC1。
- Ubantu Linux 20.04 LTS。
- GitLab 13.2.2。
- Harbor 2.0.2。
- Kubernetes 1.18.1。
- Helm v3.4.0-rc.1。
- Prometheus-Operator 0.38.1。

- Prometheus 2.22.0
- SkyWalking OAP Server 8.3.0-es7。
- SkyWalking UI 8.3.0。

本书实例所采用的集成开发工具为 IntelliJ IDEA ULTIMATE 2019.2。

姜桥

2021 年 8 月

读者服务

微信扫码回复：42267

- 获取本书配套代码
- 加入本书读者交流群，与作者互动
- 获取【百场业界大咖直播合集】（持续更新），仅需 1 元

目录

第 1 章

基础

——从 Spring Boot 单体应用到 Spring Cloud 微服务

“微服务”是近年来后端技术领域的热门话题，也是如今互联网服务端系统普遍采用的架构方式。那么，如何实施微服务？传统单体应用如何升级为微服务？微服务开发框架 Spring Boot 与 Spring Cloud 有什么样的关系？如何才能快速掌握微服务开发的技巧？

带着这些问题，让我们一起开始本书的学习吧。

本书将会通过多个完整的实战项目来介绍在实际工作场景下微服务的开发技巧，内容涵盖开发框架、编程技巧、微服务治理、Kubernetes 容器化部署及监控等。此外，本书中的每一个实战项目都是从实际业务中提炼的，通过这些实战项目除能够掌握基本的微服开发技巧外，还能直接获得特定系统的设计思路及方案。

无论是否接触过微服务开发，只要具备一定的 Java 开发基础的读者，都能通过本书快速掌握实际场景中的微服务开发技巧，并快速提升项目实战经验。

在正式开始微服务项目实战之前，本章先介绍 Spring Boot、Spring Cloud 框架的基础知识，这是目前快速落地微服务架构的主流框架，在很多互联网公司的微服务实践中它们都得到了普遍应用。本书后面所有的实战项目也都是基于 Spring Boot 及 Spring Cloud 框架来构建的。

通过本章，读者将学习到以下内容：

- 微服务的基本概念及进化史。
- Spring Boot 框架基础及核心原理。
- 快速构建 Spring Boot 应用的方法。
- Spring Cloud 微服务体系基础及运行原理。
- Spring Cloud 的技术生态圈。

1.1 微服务的概念

本节从微服务的基本概念入手，分别介绍微服务的特点、实施原因，以及主流的微服务技术栈。

1.1.1 什么是微服务

微服务从本质上来说，是一种分布式系统的架构模式。它通过将规模庞大的单体应用拆分成一组独立的“微”服务，来降低系统功能之间的耦合性，并提升系统的整体服务能力及敏捷性。

在微服务体系结构中，应用被构造为一组松耦合的、细颗粒度的服务，并通过轻量级协议（如基于 HTTP 的 RESTful API、RPC）来实现互相通信。总的来说，微服务具备如下特点：

- 松耦合。微服务需要围绕具体的业务来构建，因此从系统的边界划分上说，服务之间的依赖应该处于低耦合状态。
- 独立部署。每一个微服务都可以被独立部署，从而可以有效地避免因系统的局部变动而需要整体发布。
- 高度可维护性及可测试性。在实施微服务架构后，服务的数量会急剧上升，因此微服务体系应该具备 DevOps 的运维能力，以满足微服务高可维护性及可测试性的需求。
- 更小、更独立的团队。按照服务功能边界的划分，单体架构时代模糊的组织结构可以被拆分为更小、更独立的团队，从而实现服务迭代的敏捷性。

1.1.2 从单体应用到微服务

对于大多数互联网公司来说，在初创时期，面对的主要问题是如何将一个想法快速变成实际的软件实现。为了产品的快速上线，系统的架构一般不会设计得太复杂——服务端系统以单体应用架构为主，如图 1-1 所示。

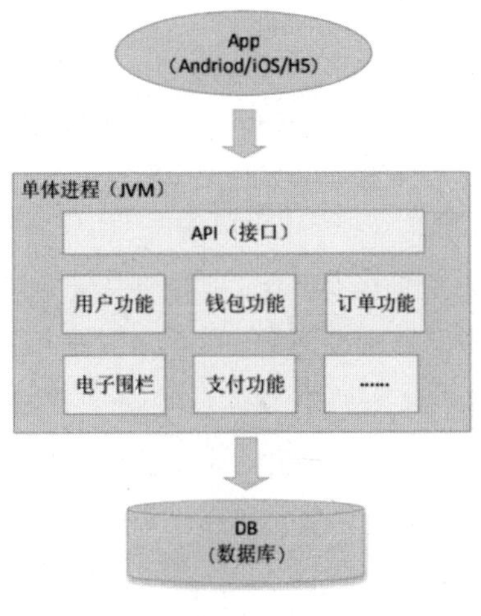

图 1-1

如图 1-1 所示，创业初期系统的功能相对简单，业务流程也不复杂，所以，功能基本都耦合在一个单体应用中。但随着业务的迅猛发展（如笔者 2017 年任职摩拜单车时业务的爆发式增长），App 的下载量、在线注册人数迅猛增长，此时整个后端服务面临的压力会陡然上升，而为了扛住流量压力，只能通过不断增加服务器来平行扩展后端服务节点数量。此时的系统架构如图 1-2 所示。

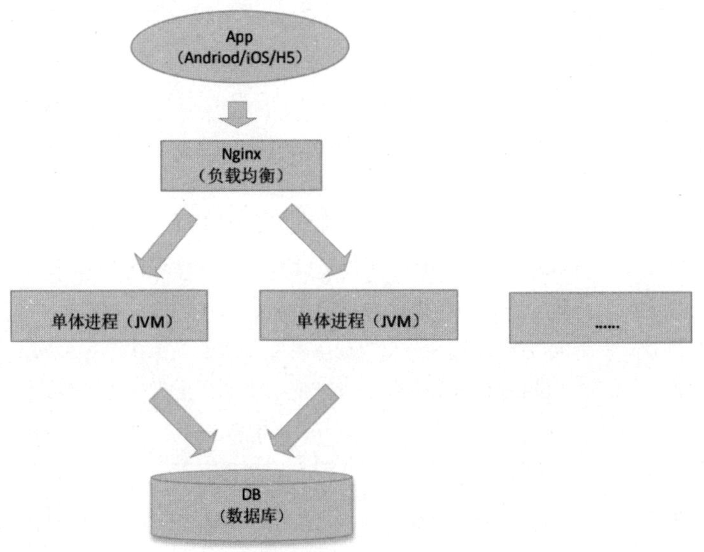

图 1-2

如图 1-2 所示，通过增加服务器的方式，虽然使得系统抗住了一波压力，但系统却并不够稳定——例如，有时会因为 API 中的某个接口的性能问题而导致整体服务的不可用。

这是因为，这些接口都在一个 JVM 进程中，虽然部署了多个节点，但底层数据库、缓存系统都是共用的，所以还是会出现“一挂全挂”的情况（作者之前就亲身经历过这样的事件）。

此外，随着业务的快速发展，之前相对简单的功能变得复杂起来——例如，满足增长策略的红包、分享拉新，以及满足变现需求的广告推荐等功能。这些复杂的功能叠加在一个单体应用中，可能导致系统的可维护性变得越来越低，反过来也会阻碍业务的发展。

另外，流量/业务的增长也意味着团队人数的增长，此时如果大家开发各自的业务功能还是共用一套后端代码，则很难想象如此规模的研发团队在同一套代码中叠加功能是一个什么样的场景。因此，如何划分业务边界、合理地进行团队配置就变成了一件非常棘手的事。

为了解决上述问题，适应业务、团队的发展，就需要实施微服务架构，按照业务边界对单体应用进行微服务拆分。而要实施微服务架构，除需要合理地划分业务边界外，还需要一套完整的技术解决方案。

1.1.3 主流的微服务技术栈

在技术方案的选择上，服务治理的框架有很多，例如早期的 WebService，近期的各种 RPC 框架（如 Dubbo、Thirft、GRPC 等）。而 Spring Cloud 是基于 Spring Boot 的一套微服务解决方案——开发友好，对服务治理相关的各类组件的支撑也非常全面。所以 Spring Cloud 成了大部分互联网公司实施微服务架构的首选技术方案。

如今以 Spring Cloud 为代表的微服务技术，在业界已经得到了普遍应用。Spring Cloud 将原本复杂的架构方式，变成了一件相对容易实施的事情，而之所以能够有这样的效果，关键就在于：以 Spring Boot、Spring Cloud 为核心的微服务技术框架，以及围绕它们所构建的开源生态，已经为我们扫平了通向微服务架构之路的绝大部分技术障碍。

虽然目前以 Istio 为代表的 Service Mesh（服务网格）也在迅猛发展，但从实施成本和普及程度来看，以 Spring Cloud 为代表的微服务技术体系仍然是大部分互联网公司实施微服务架构的首选。

本章接下来的内容，将重点讲述 Spring Boot、Spring Cloud 的基本原理及其核心组件。这将有助于读者更好地学习本书后面的项目实例。

1.2 Spring Boot 框架基础

Spring Cloud 微服务应用是基于 Spring Boot 框架来构建的，并且 Spring Cloud 微服务体系中涉及的依赖也都是遵循 Spring Boot 框架的基本逻辑来定义的。所以，在开始学习 Spring Cloud 微服务之前，需要先了解 Spring Boot 框架的相关知识。

1.2.1 Spring Boot 简介

下面从 Spring Boot 的核心能力及版本来简单介绍 Spring Boot 框架。

1. Spring Boot 是什么

随着微服务架构理念的普及，Spring Boot 已经成为 Java 服务端开发的工业级技术框架，而 Spring Boot 之所以能够迅速被接受并流行起来，关键在于：

（1）利用 Spring Boot 可以十分方便地构建生产级别的 Spring 应用。相较于早期烦琐的 Spring 应用构建方式来说，Spring Boot 框架的出现极大地改善了 Spring 应用的开发体验。

（2）Spring Boot Starter“开箱即用”的依赖集成方式，使得 Spring Boot 应用能够非常方便地集成其他技术组件。

目前在应用中经常使用的技术组件，如 Redis、MyBatis、RocketMQ 等，都对 Spring Boot 应用提供了 Spring Boot Starter 方式的接入组件。即便没有现成的依赖支持，基于 Spring Boot Starter 机制的自动配置原理，也能很方便地定制 Spring Boot Starter 组件，从而让 Spring Boot 应用无缝地接入某个技术组件。

Spring Boot 框架在 Spring 框架的基础上进行了进一步的封装（如封装了大量的注解），并延续了 Spring“约定大于配置”的思想。因此，在使用 Spring Boot 开发应用时，只需要通过几个简单的注解配置，即可快速地将应用搭建起来——这极大地提高了应用的工程构建效率，也是为什么 Spring Boot 框架能够流行起来的重要原因。

2. Spring Boot 的发布版本

Spring Boot 目前主要有两大版本：Spring Boot 1.x 系列、Spring Boot 2.x 系列。

在 Spring Boot 1.x 系列中，Spring Boot 1.5.x 是在生产中使用得比较广泛的版本。而随着 Spring Boot 2.x 的发布，目前很多公司在逐步将 Spring Boot 升级到 2.x 版本。

截止本书写作时（2021 年 8 月）目前 Spring Boot 的最新版本是 2.1.x 系列，因此本书所有的实例均采用 Spring Boot 2.1.5.RELEASE 版本来构建。

1.2.2 Spring Boot 的核心原理

Spring Boot 并不是一种全新的技术，也不是要“重复造轮子”来替代 Spring，而是基于 Spring 框架进行的一次应用开发模式的优化。在 JDK 1.5 推出注解功能后，Spring 框架后续的版本，开始大量采用定义注解来替代原有的烦琐配置，这些 Spring 注解后来被逐步用于 Spring 框架的配置管理、Bean 的依赖注入，以及 AOP 等功能的实现。

任何事物都有两面性，随着 Spring 框架中定义的注解越来越多，这些注解被大量重复地使用到 Spring 项目的各个类、方法及依赖之中，这导致应用产生了大量烦琐的注解依赖和冗余的注解代码。

到这里大家也许会想到，既然这么多 Spring 注解很烦琐，那么是不是可以组合一下呢？具体来说就是，可不可以将 Spring 的一些注解按照功能分类，通过定义一组新的注解来对相关性较强的 Spring 注解进行重新组合。这样，对于一些比较通用的场景，只需要引入这一个组合注解，就能够自动引入与之相关的其他 Spring 注解。这实际上也是 Spring Boot 框架的实质。

但要实现 Spring 注解的组合，并不只是把多个注解组合在一起就可以了，还需要解决以下两个核心问题：

- 如果将几个 Spring 注解组合成一个注解，那么该组合注解是否能够被应用在其他注解上。
- 组合注解还会遇到在 Spring 容器初始化上下文时 Bean 的注入顺序问题。例如，存在两个组件 A 和 B 需要被 Spring 容器初始化，而 A 组件的逻辑需要使用到 B 组件。如果 Spring 容器先初始化 A 组件，则可能因为缺失 B 组件而导致 A 组件创建失败，从而导致整个 Spring 容器无法正常启动。

上述问题在 Spring 框架中早已有了解决方案——组合注解及基于条件的配置。

Spring Boot 正是基于 Spring 框架提供的这种核心能力，才能够继续对 Spring 的原始注解进行组合，并通过条件配置来实现对其他技术依赖的“智能化”集成。这也正是 Spring Boot 框架的核心吸引力所在。

所以，从本质上来说，Spring Boot 更多是基于 Spring 原有的能力做了很多关于“注解”“依赖”及“自动化配置”的整合和优化，从而提高了应用的开发效率。

讲到这里大家应该对 Spring Boot 框架的本质有了一定的认识。接下来，从技术细节来分析 Spring Boot 框架的具体实现。

1. 什么是元注解

通过前面的阐述可以知道，Spring Boot 实质上就是在 Spring 框架的基础上进行了很多二次封装，其关键的特性之一是：定义了一些新的注解，来实现对部分 Spring 注解的组合。

Spring 框架提供的各种注解，其底层逻辑也基于 JDK1.5 之后推出的注解特性。

在 JDK 的注解逻辑中，如果一个注解要被其他注解所引用，则需要将该注解定义为“元注解”。

“元注解”就是可以注解到其他注解上的注解，而被注解的注解就是上面提到的组合注解。关于元注解在 JDK 中的定义，可以参考 JDK 相关的资料。

实际上在 Spring 中定义的很多注解都可以作为“元注解”，并且 Spring 本身也实现了很多组合注解。例如，经常使用的@Configuration 注解就是一个组合注解，它包含了@Component 注解。正是因为有了这样的基本条件，Spring Boot 才可以定义一些组合注解来实现特定的功能。

2. 条件注解@Conditional

接下来重点讨论条件注解@Conditional，它是 Spring 4 提供的一个具有重要意义的注解。

（1）条件注解@Conditional 的作用。

条件注解@Conditional 可以根据是否满足某一个特定条件，来控制 Spring 容器创建 Bean 的行为。

例如，某个依赖在同一个类路径下需要配置一个或多个 Bean，则可以通过条件注解@Conditional 来实现——只有当某个 Bean 被创建之后才会自动创建另外一个 Bean。这也就解决了前面提到的 Bean 注入顺序的问题。另外，也可以依据一些特定的条件来控制 Bean 的创建逻辑，这样就可以使用该特性来实现一些依赖的自动配置。

Spring Boot 之所以用起来如此便捷，核心原因就在于：它不仅本身实现了很多常用组件的自动配置，并且也支持其他组件通过自定义的自动配置来和 Spring Boot 应用快速集成。

条件注解@Conditional 是 Spring Boot 实现自动配置能力的基础。事实上，Spring Boot 之所以能够实现自动注解配置，正是基于这种能力。

（2）Spring Boot 的核心条件注解。

在 Spring Boot 框架中，以@Conditional 注解为“元注解”重新定义了一组针对不同应用场景的组合条件注解，它们是 Spring Boot 实现组件开发的重要利器。这些注解分别有：

- @ConditionalOnBean：当 Spring 容器中有指定的 Bean 时才进行实例化。
- @ConditionalOnMissingBean：功能与@ConditionalOnBean 正好相反，当 Spring 容器中没有指定的 Bean 时才进行实例化。
- @ConditionalOnClass：当 Calsspath 类路径下有指定的类时才进行实例化。
- @ConditionalOnMissingClass：该注解与@ConditionalOnClass 的功能相反，当 Classpath 类路径下没有指定的类时才进行实例化。
- @ConditionalOnWebApplication：当应用是一个 Web 应用时才进行实例化。
- @ConditionalOnNotWebApplication：当应用不是一个 Web 应用时才进行实例化。
- @ConditionalOnProperty：当指定的属性有指定的值时才进行实例化。
- @ConditionalOnExpression：基于 Spring EL 表达式的条件判断。
- @ConditionalOnJava：当 JVM 版本为指定范围内的版本时才进行实例化。
- @ConditionalOnResource：当类路径下有指定的资源时才进行实例化。
- @ConditionalOnJndi：在 JNDI 存在的条件下才进行实例化。
- @ConditionalOnSingleCandidate：当指定的 Bean 在 Spring 容器中只有一个，或者虽然有多个但是指定了首选的 Bean 时才实例化。

（3）Spring Boot 核心条件注解在自动配置类中的应用。

从 Spring Boot 的核心条件注解可以看出，基于@Conditional 元注解的条件注解占了很大的篇幅。而实际上，Spring Boot 的核心功能基本上就是基于这些条件注解来实现的。

在 Spring Boot 为实现自动配置而定义的“spring-boot-autoconfigure”项目中随意打开一些常用组件的 AutoConfiguration 文件，会发现以上条件注解被大量应用在这些自动配置类的定义中。例如，以持久层（Dao 层）开发框架 JOOQ 的自动配置类 JooqAutoConfiguration，其代码示例如下：

```
    @Configuration(proxyBeanMethods = false)
    @ConditionalOnClass(DSLContext.class)
    @ConditionalOnBean(DataSource.class)
    @AutoConfigureAfter({ DataSourceAutoConfiguration.class,
TransactionAutoConfiguration.class })
    public class JooqAutoConfiguration {
        @Bean
```

```
        @ConditionalOnMissingBean
        public DataSourceConnectionProvider
dataSourceConnectionProvider(DataSource dataSource) {
            return new DataSourceConnectionProvider(new
TransactionAwareDataSourceProxy(dataSource));
        }
        @Bean
        @ConditionalOnBean(PlatformTransactionManager.class)
        public SpringTransactionProvider
transactionProvider(PlatformTransactionManager txManager) {
            return new SpringTransactionProvider(txManager);
        }
        ...
    }
```

可以看到，在该自动配置类中涉及很多其他 Bean 的依赖注入，所以，为了确保自动配置的正确性，JOOQ 自动配置类使用了很多前面提到的 Spring Boot 条件注解来控制配置 Bean 的创建逻辑。

其他 Spring Boot 组件的自动配置类，也采用了类似的机制。正是因为有了这样的机制，所以在使用 Spring Boot 进行项目开发时，才会感觉到需要自己手工配置的地方越来越少，开发效率和体验都得到了很大的提升。

3. Spring Boot 应用的启动逻辑

接下来从 Spring Boot 最核心的组合注解@SpringBootApplicationdd 的源码角度，来介绍 Spring Boot 应用的启动逻辑。

通过梳理源码可以知道，注解@SpringBootApplication 组合了多个注解，这些注解之间的依赖关系如图 1-3 所示。

通过分析@SpringBootApplication 注解的依赖关系可以发现，除对应用开放的@ComponentScan 注解外，最核心的注解就是@EnableAutoConfiguration 了，该注解的作用是对应用开启自动配置功能，而具体实现则是通过@Import({AutoConfigurationImportSelector.class})注解导入 EnableAutoConfigurationImportSelector 类的实例来实现自动配置逻辑的。

深入分析 EnableAutoConfigurationImportSelector 类的源码，其中关键的 selectImports() 方法的代码如下：

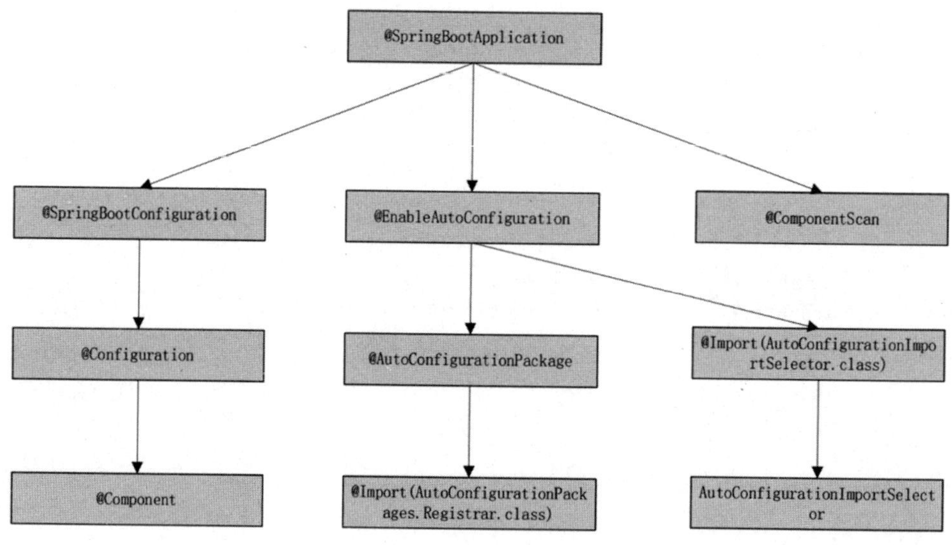

图 1-3

```
    public String[] selectImports(AnnotationMetadata annotationMetadata) {
            if (!this.isEnabled(annotationMetadata)) {
                return NO_IMPORTS;
            } else {
                AutoConfigurationMetadata autoConfigurationMetadata =
AutoConfigurationMetadataLoader.loadMetadata(this.beanClassLoader);
                AutoConfigurationImportSelector.AutoConfigurationEntry
autoConfigurationEntry = this.getAutoConfigurationEntry(autoConfigurationMetadata,
annotationMetadata);
                return
StringUtils.toStringArray(autoConfigurationEntry.getConfigurations());
            }
        }
```

在上述方法中有对 getAutoConfigurationEntry()方法的调用，该方法代码如下：

```
    protected AutoConfigurationImportSelector.AutoConfigurationEntry
getAutoConfigurationEntry(AutoConfigurationMetadata
autoConfigurationMetadata, AnnotationMetadata annotationMetadata) {
            if (!this.isEnabled(annotationMetadata)) {
                return EMPTY_ENTRY;
            } else {
                AnnotationAttributes attributes =
this.getAttributes(annotationMetadata);
                List<String> configurations =
this.getCandidateConfigurations(annotationMetadata, attributes);
                configurations = this.removeDuplicates(configurations);
```

```
            Set<String> exclusions = this.getExclusions(annotationMetadata,
attributes);
            this.checkExcludedClasses(configurations, exclusions);
            configurations.removeAll(exclusions);
            configurations = this.filter(configurations,
autoConfigurationMetadata);
            this.fireAutoConfigurationImportEvents(configurations,
exclusions);
            return new
AutoConfigurationImportSelector.AutoConfigurationEntry(configurations,
exclusions);
        }
    }
```

而顺着这个方法继续往下看，会发现其中核心的逻辑是调用 getCandidateConfigurations() 方法，该方法的代码如下：

```
   protected List<String> getCandidateConfigurations(AnnotationMetadata
metadata, AnnotationAttributes attributes) {
         List<String> configurations =
SpringFactoriesLoader.loadFactoryNames(this.getSpringFactoriesLoaderFactoryC
lass(), this.getBeanClassLoader());
         Assert.notEmpty(configurations, "No auto configuration classes found
in META-INF/spring.factories. If you are using a custom packaging, make sure that
file is correct.");
         return configurations;
   }
```

分析到这里，应该基本上能理解@EnableAutoConfiguration 注解的核心逻辑实际上就是，实现对依赖（JAR 包）中“META-INF/spring.factories”文件进行扫描，而该文件中则声明了有哪些自动配置类需要被 Spring 容器加载。利用这样的逻辑，Spring Boot 应用也就能够自动加载依赖的 Spring 配置类，从而最终完成 Spring Boot 应用配置的自动化。

上述过程对开发人员完全透明，并不需要进行额外的配置，只需要引入相应的 Starter 依赖即可实现对特定技术或业务逻辑组件的“开箱即用”。

例如，Spring Boot 官方实现自动配置的核心组件“spring-boot-autoconfigure”就在其“META-INF/spring.factories”文件中定义了 Spring Boot 应用所依赖的基础配置类（如 Spring 的容器初始化配置类及其他常用组件配置等），该文件内容示例如下：

```
   # Spring 容器初始化相关自动配置类
   org.springframework.context.ApplicationContextInitializer=\
```

```
    org.springframework.boot.autoconfigure.SharedMetadataReaderFactoryContex
tInitializer,\
    org.springframework.boot.autoconfigure.logging.ConditionEvaluationReport
LoggingListener
    # 应用监听相关自动配置类
    org.springframework.context.ApplicationListener=\
    org.springframework.boot.autoconfigure.BackgroundPreinitializer
    ...
```

上述配置文件中的配置并不会在应用启动时就被全部初始化，因为某个组件的配置是否被加载很多时候还依赖其他的条件，如果直接被初始化，则可能造成 Spring 容器启动错误。所以，这也是基于条件的注解在自动配置类中被大量使用的原因。

大部分第三方技术组件，以及在日常开发中封装某些业务逻辑的公共组件，得益于该机制，都可以通过这种“开箱即用”的 Starter 组件方式，来实现与 Spring Boot 应用的快速集成。而使用这些 Starter 组件的用户，通常只需要引入相应的依赖，而不需要进行太多额外的配置。这也是在使用 Spring Boot 进行项目开发时，为什么会感觉很多时候只需要引入一个 Starter 依赖就能够即刻生效一个组件的原因。

1.2.3 Spring Boot 的核心注解

虽然 Spring Boot 简化了应用的配置，提高了开发效率，但是对于很多初学者而言，这种方式也很容易让人产生疑惑。所以，要用好 Spring Boot 框架，一个比较好的切入点就是对其提供的各类功能注解有一个清晰的理解和认识——这样不仅可以提高开发 Spring Boot 应用的效率，还可以在开发中快速排查问题。

接下来介绍在利用 Spring Boot 开发应用时比较常见的一些注解。

1. 与 Spring Boot 密切相关的 Spring 基础注解

Spring Boot 本身是基于 Spring 框架的，所以在 Spring Boot 中有一些注解是需要与 Spring 注解搭配使用的。下面整理了在实际项目中与 Spring Boot 注解配合最为紧密的 6 个 Spring 基础注解。

（1）@Configuration。

Spring 从 3.0 版本开始通过用@Configuration 注解修饰的配置类来替换 XML 配置文件。在用@Configuration 注解修饰的类中，可以包含多个用@Bean 注解修饰的方法，这些方法会被应用容器类 AnnotationConfigApplicationContext 或 AnnotationConfigWebApplicationContext 扫描，并在 Spring 容器初始化时被构建。在下方代码中，就通过这样的方式初始化了两个类的实例 Bean。

```
@Configuration
public class TaskAutoConfiguration {
    @Bean
    @Profile("biz-electrfence-controller")
    public BizElectrfenceControllerJob bizElectrfenceControllerJob() {
        return new BizElectrfenceControllerJob();
    }
    @Bean
    @Profile("biz-consume-1-datasync")
    public BizBikeElectrFenceTradeSyncJob bizBikeElectrFenceTradeSyncJob() {
        return new BizBikeElectrFenceTradeSyncJob();
    }
}
```

（2）@ComponentScan。

基于 Spring MVC 做过 Web 开发的读者一定都使用过@Controller、@Service、@Repository 这几个注解。查看它们的源码会发现，它们有一个共同的@Component 注解。@ComponentScan 注解的作用就是将被@Component 注解修饰的类加载到 Spring 容器中。所以，被@Controller、@Service、@Repository 这些注解修饰的类也会被@ComponentScan 注解扫描到，从而被加载到 Spring 容器中。例如：

```
@ComponentScan(value = "com.user.api")
public class UserApplication {
    public static void main(String[] args) {
        SpringApplication.run(UserApplication.class, args);
    }
}
```

在 Spring Boot 应用的主类中，经常会使用@ComponentScan 注解来配置特定的扫描路径。Spring Boot 的核心注解@SpringBootApplication 也包含了注解@ComponentScan，所以，也可以基于@SpringBootApplication 来配置特定的扫描路径。例如：

```
@SpringBootApplication(scanBasePackages = {"com.user.api",
"com.user.service"})
public class UserApplication {
    public static void main(String[] args) {
        SpringApplication.run(UserApplication.class, args);
    }
}
```

（3）@Conditional。

@Conditional 是 Spring 4 新提供的注解。通过@Conditional 注解，可以根据代码中设置的条件来装载不同的 Bean。

在 1.2.2 节的“2.”小标题中提及的 Spring Boot 中的@ConditionalOnProperty、@ConditionalOnBean 等以“@Conditional*”开头的注解，都是基于@Conditional 注解来实现的。

（4）@Import。

@Import 注解可以通过导入的方式把类的实例加载到 Spring 容器中。在 1.2.2 节的“3.”小标题中介绍 Spring Boot 的启动逻辑时，在@EnableAutoConfiguration 注解的定义中就有这样的用法，例如：

```
@Target({ElementType.TYPE})
@Retention(RetentionPolicy.RUNTIME)
@Documented
@Inherited
@AutoConfigurationPackage
@Import({EnableAutoConfigurationImportSelector.class})
public @interface EnableAutoConfiguration {
    String ENABLED_OVERRIDE_PROPERTY = "spring.boot.enableautoconfiguration";
    Class<?>[] exclude() default {};
    String[] excludeName() default {};
}
```

（5）@ImportResource。

@ImportResource 注解和@Import 注解类似。但区别在于，@ImportResource 注解导入的是配置文件，而@Import 注解导入的是类。通过@ImportResource 注解可以在 Spring Boot 项目中导入额外的 XML 配置文件，例如：

```
@ImportResource("classpath:spring-redis.xml")
public class UserApplication {
    public static void main(String[] args) {
        SpringApplication.run(UserApplication.class, args);
    }
}
```

（6）@Component。

@Component 是一个元注解，可以注解其他类型的注解，如@Controller、@Service、@Repository。带此注解的类可以被看作组件，在 Spring 的类路径扫描配置中会被实例化。

其他同类型的注解，也可以被认定为是一种特殊类型的组件，比如@Controller 控制器（注入服务）、@Service 服务（注入 Dao）、@Repository（实现 Dao 访问）。

@Component 泛指组件。当组件不好归类时，可以使用这个注解进行标注，相当于 XML 配置<bean id="" class=""/>。

2. Spring Boot 框架自身提供的核心注解

接下来介绍 Spring Boot 框架自身提供的核心注解。

（1）@SpringBootApplication。

该注解是 Spring Boot 最核心的注解，用在 Spring Boot 应用的主类上标识这是一个 Spring Boot 应用，以开启 Spring Boot 应用的各项能力。

在前面讲述 Spring Boot 应用的启动逻辑时已经分析过该注解的作用，这里就不再赘述。

（2）@EnableAutoConfiguration。

开启该注解后，Spring Boot 应用就能够根据当前类路径下的包或者类来配置 Spring Bean。例如：当前类路径下有 MyBatis 的依赖时，Spring Boot 就会通过自动配置类 MybatisAutoConfiguration，来实现 MyBatis 框架的初始化。

@EnableAutoConfiguration 注解实现的关键在于引入了 AutoConfigurationImportSelector 类，其核心 selectImports()方法的逻辑大致如下：

① 从配置文件“META-INF/spring.factories”加载所有可能用到的自动配置类。

② 去重，并排除 exclude 和 excludeName 属性携带的类。

③ 过滤，返回满足条件（@Conditional）的自动配置类。

在 1.2.2 节“3.”小标题中讲述 Spring Boot 应用的启动逻辑时也提到过该方法。

（3）@SpringBootConfiguration。

该注解是@Configuration 注解的变体，用来修饰 Spring Boot 的配置，可用于 Spring Boot 后续的扩展。

（4）@ConditionalOnBean。

@ConditionalOnBean(A.class)仅在当前上下文中存在 A 对象时，才会实例化一个 Bean。

即只有当 A.class 在 spring 的 applicationContext 中存在时，当前的 Bean 才能被创建。

（5）@ConditionalOnMissingBean。

该注解和@ConditionalOnBean 注解相反，仅在当前上下文中不存在某个实例时，才会实例化一个 Bean。例如：

```
@Bean
@ConditionalOnMissingBean(RocketMQProducer.class)
public RocketMQProducer mqProducer() {
    return new RocketMQProducer();
}
```

仅在当前环境上下文中不存在 RocketMQProducer 实例时，才会实例化 RocketMQProducer 的实例 Bean。

（6）@ConditionalOnClass。

该注解仅当某些类存在于“classpath”上时才创建某个 Bean。例如：

```
@Bean
 @ConditionalOnClass(HealthIndicator.class)
 public HealthIndicator rocketMQProducerHealthIndicator(Map<String,
DefaultMQProducer> producers) {
     if (producers.size() == 1) {
          return new
RocketMQProducerHealthIndicator(producers.values().iterator().next());
     }
 }
```

当“classpath”中存在 HealthIndicator 类时，才创建 HealthIndicator Bean 对象。

（7）@ConditionalOnMissingClass。

该注解和@ConditionalOnClass 注解相反：当“classpath”中没有指定的 Class 才开启配置。

（8）@ConditionalOnWebApplication。

该注解表示在当前项目类型是 Web 项目时才开启配置。当前项目有以下 3 种类型：ANY（任何 Web 项目都开启）、SERVLET（仅基础的 Servelet 项目才开启）、REACTIVE（只有是基于响应的 Web 应用才开启）。

（9）@ConditionalOnNotWebApplication。

该注解和@ConditionalOnWebApplication 注解相反：当前项目类型不是 Web 项目才开启配置。

（10）@ConditionalOnProperty

该注解表示当指定的属性有指定的值时才开启配置。具体操作是通过其两个属性“name”及“havingValue”来实现的。其中，“name”用来从 application.properties 中读取某个属性值：

- 如果该值为空，则返回 false。
- 如果该值不为空，则将该值与“havingValue”指定的值进行比较，如果一样则返回 true，否则返回 false。

如果返回值为 false，则该配置不开启；如果返回值为 true，则该配置开启。例如：

```
@Bean
@ConditionalOnProperty(value = "rocketmq.producer.enabled", havingValue =
"true", matchIfMissing = true)
 public RocketMQProducer mqProducer() {
     return new RocketMQProducer();
 }
```

（11）@ConditionalOnExpression。

该注解表示当 SpEL 表达式为 true 时才开启配置。例如：

```
@Configuration
@ConditionalOnExpression("${enabled:false}")
public class BigpipeConfiguration {
    @Bean
    public OrderMessageMonitor orderMessageMonitor(ConfigContext
configContext) {
        return new OrderMessageMonitor(configContext);
    }
}
```

（12）@ConditionalOnResource。

该注解表示当类路径下有指定的资源时才开启配置。例如：

```
@Bean
@ConditionalOnResource(resources="classpath:shiro.ini")
protected Realm iniClasspathRealm(){
  return new Realm();
}
```

（13）@ConfigurationProperties。

Spring Boot 可使用注解的方式将自定义的 properties 文件（比如 config.properties 文件）映射到实体 Bean 中。例如：

```
@Data
```

```
@ConfigurationProperties("rocketmq.consumer")
public class RocketMQConsumerProperties extends RocketMQProperties {
    private boolean enabled = true;
    private String consumerGroup;
    private MessageModel messageModel = MessageModel.CLUSTERING;
    private ConsumeFromWhere consumeFromWhere =
ConsumeFromWhere.CONSUME_FROM_LAST_OFFSET;
    private int consumeThreadMin = 20;
    private int consumeThreadMax = 64;
    private int consumeConcurrentlyMaxSpan = 2000;
    private int pullThresholdForQueue = 1000;
    private int pullInterval = 0;
    private int consumeMessageBatchMaxSize = 1;
    private int pullBatchSize = 32;
}
```

（14）@EnableConfigurationProperties。

当@EnableConfigurationProperties 注解与@Configuration 注解配合使用时，任何被@ConfigurationProperties 注解修饰的 Bean 将自动实现 Environment 属性映射配置。例如：

```
@Configuration
@EnableConfigurationProperties({
    RocketMQProducerProperties.class,
    RocketMQConsumerProperties.class,
})
@AutoConfigureOrder
public class RocketMQAutoConfiguration {
    @Value("${spring.application.name}")
    private String applicationName;
}
```

（15）@AutoConfigureAfter。

该注解用在自动配置类上，表示该自动配置类需要在指定的自动配置类配置完之后，才可以被加载。如 MyBatis 的自动配置类，就需要在数据源自动配置类配置完成之后才可以被加载。例如：

```
@AutoConfigureAfter(DataSourceAutoConfiguration.class)
public class MybatisAutoConfiguration {
}
```

（16）@AutoConfigureBefore。

该注解和@AutoConfigureAfter 注解相反，表示该自动配置类需要在指定的自动配置类配置之前被加载。

（17）@AutoConfigureOrders

该注解是 Spring Boot 1.3.0 提供的一个注解，用于确定配置加载的优先级顺序。例如：

```
//自动配置里面的最高优先级
@AutoConfigureOrder(Ordered.HIGHEST_PRECEDENCE)
@Configuration
//仅限于 Web 应用
@ConditionalOnWebApplication
//导入内置容器的设置
@Import(BeanPostProcessorsRegistrar.class)
public class EmbeddedServletContainerAutoConfiguration {
    @Configuration
    @ConditionalOnClass({ Servlet.class, Tomcat.class })
    @ConditionalOnMissingBean(value =
EmbeddedServletContainerFactory.class, search = SearchStrategy.CURRENT)
    public static class EmbeddedTomcat {
       //...
    }
    @Configuration
    @ConditionalOnClass({ Servlet.class, Server.class, Loader.class,
WebAppContext.class })
    @ConditionalOnMissingBean(value =
EmbeddedServletContainerFactory.class, search = SearchStrategy.CURRENT)
    public static class EmbeddedJetty {
       //...
    }
}
```

以上就是 Spring Boot 中比较常见的注解。虽然在开发 Spring Boot 应用时，一些配置代码和 Starter 依赖已经有了开源的实现，只需要引入即可，但在阅读这些组件的源码时会经常遇到上述注解。

此外，如果需要自定义 Starter 组件，则在定义自动配置类时，可以根据注入逻辑使用上述注解。灵活运用 Spring Boot 相关注解，对于学习 Spring Boot 框架是非常有帮助的。

1.3 开发一个 Spring Boot 应用

通过前面内容的阐述，相信大家对 Spring Boot 应该有了一个初步的了解，接下来开发一个可以运行的 Spring Boot 应用。

1.3.1 步骤 1：创建 Spring Boot 基础工程

使用 IntelliJ IDEA 创建一个标准的 Maven 工程的项目——虽然可以直接使用 IntelliJ IDEA 创

建一个基于 Spring Boot 的项目，但是为了更清晰地理解 Spring Boot 应用的创建过程，这里将采用手工配置的方式来实现。

（1）创建一个标准的 Maven 工程。

在 IDEA 工具栏选择“File→New→Project”，选择 Maven 工程类型，如图 1-4 所示。

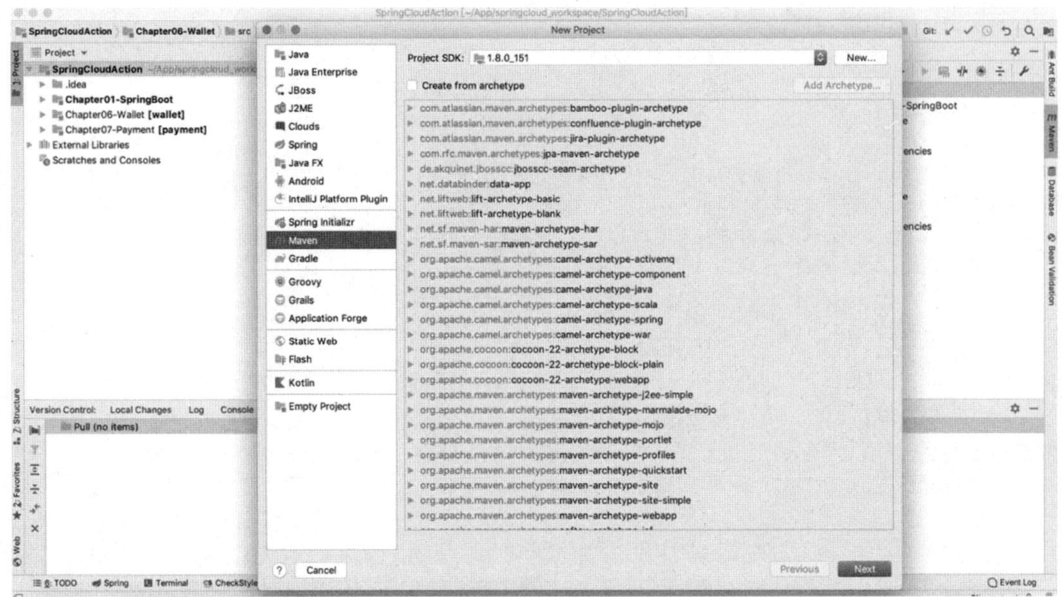

图 1-4

创建后的 Maven 工程的代码结构如图 1-5 所示。

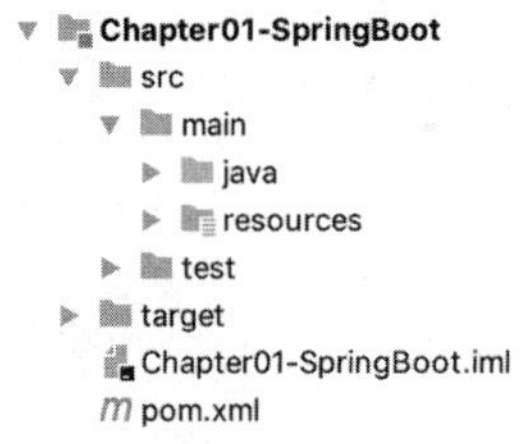

图 1-5

（2）引入 Spring Boot 的依赖。

在 Maven 项目的 pom.xml 文件中引入 Spring Boot 的依赖，这里引入的版本是 Spring Boot 2.1.5.RELEASE。代码如下：

```
<?xml version="1.0" encoding="UTF-8"?>
```

```
<project xmlns="http://maven.apache.org/POM/4.0.0"
        xmlns:xsi="http://www.w3.org/2001/XMLSchema-instance"
        xsi:schemaLocation="http://maven.apache.org/POM/4.0.0
http://maven.apache.org/xsd/maven-4.0.0.xsd">
    <modelVersion>4.0.0</modelVersion>
    <!--引入 Spring Boot 父依赖-->
    <parent>
        <groupId>org.springframework.boot</groupId>
        <artifactId>spring-boot-starter-parent</artifactId>
        <version>2.1.5.RELEASE</version>
        <relativePath/> <!-- lookup parent from repository -->
    </parent>
    <groupId>Chapter01-SpringBoot</groupId>
    <artifactId>Chapter01-SpringBoot</artifactId>
    <version>1.0-SNAPSHOT</version>
    <properties>
        <java.version>1.8</java.version>
    </properties>
    <dependencies>
        <!--引入 Spring Boot 核心 Starter 依赖-->
        <dependency>
            <groupId>org.springframework.boot</groupId>
            <artifactId>spring-boot-starter</artifactId>
        </dependency>
        <dependency>
            <groupId>org.springframework.boot</groupId>
            <artifactId>spring-boot-starter-test</artifactId>
            <scope>test</scope>
        </dependency>
    </dependencies>
    <build>
        <plugins>
            <!--引入 Spring Boot Maven 的编译插件-->
            <plugin>
                <groupId>org.springframework.boot</groupId>
                <artifactId>spring-boot-maven-plugin</artifactId>
            </plugin>
        </plugins>
    </build>
</project>
```

（3）创建 Spring Boot 应用的启动类。代码如下：

```
package com.wudimanong.demo;
import org.springframework.boot.SpringApplication;
import org.springframework.boot.autoconfigure.SpringBootApplication;
```

```
@SpringBootApplication
public class DemoApplication {
    public static void main(String[] args) {
        SpringApplication.run(DemoApplication.class, args);
    }
}
```

（4）引入 Spring Boot Web 依赖。

如果要将 Spring Boot 应用作为一个服务，并能够对外提供接口调用，则需要继续在 Spring Boot 工程的 pom.xml 文件中引入 Spring Boot Web 依赖。代码如下：

```
<!--引入 Spring Boot Web 依赖-->
  <dependency>
       <groupId>org.springframework.boot</groupId>
       <artifactId>spring-boot-starter-web</artifactId>
  </dependency>
```

在引入 spring-boot-starter-web 依赖后，因为该依赖本身也包含 spring-boot-starter 依赖，所以，为了避免重复引用，需要在项目中去掉 spring-boot-starter 的依赖。

1.3.2 步骤 2：创建项目配置文件

为了更贴近实际的项目环境，在工程的“/src/mian/resources”目录下创建配置文件 application.yml。代码如下：

```
spring:
  application:
    name: Chapter01-SpringBoot
server:
  port: 8080
```

上述配置文件定义了应用的名称及服务的端口号。

关于配置文件类型的选择，本书统一使用 YML 格式的文件。

1.3.3 步骤 3：集成 MyBatis 框架

在实际的项目中，普遍使用的数据库操作框架是 MyBatis，接下来演示如何在 Spring Boot 应用中集成 MyBatis 框架，从而在业务逻辑中实现对数据库的操作。

（1）利用 Docker 在本地快速安装一个 MySQL 数据库。

为了方便实验，笔者是在自己的 Mac 笔记本上安装了 Docker 引擎，并通过容器快速部署了 MySQL 数据库。操作命令如下：

```
docker run --name dev-mysql -p 3306:3306 -e MYSQL_ROOT_PASSWORD=123456 -d
mysql:5.7
```

执行完上述 Docker 命令后，可以通过“docker ps”命令查看 MySQL 数据库的启动情况：

```
$ docker ps
CONTAINER ID   IMAGE   COMMAND   CREATED   STATUS   PORTS   NAMES
50afedce617f   mysql:5.7  "docker-entrypoint.s…"  7 months ago  Up 2
seconds        0.0.0.0:3306->3306/tcp, 33060/tcp   dev-mysql
```

执行上述命令需要在本地安装 Docker 引擎，具体安装方式可根据自己的开发环境选择相应的安装方式。

此时就在本地环境中启动了一个 MySQL 数据库，可以通过 MySQL 连接工具创建一个名为“test”的数据库。

（2）在配置文件 application.yml 中添加数据库连接信息。代码如下：

```
spring:
  application:
    name: Chapter01-SpringBoot
  datasource:
    url: jdbc:mysql://127.0.0.1:3306/test
    username: root
    password: 123456
    type: com.alibaba.druid.pool.DruidDataSource
    driver-class-name: com.mysql.jdbc.Driver
    separator: //
server:
  port: 9090
```

在上述配置文件中，配置了数据源、数据库连接串、密码及数据库连接池等信息。

（3）引入数据库连接的相关依赖。代码如下：

```
<!--引入 Druid 连接池依赖-->
<dependency>
    <groupId>com.alibaba</groupId>
    <artifactId>druid</artifactId>
    <version>1.0.28</version>
</dependency>
```

```
<!--引入 MySQL 数据库驱动程序依赖-->
<dependency>
    <groupId>mysql</groupId>
    <artifactId>mysql-connector-java</artifactId>
    <scope>runtime</scope>
</dependency>
```

（4）引入 MyBatis 框架的 Starter 依赖。

对于 MyBatis 框架的集成非常简单，只需要引入相应的 Starter 依赖即可，代码如下：

```
<!--引入 MyBatis 的 Starter 依赖-->
<dependency>
    <groupId>org.mybatis.spring.boot</groupId>
    <artifactId>mybatis-spring-boot-starter</artifactId>
    <version>2.0.1</version>
</dependency>
```

至此，MyBatis 框架就基本配置完成了。如果此时在项目的应用代码中编写一些基于 MyBatis 的数据库持久层（Dao 层）操作类，则它们能够被 Spring Boot 自动扫描到并被 Spring 容器加载。

1.3.4 步骤 4：编写服务接口完成数据库操作

接下来编写一个模拟用户注册的接口，该接口接收参数并通过 MyBatis 完成对数据库的“写”操作。

（1）在数据库中创建一张测试表“user”。SQL 代码如下：

```
create table user(
     id bigint primary key AUTO_INCREMENT,
    username varchar(60),
    password varchar(60)
)
```

（2）在 Spring Boot 应用中开发持久层（Dao 层）代码。

在项目工程的包路径中创建一个名为“entity”的包，并创建一个 User 对象的实体类。代码如下：

```
package com.wudimanong.demo.entity;
public class User {
    private String username;
    private String password;
    public String getUsername() {
        return username;
    }
    public void setUsername(String username) {
        this.username = username;
```

```
    }
    public String getPassword() {
        return password;
    }
    public void setPassword(String password) {
        this.password = password;
    }
}
```

之后创建一个名为“dao”的包路径，然后在该包中编写 MyBatis 的持久层（Dao 层）接口。代码如下：

```
package com.wudimanong.demo.dao;
import com.wudimanong.demo.entity.User;
import org.apache.ibatis.annotations.Insert;
import org.apache.ibatis.annotations.Mapper;

@Mapper
public interface UserDao {
    String TABLE_NAME = " user ";
    String ALL_FIELDS = "username,password";
    @Insert("INSERT INTO " + TABLE_NAME + "(" + ALL_FIELDS + ") VALUES
(#{username}, #{password})")
    int addUser(User user);
}
```

这里用@Mapper 注解定义了一个 MyBatis 的数据库接口。

（3）编写 Spring MVC 服务接口。

接下来创建一个名为“controller”的包路径，并编写一个名为“addUser”的服务接口。代码如下：

```
@RestController
@RequestMapping("/user")
public class UserController {
    @Autowired
    UserDao userDao;
    @RequestMapping(value = "/addUser", method = RequestMethod.POST)
    public User getUserById(@RequestParam(value = "username") String
username,@RequestParam(value = "password") String password) {
        User user = new User();
        user.setUsername(username);
        user.setPassword(password);
        userDao.addUser(user);
        return user;
    }
```

```
}
```

在上述代码中，通过注入持久层（Dao 层）接口 UserDao 来完成对数据库的“写”操作。

（4）运行 Spring Boot 应用的 DemoApplication 主类，启动后的 Spring Boot 应用如图 1-6 所示。

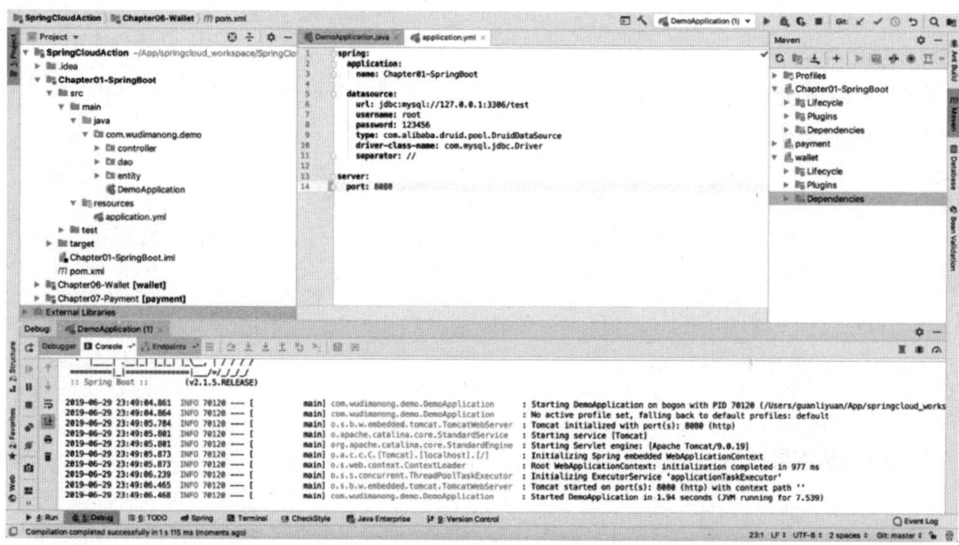

图 1-6

（5）模拟对服务接口的访问，来完成对数据库的操作。

用 HTTP 接口工具 Postman 来调用步骤（3）中编写的接口，效果如图 1-7 所示。

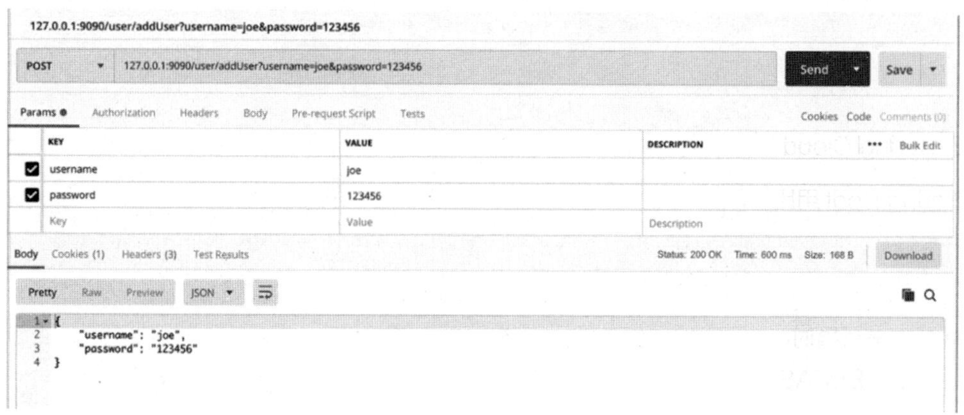

图 1-7

此时查看数据库记录，就会看到在表中插入了一条数据。至此就完成了一个 Spring Boot 基本应用的开发。

1.4　Spring Cloud 微服务体系

通过前面的学习，相信大家对 Spring Boot 框架的依赖管理及自动配置机制有了一定的认识和理解。而 Spring Boot 的这个核心能力，也正是 Spring Cloud 微服务框架得以构建的基础。接下来进一步学习 Spring Cloud 微服务体系的内容。

1.4.1　Spring Cloud 简介

接下来从基本概念及版本来简单介绍一下 Spring Cloud。

1. Spring Cloud 是什么

从技术角度来说，Spring Cloud 不是某一个具体技术的名称，也不是一种全新的技术体系。它是基于 Spring Boot 框架，将业界比较著名的、得到过实践反复验证的开源服务治理技术进行整合后的产物。所以从本质上说，Spring Cloud 并没有太多的技术创新，只是一种对微服务开发模式的优化和组合。

此外，Spring Cloud 也不是一两种技术的代名词，而是一组框架的统称。Spring Cloud 基于 Spring Boot Starter“开箱即用”的依赖集成模式，极大地简化了以往使用各类服务治理框架的烦琐步骤。

在 Spring Cloud 中，通过一个简单的注解配置，就能快速实现服务的注册与发现；通过一个简单的声明式注解，就能够实现服务的调用、负载均衡、限流、熔断等。

2. Spring Cloud 的版本

Spring Boot 用比较明确的数字编号来表示版本，而 Spring Cloud 关于版本的命名则比较特殊：它只有相应的开发代号。目前使用比较广 Spring Cloud 版本为 Edgware.SR5，对应于 Spring Boot 1.5.x.RELEASE。

截至本书写稿时（2021 年 8 月）Spring Cloud 的最新版本为 Greenwich.SR1，对应于 Spring Boot 2.1.3.RELEASE。本书所有的微服务实例采用的 Spring Cloud 版本均为 Greenwich.SR1。

1.4.2 Spring Boot 与 Spring Cloud 的关系

Spring Boot 应用可以实现与 Spring Cloud 微服务组件的无缝集成，例如，通过一些注解配置快速将 Spring Boot 应用接入 Spring Cloud 微服务体系，从而实现单体应用的微服务化转型。

因此，从某种程度上说 Spring Boot 是 Spring Cloud 的子集：Spring Boot 应用可以不使用 Spring Cloud 的微服务组件，但 Spring Cloud 微服务则必须基于 Spring Boot 来构建。

1.4.3 Spring Cloud 微服务的核心组件

在 1.4.1 节中提到 Spring Cloud 是一组框架的组合，那么构成这个“组合”的核心技术框架有哪些呢?

1. Spring Cloud 的依赖引用关系

Spring Cloud 微服务开发体系是基于 Spring Boot 框架的，如果 Spring Boot 应用要顺利实现微服务的功能，则需要定制一组针对 Spring Cloud 核心微服务组件的 Spring Boot Starter 依赖。

在构建 Spring Cloud 微服务时，一般会通过引入“spring-cloud-starter-parent”父依赖来实现 Spring Cloud 核心微服务组件的快速引入。依赖代码如下：

```
<!--引入 Spring Cloud 父依赖-->
<parent>
    <groupId>org.springframework.cloud</groupId>
    <artifactId>spring-cloud-starter-parent</artifactId>
    <version>Greenwich.SR1</version>
    <relativePath/>
</parent>
```

为了更好地理解在引入上述父依赖包后 Spring Cloud 微服务到底集成了哪些依赖，接下来分析该依赖包的引用关系，如图 1-8 所示。

从上述依赖引用关系可以看到，虽然在构建 Spring Cloud 微服务时只引入了“spring cloud parent”这一个父依赖，但该依赖向上继承了 Spring Boot 框架的父依赖“spring-boot-starter-parent”，从而间接引入了整个 Spring Boot 框架体系。

而从横向上看，Spring Cloud 的父依赖还通过引入“spring-cloud-dependencies”依赖，实现了对其他开源服务治理框架（如 Consul、ConfigServer 及 Netflix 等服务注册与发现、配置管理、限流熔断技术框架）的引入和集成。

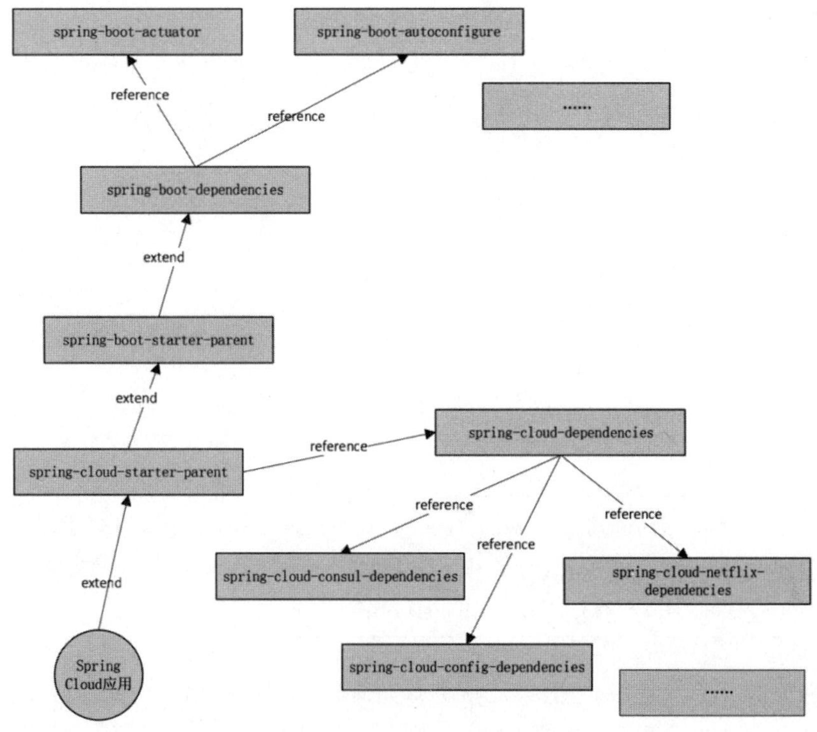

图 1-8

2. Spring Cloud 的核心组件

那么在 Spring Cloud 中具体集成了哪些核心组件来实现微服务功能呢？见表 1-1。

表 1-1　Spring Cloud 的核心组件

组件名称	功能简介
Spring Boot	构建 Spring Cloud 微服务的框架基础
Spring Cloud Starters	Spring Cloud 开箱即用的 Spring Boot 依赖管理
Consul	一个用 Go 语言编写的服务注册及配置系统，可以作为整个微服务体系的服务注册中心
Eureka	与 Consul 的功能定位类似，是一款用 Java 语言编写的服务注册中心
Feign	Feign 是一种声明式的 HTTP 客户端，主要用于简化微服务之间的通信调用方式
Ribbon	Ribbon 是一种进程间通信（远程调用）库，内置了软件负载均衡器，支持 RESTful 调用，以及各种序列化方案
Spring Cloud Gateway	服务网关组件，提供动态路由、监视、弹性及安全性等功能
Hystrix	一个延迟容错库，旨在隔离对远程系统、服务或第三方库的接口调用，防止级联故障，增强系统的弹性
Spring Cloud Config	配置管理工具包，支持将微服务的配置放到远程服务器中进行集中管理

具体说明如下。

（1）在微服务体系中最核心的组件莫过于服务注册中心，所有的微服务都需要通过它实现服务的注册与发现。在 Spring Cloud 中使用得比较普遍的注册中心有：基于 Java 语言编写的 Eureka、基于 Go 语言编写的 Consul。

考虑到系统异构问题（如为了让用 Go/Python 等语言编写的微服务能方便地与用 Java 编写的微服务进行通信），目前在生产中使用 Consul 作为服务注册中心的比较多。为了更好地贴近生产实践，本书主要使用 Consul 作为服务注册中心进行示范。

（2）在微服务通过注册中心完成服务注册后，在服务之间需要一种便捷的通信调用方式。在 Spring Cloud 中比较通用的方式是：使用 Feign 实现 HTTP 客户端的声明式服务调用。

由于所有的微服务都支持通过 Consul 或 Eureka 进行多节点集群部署，所以在客户端调用微服务时，还需要实现负载均衡等功能，而这种客户端调用负载均衡功能的实现，在 Spring Cloud 中是通过“Feign + Ribbon”组合来实现的。

（3）在微服务体系中，一个比较核心的问题是如何实现服务的限流与熔断。作为面向外部的服务，如果服务间的请求出现流量波动或阻塞现象，则需要实施限流熔断，以保证服务的可用性。Spring Cloud Gateway 及 Hystrix 可提供这样的功能。

（4）配置管理在 Spring Cloud 中是通过 Spring Cloud Config 来实现的，通过该组件可以实现 Spring Cloud 体系中微服务应用配置的集中管理，从而解决项目中配置散乱的问题，同时也可以提高敏感配置信息的安全性。

1.4.4 Spring Cloud 的核心注解

Spring Cloud 正是因为集成了一组核心的服务治理组件，所以才实现了微服务体系的基本功能。而这些功能要以友好的方式提供给开发者，则还需要通过注解的定义与封装来实现。这样，开发者只需在代码中简单地引入某个注解，就能实现相应的微服务功能。

接下来介绍 Spring Cloud 提供的核心注解，以及这些注解是怎么实现微服务功能的。

1. @EnableDiscoveryClient

微服务的注册与发现有多种实现方式，如利用注册中心 Consul 和 Eureka。

@EnableDiscoveryClient 注解定义在“spring-cloud-commons”包中，是便于开发者快速实现服务注册与发现的一个注解。

在实际项目开发中，只需要引入该注解，就能够实现微服务与服务注册中心连接。代码如下：

```
@EnableFeignClients(basePackageClasses = {PaymentClient.class})
@EnableHystrixDashboard
@EnableCircuitBreaker
@EnableDiscoveryClient
@SpringBootApplication
public class Wallet {
    public static void main(String[] args) {
        SpringApplication.run(Wallet.class, args);
    }
}
```

下面通过源码来分析@EnableDiscoveryClient 注解是如何将微服务注册到服务注册中心的。

（1）该注解通过@Import({EnableDiscoveryClientImportSelector.class})导入逻辑类。此逻辑类通过定义 isEnabled()方法来标识是否开启服务注册与发现功能。代码如下：

```
@Override
protected boolean isEnabled() {
   return getEnvironment().getProperty("spring.cloud.discovery.enabled",
     Boolean.class, Boolean.TRUE);
}
```

（2）SpringFactoryImportSelector 父类通过 selectImports()方法扫描集成的 Eureka 或 Consul 的 Spring Boot Starter 依赖。代码如下：

```
@Override
public String[] selectImports(AnnotationMetadata metadata) {
   String[] imports = super.selectImports(metadata);
   AnnotationAttributes attributes = AnnotationAttributes.fromMap(
        metadata.getAnnotationAttributes(getAnnotationClass().getName(),
true));
   boolean autoRegister = attributes.getBoolean("autoRegister");
   if (autoRegister) {
      List<String> importsList = new ArrayList<>(Arrays.asList(imports));
      importsList.add(
 "org.springframework.cloud.client.serviceregistry.AutoServiceRegistratio
nConfiguration");
      imports = importsList.toArray(new String[0]);
   }
   else {
      Environment env = getEnvironment();
      if (ConfigurableEnvironment.class.isInstance(env)) {
         ConfigurableEnvironment configEnv = (ConfigurableEnvironment) env;
         LinkedHashMap<String, Object> map = new LinkedHashMap<>();
```

```
            map.put("spring.cloud.service-registry.auto-registration.enabled",
false);
            MapPropertySource propertySource = new MapPropertySource(
                  "springCloudDiscoveryClient", map);
            configEnv.getPropertySources().addLast(propertySource);
         }
      }
      return imports;
   }
```

（3）这里以使用 Consul 作为微服务注册中心为例，在项目中引入“spring-cloud-starter-consul-discovery”依赖。代码如下：

```
<dependency>
    <groupId>org.springframework.cloud</groupId>
    <artifactId>spring-cloud-starter-consul-discovery</artifactId>
</dependency>
```

① 在该依赖包的“META-INF/spring.factories”文件中包含了一系列自动配置类，如下：

```
org.springframework.boot.autoconfigure.EnableAutoConfiguration=\
org.springframework.cloud.consul.discovery.RibbonConsulAutoConfiguration,\
org.springframework.cloud.consul.discovery.configclient.ConsulConfigServ
erAutoConfiguration,\
org.springframework.cloud.consul.serviceregistry.ConsulAutoServiceRegist
rationAutoConfiguration,\
org.springframework.cloud.consul.serviceregistry.ConsulServiceRegistryAu
toConfiguration,\
org.springframework.cloud.consul.discovery.ConsulDiscoveryClientConfigur
ation
org.springframework.cloud.bootstrap.BootstrapConfiguration=\
org.springframework.cloud.consul.discovery.configclient.ConsulDiscoveryC
lientConfigServiceBootstrapConfiguration
```

② 这些自动配置类会在应用启动时被初始化和加载，以完成微服务与 Consul 的连接。如以下这个自动配置类，会在存在配置属性“spring.cloud.config.discovery.enabled=ture”时被初始化。

```
@ConditionalOnClass(ConfigServicePropertySourceLocator.class)
@ConditionalOnProperty(value = "spring.cloud.config.discovery.enabled",
matchIfMissing = false)
@Configuration
@ImportAutoConfiguration({ ConsulAutoConfiguration.class,
      ConsulDiscoveryClientConfiguration.class })
public class ConsulDiscoveryClientConfigServiceBootstrapConfiguration {
}
```

③ 这些被初始化的组件会与项目中关于 Consul 的配置进行匹配。例如：

```
  spring:
    application:
      name: wallet
    profiles:
      active: debug
    cloud:
      consul:
        discovery:
          preferIpAddress: true
          instance-id:
${spring.application.name}:${spring.cloud.client.ipAddress}:${spring.applica
tion.instance_id:${server.port}}:@project.version@
          healthCheckPath: /actuator/health
```

通过上述分析，读者可以大致了解 Spring Cloud 微服务自动注册与发现的基本原理。

Spring Cloud 本质上还是基于 Spring Boot 框架的机制来运行的。关于微服务是如何与 Consul 注册中心连接的，感兴趣的读者可以看一下“spring-cloud-consul-discovery”这个依赖的源码。

2. @EnableFeignClients

@EnableFeignClients 注解用于生效用@FeignClient 注解定义的 Feign 客户端。在微服务消费端配置了@EnableFeignClients 注解后，就可以通过@FeignClient("wallet")注解的方式实现对其他微服务的“客户端负载均衡调用”。

如果查看@EnableFeignClients 注解的关键逻辑类 FeignClientsRegistrar，则会发现该注解默认开启了 Robbin，而 Robbin 是实现微服务客户端负载均衡的核心组件——它从 Consul 拉取服务节点信息，以轮询的策略将客户端的调用请求转发至不同的服务端节点。

3. @EnableCircuitBreaker

要在 Spring Cloud 中使用断路器，只需要加上@EnableCircuitBreaker 注解即可。该注解会引入 Hystrix 组件，其过程与本节“1.”小标题中讲解的@EnableDiscoveryClient 注解的逻辑类似。例如：

```
@Target({ElementType.TYPE})
@Retention(RetentionPolicy.RUNTIME)
@Documented
@Inherited
@Import({EnableCircuitBreakerImportSelector.class})
public @interface EnableCircuitBreaker {
}
```

@EnableCircuitBreaker 注解通过导入 EnableCircuitBreakerImportSelector 类，来开启断路器设置。代码如下：

```
    protected boolean isEnabled() {
        return ((Boolean)(new
RelaxedPropertyResolver(this.getEnvironment())).getProperty("spring.cloud.ci
rcuit.breaker.enabled", Boolean.class, Boolean.TRUE)).booleanValue();
    }
```

如果在项目中引入了“spring-cloud-starter-hystrix”依赖，那么在应用加载时就会初始化 Hystrix 的自动配置类。可以查看该依赖的“META-INF/spring.factories”文件的内容：

```
    org.springframework.boot.autoconfigure.EnableAutoConfiguration=\
    org.springframework.cloud.netflix.hystrix.HystrixAutoConfiguration,\
    org.springframework.cloud.netflix.hystrix.security.HystrixSecurityAutoCo
nfiguration
    org.springframework.cloud.client.circuitbreaker.EnableCircuitBreaker=\
    org.springframework.cloud.netflix.hystrix.HystrixCircuitBreakerConfigura
tion
```

1.4.5 Spring Cloud 的技术生态圈

事实上，实施微服务架构的复杂性较高，前面提及的 Spring Cloud 核心组件只是实现了服务的注册与发现、限流、熔断等核心功能，还有很多其他的辅助功能，如分布式链路追踪、安全等。

因此，Spring Cloud 除了一些核心功能组件，还有很多实现特定功能的组件，如 Sleuth、Turbine 等。表 1-2 中列出了一些 Spring Cloud 生态中关注度比较高的组件。

表 1-2

组件名称	功能简介
Spring Cloud Sleuth	一个面向 Spring Cloud 的分布式跟踪工具，它借鉴了 Dapper、Zipkin 及 Htrace 等框架的设计思想。但该组件在国内并不普及（目前国内比较普及的替代方案是 SkyWalking）
Spring Cloud Bus	主要用于实现微服务与轻量级消息代理的连接
Spring Cloud Security	基于 Spring Security 的安全工具包，主要用于加强微服务的安全机制
Spring Cloud CLI	基于 Spring Boot CLI，支持以命令的方式快速构建微服务组件
Turbine	发送事件流数据的工具，可以用来监控集群下 Hystrix 的 Metrics 指标情况
Spring Cloud Task	微服务任务调度、管理的组件
Archaius	配置管理 API，包含一系列的配置 API，可以提供线程安全的配置操作、轮询、回调等功能
Spring Cloud Data Flow	大数据操作工具。作为 Spring XD 的替代产品，它采用混合计算模型，结合了流数据与批量数据的处理方式
Spring Cloud Stream	数据流操作开发包，封装了与 Redis、Rabbit 及 Kafka 等组件的消息通信方式

1.5　本章小结

本章简单介绍了 Spring Boot、Spring Cloud 微服务框架的基本情况，目的是为了让大家对 Spring Cloud 微服务体系有一个基本的认识，为更好地学习本书后面的实例打下基础。

Spring Cloud 是一个比较庞大的体系——既有核心功能组件，也有针对特定场景的组件。本书将通过具体的实例来让读者掌握实际工作中的 Spring Cloud 微服务开发技巧。本书并不会一一讲解每一个组件，但会在每一个实例中，根据具体的应用场景引入相应的技术组件，这种方式可以让读者在实战中逐步掌握 Spring Cloud 微服务开发的要点。

此外，Spring Cloud 微服务只是一种架构方式，并不是“银弹”，大家要对此有一个清晰的认识。

第 2 章

【实例】用户系统——用 Spring Boot 开发应用，用 Spring Cloud 将其改为微服务架构

本章将开启本书的第一个微服务实例——基于 Spring Cloud 微服务体系完成一个用户系统，该系统就只有一个用户微服务“user”。

在真实场景下，根据复杂程度，用户系统可以被拆分为一个或多个微服务。“系统”与“微服务”这两个概念之间关系是：微服务是系统的一种实现形式。从整体上看，除系统与系统之间可以通过微服务的方式进行拆分外，在系统内部也可以根据业务的发展进一步拆分为多个微服务。

本书后续内容中关于“系统”与“微服务”的描述，均遵循此类逻辑关系，请读者提前知悉。

从业务角度来看，对于大部分互联网应用而言，用户管理都是其一项基本功能。

从技术角度来看，在 Spring Cloud 框架微服务体系中，微服务需要先通过 Spring Boot 框架进行构建，然后引入 Spring Cloud 框架的相关依赖，再通过服务注册中心与其他微服务相互连接，并利用 Spring Cloud 微服务技术栈实现服务注册/发现、负载均衡、调用等服务治理逻辑。

在本章中，除完成用户微服务的基本构建外，还将引入微服务注册中心 Consul（它是确保用户微服务能够正常运行的基础，也是整个 Spring Cloud 微服务体系的关键组件）。

通过本章，读者将学习到以下内容。

- 真实业务场景下的用户微服务设计方案。
- Spring Cloud 微服务应用构建的基本方法。
- 服务注册中心 Consul 的基本原理及部署方式。
- 将 Spring Cloud 微服务接入服务注册中心的方法。
- 微服务的开发技巧，如在应用接入 Redis 服务、使用 Docker 容器等。

2.1 功能概述

在用户微服务中，必须实现的基本功能包括用户注册、登录及登录退出等。目前国内大部分 App 的登录方式主要有：①“手机号 + 短信验证码”；②第三方账号（如微信、QQ 等）授权登录。在本实例中，将按照这两种方式进行设计。

此外，因为第三方账号授权登录需要“前端跳转到第三方界面，获取用户授权”，所以，用户微服务在进行逻辑设计时需要考虑到这一点，并在进行接口设计时兼容这两种方式在流程处理上的差异。

2.2 系统设计

下面从系统设计的角度来设计用户微服务的逻辑，主要从业务逻辑设计及数据库设计两个方面进行。

2.2.1 业务逻辑设计

目前很多互联网 App 在设计用户的登录逻辑时，考虑到用户体验的问题，都是直接支持“手机号 + 短信验证码”及第三方账号授权登录，而并没有直接显示注册入口。所以，在后端逻辑中需要兼容注册−登录一体的逻辑：如果是系统中不存在的新用户，则在注册之后自动登录；否则就是老用户，直接完成登录。

当然，并不是所有 App 的登录逻辑都这样设计的，这取决于 App 的功能性质及成本。例如，“手机号 + 短信验证码”和第三方账号授权登录都需要一定的通道成本，所以很多 App 在首次完成用户信息注册后会提示用户设置账号密码，而在下次登录时，用户可以选择使用账号密码直接登录。

接下来针对上述两种登录方式，对用户微服务的业务逻辑进行设计。

1.“手机号 + 短信验证码”登录

“手机号 + 短信验证码”登录的流程如图 2-1 所示。

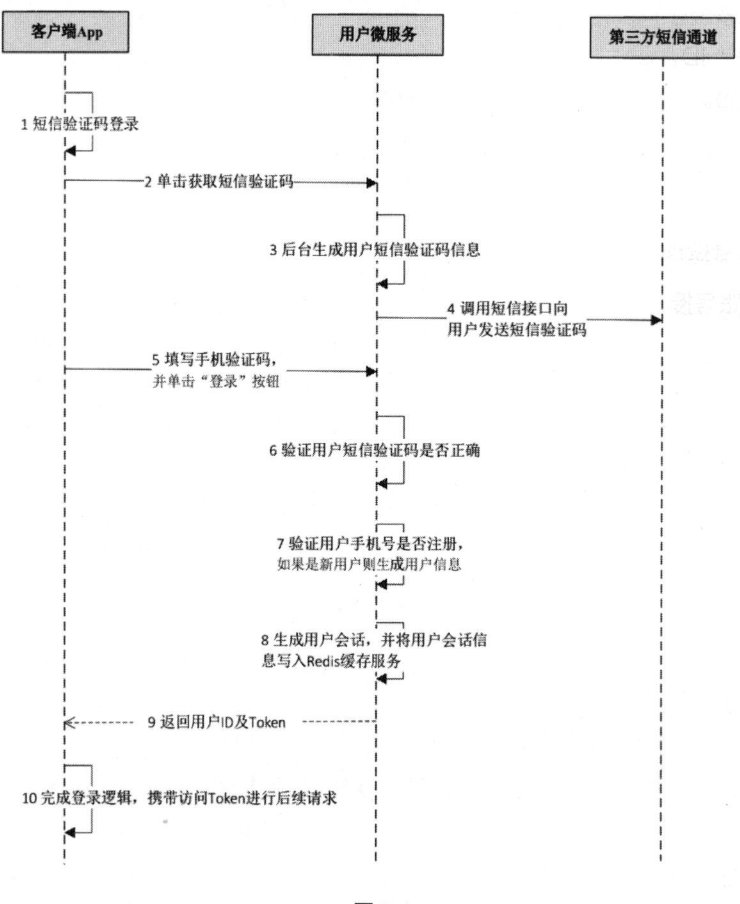

图 2-1

“手机号 + 短信验证码”登录的一般流程是：

（1）用户输入手机号并单击“获取验证码”按钮，客户端 App 调用用户微服务获取短信验证码登录接口。

（2）用户微服务在接收到请求后，会在后台随机生成该注册手机的短信验证码，并将其通过第三方短信通道发送至用户手机。

（3）用户在收到短信验证码后，在客户端 App 中输入短信验证码，单击“登录”按钮。用户微服务在收到短信验证码请求后，会先验证用户输入的短信验证码是否与之前发送的一致，如一致则表示用户验证成功。

（4）用户微服务后台逻辑判断该手机号是否已经注册：如果在系统中存在该用户手机号，则表示是老用户登录，直接生成用户会话信息（Token），并将会话信息及用户 ID 通过接口返回给客户端 App；如果在系统中不存在该用户手机号，则表示该手机号为新注册用户，此时用户微服务需要生成一条新的用户 ID，然后生成用户会话信息（Token），并将用户会话信息及新生成的用户 ID 返回至客户端 App。

此后，客户端 App 与后端服务接口的交互都需要携带用户 ID 及用户会话信息（Token），而后端服务也都需要验证用户的登录合法性。

2. 第三方账号授权登录

采用第三方账号授权登录，用户微服务的处理逻辑如图 2-2 所示。

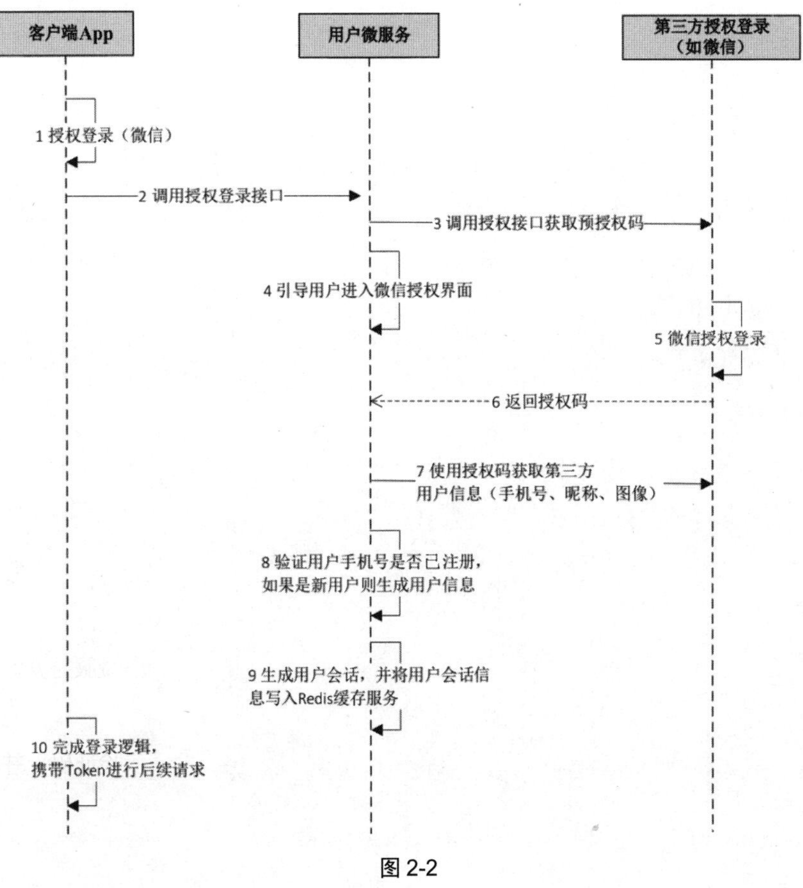

图 2-2

第三方账号授权登录的一般流程是：

（1）在用户选择使用第三方账号授权登录（如微信）后，用户微服务调用对应的第三方账号授权登录平台的开放接口。以微信登录为例，用户微服务会先调用微信开放平台的预授权登录接口获取一个预授权码。

（2）用户微服务会引导用户进入微信的授权界面（目前的流程需要用户使用微信扫描二维码），而在跳转到微信授权界面时，会将之前获取的预授权码携带过去。如果用户在微信授权界面中同意授权登录，则微信会通过 URL 回调的方式将正式的授权码返回给用户微服务对应的接口。

（3）拿到微信正式授权码后，用户微服务就可以通过该授权码再次调用微信服务获取微信用户的注册手机号、昵称、头像等信息。

而后面的流程就与“手机号 + 短信验证码”登录的逻辑基本一致了，用户微服务会校验微信授权返回的手机号是否已经注册：如未注册，则需要生成对应的用户信息，之后生成用户会话信息（Token），并将用户 ID 及 Token 返回至客户端 App，从而完成第三方账号授权登录。

2.2.2 数据库设计

接下来设计用户微服务所需的数据库表结构。

根据逻辑，先定义一张存储用户基本信息的表（user_info），然后设计一张存储短信验证码信息的表（user_sms_code）。

- 代码 这两个表在本书配置资源的“chapter02-user/src/main/resources/db.migration”目录下。

具体表结构设计如下。

1. 用户信息表

创建用户信息表的 SQL 代码（MySQL）如下：

```
create table user_info
(
   id               bigint not null auto_increment,
   user_id          varchar(10) comment '用户 ID',
   nick_name        varchar(30) comment '用户昵称',
   mobile_no        varchar(11) comment '用户注册手机号',
   password         varchar(64) comment '登录密码',
   is_login         int comment '是否登录。0-未登录；1-已登录',
   login_time       timestamp default current_timestamp comment '最近登录时间',
   is_del           int comment '是否注销。0-未注销；1-已注销',
   create_time      timestamp default current_timestamp comment '创建时间',
   primary key (id)
);
alter table user_info comment '用户信息表';
create index idx_ui_user_id on user_info(user_id);
create index idx_ui_mobile_no on user_info(mobile_no);
```

2. 短信验证码信息表

创建短信验证码信息表的 SQL 代码（MySQL）如下：

```
create table user_sms_code
(
  id              bigint not null auto_increment comment 'id',
  mobile_no       varchar(11) comment '用户注册手机号',
  sms_code        varchar(10) comment '短信验证码',
  send_time       timestamp default current_timestamp comment '短信发送信息',
  create_time     timestamp default current_timestamp comment '创建时间',
  primary key (id)
);
alter table user_sms_code comment '短信验证码表';
create index idx_usc_mobile_no on user_sms_code(mobile_no);
```

上述表结构基本满足了用户登录功能所要求的数据库存储需求。

在实际的应用场景中，用户信息的存储可能会更复杂。例如，在实名认证的业务场景中，则还需要存储用户证件信息的字段，根据具体的业务场景进行扩展即可。

2.3 步骤 1：搭建 Spring Boot 应用的工程代码

接下来以实操的形式来一步步构建基于 Spring Cloud 的用户微服务的工程代码结构，并集成系统功能开发所需要的第三方组件——数据库持久层(Dao 层)操作框架 MyBatis、缓存数据库 Redis。

2.3.1 创建 Spring Boot 应用工程

先基于 Maven 构建一个简单的 Spring Boot 工程，然后在此基础之上进行丰富。

1. 创建一个基本的 Maven 工程

（1）使用 IntelliJ IDEA 创建一个 Spring Boot 项目，选择 File→New→Project，并选择 Maven 工程类型，图 2-3 所示。

（2）创建的 Maven 工程的代码结构如图 2-4 所示。

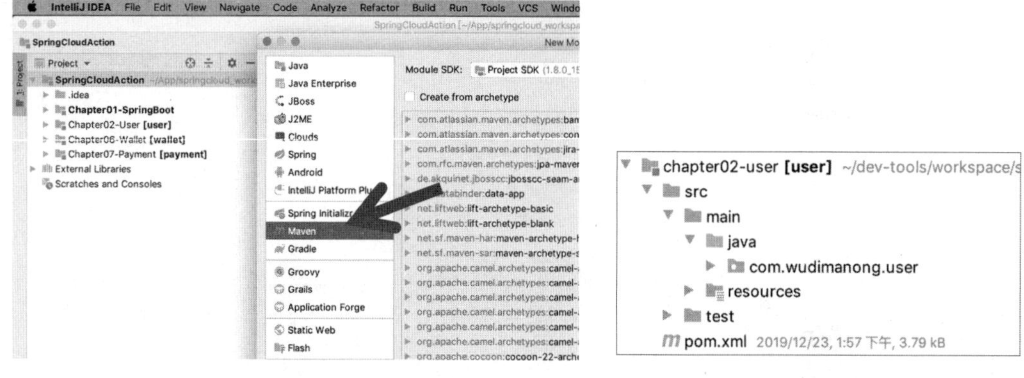

图 2-3　　　　图 2-4

此时的工程还只是一个基本的 Maven 项目结构。接下来向这个项目中添加内容，将其改造为一个能够运行并被访问的 Spring Boot 应用。

2. 将 Maven 项目改造为 Spring Boot 应用

（1）在项目的 pom.xml 文件中，引入 Spring Boot 的依赖。这里引入的版本为 Spring Boot 2.1.5.RELEASE，代码如下：

```
<?xml version="1.0" encoding="UTF-8"?>
<project xmlns="http://maven.apache.org/POM/4.0.0"
        xmlns:xsi="http://www.w3.org/2001/XMLSchema-instance"
        xsi:schemaLocation="http://maven.apache.org/POM/4.0.0
http://maven.apache.org/xsd/maven-4.0.0.xsd">
    <modelVersion>4.0.0</modelVersion>
    <groupId>com.wudimanong.user</groupId>
    <artifactId>user</artifactId>
    <version>1.0-SNAPSHOT</version>
    <!--引入 Spring Boot 的依赖-->
    <parent>
        <groupId>org.springframework.boot</groupId>
        <artifactId>spring-boot-starter-parent</artifactId>
        <version>2.1.5.RELEASE</version>
        <relativePath/> <!-- lookup parent from repository -->
    </parent>
    <dependencies>
        <!--引入 Spring Boot Starter 依赖-->
        <dependency>
            <groupId>org.springframework.boot</groupId>
            <artifactId>spring-boot-starter-web</artifactId>
        </dependency>
        <dependency>
            <groupId>org.springframework.boot</groupId>
```

```
            <artifactId>spring-boot-starter-test</artifactId>
            <scope>test</scope>
        </dependency>
    </dependencies>
    <build>
        <plugins>
            <!--引入 Spring Boot Maven 的编译插件-->
            <plugin>
                <groupId>org.springframework.boot</groupId>
                <artifactId>spring-boot-maven-plugin</artifactId>
            </plugin>
        </plugins>
    </build>
</project>
```

引入依赖后，项目就能够继承 Spring Boot 的核心组件了。

（2）创建 Spring Boot 应用的入口类，代码如下：

```
@SpringBootApplication
public class User {
    public static void main(String[] args) {
        SpringApplication.run(User.class, args);
    }
}
```

至此就构建了一个基于 Spring Boot 框架的 Web 应用了。此时，运行入口程序应用就能够以固定的端口启动一个 Web 服务了。代码的本地运行效果如图 2-5 所示。

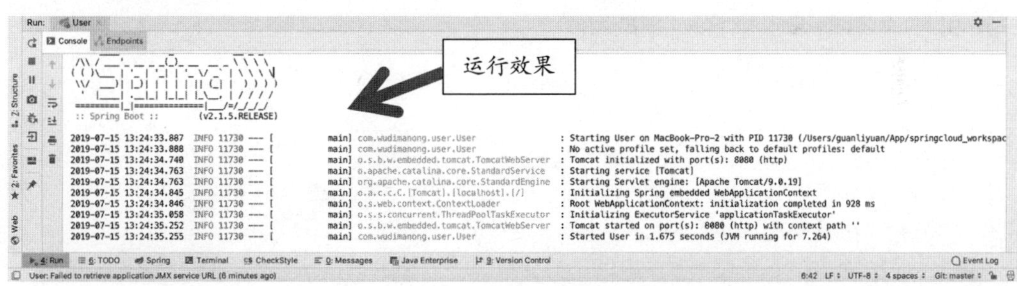

图 2-5

这里默认启动的 HTTP 端口为 8080。

2.3.2 创建应用的配置文件

在前面构建的 Spring Boot 应用中，默认启动的端口为 8080。而在实际的项目中，可以通过配置文件来指定应用的端口。通过配置文件也可以配置数据库的连接等信息。

Spring Boot 项目的配置文件主要有“*.properties”及“*.yml”两种形式。由于“*.yml”配置文件的可读性更强，所以在实际的应用中建议读者多使用“*.yml”配置文件。

这里使用的是“*.yml”形式的配置文件。在工程的“/src/mian/resources”目录下创建基本的配置文件 application.yml，代码如下：

```
spring:
  application:
    name: user
server:
  port: 9091
```

在这个配置文件中，定义了 Spring 应用的名称，以及服务的 HTTP 端口号。随着功能的增加，会在配置文件中逐渐添加数据库的连接信息、Redis 缓存服务器的连接地址等。

2.3.3 集成数据库访问框架 MyBatis

在实际的项目中，经常会用到持久层（Dao 层）框架 MyBatis 来操作数据库。对于大部分应用而言，数据库操作访问都是其业务逻辑的主要组成部分。

在第 1 章中提到，得益于 Spring Boot 框架的自动配置及依赖管理机制，Spring Boot 应用可以很轻松地集成第三方技术组件。所以，在应用中利用这个机制也可以很容易地集成 MyBatis 框架，只需要引入相应的 Starter 依赖即可，步骤如下。

1. 引入 MyBatis Spring Boot 集成组件

（1）在项目的 pom.xml 文件中，引入 MyBatis 框架的依赖（JAR 包）。如下：

```
<dependency>
  <groupId>org.mybatis.spring.boot</groupId>
  <artifactId>mybatis-spring-boot-starter</artifactId>
  <version>2.0.1</version>
</dependency>
```

（2）引入该依赖（JAR 包）后的依赖关系如图 2-6 所示。

（3）从图 2-6 中可以看到，在项目中引入 MyBatis Spring Boot Starter 依赖，也就是引入了 MyBatis 及 JDBC 相关的依赖。MyBatis 自动配置的代码在 mybatis-spring-boot-autoconfigure 中。

```
▼ org.mybatis.spring.boot:mybatis-spring-boot-starter:2.0.1
    org.springframework.boot:spring-boot-starter:2.1.5.RELEASE (omitted for duplicate)
  ▼ org.springframework.boot:spring-boot-starter-jdbc:2.1.5.RELEASE
      org.springframework.boot:spring-boot-starter:2.1.5.RELEASE (omitted for duplicate)
    ▼ com.zaxxer:HikariCP:3.2.0
        org.slf4j:slf4j-api:1.7.26 (omitted for duplicate)
    ▼ org.springframework:spring-jdbc:5.1.7.RELEASE
        org.springframework:spring-beans:5.1.7.RELEASE (omitted for duplicate)
        org.springframework:spring-core:5.1.7.RELEASE (omitted for duplicate)
      ▶ org.springframework:spring-tx:5.1.7.RELEASE
  ▼ org.mybatis.spring.boot:mybatis-spring-boot-autoconfigure:2.0.1
      org.springframework.boot:spring-boot-autoconfigure:2.1.5.RELEASE (omitted for duplicate)
    org.mybatis:mybatis:3.5.1
    org.mybatis:mybatis-spring:2.0.1
```

图 2-6

与基于 Spring Boot 的其他 Starter 组件一样，Spring Boot 应用在启动时会自动扫描在该依赖（JAR 包）中“META-INF/spring.factories”文件中定义的自动配置类，内容如下：

```
# Auto Configure
org.springframework.boot.autoconfigure.EnableAutoConfiguration=\
org.mybatis.spring.boot.autoconfigure.MybatisAutoConfiguration
```

可以看到，在该文件中定义了 MyBatis 的自动配置类 MybatisAutoConfiguration。

如果查看定义的自动配置类的源码，则会发现其中的逻辑实际上与第 1 章中 Spring Boot 自动配置的原理是一致的，例如部分代码如下：

```
    @org.springframework.context.annotation.Configuration
    @ConditionalOnClass({ SqlSessionFactory.class,
SqlSessionFactoryBean.class })
    @ConditionalOnSingleCandidate(DataSource.class)
    @EnableConfigurationProperties(MybatisProperties.class)
    @AutoConfigureAfter(DataSourceAutoConfiguration.class)
    public class MybatisAutoConfiguration implements InitializingBean {
      private static final Logger logger =
LoggerFactory.getLogger(MybatisAutoConfiguration.class);
      private final MybatisProperties properties;
      private final Interceptor[] interceptors;
      private final ResourceLoader resourceLoader;
      private final DatabaseIdProvider databaseIdProvider;
      private final List<ConfigurationCustomizer> configurationCustomizers;
      public MybatisAutoConfiguration(MybatisProperties properties,
          ObjectProvider<Interceptor[]> interceptorsProvider,
          ResourceLoader resourceLoader,
          ObjectProvider<DatabaseIdProvider> databaseIdProvider,
          ObjectProvider<List<ConfigurationCustomizer>>
configurationCustomizersProvider) {
        this.properties = properties;
        this.interceptors = interceptorsProvider.getIfAvailable();
```

```
        this.resourceLoader = resourceLoader;
        this.databaseIdProvider = databaseIdProvider.getIfAvailable();
        this.configurationCustomizers =
configurationCustomizersProvider.getIfAvailable();
      }
      @Override
      public void afterPropertiesSet() {
        checkConfigFileExists();
      }
      private void checkConfigFileExists() {
        if (this.properties.isCheckConfigLocation() && StringUtils.hasText
(this.properties.getConfigLocation())) {
          Resource resource =
this.resourceLoader.getResource(this.properties.getConfigLocation());
          Assert.state(resource.exists(), "Cannot find config location: " +
resource
              + " (please add config file or check your Mybatis configuration)");
        }
      }
    ...
```

关于 MyBatis 自动配置类的细节逻辑这里就不再赘述了。大家只需要理解 MyBatis 框架能够被快速集成的基本原理即可。如果要了解更多的细节逻辑，则需要仔细阅读 MybatisAutoConfiguration 类的源码。

2. 引入“数据库驱动程序 + 连接池”依赖

实际上 MyBatis 框架只是一款数据库操作框架，而对于数据库的基本连接及数据库连接池的管理，则需要在项目中引入相应的数据库连接池依赖，以及针对特定数据库的 JDBC 驱动程序。

在数据库连接池的选择上，这里选择的是业界使用得比较广泛的 Druid 连接池。对于 JDBC 驱动程序，因为本书中所有的实例都使用的是 MySQL 数据库，所以引入 MySQL 的 JDBC 驱动程序即可。代码如下：

```
    <!--引入 Druid 连接池的依赖-->
    <dependency>
       <groupId>com.alibaba</groupId>
       <artifactId>druid</artifactId>
       <version>1.0.28</version>
    </dependency>
    <!--引入 MySQL 的 JDBC 驱动程序-->
    <dependency>
       <groupId>mysql</groupId>
       <artifactId>mysql-connector-java</artifactId>
       <scope>runtime</scope>
```

```
    </dependency>
```

3. 配置项目 MySQL 数据库连接信息

在引入 MyBatis 依赖后，在应用启动时，MyBatis 自动配置类需要初始化数据源。而在这个过程中，需要读取项目中的数据库连接信息，从而初始化数据库连接，否则 Spring Boot 应用将无法启动。

在项目的配置文件 application.yml 中，添加数据库连接信息，具体代码如下：

```
spring:
  application:
    name: user
  datasource:
    url: jdbc:mysql://127.0.0.1:3306/user
    username: root
    password: 123456
    type: com.alibaba.druid.pool.DruidDataSource
    driver-class-name: com.mysql.jdbc.Driver
    separator: //
server:
  port: 9091
```

这里利用第 1 章中给大家演示过的“用 Docker 安装 MySQL 数据库”方式，在本地环境中启动了一个 MySQL 数据库，并创建了一个名为“user”的数据库实例。

至此，完成了在 Spring Boot 应用中集成 MyBatis 框架。在后面业务逻辑开发章节中，就可以使用 MyBatis 框架完成业务逻辑中对数据库的 DML 操作了。

2.3.4 集成缓存数据库 Redis

Redis 是一款应用非常广泛的 NoSQL 缓存数据库，用于提高系统的并发处理性能。在用户微服务中,主要使用 Redis 来缓存用户的会话信息。接下来演示在 Spring Boot 应用中具体集成 Redis 的方法。

一般在公司中会有专人维护一组 Redis 集群服务来供不同的业务线使用。这里出于测试的目的，暂且用 Docker 启动一个本地 Redis 环境。

1. 用 Docker 启动一个本地 Redis 环境

具体的 Docker 命令如下：

```
    docker run -p 6379:6379 -v $PWD/data:/data  -d redis:3.2 -server
--requirepass "123456"  --appendonly yes
```

命令说明如下。

- −p 6379:6379：将容器的 6379 端口映射到主机的 6379 端口。
- −v $PWD/data:/data：将主机当前目录下的 data 目录挂载到容器的“/data”目录中。
- redis−server −−requirepass "123456" −−appendonly yes：在容器中执行 Redis Server 启动命令，打开 Redis 持久化配置并设置访问密码为“123456”。

之后，通过“docker ps”命令查看 Redis 服务的 Docker 容器信息，如图 2−7 所示。

```
bogon:1.0.10-SNAPSHOT guanliyuan$ docker ps
CONTAINER ID    IMAGE       COMMAND                  CREATED         STATUS          PORTS                     NAMES
20206f811aad    redis:3.2   "docker-entrypoint.s…"   9 minutes ago   Up 9 minutes    0.0.0.0:6379->6379/tcp    silly_ganguly
```

图 2-7

也可以通过 Redis 客户端工具来访问 Docker 容器中的 Redis 服务，命令如下：

```
./redis-cli -h 127.0.0.1 -p 6379 -a 123456
```

在连接上后，为了测试 Redis 服务的可用性，可以通过 set/get 命令进行赋/取值的操作，命令如下：

```
127.0.0.1:6379> set a 123
OK
127.0.0.1:6379> get a
"123"
```

在上述命令中，通过 set 命令设置了属性 a 的值，并通过 get 命令获取属性 a 的值。这说明基于 Docker 运行的 Redis 服务是可用的，可以基本满足开发需求了。

2. 集成 Spring Boot 访问组件来实现对 Redis 的访问操作

接下来看看在 Spring Boot 应用中如何通过集成 Redis 访问组件，以实现应用对 Redis 服务的访问操作。

（1）引入 Spring Boot 的 Redis 访问组件。

与在 Spring Boot 应用中集成 MyBatis 框架一样，Spring Boot 针对 Redis 服务的访问也已经提供了现成的 Starter 依赖，只需要引入它即可，代码如下：

```
<dependency>
    <groupId>org.springframework.boot</groupId>
    <artifactId>spring-boot-starter-data-redis</artifactId>
</dependency>
```

引入该组件后，Spring Boot 应用就具备了访问和操作 Redis 的能力。而其基本原理与前面讲述的 MyBatis 框架一致，也是利用了 Spring Boot 框架提供的自动配置能力。具体细节大家可以阅读 Starter 依赖（JAR 包）的源码。

（2）配置 Redis 服务的连接信息。

当然，此时 Redis 的访问并不会生效，还需要在 Spring Boot 的应用配置文件中配置 Redis 服务的连接信息，配置如下：

```
spring:
  application:
    name: user
  datasource:
    url: jdbc:mysql://127.0.0.1:3306/user
    username: root
    password: 123456
    type: com.alibaba.druid.pool.DruidDataSource
    driver-class-name: com.mysql.jdbc.Driver
    separator: //
  redis:
    host: 127.0.0.1
    port: 6379
    password: 123456
server:
  port: 9091
```

完成后，Spring Boot 应用就可以通过 RedisTemplate 很方便地操作和使用 Redis 服务了。关于具体的使用方法，会在用户微服务业务逻辑开发的过程中演示。

2.4 步骤 2：用 Spring Boot 实现业务逻辑

在前面的章节中，设计了业务流程及数据库结构，并且完成了项目开发所需的工程环境。接下来将具体实现用户微服务的业务逻辑。

按照基本的软件分层思想，将分别从服务接口层（Controller 层）、业务层（Service 层）及持久层（Dao 层）逐层进行开发。

2.4.1 定义用户微服务服务接口层（Controller 层）

在本实例中，Spring Cloud 微服务之间的通信协议采用的是 HTTP 协议，所以，微服务对外接口的定义与基于 Spring MVC 的接口编写方式是一致的。

按照本章前面关于用户微服务需求的概述，整个用户微服务所需要的接口见表 2-1。

表 2-1 实现用户微服务所需要的接口

接　口	功　能
/user/getSmsCode	获取短信验证码
/user/loginByMobile	短信验证码登录
/user/loginExit	登录退出

根据上述接口定义，用 Spring MVC 分别予以定义。

1. 编写“获取短信验证码”接口

“获取短信验证码”接口（/user/getSmsCode）的代码如下：

```
@RestController
@RequestMapping("/user")
public class UserController {
    @Autowired
    UserService userServiceImpl;
    @RequestMapping(value = "getSmsCode", method = RequestMethod.POST)
    public Boolean getSmsCode(@RequestParam("reqId") String reqId,
            @RequestParam("mobileNo") String mobileNo) {
        GetSmsCodeReqVo getSmsCodeReqVo =
GetSmsCodeReqVo.builder().reqId(reqId).mobileNo(mobileNo).build();
        boolean result = userServiceImpl.getSmsCode(getSmsCodeReqVo);
        return result;
    }
}
```

在上述代码中，定义了获取短信验证码的 Spring MVC 的服务接口层（Controller 层），该接口并不做具体的业务逻辑，只是将请求封装为对象后交给业务层（Service 层）的方法。这里先定义 UserService 业务层（Service 层）接口，具体的实现在 2.4.2 节实现。

2. 编写“短信验证码登录”接口

下面编写“短信验证码登录”接口（/user/loginByMobile）。继续在 UserController 类中添加 loginByMobile()方法。代码如下：

```
@RequestMapping(value = "loginByMobile", method = RequestMethod.POST)
    public ApiResponse loginByMobile(@RequestParam("reqId") String reqId,
            @RequestParam("mobileNo") String mobileNo,
@RequestParam("smsCode") String smsCode) {
        LoginByMobileReqVo loginByMobileReqVo =
LoginByMobileReqVo.builder().reqId(reqId).mobileNo(mobileNo)
                .smsCode(smsCode).build();
        LoginByMobileResVo loginByMobileResVo =
userServiceImpl.loginByMobile(loginByMobileReqVo);
        return ApiResponse.success(ResultCode.SUCCESS.getCode(),
ResultCode.SUCCESS.getDesc(), loginByMobileResVo);
    }
```

与“1.”小标题中类似，在将请求参数封装为具体对象后，调用业务层（Service 层）接口 UserService 对应的方法，而业务层（Service 层）定义了返回对象类 LoginByMobileResVo，代码如下：

```
@Data
@Builder
public class LoginByMobileResVo implements Serializable {
    private String userId;
    private String accessToken;
}
```

针对上述接口的定义，这里有一些辅助的依赖及公共类需要定义。

（1）引入 lombok 工具包依赖。

在 LoginByMobileResVo 实体类的定义中使用了@Data、@Builder 注解。这两个注解来自目前使用比较广泛的开源工具包 lombok，以减少编写“setter/getter”这样的重复代码。

要使用 lombok 工具包所带来的功能，只需要在项目的 pom.xml 文件中引入其依赖即可，代码如下：

```
<dependency>
    <groupId>org.projectlombok</groupId>
    <artifactId>lombok</artifactId>
</dependency>
```

（2）定义接口层统一 JSON 响应报文对象。

由于 Controller 层需要向网络调用方返回具体的报文，而目前大部分业务接口协议采用 JSON 格式比较多，所以在这个接口中，为了让返回报文更加规范，定义了统一的 JSON 响应报文对象 ApiResponse 类。

该对象的主要作用就是封装接口响应报文信息。其代码如下：

```
@Data
@Builder
public class ApiResponse<T> {
    private String code;
    private String message;
    private T data;
    public ApiResponse() {
    }
    public ApiResponse(String code, String message, T data) {
        this.code = code;
        this.message = message;
        this.data = data;
    }
```

```
    public static <T> ApiResponse<T> success(String code, String message, T
data) {
        ApiResponse<T> response = new ApiResponse<>();
        response.setCode(code);
        response.setMessage(message);
        response.setData(data);
        return response;
    }
}
```

该对象定义了返回报文的响应码、响应信息及具体的返回数据，而最终的响应报文会被 Spring MVC 转化为 JSON 格式展示给调用端。

3. 编写“登录退出”接口

用户登录退出也是所有客户端 App 必须具备的功能。下面定义“登录退出”接口（/user/loginExit）。代码如下：

```
@RequestMapping(value = "loginExit", method = RequestMethod.POST)
public Boolean loginExit(@RequestParam("userId") String userId,
        @RequestParam("accessToken") String accessToken) {
    LoginExitReqVo loginExitReqVo =
LoginExitReqVo.builder().userId(userId).accessToken(accessToken).build();
    boolean result = userServiceImpl.loginExit(loginExitReqVo);
    return result;
}
```

2.4.2 开发用户微服务业务层（Service 层）代码

2.4.1 节定义了用户微服务服务接口的 Controller 层。一般情况下，Controller 层不会承担太多的业务逻辑，所以，真正实现业务功能的还是业务层（Service 层）。

而在上述的接口中，已经定义了业务层（Service 层）的业务接口 UserService，并在该接口中定义了具体的业务层（Service 层）方法，代码如下：

```
public interface UserService {
    //获取短信验证码
    boolean getSmsCode(GetSmsCodeReqVo getSmsCodeReqVo);
    //短信登录
    LoginByMobileResVo loginByMobile(LoginByMobileReqVo
loginByMobileReqVo);
    //登录退出
    boolean loginExit(LoginExitReqVo loginExitReqVo);
}
```

接下来，具体实现以上业务层（Service 层）的方法：创建 UserServiceImpl 实现类，并逐一实行上述方法。

1. 实现“获取短信验证码”接口的业务层（Service 层）方法

实现“获取短信验证码”接口的业务层（Service 层）方法：

```
@Service
@Slf4j
public class UserServiceImpl implements UserService {
@Autowired
    UserSmsCodeDao userSmsCodeDao;
    @Override
    public boolean getSmsCode(GetSmsCodeReqVo getSmsCodeReqVo) {
        //随机生成 6 位短信验证码
        String smsCode = String.valueOf((int) (Math.random() * 100000 + 1));
        //真实场景中，这里需要调用短信平台接口
        //存储用户短信验证码信息至短信验证码信息表
        UserSmsCode userSmsCode =
UserSmsCode.builder().mobileNo(getSmsCodeReqVo.getMobileNo()).smsCode(smsCod
e)
                .sendTime(new Timestamp(new
Date().getTime())).createTime(new Timestamp(new Date().getTime()))
                .build();
        userSmsCodeDao.insert(userSmsCode);
        return true;
    }
}
```

由于业务层（Service 层）方法需要操作数据库，所以先提前定义了用户短信信息表在持久层（Dao 层）的接口（参见 2.4.3 节）。

而实际的逻辑是：系统随机生成 6 位短信验证码，并通过短信通道发送给用户手机；之后，系统本地存储短信验证码信息，用于用户登录时的校验比对。

2. 实现“短信验证码登录”接口的业务层（Service 层）方法

继续根据 2.2.1 节中业务逻辑的设计，实现“短信验证码登录”接口的业务层（Service 层）方法。代码如下：

```
@Service
@Slf4j
public class UserServiceImpl implements UserService {
    @Autowired
    UserSmsCodeDao userSmsCodeDao;
    @Autowired
    UserInfoDao userInfoDao;
    @Autowired
```

```
    RedisTemplate redisTemplate;
    @Override
    public LoginByMobileResVo loginByMobile(LoginByMobileReqVo
loginByMobileReqVo) throws BizException {
        //1 验证短信验证码是否正确
        UserSmsCode userSmsCode =
userSmsCodeDao.selectByMobileNo(loginByMobileReqVo.getMobileNo());
        if (userSmsCode == null) {
            throw new BizException(-1, "验证码输入错误");
        } else if
(!userSmsCode.getSmsCode().equals(loginByMobileReqVo.getSmsCode())) {
            throw new BizException(-1, "验证码输入错误");
        }
        //2 判断用户是否已注册
        UserInfo userInfo =
userInfoDao.selectByMobileNo(loginByMobileReqVo.getMobileNo());
        if (userInfo == null) {
            //随机生成用户 ID
            String userId = String.valueOf((int) (Math.random() * 100000 + 1));
            userInfo =
UserInfo.builder().userId(userId).mobileNo(loginByMobileReqVo.getMobileNo())
.isLogin("1")
                    .loginTime(new Timestamp(new
Date().getTime())).createTime(new Timestamp(new Date().getTime()))
                    .build();
            //完成系统默认注册流程
            userInfoDao.insert(userInfo);
        } else {
            userInfo.setIsLogin("1");
            userInfo.setLoginTime(new Timestamp(new Date().getTime()));
            userInfoDao.updateById(userInfo);
        }
        //3 生成用户会话信息
        String accessToken =
UUID.randomUUID().toString().toUpperCase().replaceAll("-", "");
        //将用户会话信息存储至 Redis 服务
        redisTemplate.opsForValue().set("accessToken", userInfo, 30,
TimeUnit.DAYS);
        //4 封装响应参数
        LoginByMobileResVo loginByMobileResVo =
LoginByMobileResVo.builder().userId(userInfo.getUserId())
                .accessToken(accessToken).build();
        return loginByMobileResVo;
    }
}
```

登录验证及注册总体上分为 4 步：

（1）判断短信验证码是否正确。如果正确，则进行后续逻辑；如果不正确，则抛出异常，提示用户短信验证码输入错误。

（2）判断用户是否已经注册。如果未注册，则生成用户信息并保存。

（3）将用户信息及随机生成的用户会话信息（Token）存入 Redis 缓存服务器，并设置有效时间为 30 天。

（4）将会话信息及用户 ID 返回至调用端，调用端后续通过 Token 与服务端进行交互。

3. 实现“登录退出”接口的业务层（Service 层）方法

下面继续实现“登录退出”接口的业务层（Service 层）方法。代码如下：

```
@Service
@Slf4j
public class UserServiceImpl implements UserService {
    @Autowired
    RedisTemplate redisTemplate;
    @Override
    public boolean loginExit(LoginExitReqVo loginExitReqVo) {
        try {
            redisTemplate.delete(loginExitReqVo.getAccessToken());
            return true;
        } catch (Exception e) {
            log.error(e.toString() + "_" + e);
            return false;
        }
    }
}
```

会话登录退出比较简单——根据 Token 删除存储在 Redis 中的用户信息。

2.4.3 开发 MyBatis 持久层（Dao 层）组件

在业务层（Service 层）方法的实现过程中，需要使用数据库操作组件。本节就针对不同表的操作，编写基于 MyBatis 的数据库操作代码。

1. 开发用户短信信息表的持久层（Dao 层）组件 UserSmsCodeDao

在“获取短信验证码”接口的业务层（Service 层）实现中，需要将短信验证码信息进行存储。所以，需要在 UserSmsCodeDao 接口中实现一个 insert()方法。另外，在用户登录时，需要根据手机号获取最新的短信验证码信息进行校验，所以还需要一个 selectByMobileNo()方法。步骤如下：

（1）定义持久层（Dao 层）接口类。

先在代码包路径中创建一个名称为“dao”的包，然后在该包路径中创建 MyBatis 持久层（Dao 层）接口 UserSmsCodeDao，并定义相应接口方法。代码如下：

```
@Mapper
public interface UserSmsCodeDao {

    int insert(UserSmsCode userSmsCode);
    UserSmsCode selectByMobileNo(String mobileNo);
}
```

（2）编写对应的 MyBatis SQL 映射文件。

定义 MyBatis SQL 映射的 XML 文件，在工程代码 resources 目录中创建 mybatis 目录作为存放 SQL 映射文件专用目录，并创建针对用户短信信息表的 MyBatis SQL 映射 XML 文件 user_sms_code.xml，代码如下：

```
<?xml version="1.0" encoding="UTF-8" ?>
<!DOCTYPE mapper PUBLIC "-//mybatis.org//DTD Mapper 3.0//EN"
"http://mybatis.org/dtd/mybatis-3-mapper.dtd" >
<mapper namespace="com.wudimanong.user.dao.UserSmsCodeDao">
    <sql id="Base_Column_List">
        id, mobile_no, sms_code, send_time, create_time
    </sql>
    <insert id="insert"
parameterType="com.wudimanong.user.entity.UserSmsCode">
        insert into user_sms_code(mobile_no, sms_code, send_time,
create_time)
        values (#{mobileNo}, #{smsCode}, #{sendTime}, #{createTime})
    </insert>
    <select id="selectByMobileNo"
resultType="com.wudimanong.user.entity.UserSmsCode">
        SELECT
        <include refid="Base_Column_List"/>
        FROM user_sms_code
        WHERE mobile_no=#{mobileNo} order by create_time desc limit 1
    </select>
</mapper>
```

（3）在 Maven 项目文件 pom.xml 中增加对 XML 文件的包含。

在项目中新增加 XML 文件，需要在 pom.xml 中增加对 XML 文件的包含，代码如下：

```
<build>
    <plugins>
        <!--引入 Spring Boot Maven 编译插件-->
```

```
        <plugin>
            <groupId>org.springframework.boot</groupId>
            <artifactId>spring-boot-maven-plugin</artifactId>
        </plugin>
    </plugins>
    <resources>
        <resource>
            <directory>${basedir}/src/main/resources</directory>
            <filtering>true</filtering>
            <includes>
                <include>**/application*.yml</include>
                <include>**/application*.yaml</include>
                <include>**/bootstrap.yml</include>
                <include>**/bootstrap.yaml</include>
                <include>**/*.xml</include>
            </includes>
        </resource>
    </resources>
</build>
```

2. 开发用户信息表的持久层（Dao 层）组件 UserInfoDao

本 MyBatis 持久层（Dao 层）接口涉及 selectByMobileNo()、insert()及 updateById()这 3 个数据库操作方法。步骤如下。

（1）定义持久层（Dao 层）接口类。

与“1.”小标题中类似，在对应的工程代码目录中创建 UserInfoDao 接口。代码如下：

```
@Mapper
public interface UserInfoDao {
    UserInfo selectByMobileNo(String mobikeNo);
    int insert(UserInfo userInfo);
    int updateById(UserInfo userInfo);
}
```

（2）编写对应的 MyBatis SQL 映射文件。

继续实现 MyBatis SQL 映射 XML 文件：在“resources/mybatis”目录中创建 user_info.xml 文件，并具体实现 SQL 映射方法，代码如下：

```
<?xml version="1.0" encoding="UTF-8" ?>
<!DOCTYPE mapper PUBLIC "-//mybatis.org//DTD Mapper 3.0//EN"
"http://mybatis.org/dtd/mybatis-3-mapper.dtd" >
<mapper namespace="com.wudimanong.user.dao.UserInfoDao">
    <sql id="Base_Column_List">
```

```
            id, user_id, nick_name, mobile_no,
password,is_login,login_time,is_del,create_time
        </sql>
        <insert id="insert"
parameterType="com.wudimanong.user.entity.UserInfo">
            insert into user_info(user_id, nick_name, mobile_no,
password,is_login,login_time,is_del,create_time)
            values (#{userId}, #{nickName}, #{mobileNo},
#{password},#{isLogin},#{loginTime},#{isDel},#{createTime})
        </insert>
        <select id="selectByMobileNo"
resultType="com.wudimanong.user.entity.UserInfo">
            SELECT
            <include refid="Base_Column_List"/>
            FROM user_info
            WHERE mobile_no=#{mobileNo}
        </select>
        <update id="updateById"
parameterType="com.wudimanong.user.entity.UserInfo">
            update user_info
            <set>
                <if test="userId != null">
                    user_id = #{userId,jdbcType=VARCHAR},
                </if>
                <if test="nickName != null">
                    nick_name = #{nickName,jdbcType=VARCHAR},
                </if>
                <if test="mobileNo != null">
                    mobile_no = #{mobileNo,jdbcType=VARCHAR},
                </if>
                <if test="password != null">
                    password = #{password,jdbcType=VARCHAR},
                </if>
                <if test="isLogin != null">
                    is_login = #{isLogin,jdbcType=VARCHAR},
                </if>
                <if test="loginTime != null">
                    login_time = #{loginTime,jdbcType=TIMESTAMP},
                </if>
                <if test="isDel != null">
                    is_del = #{isDel,jdbcType=VARCHAR},
                </if>
                <if test="createTime != null">
                    create_time = #{createTime,jdbcType=TIMESTAMP},
                </if>
```

```
        </set>
        where id = #{id,jdbcType=BIGINT}
    </update>
</mapper>
```

至此就完成了基于 Spring Boot 框架的用户微服务业务功能开发。此时可以通过 Postman 进行接口调用，来验证逻辑的正确性。

2.5 步骤 3：将 Spring Boot 应用升级为 Spring Cloud 微服务

前面实际上已经基于 Spring Boot 框架完成了用户微服务的主要业务功能。但目前，该应用还不是一个真正意义上的微服务，而只是一个普通的 Spring Boot 应用。

下面将该 Spring Boot 应用升级为基于 Spring Cloud 的微服务。

2.5.1 部署服务注册中心 Consul

在对 Spring Boot 应用进行 Spring Cloud 微服务化改造之前，先来认识下“服务注册中心”，它是 Spring Cloud 微服务体系中最重要的组件。

1. 了解“服务注册中心”组件

目前 Spring Cloud 微服务体系中比较主流的服务注册中心组件有 Consul、Eureka 和 Nacos。

现在使用 Spring Cloud 微服务技术栈优先选择的注册中心组件是 Consul 或 Nacos。本书以 Consul 为例。

Consul 是一款开源的、使用 Go 语言编写的服务注册中心，它内置了服务注册/发现、分布式一致性协议实现、健康性检查、Key/Value 存储、多数据中心等多个方案。

在使用 Consul 作为微服务注册中心的架构中，一旦作为关键组件的 Consul 宕机，则可能造成所有微服务系统的宕机。所以在生产环境中，Consul 都会以集群部署的方式实现高可用。

2. 部署一个单机版的 Consul 服务

为了便于开发和测试，这里先以 Docker 的方式部署一个单机版的 Consul 服务。

（1）通过 Docker 部署 Consul，命令如下：

```
//创建映射目录
mkdir -p /tmp/consul/{conf,data}
//Docker 启动 Consul 容器
```

```
docker run --name consul -p 8500:8500 -v /tmp/consul/conf/:/consul/conf/ -v
/tmp/consul/data/:/consul/data/ -d consul
```

（2）通过 Docker 命令查看容器运行情况，命令如下：

```
docker ps
```

（3）演示 Consul 控制台效果。

如果 Consul 容器启动成功，则可以打开控制台地址进行查看。在浏览器中输入地址“http://127.0.0.1:8500”。Consul 部署成功后的控制台界面如图 2-8 所示。

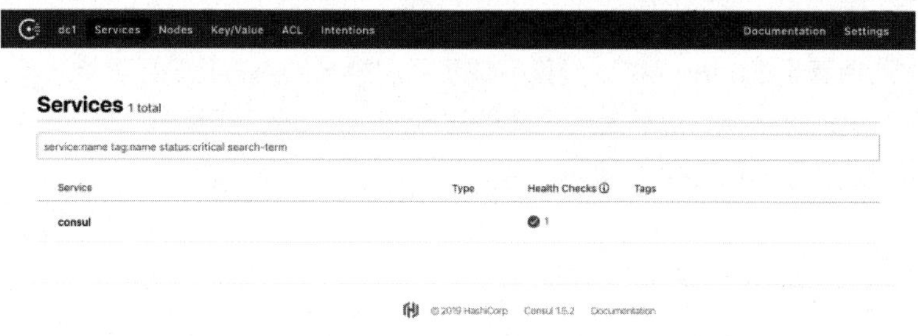

图 2-8

2.5.2 对 Spring Boot 应用进行微服务改造

Spring Cloud 是基于 Spring Boot 框架的，所以，要将一个 Spring Boot 应用改造为基于 Spring Cloud 的微服务应用是一件相对简单的事情。

1. 引入 Spring Cloud 父依赖

将 Spring Boot 应用的 Maven 父依赖改成 Spring Cloud 父依赖。代码如下：

```
<!--引入 Spring Cloud 父依赖-->
<parent>
    <groupId>org.springframework.cloud</groupId>
    <artifactId>spring-cloud-starter-parent</artifactId>
    <version>Greenwich.SR1</version>
    <relativePath/>
</parent>
```

这里将原先 Spring Boot 工程中基于 spring-boot-starter-parent 父依赖改为 spring-cloud-starter-parent 父依赖。

在第 1 章的内容中分析过 Spring Cloud 框架是继承 Spring Boot 的，所以，引入 Spring Cloud 框架的父依赖，实际是引入 Spring Boot 相关的依赖。

在 Spring Cloud 的版本选择上，本书使用的是 Greenwich.SR1，它是基于 Spring Boot 2.1.3.RELEASE 版本的。

2. 引入 Spring Cloud 微服务的核心依赖

接下来，在项目 pom.xml 文件中引入构建 Spring Cloud 微服务的核心依赖。代码如下：

```
<!--Spring Cloud 核心依赖-->
<!--微服务健康性检查-->
<dependency>
    <groupId>org.springframework.boot</groupId>
    <artifactId>spring-boot-starter-actuator</artifactId>
</dependency>
<!--基于 Consul 的服务注册/发现依赖-->
<dependency>
    <groupId>org.springframework.cloud</groupId>
    <artifactId>spring-cloud-starter-consul-discovery</artifactId>
</dependency>
<!--基于 Hystrix 的服务限流熔断依赖-->
<dependency>
    <groupId>org.springframework.cloud</groupId>
    <artifactId>spring-cloud-starter-netflix-hystrix</artifactId>
</dependency>
<!--Spring Cloud 公共代码-->
<dependency>
    <groupId>org.springframework.cloud</groupId>
    <artifactId>spring-cloud-commons</artifactId>
</dependency>
```

在上方的代码中，除引入 Spring Boot 的核心依赖外，还引入了 Spring Cloud 微服务的几个核心依赖，主要包括微服务健康性检查、Consul 的服务注册/发现依赖、Hystrix 的服务限流熔断依赖，以及 Spring Cloud 公共代码。

3. 新增 Spring Cloud 微服务的相关配置

在项目代码的 resources 目录中，新建一个基础性配置文件 bootstrap.yml，该配置文件的功能与 application.yml 类似——用于在 Spring Cloud 微服务架构中配置一些与微服务相关的基础性信息，例如服务注册中心相关的配置，以及 Hystrix 熔断降级等。

新建配置文件 bootstrap.yml 不是必须的步骤，也可以将微服务的基本信息配置到 application.yml 中。只是在 Spring Cloud 微服务实践中，大家比较倾向于使用 bootstrap.yml 来配置应用的不区分环境的基础性信息，使用 application.yml 来配置区分环境的信息。

具体步骤如下。

（1）创建 bootstrap.yml 配置文件。

创建基于 YAML 格式的配置文件，代码如下：

```
spring:
  application:
    name: user
  profiles:
    active: debug
  cloud:
    consul:
      discovery:
        preferIpAddress: true
        instance-id:
${spring.application.name}:${spring.cloud.client.ipAddress}:${spring.applica
tion.instance_id:${server.port}}:@project.version@
        healthCheckPath: /actuator/health
server:
  port: 9090
```

在 bootstrap.yml 文件中，将原先在 application.yml 中的应用名称及 HTTP 端口信息移过来了，并且配置了 Consul 服务注册的相关信息，例如 instance-id 及 healthCheckPath 服务健康性检查的地址。

Spring Boot 会默认加载 application.yml 配置文件，但不会加载 bootstrap.yml 配置文件。所以，还需要在 pom.xml 中添加 Maven 资源相关的配置。

（2）在 pom.xml 中添加 Maven 资源相关的配置。

代码如下：

```
<build>
    <plugins>
        <!--引入 Spring Boot Maven 编译插件-->
        <plugin>
            <groupId>org.springframework.boot</groupId>
            <artifactId>spring-boot-maven-plugin</artifactId>
        </plugin>
    </plugins>
    <resources>
        <resource>
            <directory>${basedir}/src/main/resources</directory>
```

```
            <filtering>true</filtering>
            <includes>
                <include>**/application*.yml</include>
                <include>**/application*.yaml</include>
                <include>**/bootstrap.yml</include>
                <include>**/bootstrap.yaml</include>
            </includes>
        </resource>
    </resources>
</build>
```

此时，bootstrap.yml 文件中的配置就能够被 Spring Boot 应用读取了。

4. 修改 Spring Boot 应用入口类，开启微服务自动注册/发现机制

经过以上几个步骤的操作，此时 Spring Boot 应用已经具备成为 Spring Cloud 微服务的必要前提条件了。

接下来，在入口类上通过使用 Spring Cloud 提供的注解@EnableDiscoveryClient，正式将 Spring Boot 应用升级成为 Spring Cloud 微服务。代码如下：

```
@EnableDiscoveryClient
@SpringBootApplication
public class User {
    public static void main(String[] args) {
        SpringApplication.run(User.class, args);
    }
}
```

通过@EnableDiscoveryClient 注解，微服务会自动根据在前面引入的“spring-cloud-starter-consul-discovery”依赖，将微服务注册到在应用启动时指定的 Consul 中去。

2.5.3 将 Spring Cloud 微服务注入服务注册中心 Consul

经过前面的准备，现在就可以将用户微服务正式注入服务注册中心 Consul，实现真正意义上的微服务架构了。

1. 连接本地 Consul 服务

为了方便开发测试，可以通过 IDEA 工具启动微服务，并将其注册到 2.5.1 节所部署的本地 Consul 中去。步骤如下：

（1）通过 IDEA 运行用户微服务。

在开发集成环境中运行用户微服务，观察启动日志。可以看到，应用已经成功注册到服务注册中心 Consul 中了，如图 2-9 所示。

```
21:39.220  INFO 13348 --- [           main] c.netflix.config.DynamicPropertyFactory  : DynamicPropertyFactory is initialized with configuration sources: com.netflix.config.ConcurrentComposit
21:40.157  WARN 13348 --- [           main] c.n.c.sources.URLConfigurationSource     : No URLs will be polled as dynamic configuration sources.
21:40.158  INFO 13348 --- [           main] c.n.c.sources.URLConfigurationSource     : To enable URLs as dynamic configuration sources, define System property archaius.configurationSource.ac
21:40.348  INFO 13348 --- [           main] o.s.s.concurrent.ThreadPoolTaskExecutor  : Initializing ExecutorService 'applicationTaskExecutor'
21:40.773  INFO 13348 --- [           main] o.s.s.c.ThreadPoolTaskScheduler          : Initializing ExecutorService 'catalogWatchTaskScheduler'
21:40.824  WARN 13348 --- [           main] o.s.c.n.core.CoreAutoConfiguration       : This module is deprecated. It will be removed in the next  . Please use spring-cloud-netfli
21:40.830  INFO 13348 --- [           main] o.s.b.a.e.web.EndpointLinksResolver      : Exposing 2 endpoint(s) beneath base path '/actuator'
21:40.937  INFO 13348 --- [           main] o.s.b.w.embedded.tomcat.TomcatWebServer  : Tomcat started on port(s): 9090 (http) with context pa
21:40.946  INFO 13348 --- [           main] o.s.c.c.s.ConsulServiceRegistry          : Registering service with consul: NewService{id='user-spring-   -client-ipAddress-9090-1-0-SNAPSHOT',
21:40.966  INFO 13348 --- [           main] com.wudimanong.user.User                 : Started User in 3.722 seconds (JVM running for 9.209)
21:41.454  INFO 13348 --- [-192.168.50.178] o.a.c.c.C.[Tomcat].[localhost].[/]       : Initializing Spring DispatcherServlet 'dispatcherServlet'
21:41.454  INFO 13348 --- [-192.168.50.178] o.s.web.servlet.DispatcherServlet        : Initializing Servlet 'dispatcherServlet'
`com.mysql.jdbc.Driver'. This is deprecated. The new driver class is `com.mysql.cj.jdbc.Driver'. The driver is automatically registered via the SPI and manual loading of the driver class is
21:41.462  INFO 13348 --- [-192.168.50.178] o.s.web.servlet.DispatcherServlet        : Completed initialization in 8 ms
21:41.484  INFO 13348 --- [-192.168.50.178] com.alibaba.druid.pool.DruidDataSource   : {dataSource-1} inited
21:42.008  INFO 13348 --- [nio-9090-exec-1] io.lettuce.core.EpollProvider            : Starting without optional epoll library
21:42.009  INFO 13348 --- [nio-9090-exec-1] io.lettuce.core.KqueueProvider           : Starting without optional kqueue library
```

图 2-9

（2）通过 Consul 控制台观察微服务注册信息。

打开 2.5.1 节部署的 Consul 控制台，可以看到，Consul 控制台已经能够显示 User 微服务信息了，如图 2-10 所示。

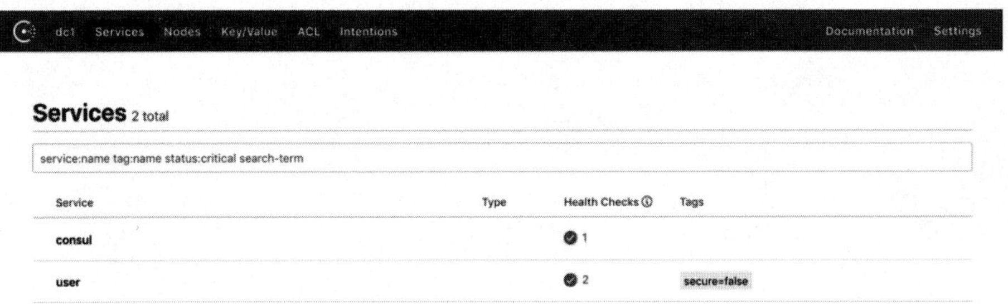

图 2-10

这样，其他微服务就可以通过 Consul 对 User 微服务实现发现，并通过 FeignClient 实现接口访问了。

2. 连接远程 Consul 服务

至此可能有读者会有疑惑，在程序运行的过程中并没有具体指定 Consul 服务的地址，那么应用是怎么连接上 Consul 的呢?

实际上，在没有配置任何 Consul 地址的情况下，Spring Cloud 默认连接本地 IP 地址的 8500 端口。如果需要连接远程 Consul 服务，则可以通过参数进行指定，例如：

```
spring:
  application:
    name: user
  profiles:
    active: debug
```

```
      cloud:
        consul:
          host: 127.0.0.1
          port: 8500
          discovery:
            preferIpAddress: true
            instance-id:
${spring.application.name}:${spring.cloud.client.ipAddress}:${spring.applica
tion.instance_id:${server.port}}:@project.version@
            healthCheckPath: /actuator/health
    server:
      port: 9090
```

在实际的生产部署实践中，出于灵活性的考虑，一般不在代码中固定配置 Consul 服务的地址，而是在启动时根据系统环境来动态配置。

2.6 本章小结

本章以实际的业务场景为例完成了一个用户微服务的实例，介绍了如何构建基本的 Spring Boot 应用，以及如何将 Spring Boot 应用升级为 Spring Cloud 微服务。在这个过程中，演示了如何集成 MyBatis、Redis 等常用的组件。这些组件的集成方式，与第 1 章中 Spring Boot 框架的集成方法是完全一致的。

本章还介绍了 Spring Cloud 微服务体系中“服务注册中心”这个最重要的依赖，并以主流的 Consul 为例，通过 Docker 的方式完成了一个开发版本的部署，最终将用户微服务注册到 Consul 服务中心中。

第 3 章

【实例】SSO 授权认证系统

——用“Spring Security + Spring Cloud Gateway”构建 OAuth 2.0 授权认证服务

大家应该都有过这样的体验：在登录一些不常用的网站时，可以使用常用的 QQ、微信等第三方账号进行授权登录。在公司内部一般会存在多种不同的软件系统（例如 GitLab 代码管理平台、财务报销系统、人事系统等），也可以提供类似的授权登录功能，从而实现员工利用账号登录一次就能访问多个系统的效果。要实现这样的效果，需要一套独立的账号授权认证系统——SSO（SingleSignOn，单点登录）授权认证系统。

本章将以 OAuth 2.0 授权认证流程为基础，实现一套基于 Spring Cloud 微服务体系的 SSO 授权认证系统。

因为考虑到 SSO 授权认证系统在某些情况下会涉及外部服务对内部微服务的访问，所以，从微服务访问安全的角度考虑，本章将引入 Spring Cloud 微服务体系中另一个比较重要的组件——Spring Cloud Gateway 微服务网关。

此外，由于在网关服务与具体的微服务（如资源微服务）之间还存在着调用关系，所以，为了实现 Spring Cloud 微服务的远程服务访问，还将引入“Feign + Ribbon”组合，以实现 Spring Cloud 微服务远程通信、负载均衡调用等功能。

通过本章，读者将学习到以下内容：

- OAuth 2.0 协议的授权认证流程。
- 引入 Spring Security Oauth 2 组件，实现 OAuth 2.0 授权认证服务。

- 编写 FeignClient 微服务调用接口。
- 构建基于 Spring Cloud Gateway 的微服务网关。
- 使用“Feign + Ribbon”组件，实现微服务的远程负载均衡调用。

3.1 功能概述

为了实现企业内部网络信息的安全访问、打通内部信息孤岛，如今稍有规模的 IT 企业都会实行内部员工账号的集中认证、访问授权，以及访问日志定期审核。具体来说，就是提供一套独立的 SSO 授权认证系统，为内部各业务系统提供集中认证、授权登录的服务。

本章要实现的系统主要包括以下几个方面功能。

（1）集中式的账户系统：实现员工登录账号的集中管理，确保公司内部不同系统之间账户信息及时、准确地同步，为获得授权的内部业务系统提供账户信息查询的服务。

（2）统一认证服务：提供基于 OAuth 2.0 协议的统一授权认证服务。

（3）授权管理：实现应用级别的授权管理，提供门户入口，便于应用管理者注册应用，以使用 SSO 服务。另外，支持对权限及角色的编辑管理，可以对用户进行角色指派、编组等，从而实现应用级别的授权管理。

（4）审计服务：记录授权认证的日志，实现用户访问记录的快速查询，从而为定期的 IT 安全审计提供支撑。

3.2 系统设计

接下来从系统设计的角度，对 SSO 授权认证系统的逻辑进行设计，主要从 OAuth 2.0 授权认证流程、系统结构设计和数据库设计这 3 个方面进行阐述。

3.2.1 OAuth 2.0 授权认证流程

OAuth 2.0 是一种允许第三方 App 使用资源所有者颁发的令牌，对资源实现有限访问的一种授权认证协议。例如，通过 QQ、微信授权登录某个 App，相当于 QQ、微信允许该 App 在经过用户授权后，通过 QQ、微信授权认证服务颁发的令牌有限地访问 QQ、微信的用户信息。

在 OAuth 2.0 协议中，第三方 App 获取的令牌就是一个表示特定范围、生存周期，以及访问权限的字符串令牌，即通常所说的 Token 信息。

1. 四种角色

在 OAuth 2.0 协议中定义了以下四种角色。

（1）Resource owner（资源拥有者）。

资源拥有者即有权对受保护资源授予访问权限的实体。例如，在通过微信账号登录其他第三方 App 的过程中，微信用户就是资源拥有者（也被称为最终用户）。在这个过程中一般需要微信用户进行手动确认。

（2）Resource Server（资源服务）。

资源服务即承载受保护资源的服务。它接收第三方 App 应用通过访问令牌对受保护资源发起的请求，并予以响应。

在实际场景中，资源服务与授权认证服务可以是同一服务，也可以是不同服务，这取决于 SSO 授权认证系统的架构设计。以微信账号授权登陆第三方 App 为例，资源服务就是存储微信用户信息的服务器。

（3）Client（客户端）。

客户端即需要被授权登录的第三方 App。在通过微信账号登录第三方 App 的过程中，第三方 App 就是客户端。

（4）Authorization Server（授权认证服务）。

在整个 OAuth 2.0 授权认证协议中，授权认证服务用于处理授权认证请求，从而将第三方 App 与受保护资源进行隔离。

2. 授权认证过程

前面基本介绍了 OAuth 2.0 认证流程中的不同角色，那么这些角色是如何实现一次完整的授权认证过程的呢?

在 OAuth 2.0 标准协议中，定义了四种不同的客户端授权模式：①授权码模式（Authorization Code）；②简化模式（Implicit）；③密码模式（Resource Owner Password Credentials）；④客户端模式（Client Credentials）。

不同客户端授权模式的处理流程也不同。但在实践中，应用最广泛的是授权码模式。授权码模式也是这四种模式中功能最完整、流程最严密的授权认证模式。

在授权码模式中，授权码并不是客户端直接从资源服务获取的，而是从授权认证服务获取的。授权认证的过程是：资源所有者直接通过授权认证服务进行身份认证。这样就避免了资源所有者与客户端共享身份凭证。

本实例将使用授权码模式作为 OAuth 2.0 授权认证服务的实现方式。

以使用微信授权登录微博为例，OAuth 2.0 授权码模式系统的运行流程如图 3-1 所示。

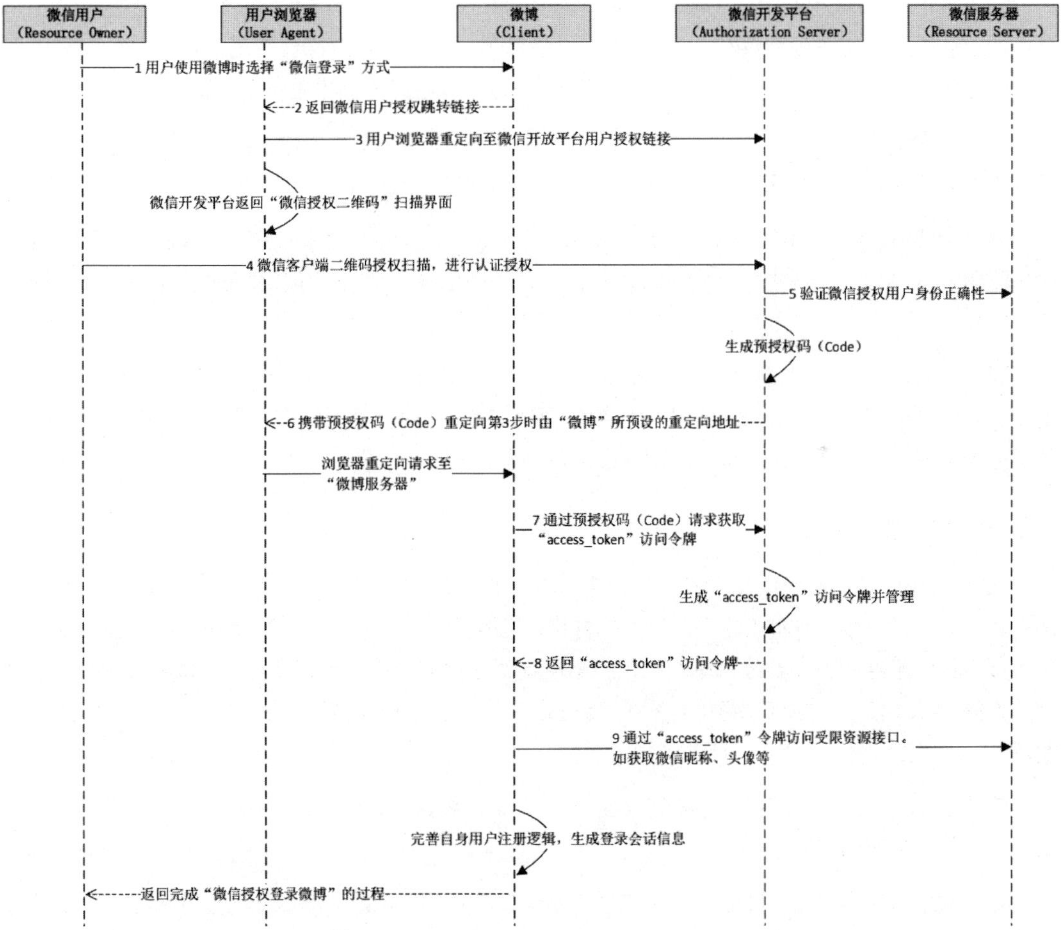

图 3-1

具体说明如下：

（1）用户在登录微博 App 时选择“微信登录”按钮，之后微博 App 会将用户请求以 URL 重定向的方式跳转至微信开放平台的第三方账号授权登录界面。

（2）用户打开微信客户端，使用“扫一扫”功能扫描此时浏览器展示的微信授权二维码，完成微信用户授权登录操作。

具体的用户授权形式取决于授权认证服务具体的实现，例如 QQ 授权会要求用户输入用户账号及密码。

（3）微信服务端验证用户身份信息的正确性，如正确，则用户授权登录微博 App 成功。此时，微信授权认证服务生成预授权码（Code），并携带该预授权码将浏览器重定向至第（1）步——微博 App 在生成“微信开放平台”授权跳转链接时设置的本系统回调地址（callBackUrl）。

（4）用户浏览器携带该预授权码再次请求微博 App。在接收到此预授权码后，微博 App 后台会再调用微信开放平台的授权认证服务接口，换取正式的访问令牌（access_token）。

（5）微博 App 在获取微信授权访问令牌（access_toke）后，用户的授权认证流程即完成。之后，微博 App 会通过该令牌（access_token）调用微信开放平台的授权认证服务接口，获取用户的微信头像、昵称、手机号等信息，并完成自身的用户注册逻辑及会话信息。

回到本实例——SSO 授权认证系统，在企业内部实现 SSO 单点登录时，并不需要各个业务系统（Client 端系统）都建立自己的用户体系，而是实行账号集中管理，统一授权，各个业务系统只需要记录用户访问凭证，实现统一的用户会话逻辑即可。

3.2.2 系统结构设计

根据 OAuth 2.0 授权认证系统的流程特点，本实例将“授权认证服务”与“资源服务”进行分离设计。这样，在系统结构上也更加清晰，并且从 OAuth 2.0 的交互细节上看——Client 端系统在接入授权认证服务时，需要经过多次的用户浏览器页面跳转。由于授权认证服务本身是具备安全控制能力的，所以 Client 端系统可以更安全地通过 HTTP 直接访问授权认证服务，实现用户授权认证逻辑。

除此之外，Client 端系统在完成授权认证获取访问令牌（access_token）后，会访问“资源服务”以获取授权用户的信息。

但从微服务架构来看，Client 端系统很可能是来自外部的系统，所以 Client 端系统不能直接访问资源微服务，而是通过服务网关（Api Gateway）来访问资源微服务，并由服务网关实现微服务入口请求的统一管控。

OAuth 2.0 授权认证系统的结构如图 3-2 所示。

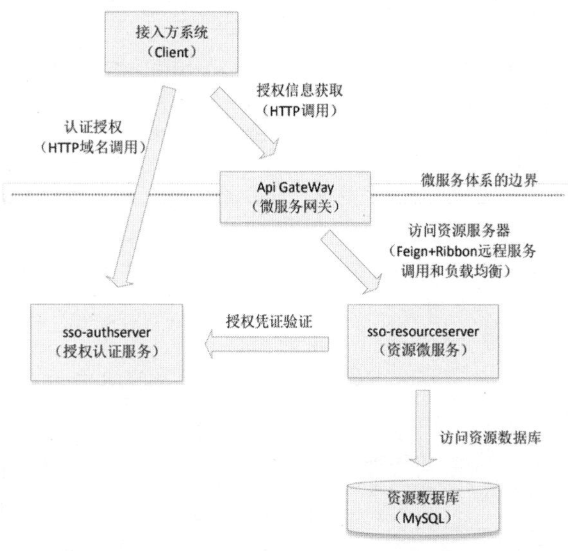

图 3-2

根据 OAuth 2.0 授权认证流程，在图 3-2 中将 SSO 授权认证系统拆分为：①授权认证微服务（sso-authsever）；②资源微服务（sso-resourceserver）。微服务之间的交互说明如下：

（1）Client 端系统在通过“授权认证微服务”完成授权认证动作，之后携带令牌调用资源微服务的接口以获取用户授权的信息。

在这个过程中，资源微服务也会调用授权认证微服务，以验证 Client 端系统携带令牌的合法性。

（2）出于微服务访问安全的考虑，Client 端系统在拿到授权令牌后，需要通过微服务网关才能访问资源微服务。因此，需要在 Client 端系统与授权认证微服务体系的边界架设微服务网关。

（3）在服务网关与资源微服务之间，通过 Spring Cloud 微服务通信组件“Feign + Ribbon”完成远程服务调用和负载均衡功能。

3.2.3 数据库设计

本实例涉及两个数据库：①授权认证数据库；②用户资源数据库。

1. 授权认证数据库

以下 6 个表结构主要用于授权认证服务，执行在授权认证时所需的数据库存取操作。其表结构与“Spring Security OAuth 2.0”开源组件所提供的表结构是一致的。

- 代码 以下表在本书配置资源的“chapter03-sso-authserver/src/main/resources/db.migration/”目录下。

以 MySQL 数据库为例，具体表结构设计如下。

（1）Client 配置信息表。

Client 配置信息表，用于存储 Client 端系统的身份信息。例如，要接入 SSO 授权认证系统的 Client 端系统，则在申请接入后会在此表中生成一条配置记录。具体的 SQL 代码如下：

```
create table oauth_client_details
(
    client_id varchar(256) primary key comment '用于标识客户端，类似于appKey',
    resource_ids varchar(256) comment '客户端能访问的资源ID集合，以逗号分隔，例如order-resource,pay-resource',
    client_secret varchar(256) comment '客户端访问密钥类似于appSecret，必须要有前缀代表加密方式，例如：
{bcrypt}10gY/Hauph1tqvVWiH4atxteSH8sRX03IDXRIQi03DVTFGzKfz8ZtGi',
    scope varchar(256) comment '用于指定客户端的权限范围，如读写权限、移动端或Web端等，例如：read,write/web,mobile',
    authorized_grant_types varchar(256) comment '可选值，如授权码模式:authorization_code；密码模式:password；刷新token:refresh_token；隐式模式:implicit；客户端模式:client_credentials，支持多种方式以逗号分隔，例如：password,refresh_token',
    web_server_redirect_uri varchar(256) comment '客户端重定向URL，在authorization_code和implicit模式时需要该值进行校验',
    authorities varchar(256) comment '可为空，指定用户的权限范围，如果授权认证过程需要用户登录，该字段不生效，在implicit和client_credentials模式时需要；例如：ROLE_ADMIN,ROLE_USER',
    access_token_validity integer comment '可为空，设置access_token的有效时间(秒)，默认为12小时，例如：3600',
    refresh_token_validity integer comment '可为空，设置refresh_token有效期(秒)，默认30天，例如7200',
    additional_information varchar(4096) comment '可为空，值必须是JSON格式，例如{"key", "value"}',
    autoapprove varchar(256) comment '默认false，适用于authorization_code模式，设置用户是否自动approval操作，如果设置为true，则跳过用户确认授权操作页面，直接跳到redirect_uri'
);
```

（2）Client 授权信息表。

Client 授权信息表，用于存储 Client 端系统的 Token。具体的 SQL 代码如下：

```
create table oauth_client_token
(
    token_id varchar(256) comment '从服务器端获取的access_token的值',
    token blob comment '这是一个二进制的字段，存储的数据是OAuth2AccessToken.java对象序列化后生成的二进制数据',
    authentication_id varchar(256) primary key comment '该字段具有唯一性，是根据当前的username（如果有）、client_id与scope通过MD5加密生成的（可参考DefaultClientKeyGenerator.java类）',
    user_name varchar(256) comment '登录用户账号',
    client_id varchar(256) comment '客户端ID'
);
```

（3）预授权码信息表。

预授权码信息表，用于记录授权认证服务颁发的预授权码信息。具体的 SQL 代码如下：

```
create table oauth_code
(
    code varchar(256) comment '存储由服务端系统生成的code的值（未加密）',
    authentication blob comment '存储将AuthorizationRequestHolder.java对象序列化后生成的二进制数据'
);
```

（4）授权 Token 信息表。

授权 Token 信息表，用于存储授权认证服务颁发的正式 access_token 信息。具体的 SQL 代码如下：

```
create table oauth_access_token
(
    token_id varchar(256) comment '该字段的值是将access_token的值通过MD5加密后生成的',
    token blob comment '存储将OAuth2AccessToken.java对象序列化后生成的二进制数据，是真实的AccessToken的数据值',
    authentication_id varchar(256) primary key comment '该字段具有唯一性，是根据当前的username（如果有）、client_id与scope通过MD5加密生成的（可参考DefaultClientKeyGenerator.java类）',
    user_name varchar(256) comment '登录时的用户账号，若客户端没有用户账号(如grant_type="client_credentials")，则该值为client_id',
    client_id varchar(256) comment '客户端ID',
    authentication blob comment '存储将OAuth2Authentication.java对象序列化后生成的二进制数据',
```

```
    refresh_token varchar(256) comment '该字段的值是将 refresh_token 的值通过 MD5 加密后生成的'
);
```

（5）授权 Token 刷新记录表。

授权 Token 刷新记录表，主要存储在 access_token 过期后重新获取的 access_token 记录信息。具体的 SQL 代码如下：

```
create table oauth_refresh_token
(
    token_id varchar(256) comment '该字段的值是将 refresh_token 的值通过 MD5 加密后生成的',
    token blob comment '存储将 OAuth2RefreshToken.java 对象序列化后生成的二进制数据',
    authentication binary comment '存储将 OAuth2Authentication.java 对象序列化后生成的二进制数据'
);
```

（6）用户授权历史表。

用户授权历史表，用于记录用户（资源所有者）授权访问的操作日志。具体的 SQL 代码如下：

```
create table oauth_approvals
(
    userId varchar(256) comment '授权用户 ID',
    clientId varchar(256) comment '客户端 ID',
    scope varchar(256) comment '授权访问',
    status varchar(10) comment '授权状态，例如：APPROVED',
    expiresAt timestamp comment '授权失效时间',
    lastModifiedAt timestamp default current_timestamp comment '最后修改时间'
);
```

2. **用户资源数据库**

用户资源数据库用于存储具体的用户信息，如用户的账户名/密码、姓名、昵称等。

- 代码 该表在本书配置资源的“chapter03-sso-resourceserver/src/main/resources/db.migration”目录下。

该库所存储的用户信息主要被“资源服务”读取。在本实例中，只设计了一张用户基本信息表，在实际场景中可根据业务需要进行扩展。

具体的 SQL 代码如下：

```
create table oauth_user_details
(
    user_name varchar(200) not null comment '用户 ID',
```

```
    password varchar(256) not null default '' comment '用户密码' ,
    salt varchar(256) not null default '' comment '生成用户密码的 MD5 密钥',
    nick_name  varchar(128) not null default '' comment '用户昵称',
    mobile  varchar(11) not null default '' comment '用户手机号' ,
    gender  int not null default 3 comment '性别。1-女; 2-男; 3-未知' ,
    authorities  varchar(256) not null default 'all' comment '用户权限，使用半角逗号分隔',
    non_expired  boolean default true comment '用户账号是否过期，boolean 值。1-表示 true; 0-表示 false',
    non_locked  boolean default true comment '用户账号是否锁定，boolean 值。1-表示 true; 0-表示 false',
    credentials_non_expired  boolean default true comment '用户密码是否过期，boolean 值。1-表示 true; 0-表示 false',
    enabled  boolean default true comment '账号是否生效。1-表示生效; 0-表示 false',
    create_time  timestamp not null default current_timestamp comment '用户账号创建时间',
    create_by  varchar(100) not null default 'system' comment '创建者',
    update_time  timestamp not null default current_timestamp comment '最后更新时间',
    update_by  varchar(100) not null default '' comment '最后更新人',
   primary key (user_name)
  ) engine=innodb default charset=utf8 comment '外部用户详细信息表';
```

3.3 步骤 1：构建 Spring Cloud 授权认证微服务

按照 3.2.2 节中设计的系统结构，本节将构建 Spring Cloud“授权认证微服务”。

3.3.1 创建 Spring Cloud 微服务工程

接下来创建授权认证微服务所需要的 Spring Cloud 微服务工程。

1. 创建一个基本的 Maven 工程

利用 2.3.1 节介绍的方法创建一个 Maven 工程。创建后的工程代码结构如图 3-3 所示。

```
▼ chapter03-sso-authserver [sso-authserver] ~/dev-tools/workspace/springcloud_works
  ▶ .idea
  ▼ src
    ▼ main
      ▶ java
      ▶ resources
    ▶ test
  ▶ target
    pom.xml 2020/2/6, 11:43 上午, 6.34 kB
    sso-authserver.iml 2020/5/15, 9:39 上午, 21.2 kB
```

图 3-3

2. 引入 Spring Cloud 依赖，将 Maven 工程改造为微服务项目

（1）引入 Spring Cloud 微服务的核心依赖。

这里可以参考 2.5.2 节中的具体步骤。

（2）在工程代码的 resources 目录下，新建一个基础性配置文件（bootstrap.yml）。

其中的代码如下：

```
spring:
  application:
    name: sso-authserver
  profiles:
    active: debug
  cloud:
    consul:
      discovery:
        preferIpAddress: true
        instance-id:
${spring.application.name}:${spring.cloud.client.ipAddress}:${spring.applica
tion.instance_id:${server.port}}:@project.version@
        healthCheckPath: /actuator/health
server:
  port: 9090
```

（3）在 2.5.2 节提到过，Spring Boot 并不会默认加载 bootstrap.yml 这个文件，所以需要在 pom.xml 中添加 Maven 资源的相关配置，具体参考 2.5.2 节内容。

（4）创建授权认证微服务的入口程序类。

代码如下：

```
package com.wudimanong.authserver;
import org.springframework.boot.SpringApplication;
import org.springframework.boot.autoconfigure.SpringBootApplication;
import org.springframework.cloud.client.discovery.EnableDiscoveryClient;
@EnableDiscoveryClient
@SpringBootApplication
public class AuthServer {
    public static void main(String[] args) {
        SpringApplication.run(AuthServer.class, args);
    }
}
```

至此，Spring Cloud 授权认证微服务就基本构建出来了。

3.3.2 将 Spring Cloud 微服务注入服务注册中心 Consul

参考 2.5.1 节、2.5.3 节的内容，将“sso-authserver”微服务注入服务注册中心 Consul 中。然后运行所构建的“sso-authserver”微服务工程，可以看到该服务已经注册到 Consul 中了，如图 3-4 所示。

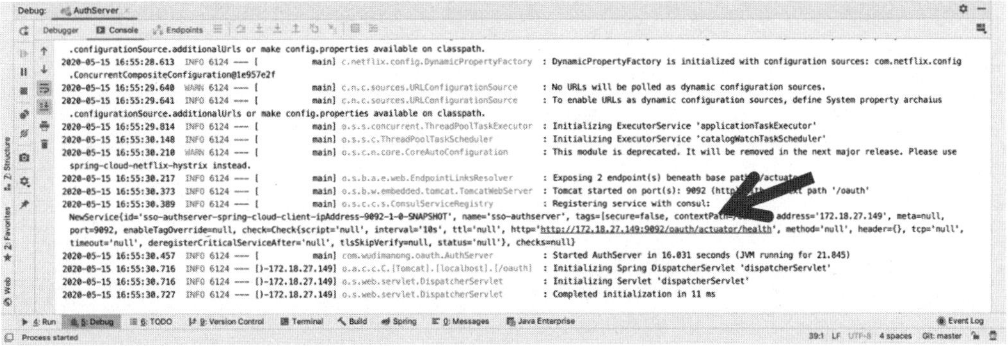

图 3-4

打开 Consul 控制台，“sso-authserver”微服务被注册到 Consul 中的效果如图 3-5 所示。

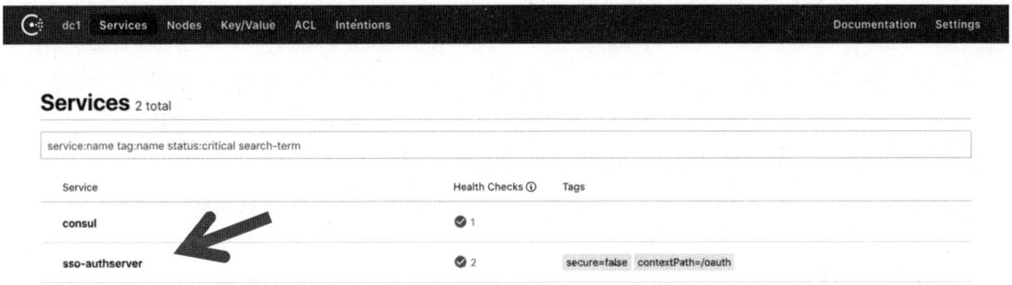

图 3-5

至此，从技术层面完成了 Spring Cloud 微服务的搭建。接下来将从业务功能层面完善授权认证微服务的其他逻辑。

3.3.3 集成 JDBC 数据源，以访问 MySQL 数据库

本节将通过“Spring Security”开源组件，实现基于 OAuth 2.0 协议的授权认证微服务。在该服务中，实现了基于数据库的授权认证信息存储逻辑。

具体的数据库表操作，已经封装在该开源依赖（JAR 包）中。下面只需要在工程中集成访问 MySQL 数据库的数据源组件，具体步骤如下。

1. 引入数据库连接池及 MySQL 驱动程序

在数据库连接池的选择上，这里选择 Druid 连接池；而 JDBC 驱动程序，因为本章实例使用的是 MySQL 数据库，所以引入的是 MySQL 的 JDBC 驱动程序。代码如下：

```
<!--引入 Druid 连接池的依赖-->
<dependency>
    <groupId>com.alibaba</groupId>
    <artifactId>druid</artifactId>
    <version>1.0.28</version>
</dependency>
<!--引入 MySQL 的 JDBC 驱动程序-->
<dependency>
    <groupId>mysql</groupId>
    <artifactId>mysql-connector-java</artifactId>
    <scope>runtime</scope>
</dependency>
<!--引入 Spring Boot 的数据源自动配置组件-->
<dependency>
    <groupId>org.springframework.boot</groupId>
    <artifactId>spring-boot-starter-jdbc</artifactId>
</dependency>
```

2. 配置项目数据库连接信息

引入数据源依赖后，在应用启动时，Spring Boot 配置类会自动初始化数据源，而在这个过程中需要读取项目配置中的数据库连接信息，所以需要在工程 resources 目录中创建一个用于配置数据库连接信息的配置文件——application.yml。其具体代码如下：

```
spring:
  datasource:
    #MySQL 连接信息
    url:
jdbc:mysql://127.0.0.1:3306/auth?zeroDateTimeBehavior=convertToNull&useUnicode=true&useUnicode=true&characterEncoding=utf-8
    username: root
    password: 123456
    type: com.alibaba.druid.pool.DruidDataSource
    driver-class-name: com.mysql.jdbc.Driver
    separator: //
```

上述配置过程涉及的数据库信息，可以参考 1.3.3 节通过 Docker 部署本地 MySQL 的步骤——创建一个名为 auth 的数据库，并执行 3.2.3 节中与授权认证数据库相关的 SQL 脚本。

3.3.4 构建 OAuth 2.0 授权认证微服务

OAuth 2.0 是一个标准的授权认证协议，业界有许多开源的实现组件。在 Spring Cloud 微服务体系中，提供了对 Spring Security OAuth 2.0 开源组件的快速集成方案。

本节将基于 Spring Security OAuth 2.0 开源组件来实现 SSO 授权认证微服务。步骤如下。

1. 引入 Spring Security OAuth 2.0 的依赖

在 Spring Cloud 中已经提供了对 OAuth 2.0 开源组件的 Starter 集成依赖。所以，这里只需要在代码工程 pom.xml 中引入如下依赖：

```
<!--引入 OAuth 2.0 的 Spring Cloud Starter 依赖-->
<dependency>
    <groupId>org.springframework.cloud</groupId>
    <artifactId>spring-cloud-starter-oauth2</artifactId>
</dependency>
<dependency>
    <groupId>org.springframework.boot</groupId>
    <artifactId>spring-boot-starter-security</artifactId>
</dependency>
```

引入的依赖（JAR 包）的版本号，默认与 Spring Cloud 父依赖中 Spring Boot 的版本号一致。

2. 创建授权认证微服务配置类

在构建基于 Spring Security OAuth 2.0 组件的授权认证微服务过程中，需要进行一些配置。具体步骤如下：

（1）创建 Spring 配置类。

具体代码如下：

```
package com.wudimanong.authserver.config;
import java.util.concurrent.TimeUnit;
import javax.sql.DataSource;
...
import org.springframework.security.oauth2.provider.token.store.
KeyStoreKeyFactory;
@Configuration
@EnableAuthorizationServer
public class AuthServerConfiguration extends
AuthorizationServerConfigurerAdapter {
    /**
     * JDBC 数据源的依赖
     */
    @Autowired
```

```
        private DataSource dataSource;
        /**
         * 授权认证管理接口
         */
        AuthenticationManager authenticationManager;

        /**
         * 构造方法
         */
        public AuthServerConfiguration(AuthenticationConfiguration
authenticationConfiguration) throws Exception {
            this.authenticationManager =
authenticationConfiguration.getAuthenticationManager();
        }
        /**
         * 通过 JDBC 操作数据库，实现对客户端信息的管理
         */
        @Override
        public void configure(ClientDetailsServiceConfigurer clients) throws
Exception {
            clients.withClientDetails(new
JdbcClientDetailsService(dataSource));
        }
        /**
         * 配置授权认证微服务相关的服务端点
         */
        @Override
        public void configure(AuthorizationServerEndpointsConfigurer endpoints) {
            //配置 TokenService 参数
            DefaultTokenServices tokenServices = new DefaultTokenServices();
            tokenServices.setTokenStore(getJdbcTokenStore());
            //支持访问令牌的刷新
            tokenServices.setSupportRefreshToken(true);
            tokenServices.setReuseRefreshToken(false);
            //设置 accessToken 的有效时间，这里设置为 30 天
            tokenServices.setAccessTokenValiditySeconds((int)
TimeUnit.DAYS.toSeconds(30));
            //设置 refreshToken 的有效时间，这里设置为 15 天
            tokenServices.setRefreshTokenValiditySeconds((int)
TimeUnit.DAYS.toSeconds(15));
            tokenServices.setClientDetailsService
(getJdbcClientDetailsService());
            // 数据库管理授权信息
            endpoints.authenticationManager(this.authenticationManager).
accessTokenConverter(jwtAccessTokenConverter())
```

```
                    .tokenStore(getJdbcTokenStore()).tokenServices(tokenServices)
                    .authorizationCodeServices(getJdbcAuthorizationCodeServices
()).approvalStore(getJdbcApprovalStore());
        }
        /**
         * 安全约束配置
         */
        @Override
        public void configure(AuthorizationServerSecurityConfigurer security) {
            security.tokenKeyAccess("permitAll()").checkTokenAccess
("hasAuthority('ROLE_TRUSTED_CLIENT')")
                    .allowFormAuthenticationForClients();
        }
        /**
         * 数据库管理的 Token 实例
         */
        @Bean
        public JdbcTokenStore getJdbcTokenStore() {
            return new JdbcTokenStore(dataSource);
        }
        /**
         * 数据库管理的客户端信息
         */
        @Bean
        public ClientDetailsService getJdbcClientDetailsService() {
            return new JdbcClientDetailsService(dataSource);
        }
        /**
         * 数据库管理的授权码信息
         */
        @Bean
        public AuthorizationCodeServices getJdbcAuthorizationCodeServices() {
            return new JdbcAuthorizationCodeServices(dataSource);
        }
        /**
         * 数据库管理的用户授权确认记录
         */
        @Bean
        public ApprovalStore getJdbcApprovalStore() {
            return new JdbcApprovalStore(dataSource);
        }
        /**
         * AccessToken 颁发管理（使用非对称加密算法来对 Token 进行签名）
         */
        @Bean
```

```
        public JwtAccessTokenConverter jwtAccessTokenConverter() {
            final JwtAccessTokenConverter converter = new
JwtAccessTokenConverter();
            // 导入证书
            KeyStoreKeyFactory keyStoreKeyFactory = new KeyStoreKeyFactory(new
ClassPathResource("keystore.jks"),
                    "mypass".toCharArray());
            converter.setKeyPair(keyStoreKeyFactory.getKeyPair("mytest"));
            return converter;
        }
    }
```

上述代码涉及的配置较多。其中，通过@EnableAuthorizationServer 注解开启授权认证微服务的相关功能；继承 AuthorizationServerConfigurerAdapter 类并重写 configure()方法，则实现了授权认证信息的数据库存储管理（包括客户端管理、授权码管理、访问 Token 令牌管理等）。

（2）创建用于生成访问令牌的加密证书。

在步骤（1）中，在生成访问令牌的过程中需要用到加密算法。加密算法涉及的证书可通过如下命令创建：

```
    $ keytool -genkeypair -alias mytest -keyalg RSA -keypass mypass -keystore
keystore.jks -storepass mypass
    您的名字与姓氏是什么？
      [Unknown]:  wudimanong
    您的组织单位名称是什么？
      [Unknown]:  wudimanong
    您的组织名称是什么？
      [Unknown]:  wudimanong
    您所在的城市或区域名称是什么？
      [Unknown]:  beijing
    您所在的省/市/自治区名称是什么？
      [Unknown]:  beijing
    该单位的双字母国家/地区代码是什么？
      [Unknown]:  CH
    CN=wudimanong, OU=wudimanong, O=wudimanong, L=beijing, ST=beijing, C=CH 是
否正确？
      [否]:  Y
    Warning:
    JKS 密钥库使用专用格式。建议使用“keytool -importkeystore -srckeystore
keystore.jks -destkeystore keystore.jks -deststoretype pkcs12”迁移到行业标准格
式 PKCS12。
```

（3）将生成的加密证书复制到工程资源目录下，并配置 Maven 资源文件的加载。

将步骤（2）中生成的“.jks”加密证书文件复制至工程目录“/src/main/resources”下。要

使该证书文件能够被 Maven 项目加载，还需要在 pom.xml 文件的资源加载配置<build>/<resources>标签下添加如下配置：

```
<resource>
    <directory>src/main/resources</directory>
    <filtering>false</filtering>
    <includes>
        <include>**/*.jks</include>
        <include>**/*.ftl</include>
        <include>/static/**</include>
    </includes>
    <excludes>
        <exclude>**/*.yml</exclude>
    </excludes>
</resource>
```

在上面配置 Maven 加密证书资源文件的加载时，也一并配置了加载“/static/**”目录下的资源（该目录在后面将被用于存放前端的静态资源）。

3. 创建 Spring Security 的安全配置类

授权认证微服务的相关接口是受安全认证保护的。但在实现授权认证微服务的具体逻辑中，有些服务却是要被开放的，所以需要进行一些安全相关的配置。

（1）创建 Spring 配置类。

具体代码如下：

```
package com.wudimanong.authserver.config;
import com.wudimanong.authserver.config.provider.
UserNameAuthenticationProvider;
...
import org.springframework.security.crypto.bcrypt.BCryptPasswordEncoder;
@Configuration
public class WebSecurityConfig extends WebSecurityConfigurerAdapter {
    /**
     * 处理授权用户信息的 Service 类
     */
    @Autowired
    UserDetailsService baseUserDetailService;
    /**
     * 安全路径过滤
     */
    @Override
```

```
    public void configure(WebSecurity web) throws Exception {
        web.ignoring().antMatchers("/css/**", "/js/**", "/fonts/**",
"/icon/**", "/images/**", "/favicon.ico");
    }
    /**
     * 放开部分授权认证入口服务的访问限制
     */
    @Override
    protected void configure(HttpSecurity http) throws Exception {
        http.requestMatchers().antMatchers("/login", "/oauth/authorize",
"/oauth/check_token").and().authorizeRequests()
                .anyRequest().authenticated().and().formLogin().loginPage("
/login").failureUrl("/login-error")
                .permitAll();
        http.csrf().disable();
    }
    /**
     * 授权认证管理配置
     */
    @Override
    public void configure(AuthenticationManagerBuilder auth) {
        auth.authenticationProvider(daoAuthenticationProvider());
    }
    /**
     * 授权用户信息数据库提供者的对象配置
     */
    @Bean
    public AbstractUserDetailsAuthenticationProvider
daoAuthenticationProvider() {
        UserNameAuthenticationProvider authProvider = new
UserNameAuthenticationProvider();
        // 设置 userDetailsService
        authProvider.setUserDetailsService(baseUserDetailService);
        // 禁止隐藏未被发现的异常
        authProvider.setHideUserNotFoundExceptions(false);
        // 使用 BCrypt 进行密码的 Hash 运算
        authProvider.setPasswordEncoder(new BCryptPasswordEncoder(6));
        return authProvider;
    }
}
```

上面的配置类的主要包括：①设置忽略静态资源的安全路径过滤；②设置自定义的用户管理实例——baseUserDetailService，以实现对授权用户信息的管理。

（2）开发 baseUserDetailService 实例所对应的类的代码。

具体代码如下：

```
    package com.wudimanong.authserver.service;
    import com.wudimanong.authserver.client.ResourceServerClient;
    ...
    import org.springframework.stereotype.Service;
    @Service
    public class BaseUserDetailService implements UserDetailsService {
        /**
         * 将资源微服务的 FeignClient 接口注入本实例
         */
        @Autowired
        ResourceServerClient resourceServerClient;
        @Override
        public UserDetails loadUserByUsername(String username) throws
UsernameNotFoundException {
            CheckPassWordDTO checkPassWordDTO =
CheckPassWordDTO.builder().userName(username).build();
            ResponseResult<CheckPassWordBO> responseResult =
resourceServerClient.checkPassWord(checkPassWordDTO);
            CheckPassWordBO checkPassWordBO = responseResult.getData();
            List<GrantedAuthority> authorities = new ArrayList<>();
            // 返回带有用户权限信息的 User
            User user = new User(checkPassWordBO.getUserName(),
                    checkPassWordBO.getPassWord() + "," +
checkPassWordBO.getSalt(), true, true, true, true, authorities);
            return user;
        }
    }
```

该类实现了 Spring Security OAuth 2.0 组件中的 UserDetailsService 接口，并通过实现其 loadUserByUsername()方法来加载需要授权认证的用户信息。

在 loadUserByUsername()方法中，通过 Spring Cloud 微服务的通信方式来获取 OAuth 2.0 资源微服务中的用户信息。

（3）实现用户账号/密码验证的功能。

在步骤（2）的代码中，只是根据用户登录名称获取了用户的基本信息。而具体的身份验证逻辑，则是在安全配置类 WebSecurityConfig 中用通过 @Bean 注解所修饰的 daoAuthenticationProvider()方法来设置的。

而 daoAuthenticationProvider()方法中的 UserNameAuthenticationProvider 类，则是继承了 Spring Security OAuth 2.0 组件中的抽象类 AbstractUserDetailsAuthenticationProvider（主要用于实现用户账号/密码验证的功能），其部分代码如下：

```
    package com.wudimanong.authserver.config.provider;
    import com.wudimanong.authserver.utils.Md5Utils;
    ...
    import org.springframework.util.Assert;
    /**
     * @描述：自定义实现用户账号/密码验证的功能类
     */
    @Data
    public class UserNameAuthenticationProvider extends
AbstractUserDetailsAuthenticationProvider {
        private static final String USER_NOT_FOUND_PASSWORD =
"userNotFoundPassword";
        private PasswordEncoder passwordEncoder;
        private volatile String userNotFoundEncodedPassword;
        private UserDetailsService userDetailsService;
        private UserDetailsPasswordService userDetailsPasswordService;
        public UserNameAuthenticationProvider() {
            this.setPasswordEncoder
(PasswordEncoderFactories.createDelegatingPasswordEncoder());
        }

        /**
         * 重写授权认证的检查方法，实现通过用户账号和密码进行登录验证的功能
         */
        @Override
        protected void additionalAuthenticationChecks(UserDetails userDetails,
                UsernamePasswordAuthenticationToken authentication) throws
AuthenticationException {
            if (authentication.getCredentials() == null) {
                this.logger.debug("Authentication failed: no credentials
provided");
                throw new BadCredentialsException(this.messages
                        .getMessage("AbstractUserDetailsAuthenticationProvider.b
adCredentials", "Bad credentials"));
            } else {
                //获取用户输入的密码
                String presentedPassword =
authentication.getCredentials().toString();
                //约定输入密码信息，拆分加密值
                String[] strArray = userDetails.getPassword().split(",");
                String userPasswordEncodeValue = strArray[0];
                String presentedPasswordEncodeValue =
Md5Utils.md5Hex(presentedPassword + "&" + strArray[1], "UTF-8");
                if
(!userPasswordEncodeValue.equals(presentedPasswordEncodeValue)) {
```

```
                this.logger.debug("Authentication failed: password does not
match stored value");
                throw new BadCredentialsException(this.messages
                        .getMessage("AbstractUserDetailsAuthenticationProvid
er.badCredentials", "Bad credentials"));
            }
        }
    }
    ...
}
```

- 代码 由于篇幅关系，这里只给出了 UserNameAuthenticationProvider 类的部分关键代码。其完整代码在本书配置资源的“chapter03-sso-authserver/src/main/java/com/wudimanong/authserver/config/provider/” 目录下。

UserNameAuthenticationProvider 类的主要逻辑，与 Spring Security OAuth 2.0 开源组件中默认提供的 DaoAuthenticationProvider 类的主要逻辑基本一致。

但 UserNameAuthenticationProvider 类重写了 additionalAuthenticationChecks()方法，重写后的逻辑是：通过 MD5 的方式对输入的密码进行 Hash 运算，然后将运算结果与“通过 BaseUserDetailService 获取的用户密码的 Hash 值”进行比较，从而降低用户密码被泄露的风险。

（4）定义 MD5 工具类的代码。

步骤（3）中涉及的 MD5 工具类的代码如下：

```
package com.wudimanong.authserver.utils;
import java.security.MessageDigest;
import java.util.UUID;
public class Md5Utils {
    private static final String hexDigits[] = {"0", "1", "2", "3", "4", "5",
"6", "7", "8", "9","a", "b", "c", "d", "e", "f"};
    /**
     * 获取MD5哈希值的方法
     */
    public static String md5Hex(String origin, String charsetname) {
        String resultString = null;
        try {
            resultString = new String(origin);
            MessageDigest md = MessageDigest.getInstance("MD5");
            if (charsetname == null || "".equals(charsetname)) {
                resultString =
byteArrayToHexString(md.digest(resultString.getBytes()));
            } else {
```

```
                    resultString =
byteArrayToHexString(md.digest(resultString.getBytes(charsetname)));
                }
            } catch (Exception exception) {
            }
            return resultString;
        }
        private static String byteArrayToHexString(byte b[]) {
            StringBuffer resultSb = new StringBuffer();
            for (int i = 0; i < b.length; i++) {
                resultSb.append(byteToHexString(b[i]));
            }
            return resultSb.toString();
        }
        private static String byteToHexString(byte b) {
            int n = b;
            if (n < 0) {
                n += 256;
            }
            int d1 = n / 16;
            int d2 = n % 16;
            return hexDigits[d1] + hexDigits[d2];
        }
    }
```

以上基于 Spring Security OAuth 2.0 开源组件，完成了 OAuth 2.0 授权认证微服务的代码实现。在授权认证微服务调用资源微服务的代码中涉及资源微服务的 FeignClient 接口，该接口的定义可以参考 3.3.5 节内容。

3.3.5 开发调用资源微服务的 FeignClient 代码

在构建授权认证微服务的过程中，对于授权用户信息的获取，授权认证微服务是通过 Spring Cloud 微服务调用的方式从资源微服务获取的（OAuth 2.0 资源微服务的构建将在 3.4 节介绍）。主要方式是：通过集成“Feign + Ribbon”组件实现微服务的远程 HTTP 通信，以及客户端负载均衡。具体步骤如下。

1. 引入 Feign 的依赖

在工程 pom.xml 文件中，引入 Feign 的依赖，代码如下：

```
<!—引入 Feign 的依赖-->
<dependency>
    <groupId>org.springframework.cloud</groupId>
    <artifactId>spring-cloud-starter-openfeign</artifactId>
</dependency>
```

2. 编写 Feign 远程通信客户端接口

（1）通过@FeignClient 注解定义微服务 sso-resourceserver 的远程访问接口。代码如下：

```
package com.wudimanong.authserver.client;
import com.wudimanong.authserver.client.bo.CheckPassWordBO;
...
import org.springframework.web.bind.annotation.PostMapping;
@FeignClient(value = "sso-resourceserver", configuration =
ResourceServerConfiguration.class, fallbackFactory =
ResourceServerFallbackFactory.class)
public interface ResourceServerClient {
    /**
     * "登录密码验证"接口
     */
    @PostMapping("/auth/checkPassWord")
    public ResponseResult<CheckPassWordBO> checkPassWord(CheckPassWordDTO
checkPassWordDTO);
}
```

（2）定义 checkPassWord()方法的请求参数对象。代码如下：

```
package com.wudimanong.authserver.client.dto;
import lombok.Builder;
import lombok.Data;
@Data
@Builder
public class CheckPassWordDTO {
    /**
     * 登录账号
     */
    private String userName;
}
```

（3）定义 checkPassWord()方法的返回参数对象。代码如下：

```
package com.wudimanong.authserver.client.bo;
import lombok.Builder;
import lombok.Data;
@Data
@Builder
public class CheckPassWordBO {
    /**
     * 用户账号
     */
    private String userName;
    /**
```

```
     * 密码
     */
    private String passWord;
    /**
     * 密码加密密钥
     */
    private String salt;
    /**
     * 用户权限
     */
    private String authorities;
}
```

（4）为了返回统一的报文格式，资源微服务通过定义 ResponseResult 类对返回的数据进行统一的包装。代码如下：

```
package com.wudimanong.authserver.entity;
import com.fasterxml.jackson.annotation.JsonInclude;
...
import lombok.Data;
import lombok.NoArgsConstructor;
@NoArgsConstructor
@AllArgsConstructor
@Builder
@Data
@JsonPropertyOrder({"code", "message", "data"})
public class ResponseResult<T> implements Serializable {
    private static final long serialVersionUID = 1L;
    /**
     * 返回的对象
     */
    @JsonInclude(JsonInclude.Include.NON_NULL)
    private T data;
    /**
     * 返回的编码
     */
    private Integer code;
    /**
     * 返回的描述信息
     */
    private String message;
    /**
     * 返回成功响应码
     *
     * @return 响应结果
     */
```

```
        public static ResponseResult<String> OK() {
            return packageObject("", GlobalCodeEnum.GL_SUCC_0000);
        }
        /**
         * 返回响应数据
         */
        public static <T> ResponseResult<T> OK(T data) {
            return packageObject(data, GlobalCodeEnum.GL_SUCC_0000);
        }
        /**
         * 对返回的数据进行包装
         */
        public static <T> ResponseResult<T> packageObject(T data, GlobalCodeEnum
globalCodeEnum) {
            ResponseResult<T> responseResult = new ResponseResult<>();
            responseResult.setCode(globalCodeEnum.getCode());
            responseResult.setMessage(globalCodeEnum.getDesc());
            responseResult.setData(data);
            return responseResult;
        }
        /**
         * 在系统发生异常不可用时返回
         */
        public static <T> ResponseResult<T> systemException() {
            return packageObject(null, GlobalCodeEnum.GL_FAIL_9999);
        }
        /**
         * 在发现可感知的系统异常时返回
         */
        public static <T> ResponseResult<T> systemException(GlobalCodeEnum
globalCodeEnum) {
            return packageObject(null, globalCodeEnum);
        }
    }
```

3. 开发微服务调用降级代码

由于网络、服务本身的原因，有时微服务调用会失败。在这种情况下，在定义 FeignClient 远程接口时可以指定服务降级代码。具体步骤如下：

（1）定义降级逻辑的代码。

```
    package com.wudimanong.authserver.client;
    import com.wudimanong.authserver.entity.ResponseResult;
    import feign.hystrix.FallbackFactory;
    import lombok.extern.slf4j.Slf4j;
```

```
@Slf4j
public class ResourceServerFallbackFactory implements 
FallbackFactory<ResourceServerClient> {
    @Override
    public ResourceServerClient create(Throwable cause) {
        return checkPassWordDTO -> {
            log.info("资源微服务调用降级逻辑处理...");
            log.error(cause.getMessage());
            return ResponseResult.systemException();
        };
    }
}
```

（2）定义实例化降级代码的配置类。

对步骤（1）中降级逻辑类的实例化，是在“2.”小标题中编写 Feign 客户端接口时，通过 @FeignClient 注解中 configuration 属性指定的配置类来实现的。该配置类的代码如下：

```
package com.wudimanong.authserver.client;
import org.springframework.context.annotation.Bean;
import org.springframework.context.annotation.Configuration;
@Configuration
public class ResourceServerConfiguration {
    @Bean
    ResourceServerFallbackFactory resourceServerFallbackFactory() {
        return new ResourceServerFallbackFactory();
    }
}
```

至此，完成了授权认证微服务调用资源微服务所需的远程 FeignClient 接口的代码编写。

（3）在运行类上开启对 FeignClient 的支持。

为了在微服务中使 FeignClient 通信组件生效，需要在服务入口类中通过 @EnableFeignClients 注解进行开启。代码如下：

```
package com.wudimanong.authserver;
import com.wudimanong.authserver.client.ResourceServerClient;
...
import org.springframework.web.bind.annotation.SessionAttributes;
@EnableDiscoveryClient
@SpringBootApplication
@SessionAttributes("authorizationRequest")
@EnableFeignClients(basePackageClasses = ResourceServerClient.class)
public class AuthServer {
    public static void main(String[] args) {
        SpringApplication.run(AuthServer.class, args);
```

```
    }
}
```

3.3.6 开发授权认证的自定义登录界面

在 Spring Security OAuth 2.0 组件中内嵌了简单的用户登录授权认证界面。但一般情况下，都会根据实际需要自定义登录界面系统。

1. 自定义登录界面效果

下面通过 Freemarker 模板引擎来实现一个自定义的登录界面。自定义的登录界面如图 3-6 所示。

图 3-6

2. 使用 Freemarker 实现自定义登录界面

在“1.”小标题中展示的自定义登录界面将在“授权认证微服务”中通过嵌入 Freemarker 模板来实现。步骤如下。

（1）在项目 pom.xml 文件中，引入 Freemarker 的相关依赖。具体代码如下：

```
<!--引入 Freemarker 的相关依赖-->
<dependency>
    <groupId>org.springframework.boot</groupId>
    <artifactId>spring-boot-starter-freemarker</artifactId>
</dependency>
<dependency>
    <groupId>org.webjars</groupId>
    <artifactId>Semantic-UI</artifactId>
    <version>2.2.10</version>
</dependency>
<dependency>
    <groupId>org.webjars</groupId>
    <artifactId>jquery</artifactId>
```

```
        <version>3.2.1</version>
    </dependency>
```

（2）在工程目录“/src/resources/templates”下定义 login.ftl 模板文件。具体代码如下：

```
    <!DOCTYPE html>
    <html lang="en">
    <head>
        <meta charset="UTF-8">
        <title>OAuth 2.0 统一授权认证中心</title>
        <meta name="viewport" content="width=device-width, initial-scale=1,
maximum-scale=1, user-scalable=no">
        <link rel="stylesheet" type="text/css"
href="${request.contextPath}/css/login.css">
    </head>
    <body>
    <div class="authcenter" id="J-authcenter">
        <!--头部代码定义-->
        <div class="authcenter-head">
            <div class="container fn-clear">
                <ul class="container-left">
                    <li class="container-left-item container-left-first">
                        <a href="#" class="on" target="_blank">SSO 统一授权登录中心
</a>
                    </li>
                </ul>
                <ul class="container-right">
                    <li class="container-right-item"><a href="https://
www.baidu.com/" target="_blank" title="统一授权中心首页">统一授权中心首页</a></li>
                </ul>
            </div>
        </div>
        <!--登录代码—form 表单的定义-->
        <div class="authcenter-body fn-clear">
            <div class="authcenter-body-login">
                <ul class="ui-nav" id="J-loginMethod-tabs">
                    <li class="active" data-status="show_login">账号登录</li>
                </ul>
                <div class="login login-modern " id="J-login">
                    <form name="loginForm" id="login" action="/login"
method="post" class="ui-form"
                          novalidate="novalidate"
                          data-widget-cid="widget-3" data-qrcode="false"><input
type="hidden" name="ua" id="UA_InputId" value="">
                        <fieldset>
                            <div class="ui-form-item" id="J-username">
```

```
                        <label id="J-label-user" class="ui-label">
                            <span class="ui-icon ui-icon-userDEF"><i
class="iconauth-men"></i></span>
                        </label>
                        <input type="text" id="J-input-user"
class="ui-input ui-input-normal" name="username"
                               tabindex="1" value="" autocomplete="off"
maxlength="100" placeholder="账号">
                        <div class="ui-form-explain"></div>
                    </div>
                    <div class="ui-form-item ui-form-item-20pd"
id="J-password">
                        <label id="J-label-editer" class="ui-label"
data-desc="登录密码">
                            <span class="ui-icon ui-icon-securityON"
id="safeSignCheck"><i class="iconauth-lock"></i></span>
                        </label>
                        <span class="alieditContainer"
id="password_container">
                            <input type="password" tabindex="2"
id="password_rsainput" name="password"
                                   class="ui-input i-text" value=""
placeholder="密码"></span>
                        <p class="ui-form-other ui-form-other-fg">
                            <a class="textlink forgot"
href="javascript:void(0)" tabindex="5"></a>
                        </p>
                        <#if _csrf??>
                            <input type="hidden"
name="${_csrf.parameterName}" value="${_csrf.token}"/>
                        </#if>
                        <div class="ui-form-explain"></div>
                    </div>
                    <!--登录按钮-->
                    <div class="ui-form-item ui-form-item-30pd"
id="J-submit">
                        <input type="submit" value="登 录"
class="ui-button" id="J-login-btn">
                        <p class="ui-form-other">
                            <a class="register" href="javascript:void(0)"
target="_blank">免费注册</a>
                        </p>
                    </div>
                </fieldset>
            </form>
```

```
            </div>
        </div>
    </div>
</div>
<!--页面尾部的定义-->
<div class="authcenter-foot" id="J-authcenter-foot">
    <div class="authcenter-foot-container">
        <p class="authcenter-foot-link">
            <a
href="https://mp.weixin.qq.com/mp/profile_ext?action=home&__biz=MzU3NDY4NzQw
NQ==&scene=124#wechat_redirect">关于无敌码农</a>
        </p>
        <div class="copyright">
            <a href="" target="_blank">Copyright©  微信公众号(无敌码
农)</a>
        </div>
    </div>
</div>
</body>
</html>
```

上面用到的样式及图片静态资源在本书配套资源“chapter03-sso-authserver/src/main/resources/static/”目录下，实操时复制相应文件即可。

此外，静态资源文件的加载配置已经在 3.3.4 节中提前配置了。

3.4 步骤 2：构建 Spring Cloud 资源微服务

在 3.3 中节中构建了 SSO 授权认证系统中的授权认证微服务。本节将构建该系统的另外一个重要组成部分——基于 Spring Cloud 体系的资源微服务。

3.4.1 创建 Spring Cloud 微服务工程

与 3.3.1 节一样，接下来创建资源微服务的 Spring Cloud 工程。

1. 创建一个基本的 Maven 工程

参考 2.3.1 节介绍的方法，创建后的工程代码结构如图 3-7 所示。

```
▼ chapter03-sso-resourceserver [sso-resourceserver] ~/dev-tools/workspace/springclou
  ▼ src
    ▼ main
      ▶ java
      ▶ resources
    ▼ test
        java
  ▶ target
    pom.xml 2020/5/18, 3:54 下午, 3.91 kB
```

图 3-7

2. 引入 Spring Cloud 依赖，将其改造为微服务项目

（1）引入 Spring Cloud 微服务的核心依赖。

参考 2.5.2 节中的具体方法。

（2）在工程代码的 resources 目录下，创建基础配置文件——bootstrap.yml。代码如下：

```
spring:
  application:
    name: sso-resourceserver
  profiles:
    active: debug
  cloud:
    consul:
      discovery:
        preferIpAddress: true
        instance-id:
${spring.application.name}:${spring.cloud.client.ipAddress}:${spring.applica
tion.instance_id:${server.port}}:@project.version@
        healthCheckPath: /actuator/health
server:
  port: 9091
  use-forward-headers: true
```

（3）因为 Spring Boot 不会默认加载 bootstrap.yml 这个文件，所以还需要在 pom.xml 中添加 Maven 资源相关的配置，具体参考 2.5.2 节内容。

（4）创建 SSO 资源微服务的入口程序类。代码如下：

```
package com.wudimanong.resourceserver;
import org.springframework.boot.SpringApplication;
import org.springframework.boot.autoconfigure.SpringBootApplication;
import org.springframework.cloud.client.discovery.EnableDiscoveryClient;
@EnableDiscoveryClient
@SpringBootApplication
public class ResourceServer {
    public static void main(String[] args) {
```

```
        SpringApplication.run(ResourceServer.class, args);
    }
}
```

3.4.2 将 Spring Cloud 微服务注入 Consul

将构建的资源微服务接入 Consul 的过程可以参考 3.3.2 节内容。

3.4.3 集成 MyBatis 框架，以访问 MySQL 数据库

在资源微服务中，将基于 MyBatis 访问 MySQL 数据库。集成 MyBatis 的具体步骤如下。

1. 引入 MyBatis 框架的依赖，以及 MySQL 的驱动程序

具体步骤可以参考 2.3.3 节的内容。

2. 配置项目数据库连接信息，以及 MyBatis 的配置信息

在项目中创建一个新的配置文件 application.yml，添加 MySQL 数据库的连接信息如下：

```
spring:
  datasource:
    #防止乱码添加字符集
    url: jdbc:mysql://127.0.0.1:3306/resource?zeroDateTimeBehavior=
convertToNull&useUnicode=true&useUnicode=true&characterEncoding=utf-8
    username: root
    password: 123456
    type: com.alibaba.druid.pool.DruidDataSource
    driver-class-name: com.mysql.jdbc.Driver
    separator: //
```

在配置文件中添加 MyBatis SQL 映射文件的路径，具体如下：

```
#添加 MyBatis SQL 映射文件的路径
mybatis:
  mapper-locations: classpath:mybatis/*.xml
  configuration:
    map-underscore-to-camel-case: true
```

上述配置中涉及的数据库信息，可以参考 1.3.3 节通过 Docker 部署本地 MySQL 的步骤——创建一个名为 resource 的数据库，并执行 3.2.3 节中与用户资源数据库相关的 SQL 脚本。

3.4.4 构建 OAuth 2.0 资源微服务

在 OAuth 2.0 协议中，资源服务主要提供“用户受保护信息查询”接口。

以微信授权登录某个网站为例，在该网站通过微信的授权认证服务获得用户的授权后，授权认证服务会给该网站颁发一个令牌。之后，该网站就可以通过该令牌去查询微信用户的头像、昵称及手机号等受保护的用户信息了，而提供这些信息查询的服务就是资源服务。

本节关于资源微服务的实现，将基于 Spring Security OAuth 2.0 开源组件来实现。

1. 引入 Spring Security OAuth 2.0 的依赖

在项目工程的 pom.xml 文件中，引入构建资源微服务所需的依赖，具体代码如下：

```
<!—引入 Spring-Security OAuth 2.0 的依赖-->
<dependency>
    <groupId>org.springframework.boot</groupId>
    <artifactId>spring-boot-starter-security</artifactId>
</dependency>
<dependency>
    <groupId>org.springframework.security.oauth.boot</groupId>
    <artifactId>spring-security-oauth2-autoconfigure</artifactId>
</dependency>
```

2. 创建 OAuth 2.0 资源微服务的配置类

资源微服务在接到“用户受保护信息查询”请求后，会对请求所携带的令牌（access_token）向授权认证微服务发起校验请求。只有在授权认证微服务通过对令牌的合法性检查后，资源微服务才会对该查询请求进行响应。

（1）创建资源微服务访问授权认证微服务的配置类。具体代码如下：

```
package com.wudimanong.resourceserver.config;
import org.springframework.beans.factory.annotation.Value;
...
import
org.springframework.security.oauth2.provider.token.RemoteTokenServices;
@Configuration
@EnableResourceServer
public class ResourceServerConfiguration extends
ResourceServerConfigurerAdapter {
    /**
     * 授权认证微服务的“令牌验证”接口的地址
     */
    @Value("${security.oauth2.checkTokenUrl}")
    private String checkTokenUrl;
    /**
```

```
     * 在授权认证微服务中为资源微服务配置的客户端 ID
     */
    @Value("${security.oauth2.clientId}")
    private String clientId;
    /**
     * 在授权认证微服务中为资源微服务配置的客户端密钥
     */
    @Value("${security.oauth2.clientSecret}")
    private String clientSecret;
    @Override
    public void configure(ResourceServerSecurityConfigurer resources) {
        RemoteTokenServices tokenService = new RemoteTokenServices();
        tokenService.setCheckTokenEndpointUrl(checkTokenUrl);
        tokenService.setClientId(clientId);
        tokenService.setClientSecret(clientSecret);
        resources.tokenServices(tokenService);
    }
}
```

这个配置类通过@EnableResourceServer 注解开启了 OAuth 2.0 资源微服务的相关功能，并通过重写 configure()方法实现了对授权认证微服务访问的配置。

（2）配置访问授权认证微服务的 URL、客户端 ID 及密钥信息。

在项目配置文件 application.yml 中添加如下配置：

```
#配置资源微服务访问授权认证微服务的信息
security:
  oauth2:
    checkTokenUrl: http://localhost:9092/oauth/check_token
    clientId: resourceClient
    clientSecret: 123456
```

在该配置中，checkTokenUrl 属性配置了授权认证微服务中“令牌验证”接口的地址；而 clientId 和 clientSecret 属性则配置了资源微服务接入授权认证微服务需要的接入 ID 及密钥信息。

接入 ID 及密钥信息可以通过在授权认证微服务的数据库（auth）中进行配置，具体 SQL 语句如下：

```
#为资源微服务配置的 Client 信息
insert into `auth`.`oauth_client_details`(`client_id`, `resource_ids`,
`client_secret`, `scope`, `authorized_grant_types`, `web_server_redirect_uri`,
`authorities`, `access_token_validity`, `refresh_token_validity`,
`additional_information`, `autoapprove`) values ('resourceclient', null,
'{noop}123456', 'all,read,write', 'authorization_code,refresh_token,password',
'http://www.baidu.com', 'role_trusted_client', 7200, 7200, null, 'true');
```

3. 创建 Spring Security 配置类

为保证资源微服务中用户信息的安全，创建一个实现 Web 安全的配置类，具体代码如下：

```
package com.wudimanong.resourceserver.config;
import
org.springframework.security.config.annotation.web.builders.HttpSecurity;
...
import
org.springframework.security.config.annotation.web.configuration.WebSecurity
ConfigurerAdapter;
@EnableWebSecurity
public class ResourceServerSecurityConfiguration extends
WebSecurityConfigurerAdapter {
    /**
     * 配置受保护资源的接口路径
     *
     * @param http
     * @throws Exception
     */
    @Override
    protected void configure(HttpSecurity http) throws Exception {
        http.authorizeRequests().antMatchers("/user/**").authenticated();
        http.csrf().disable();
    }
    /**
     * 配置需要忽略安全控制的接口路径
     *
     * @param web
     * @throws Exception
     */
    @Override
    public void configure(WebSecurity web) throws Exception {
        web.ignoring().antMatchers("/auth/**", "/actuator/health");
    }
}
```

在上述代码中，通过继承 WebSecurityConfigurerAdapter 类实现了 Web 安全配置。其中，配置的路径说明如下。

- 路径“/user/**”：受保护的用户资源，需要授权认证后才能访问。
- 路径“/auth/**”：专门为授权认证微服务提供的，用来获取用户身份信息的接口。
- 路径“/actuator/health”：Spring Cloud 微服务用于进行健康性检查的接口。

3.4.5 实现“用户受保护信息查询”的业务逻辑

在 OAuth 2.0 中，资源服务会提供“用户受保护信息查询”接口。但这并不意味着所有受保护用户信息都存储在资源服务中。

以微信授权登录某网站为例，在 Client 端系统通过获取的访问令牌向腾讯 OAuth 2.0 资源服务获取微信用户信息时，微信用户的信息并不一定会存储在该资源服务中，但资源服务会保证它所提供的查询接口可以通过访问其他内部服务（如微信用户服务）获得相关的信息。

在本实例中，SSO 授权认证系统所涉及的用户信息都直接存储在资源微服务中，并通过资源微服务对外提供“用户受保护信息查询”接口。

本节将实现资源微服务对外提供的“用户受保护信息查询”接口的业务逻辑。具体步骤如下。

1. 定义服务接口层（Controller 层）

Controller 层是服务的入口，它接收请求数据，将请求数据转换为 Java 对象，并对请求数据的合法性进行校验，在完成业务逻辑的处理后返回统一的响应数据。

（1）接口数据格式的约定。

关于接口的请求方式及报文协议，这里采用实际项目中的普遍约定：对于无数据变更的查询类接口，采用“form 表单格式 + Get 请求方式”进行提交；对于存在数据变更的事务型接口，采用“JSON 格式 + Post 请求方式”进行提交；所有接口的返回报文格式统一为 JSON 格式。

约定的接口会返回报文格式的数据对象，代码如下：

```
package com.wudimanong.resourceserver.entity;
import com.fasterxml.jackson.annotation.JsonInclude;
...
import lombok.NoArgsConstructor;
@NoArgsConstructor
@AllArgsConstructor
@Builder
@Data
@JsonPropertyOrder({"code", "message", "data"})
public class ResponseResult<T> implements Serializable {
    private static final long serialVersionUID = 1L;
    /**
     * 返回的业务数据对象
     */
    @JsonInclude(JsonInclude.Include.NON_NULL)
    private T data;
    /**
```

```
     * 返回的响应编码
     */
    private Integer code;
    /**
     * 返回的响应信息
     */
    private String message;
    /**
     * 返回的成功响应码
     *
     * @return 响应结果
     */
    public static ResponseResult<String> OK() {
        return packageObject("", GlobalCodeEnum.GL_SUCC_0000);
    }
    /**
     * 返回的成功响应数据
     *
     * @param data 返回的数据
     * @param <T>  返回的数据类型
     * @return 响应结果
     */
    public static <T> ResponseResult<T> OK(T data) {
        return packageObject(data, GlobalCodeEnum.GL_SUCC_0000);
    }
    /**
     * 对返回的消息进行包装的方法
     *
     * @param data             返回的数据对象
     * @param globalCodeEnum 自定义的返回码枚举类型
     * @param <T>              返回的数据类型
     * @return 响应结果
     */
    public static <T> ResponseResult<T> packageObject(T data, GlobalCodeEnum
globalCodeEnum) {
        ResponseResult<T> responseResult = new ResponseResult<>();
        responseResult.setCode(globalCodeEnum.getCode());
        responseResult.setMessage(globalCodeEnum.getDesc());
        responseResult.setData(data);
        return responseResult;
    }
    /**
     * 在系统发生异常不可用时返回的信息
     *
     * @param <T> 返回的数据类型
```

```
     * @return 响应结果
     */
    public static <T> ResponseResult<T> systemException() {
        return packageObject(null, GlobalCodeEnum.GL_FAIL_9999);
    }
    /**
     * 在发生可感知的系统异常时返回的信息
     *
     * @param globalCodeEnum
     * @param <T>
     * @return
     */
    public static <T> ResponseResult<T> systemException(GlobalCodeEnum
globalCodeEnum) {
        return packageObject(null, globalCodeEnum);
    }
}
```

在上述代码中，定义了统一的返回报文的包装类，并定义了处理成功和失败逻辑的响应方法。后面在业务层（Service 层）定义的返回数据对象都通过此类进行包装。

在 ResponseResult 包装类中涉及响应码枚举类的定义，代码如下：

```
package com.wudimanong.resourceserver.entity;
public enum GlobalCodeEnum {
    /**
     * 全局返回码的定义
     */
    GL_SUCC_0000(0, "成功"),
    GL_FAIL_9996(996, "不支持的 HttpMethod"),
    GL_FAIL_9997(997, "HTTP 错误"),
    GL_FAIL_9998(998, "参数错误"),
    GL_FAIL_9999(999, "系统异常"),
    /**
     * 业务逻辑异常码的定义
     */
    BUSI_USER_NOT_EXIST(1001, "用户信息不存在");
    /**
     * 编码
     */
    private Integer code;
    /**
     * 描述
     */
    private String desc;
    GlobalCodeEnum(Integer code, String desc) {
```

```
        this.code = code;
        this.desc = desc;
    }
    /**
     * 根据编码获取枚举类型的方法
     *
     * @param code 编码
     * @return
     */
    public static GlobalCodeEnum getByCode(String code) {
        //判断编码是否为空
        if (code == null) {
            return null;
        }
        //循环处理
        GlobalCodeEnum[] values = GlobalCodeEnum.values();
        for (GlobalCodeEnum value : values) {
            if (value.getCode().equals(code)) {
                return value;
            }
        }
        return null;
    }
    public Integer getCode() {
        return code;
    }
    public String getDesc() {
        return desc;
    }
}
```

（2）“登录密码验证”接口的定义。

在授权认证微服务中会提供统一的“用户账号 + 密码”的登录界面。但由于用户账号信息是存储在资源微服务中的，所以，为了实现授权认证微服务中的密码验证功能，需要在资源微服务中定义一个“登录密码验证”接口。具体代码如下：

```
package com.wudimanong.resourceserver.controller;
import com.wudimanong.resourceserver.entity.ResponseResult;
...
import org.springframework.web.bind.annotation.RestController;
@RestController
@RequestMapping("/auth")
public class UserAuthController {
    /**
     * 业务层（Service 层）的依赖
```

```
     */
    @Autowired
    UserAuthService userAuthServiceImpl;
    /**
     * 定义“登录密码验证”接口
     *
     * @param checkPassWordDTO
     * @return
     */
    @PostMapping("/checkPassWord")
    public ResponseResult<CheckPassWordBO> checkPassWord(@RequestBody
@Validated CheckPassWordDTO checkPassWordDTO) {
        return ResponseResult.OK(userAuthServiceImpl.
checkPassWord(checkPassWordDTO));
    }
}
```

定义该接口请求参数对象的代码如下:

```
package com.wudimanong.resourceserver.entity.dto;
import lombok.Data;
@Data
public class CheckPassWordDTO {
    /**
     * 登录账号
     */
    private String userName;
}
```

定义该接口返回参数对象的代码如下:

```
package com.wudimanong.resourceserver.entity.bo;
import lombok.Builder;
import lombok.Data;
@Data
@Builder
public class CheckPassWordBO {
    /**
     * 用户账号
     */
    private String userName;
    /**
     * 密码
     */
    private String passWord;
    /**
     * 密码加密密钥
```

```
     */
    private String salt;
    /**
     * 用户权限
     */
    private String authorities;
}
```

在 UserAuthController 类中涉及的业务层（Service 层）的依赖接口，可以参见下方“2.开发业务层（Service 层）代码”小标题中的内容。

（3）定义“用户受保护信息查询”接口。

Client 端系统在完成授权认证后，会通过获得的访问令牌来查询受保护的用户信息（例如用户昵称、手机号、性别等信息）。而 Client 端系统据此来完善自身的用户注册及登录会话的逻辑。

定义向 Client 端系统暴露的“用户受保护信息查询”接口的代码如下：

```
package com.wudimanong.resourceserver.controller;
import com.wudimanong.resourceserver.entity.ResponseResult;
...
import org.springframework.web.bind.annotation.RestController;
@RestController
@RequestMapping("/user")
public class UserResourcesController {
    /**
     * 注入业务层（Service 层）的依赖
     */
    @Autowired
    UserResourcesService userResourcesServiceImpl;
    /**
     *定义的“用户受保护信息查询”接口
     *
     * @param getUserInfoDTO
     * @return
     */
    @GetMapping("/getUserInfo")
    public ResponseResult<GetUserInfoBO> getUserInfo(@Validated
GetUserInfoDTO getUserInfoDTO) {
        return
ResponseResult.OK(userResourcesServiceImpl.getUserInfo(getUserInfoDTO));
    }
}
```

定义该接口请求参数对象的代码如下：

```
package com.wudimanong.resourceserver.entity.dto;
```

```
import lombok.Data;
@Data
public class GetUserInfoDTO {
    /**
     * 登录账号
     */
    private String userName;
}
```

定义该接口返回参数对象的代码如下：

```
package com.wudimanong.resourceserver.entity.bo;
import lombok.Builder;
import lombok.Data;
@Data
@Builder
public class GetUserInfoBO {
    /**
     * 用户昵称
     */
    private String nickName;
    /**
     * 用户手机号
     */
    private String mobileNo;
    /**
     * 用户性别。1-女；2-男；3-未知
     */
    private Integer gender;
    /**
     * 用户描述
     */
    private String desc;
}
```

2. 开发业务层（Service 层）代码

在服务接口的 Controller 层中所依赖的业务层（Service 层）组件将在这里实现。

在开发业务层（Service 层）代码时，除编写正常逻辑外，一般还需要处理异常逻辑：一旦数据或者条件逻辑不满足程序正常运行的要求，则中断逻辑的处理，并通过 Controller 层向调用方返回响应的业务异常信息。

所以，业务层（Service 层）需要封装业务异常，而 Controller 层则需要将业务异常转换为接口返回数据。为了减少代码量，在实践中可以通过 Spring 提供的全局异常处理机制来实现对业务异常的统一处理，步骤如下。

（1）通用的业务层（Service 层）业务异常基类的定义：

```
package com.wudimanong.resourceserver.exception;
public class ServiceException extends RuntimeException {
    private final Integer code;
    public ServiceException(Integer code, String message) {
        super(message);
        this.code = code;
    }
    public ServiceException(Integer code, String message, Throwable e) {
        super(message, e);
        this.code = code;
    }
    public Integer getCode() {
        return code;
    }
}
```

这是一个继承了运行时异常的业务层（Service 层）异常基类，业务层（Service 层）通过它来实现统一的业务异常处理。

（2）通过@ControllerAdvice 注解实现系统全局异常处理类：

```
package com.wudimanong.resourceserver.exception;
import com.wudimanong.resourceserver.entity.GlobalCodeEnum;
...
import org.springframework.web.bind.annotation.ResponseBody;
@Slf4j
@ControllerAdvice
public class GlobalExceptionHandler {
    /**
     * 业务异常处理的方法
     */
    @ExceptionHandler(ServiceException.class)
    @ResponseBody
    public ResponseResult<?> processServiceException(
            HttpServletResponse response, ServiceException e) {
        response.setStatus(HttpStatus.OK.value());
        response.setContentType("application/json;charset=UTF-8");
        ResponseResult result = new ResponseResult();
        result.setCode(e.getCode());
        result.setMessage(e.getMessage());
        log.error(e.toString() + "_" + e.getMessage(), e);
        return result;
    }
    /**
```

```
     * 参数校验错误异常的处理方法 1
     *
     * @param response
     * @param e
     * @return
     */
    @ExceptionHandler(MethodArgumentNotValidException.class)
    @ResponseBody
    public ResponseResult<?> processValidException(HttpServletResponse
response, MethodArgumentNotValidException e) {
        response.setStatus(HttpStatus.INTERNAL_SERVER_ERROR.value());
        List<String> errorStringList = e.getBindingResult().getAllErrors()
                .stream().map(ObjectError::getDefaultMessage).collect(Colle
ctors.toList());
        String errorMessage = String.join("; ", errorStringList);
        response.setContentType("application/json;charset=UTF-8");
        log.error(e.toString() + "_" + e.getMessage(), e);
        return
ResponseResult.systemException(GlobalCodeEnum.GL_FAIL_9998);
    }
    /**
     * 参数校验错误异常的处理方法 2
     *
     * @param response
     * @param e
     * @return
     */
    @ExceptionHandler(BindException.class)
    @ResponseBody
    public ResponseResult<?> processValidException(HttpServletResponse
response, BindException e) {
        response.setStatus(HttpStatus.INTERNAL_SERVER_ERROR.value());
        List<String> errorStringList = e.getBindingResult().getAllErrors()
                .stream().map(ObjectError::getDefaultMessage).collect(Colle
ctors.toList());
        String errorMessage = String.join("; ", errorStringList);
        response.setContentType("application/json;charset=UTF-8");
        log.error(e.toString() + "_" + e.getMessage(), e);
        return
ResponseResult.systemException(GlobalCodeEnum.GL_FAIL_9998);
    }
    /**
     * 参数校验错误异常的处理方法 3
     *
     * @param response
```

```
         * @param e
         * @return
         */
        @ExceptionHandler(HttpRequestMethodNotSupportedException.class)
        @ResponseBody
        public ResponseResult<?> processValidException(HttpServletResponse
response,
                HttpRequestMethodNotSupportedException e) {
            response.setStatus(HttpStatus.INTERNAL_SERVER_ERROR.value());
            String[] supportedMethods = e.getSupportedMethods();
            String errorMessage = "此接口不支持" + e.getMethod();
            if (!ArrayUtils.isEmpty(supportedMethods)) {
                errorMessage += "(仅支持" + String.join(",", supportedMethods) +
")";
            }
            response.setContentType("application/json;charset=UTF-8");
            log.error(e.toString() + "_" + e.getMessage(), e);
            return
ResponseResult.systemException(GlobalCodeEnum.GL_FAIL_9996);
        }
        /**
         * 未知系统异常的处理方法
         */
        @ExceptionHandler(Exception.class)
        @ResponseBody
        public ResponseResult<?> processDefaultException(HttpServletResponse
response, Exception e) {
            response.setStatus(HttpStatus.INTERNAL_SERVER_ERROR.value());
            response.setContentType("application/json;charset=UTF-8");
            log.error(e.toString() + "_" + e.getMessage(), e);
            return ResponseResult.systemException();
        }
    }
```

在全局异常处理类中，定义了业务异常的统一处理方法。这样业务层（Service 层）在处理业务错误时就可以直接抛出定义的业务异常类型，而无须进行额外的处理。

此外，在全局异常处理类中，还定义了针对数据校验、HTTP 请求方式，以及未知系统异常处理的方法，这样 Controller 层、Service 层只需专注于处理正常的业务逻辑即可。

（3）实现“登录密码验证”接口的业务层（Service 层）逻辑。

定义业务层（Service 层）方法的代码如下：

```
    package com.wudimanong.resourceserver.service;
    import com.wudimanong.resourceserver.entity.bo.CheckPassWordBO;
```

```
import com.wudimanong.resourceserver.entity.dto.CheckPassWordDTO;
public interface UserAuthService {
    /**
     * 定义“登录密码验证”接口的业务层（Service 层）方法
     *
     * @param checkPassWordDTO
     * @return
     */
    CheckPassWordBO checkPassWord(CheckPassWordDTO checkPassWordDTO);
}
```

具体实现类的代码如下：

```
package com.wudimanong.resourceserver.service.impl;
import com.wudimanong.resourceserver.dao.mapper.OauthUserDetailsDao;
...
import org.springframework.stereotype.Service;
@Service
public class UserAuthServiceImpl implements UserAuthService {
    /**
     * 持久层（Dao 层）的依赖
     */
    @Autowired
    OauthUserDetailsDao oauthUserDetailsDao;
    @Override
    public CheckPassWordBO checkPassWord(CheckPassWordDTO checkPassWordDTO) {
        //获取用户信息
        OauthUserDetailsPO oauthUserDetailsPO =
oauthUserDetailsDao.getUserDetails(checkPassWordDTO.getUserName());
        if (oauthUserDetailsPO == null) {
            throw new
ServiceException(GlobalCodeEnum.BUSI_USER_NOT_EXIST.getCode(),
                    GlobalCodeEnum.BUSI_USER_NOT_EXIST.getDesc());
        }
        //返回密码验证信息
        return
CheckPassWordBO.builder().userName(oauthUserDetailsPO.getUserName())
                .passWord(oauthUserDetailsPO.getPassword()).salt(oauthUserD
etailsPO.getSalt())
                .authorities(oauthUserDetailsPO.getAuthorities()).build();
    }
}
```

该实现类的主要逻辑是：通过 MyBatis 持久层（Dao 层）组件来查询用户账号和密码等信息，并将这些信息返回。而持久层（Dao 层）的依赖可以参考持久层（Dao 层）代码的实现。

（4）开发“用户受保护信息查询”接口的业务层（Service 层）代码。

定义“用户受保护信息查询”接口的业务层（Service 层）方法的代码如下：

```
package com.wudimanong.resourceserver.service;
import com.wudimanong.resourceserver.entity.bo.GetUserInfoBO;
import com.wudimanong.resourceserver.entity.dto.GetUserInfoDTO;
public interface UserResourcesService {
    /**
     * 定义“用户受保护信息查询”接口的业务层（Service 层）方法
     *
     * @param getUserInfoDTO
     * @return
     */
    GetUserInfoBO getUserInfo(GetUserInfoDTO getUserInfoDTO);
}
```

具体实现类的代码如下：

```
package com.wudimanong.resourceserver.service.impl;
import com.wudimanong.resourceserver.dao.mapper.OauthUserDetailsDao;
...
import org.springframework.stereotype.Service;
@Service
public class UserResourcesServiceImpl implements UserResourcesService {
    /**
     * 持久层（Dao 层）的依赖
     */
    @Autowired
    OauthUserDetailsDao oauthUserDetailsDao;
    @Override
    public GetUserInfoBO getUserInfo(GetUserInfoDTO getUserInfoDTO) {
        //查询用户信息
        OauthUserDetailsPO oauthUserDetailsPO =
oauthUserDetailsDao.getUserDetails(getUserInfoDTO.getUserName());
        if (oauthUserDetailsPO == null) {
            throw new
ServiceException(GlobalCodeEnum.BUSI_USER_NOT_EXIST.getCode(),
                    GlobalCodeEnum.BUSI_USER_NOT_EXIST.getDesc());
        }
        return GetUserInfoBO.builder().nickName
(oauthUserDetailsPO.getNickName())
                .mobileNo(oauthUserDetailsPO.getMobile()).gender(oauthUserD
etailsPO.getGender()).build();
    }
}
```

3. 开发 MyBatis 持久层（Dao 层）组件

在开发业务层（Service 层）代码过程中，需要用到基于 MyBatis 持久层（Dao 层）的依赖。具体步骤如下：

（1）定义持久层（Dao 层）的接口。代码如下：

```
package com.wudimanong.resourceserver.dao.mapper;
import com.wudimanong.resourceserver.dao.model.OauthUserDetailsPO;
import org.apache.ibatis.annotations.Mapper;
@Mapper
public interface OauthUserDetailsDao {
    /**
     * 根据用户账号获取用户的详情信息
     *
     * @param userName
     * @return
     */
    OauthUserDetailsPO getUserDetails(String userName);
}
```

定义该持久层（Dao 层）接口涉及的数据库实体对象（OauthUserDetailsPO）的代码如下：

```
package com.wudimanong.resourceserver.dao.model;
import java.sql.Timestamp;
import lombok.Data;
@Data
public class OauthUserDetailsPO {
    private String userName;
    private String password;
    private String salt;
    private String nickName;
    private String mobile;
    private Integer gender;
    private String authorities;
    private Boolean nonExpired;
    private Boolean nonLocked;
    private Boolean credentialsNonExpired;
    private Boolean enabled;
    private Timestamp createTime;
    private String createBy;
    private Timestamp updateTime;
    private String updateBy;
}
```

该实体对象映射的数据库表为“oauth_user_details”。

（2）定义持久层（Dao 层）接口的 MyBatis 映射 XML 文件。

在步骤（1）中涉及的数据库操作，需要在 MyBatis 的 SQL 映射 XML 文件中定义。在工程的“/src/resources/mybatis”目录下创建名为“OauthUserDetailsDao.xml”的文件，代码如下：

```
<?xml version="1.0" encoding="UTF-8" ?>
<!DOCTYPE mapper PUBLIC "-//mybatis.org//DTD Mapper 3.0//EN"
"http://mybatis.org/dtd/mybatis-3-mapper.dtd" >
<mapper
namespace="com.wudimanong.resourceserver.dao.mapper.OauthUserDetailsDao">
    <sql id="Base_Column_List">
        user_name, password, salt, nick_name,
mobile,gender,authorities,non_expired,non_locked,credentials_non_expired,ena
bled,create_time,create_by,update_time,update_by
    </sql>
    <select id="getUserDetails"
resultType="com.wudimanong.resourceserver.dao.model.OauthUserDetailsPO">
        SELECT
        <include refid="Base_Column_List"/>
        FROM oauth_user_details
        WHERE user_name=#{userName}
    </select>
</mapper>
```

至此，完成了构建 Spring Cloud 资源微服务的全部代码。具体的微服务调用将在 3.6 节中演示。

3.5 步骤 3：搭建基于 Spring Cloud Gateway 的服务网关

服务网关处于 Spring Cloud 微服务体系的边界，主要用于在内部微服务向外暴露服务时提供统一的安全认证及路由功能。

在本章实例中，资源微服务需要向外部的 Client 端系统提供“用户受保护信息查询”服务，因此需要将资源微服务的内部接口直接暴露在 Spring Cloud 体系之外。为了保证 SSO 授权认证系统内部微服务的访问安全，本节将介绍 Spring Cloud 微服务中的另一个重要组件——Spring Cloud Gateway。

3.5.1 认识微服务网关

一般来说，Spring Cloud 体系中的微服务会通过服务注册中心发现彼此，从而实现微服务之间的信任调用。

但有时也需要将一些内部微服务的接口直接暴露在微服务体系之外，因此就需要实现一定的接口安全性——例如数字签名、报文加解密等。这就意味着，很多微服务都需要考虑接口的安全性，这不仅会造成重复开发，也会增加微服务系统实施的复杂性。因此通常的做法是——在微服务体系的边界架设一个网关服务。这样，微服务就不再将内部服务地址直接暴露给外部，而是由服务网关作为请求的入口。外部的请求先被发送到服务网关，服务网关在进行了统一的安全认证（如签名验证、解密、登录会话验证等）后，再将请求路由到具体微服务的接口。

服务网关在 Spring Cloud 微服务体系中的位置如图 3-8 所示。

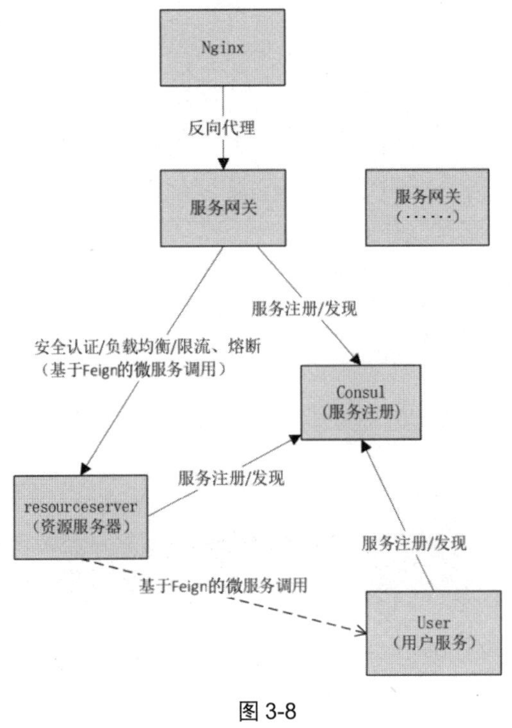

图 3-8

3.5.2 了解常见的服务网关组件

在 Spring Cloud 微服务的技术栈中，服务网关的技术组件主要有 Netflix 开源的 Zuul，以及 Spring Cloud 官方开源的 Spring Cloud Gateway 这两种。

在 Spring Boot 2.0 版本之前，Spring Cloud 默认支持的服务网关组件是 Zuul，只不过彼时还是基于 BIO 模型的 Zuul 1。Spring Cloud 在基于 Spring Boot 2.0 的版本中推出了基于 NIO 模型的服务网关组件——Spring Cloud Gateway。

从性能上说，Spring Cloud Gateway 要优于 Zuul 1 版本。虽然后来 Zuul 在 1 版本的基础上实现了基于 NIO 模型的 Zuul 2，但是由于发布时间太晚，所以 Spring Cloud 之后的官方版本已经不再默认支持 Zuul 了。

在基于 Spring Boot 2.0 的 Spring Cloud 微服务版本中，主流的微服务网关组件已经变成 Spring Cloud Gateway，所以本实例用 Spring Cloud Gateway 来实现。

3.5.3 服务网关的具体构建

Spring Cloud Gateway 作为 Spring Cloud 的官方子项目，相较于 Zuul 而言具有一定的后发优势。它提供了统一的路由功能，并可以基于 Filter 过滤链来实现接口安全、限流及监控指标上报等功能。

接下来介绍基于 Spring Cloud Gateway 搭建 Spring Cloud 微服务网关的具体步骤。

1. 构建 Spring Cloud Gateway 工程结构

参考 2.3.1、2.5.2 节中的具体步骤，来构建服务网关所需的微服务工程。

2. 引入 Spring Cloud Gateway 的依赖

在 pom.xml 文件中，引入 Spring Cloud Gateway 的依赖，代码如下：

```
<!--Spring Cloud Gateway 的依赖-->
<dependency>
    <groupId>org.springframework.cloud</groupId>
    <artifactId>spring-cloud-starter-gateway</artifactId>
</dependency>
```

需要说明的是，Spring Cloud Gateway 使用 WebFlux 作为 Web 框架，因此并不需要引入“spring-boot-starter-web”组件。

3. 创建 Spring Cloud Gateway 项目的配置

在项目 resources 目录中，创建一个基础配置文件 bootstrap.yml。代码如下：

```
spring:
  application:
    name: gateway
  profiles:
    active: debug
  cloud:
    consul:
      host: 127.0.0.1
```

```
      port: 8500
      discovery:
        preferIpAddress: true
        instance-id:
${spring.application.name}:${spring.cloud.client.ipAddress}:${spring.applica
tion.instance_id:${server.port}}:@project.version@
        healthCheckPath: /actuator/health
server:
  port: 9090
```

4. 配置 Maven 资源文件的加载支持

可以参考 2.5.2 节中的具体步骤。

5. 编写服务网关的入口程序类

```
@EnableDiscoveryClient
@SpringBootApplication
public class GatewayApplication {
    public static void main(String[] args) {
        SpringApplication.run(GatewayApplication.class, args);
    }
}
```

在成功运行服务网关的微服务工程后，打开 Consul 控制台，"gateway"微服务被注册到 Consul 中的效果如图 3-9 所示。

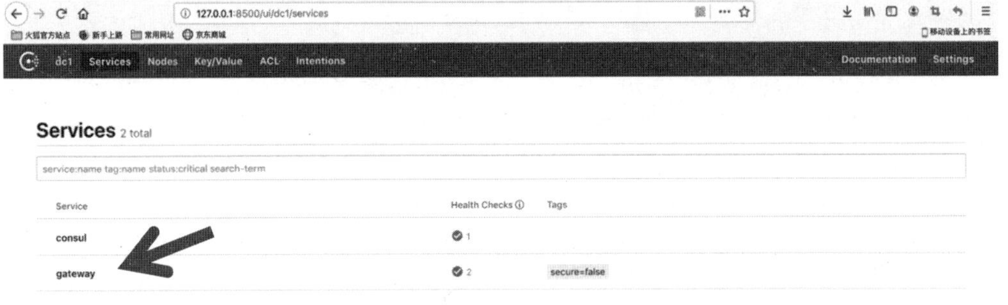

图 3-9

3.5.4 添加安全认证机制

Spring Cloud 微服务网关作为边界系统，需要对外部请求进行统一的路由、安全控制、限流及 URL 过滤。这些功能在 Spring Cloud Gateway 中是通过路由规则配置和自定义过滤器来实现的。

在 Spring Cloud Gateway 中，提供了很多现成的路由匹配规则和实现特定功能的过滤器。本节通过定义一个全局过滤器，来实现 URL 过滤及接口访问认证功能。

1. 配置微服务网关的路由规则

在工程的 resources 目录中创建配置文件 application.yml，并定义从微服务网关到资源微服务的路由规则，代码如下：

```
spring:
  cloud:
    gateway:
      #开启服务网关的注册/发现机制
      discovery:
        locator:
          enabled: true
      #路由配置（规则由 ID、目标 URL、一组 predicates 及一组 filters 构成）
      routes:
        - id: sso-resourceserver
          #lb 代表从注册中心获取服务，格式为 lb://$(注册服务的名字)
          uri: lb://sso-resourceserver
          predicates:
            #通过路径进行匹配
            - Path=/resources/**
          filters:
            - StripPrefix=1
```

在引入 spring-cloud-starter-gateway 服务网关的依赖后，在默认情况下其是关闭的，需要在配置中将其打开。而路由规则的配置则是通过"谓词匹配"来实现的。

Spring Cloud Gateway 对资源微服务的路由过程，是基于 Consul 服务发现，以客户端负载均衡的方式实现的。

2. 创建实现微服务安全认证的全局过滤器

通过"1."小标题中配置的路由规则，实现了 Spring Cloud 微服务网关与资源微服务之间的路由功能。而要实现服务网关的其他扩展功能，则可以通过创建一组全局（或者局部）的过滤器来实现。

以下定义了一个针对微服务接口安全认证的全局过滤器：

```
package com.wudimanong.gateway.filter;
import org.springframework.cloud.gateway.filter.GatewayFilterChain;
...
import reactor.core.publisher.Mono;
@Configuration
public class AuthSignatureGlobalFilter implements GlobalFilter, Ordered {
    @Override
    public Mono<Void> filter(ServerWebExchange exchange, GatewayFilterChain
chain) {
```

```
        String requestPath = exchange.getRequest().getPath().value();
        //判断接口的URL路径，如果为内部服务接口，则拦截它
        if (requestPath.contains("internal")) {
            exchange.getResponse().setStatusCode(HttpStatus.UNAUTHORIZED);
            return exchange.getResponse().setComplete();
        }
        //验证accessToken的有效性（这里只是简单判断编码是否为空，可以根据实际的业务
场景进行扩展）
        String accessToken =
exchange.getRequest().getHeaders().getFirst("access_token");
        if (accessToken == null) {
            exchange.getResponse().setStatusCode(HttpStatus.UNAUTHORIZED);
            return exchange.getResponse().setComplete();
        }
        //正常进行返回
        return chain.filter(exchange);
    }
    @Override
    public int getOrder() {
        return -1;
    }
}
```

以上代码定义了一个“Spring Cloud Gateway”全局过滤器，实现了内部服务接口 URL 路径过滤，以及接口访问权限认证功能。

如果定义多个全局过滤器，则可以通过 getOrder()方法指定它们执行的优先级，值越大优先级越低。

3.6 步骤 4：演示 OAuth 2.0 授权认证流程

经过前面的步骤已经构建出了授权认证微服务、资源微服务，以及基于 Spring Cloud Gateway 的微服务网关。接下来从 Client 端系统实际接入的角度，来演示授权认证微服务体系的运行过程。

3.6.1 编写注册 Client 端系统的 SQL 语句

在实际场景中，可以开发一个管理系统来实现对 Client 端系统接入的管理。在本实例中可以先通过 SQL 语句的方式配置用于测试的 Client 端系统的信息：

```
#配置用于测试的Client端系统的信息
```

```
    insert intO `auth`.`oauth_client_details`(`client_id`, `resource_ids`,
`client_secret`, `scope`, `authorized_grant_types`, `web_server_redirect_uri`,
`authorities`, `access_token_validity`, `refresh_token_validity`,
`additional_information`, `autoapprove`) VALUES ('accessDemo', NULL,
'{noop}123456', 'all,read,write', 'authorization_code,refresh_token,password',
'http://www.baidu.com', 'ROLE_TRUSTED_CLIENT', 7200, 7200, NULL, 'true');
```

该 SQL 语句需要在授权认证微服务的数据库（auth）中执行。另外，为了演示用户授权认证登录过程，还需要在资源微服务的数据库（resource）中配置一条用于测试的用户信息，具体 SQL 语句如下：

```
    #配置一条用于测试的用户信息
    insert into `resource`.`oauth_user_details`(`user_name`, `password`, `salt`,
`nick_name`, `mobile`, `gender`, `authorities`, `non_expired`, `non_locked`,
`credentials_non_expired`, `enabled`, `create_time`, `create_by`,
`update_time`, `update_by`) VALUES ('wudimanong',
'7b952cb6a19dad78e50cbff9dde121ef', 'e7909fd872764f0fa286a93c73441e71', '无敌
码农', '18610380625', 3, 'all', 1, 1, 1, 1, '2020-05-21 05:59:07', 'system',
'2020-05-21 05:59:07', '');
```

3.6.2 演示用户授权认证登录的过程

在配置完相关信息后，通过 IDEA 启动授权认证微服务、资源微服务，以及微服务网关。如果运行正常，则服务注册中心 Consul 的效果如图 3-10 所示。

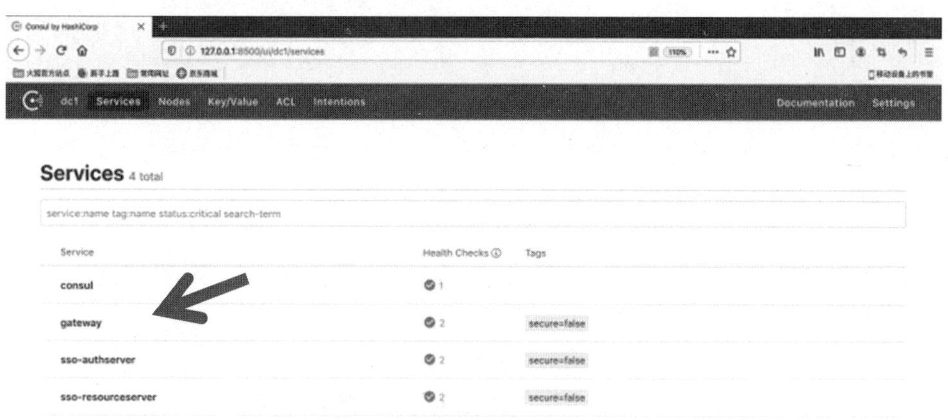

图 3-10

将 Client 端系统接入 SSO 授权认证微服务的步骤如下。

1. 访问授权认证微服务，获取预授权码

在这个步骤中，Client 端系统通过浏览器跳转的方式将授权登录的请求重定向至授权认证微服

务的“/oauth/authorize”接口。

授权认证微服务根据传入的 Client 端系统信息及授权验证类型，将用户界面重定向至验证用户身份的界面（如登录界面）。请求 URL 及参数如下：

```
http://127.0.0.1:9092/oauth/authorize?response_type=code&client_id=accessDemo&redirect_uri=http://www.baidu.com
```

参数“response_type=code”表示的是授权码模式，client_id 是在 3.6.1 节中配置的 Client 端系统接入信息，redirect_uri 是重定向到 Client 端系统的地址（必须与在 3.6.1 节中配置的 Client 端系统重定向地址一致）。

在浏览器输入授权认证微服务的请求 URL 后，跳转到的用户登录界面如图 3-11 所示。

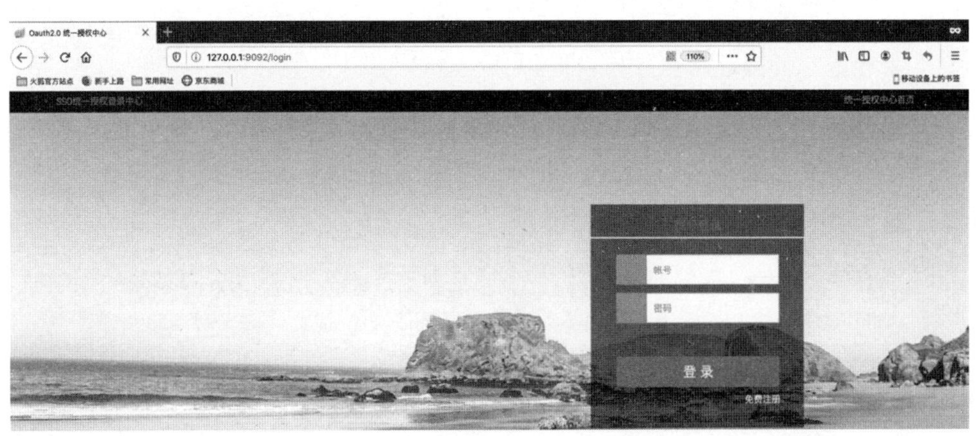

图 3-11

输入在 3.6.1 节中通过 SQL 语句配置的用户账号及密码，授权认证微服务会在用户帐号及密码验证通过后携带生成的预授权码重定向至 redirect_uri 所指的链接，如图 3-12 所示。

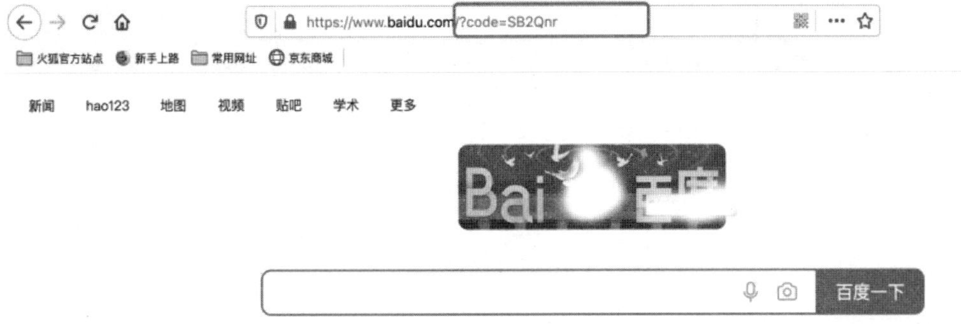

图 3-12

2. 通过预授权码向授权认证微服务获取访问令牌（access_token）

Client 端系统在获得授权认证微服务返回的预授权码后，继续访问授权认证微服务的“/oauth/token”接口，以换取正式的访问令牌（access_token）。通过 Postman 接口工具获取的 access_token 如图 3-13 所示。

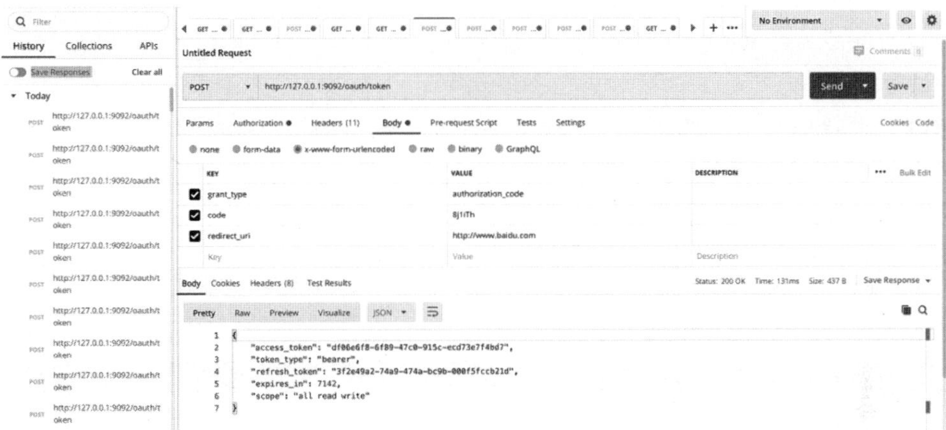

图 3-13

需要说明的是，向授权认证微服务获取正式的令牌，需要采取“POST/x-www-form-urlencoded”的请求方式。请求参数“grant_type=authorization_code”表示授权码模式。

此外，在获取访问令牌时，需要以“Basic Auth”的方式认证 Client 端系统的身份。在使用 Postman 进行请求时，可以通过设置 Authorization 信息来实现，如图 3-14 所示。

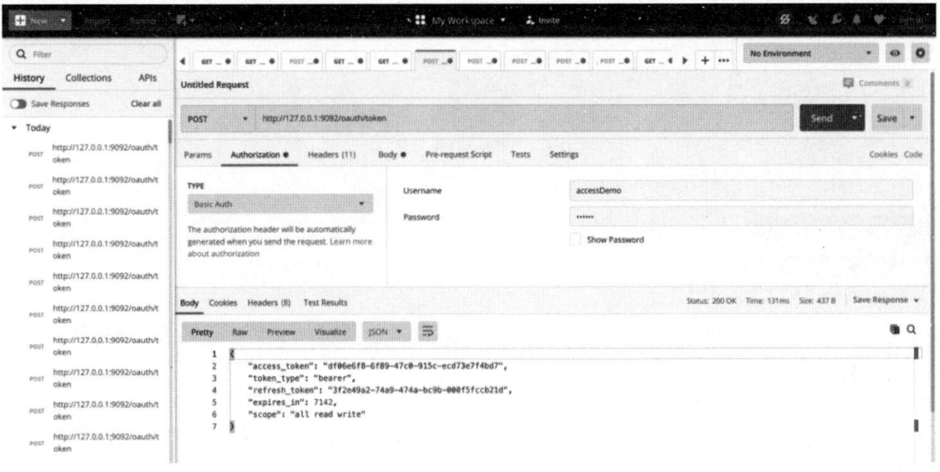

图 3-14

在得到授权认证微服务颁发的正式访问令牌后，Client 端系统就可以通过该令牌访问受保护的用户信息了。

3.6.3 通过微服务网关访问 OAuth 资源微服务

在获取正式访问令牌后，Client 端系统在访问资源微服务时，需要通过微服务网关的入口，以求在 Spring Cloud 微服务体系的边界对外部请求进行统一的身份验证及限流等操作。

通过调用 Spring Cloud Gateway 访问 OAuth 2.0 资源微服务的 URL 如下：

```
http://127.0.0.1:9090/resources/user/getUserInfo?access_token=df06e6f8-6f89-47c0-915c-ecd73e7f4bd7&userName=wudimanong
```

其中，“/resources/”是在 3.5.4 节中所设置的路由转发规则，而“/user/getUserInfo”则是资源微服务所提供的“用户受保护信息查询”接口。

访问资源微服务的效果如图 3-15 所示。

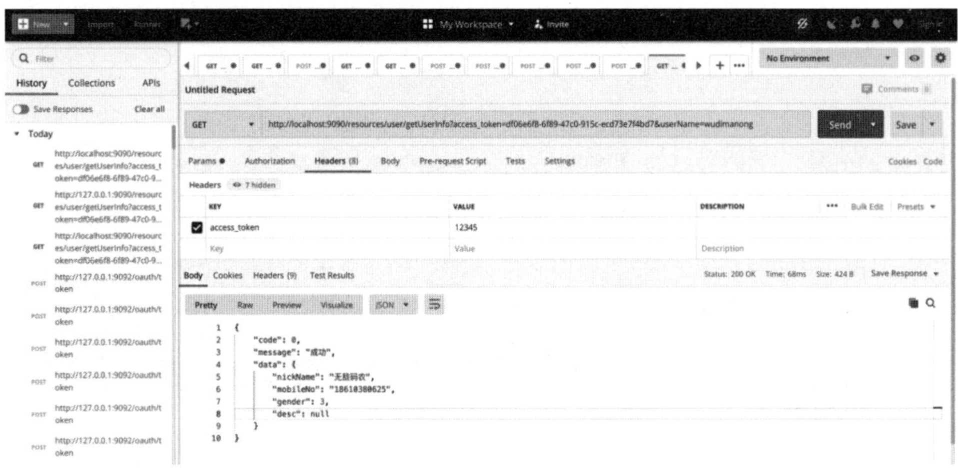

图 3-15

3.7 本章小结

本章基于 Spring Security OAuth 2.0 组件完成了一个 SSO 授权认证系统，介绍了基于 OAuth 2.0 协议的授权认证体系的实现方式。

另外，还针对 Spring Cloud 微服务系统边界的问题介绍了一个重要的概念——微服务网关，并演示了基于 Spring Cloud Gateway 构建微服务网关的具体过程。

第 4 章

【实例】车辆电子围栏系统

——用“PostgreSQL + PostGis”实现电子围栏服务，并利用配置中心管理微服务的多环境配置信息

近年来，以共享单车为代表的出行模式蓬勃发展，这在方便日常生活的同时，也带来了很多负面影响，例如乱停乱放现象就严重影响了交通秩序。为了规范用户使用共享单车的行为，越来越多的共享单车企业开始注重精细化运营，一个很重要的手段就是采用电子围栏技术对用户的停车行为进行约束，例如对违停行为进行罚款，以减少乱停乱放现象的发生。

本实例将基于 Spring Cloud 微服务体系，利用地理信息系统（GIS）的相关技术手段实现一个简单的车辆电子围栏系统，该系统就只有一个电子围栏微服务“efence”。

在本实例的实现过程中，将利用 Spring Cloud 微服务配置中心组件——Spring Cloud ConfigServer 来管理电子围栏微服务的多环境配置信息。

通过本章，读者将学习到以下内容：

- 车辆电子围栏系统的基本概念。
- PostgreSQL 数据库的使用。
- PostGIS 空间数据库的使用。
- Spring Cloud 微服务的构建方式。
- 使用 MyBatis-Plus 插件简化 MyBatis 的数据库操作。
- 利用配置中心管理微服务的多环境配置信息。

4.1 功能概述

从功能上看，车辆电子围栏系统主要是通过定义车辆运营区域及停车点的地图坐标，来实现对车辆电子围栏地理范围的划分。

一般来说，运营人员会根据平台的实际运营情况来定义“运营区域”，每个运营区域内的运营人员根据实际的街道情况来划定“停车围栏”，因此在实际环境中使用的车辆电子围栏系统会提供完善的可视化操作平台。

由于篇幅的关系，本章实例主要从“后端服务接口 + 数据库操作”的角度来实现简单的电子围栏微服务。其核心功能有：①批量导入围栏数据；②单个围栏数据的地图展示；③违停行为判断。

本章后面的内容将围绕实现这几个核心功能来展开。

4.2 系统设计

在设计电子围栏微服务时，需要考虑对地理位置这样的空间数据进行存储和计算的能力。而对于空间数据的存储和计算，在实践中采用得比较广泛的技术方案是：利用“PostgreSQL + PostGIS”来存储地理位置信息，利用 PostGIS 的空间数据处理能力对围栏数据进行计算和判断。

因此在数据库设计中，将利用 PostGIS 来定义位置信息的存储字段。

4.2.1 系统结构设计

在系统实现上将采用经典的 MVC 分层模式，电子围栏微服务的结构如图 4-1 所示。

系统结构主要分为 3 层：①服务接口层（Controller 层）；②业务层（Service 层）；③持久层（Dao 层）。其中，服务接口层（Controller 层）用于定义服务接口，业务层（Service 层）用于处理业务逻辑，持久层（Dao 层）用于封装针对 PostgreSQL 数据库的操作。

在微服务的调用上，内部微服务之间的调用采用 FeignClient 来实现，而外部服务对内部微服务的访问则通过 3.5 节介绍的微服务网关来进行隔离。

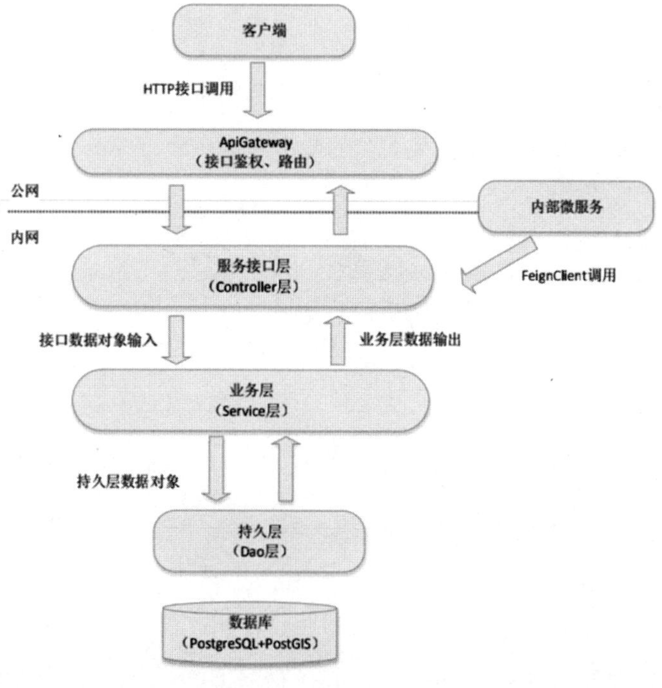

图 4-1

4.2.2 数据库设计

在本实例的数据库设计中，会使用到 PostGIS 所支持的空间数据存储字段，如“Geometry”类型——这是一种既支持平面对象又支持空间对象的数据存储类型。

- 代码 以下表在本书配置资源的“chapter04-efence/src/main/resources/db.migration”目录下。

利用“PostgreSQL + PostGIS”数据库的相关函数，可以方便地进行与空间位置相关的计算。具体表结构设计如下（SQL 语法遵循 PostgreSQL 及 PostGIS 数据库的约定）。

1. 电子围栏图层信息表

电子围栏图层信息表主要用于约定电子围栏的用途、类型等信息。具体的 SQL 代码如下：

```
--创建图层 ID 自增序列
create sequence fence_geo_layer_id_seq;
--创建电子围栏图层信息表
create table fence_geo_layer
(
```

```
        id integer not null default nextval('fence_geo_layer_id_seq'::regclass),
        code character varying(16) collate pg_catalog."default" not null default
''::character varying,
        name character varying(32) collate pg_catalog."default" not null default
''::character varying,
        explain character varying(100) collate pg_catalog."default" not null
default ''::character varying,
        check_city boolean not null default false,
        city_code character varying(16) collate pg_catalog."default",
        state smallint not null default 0,
        type smallint  not null default 0,
        create_time timestamp with time zone not null default now(),
        update_time timestamp with time zone not null default now(),
        create_user character varying(32) collate pg_catalog."default" not null
default ''::character varying,
        update_user character varying(32) collate pg_catalog."default" not null
default ''::character varying,
        constraint fence_geo_layer_pkey primary key (id)
    )
    with (
        oids = false
    );
    comment on table fence_geo_layer is '电子围栏图层信息表';
    comment on column fence_geo_layer.code is '图层编码';
    comment on column fence_geo_layer.name is '图层名称';
    comment on column fence_geo_layer.explain is '图层说明';
    comment on column fence_geo_layer.check_city is '是否检查城市，配合 city_code
字段使用';
    comment on column fence_geo_layer.city_code is '所属城市编码';
    comment on column fence_geo_layer.state is '图层状态。0-有效（默认）；1-删除';
    comment on column fence_geo_layer.type is '图层类型。0-未知分类；1-干预；2-调
度；3-停车围栏；4-运营范围；5-技术定义';
    comment on column fence_geo_layer.create_time is '创建时间';
    comment on column fence_geo_layer.update_time is '修改时间';
    comment on column fence_geo_layer.create_user is '创建用户';
    comment on column fence_geo_layer.update_user is '修改用户';
```

2. 电子围栏信息表

电子围栏信息表主要用于存储根据运营城市、区域和街道而划定的电子围栏坐标信息。具体的 SQL 代码如下：

```
    --创建围栏 ID 自增序列
    create sequence fence_geo_id_seq;
    --创建电子围栏信息表
    create table fence_geo_info
```

```
(
    id bigint not null default nextval('fence_geo_id_seq'::regclass),
    name character varying(254) collate pg_catalog."default" not null,
    explain character varying(200) collate pg_catalog."default" default
''::character varying,
    city_code character varying(16) collate pg_catalog."default" not null
default ''::character varying,
    ad_code character varying(16) collate pg_catalog."default",
    layer_code character varying(16) collate pg_catalog."default" not null,
    region geometry(geometry,4326) not null,
    centre geometry(point,4326),
    area numeric(16,2),
    custom_info jsonb,
    batch_id bigint,
    from_id bigint,
    geo_json text collate pg_catalog."default",
    geo_hash character varying(16)[] collate pg_catalog."default",
    date_range tstzrange,
    time_bucket int4range[],
    state smallint,
    update_time timestamp with time zone,
    create_time timestamp with time zone,
    update_user character varying(32) collate pg_catalog."default" default
''::character varying,
    create_user character varying(32) collate pg_catalog."default" default
''::character varying,
    constraint fence_geo_pkey primary key (id)
)
with (
    oids = false
);
--添加字段备注
comment on table fence_geo_info is '电子围栏信息表';
comment on column fence_geo_info.id is '围栏ID';
comment on column fence_geo_info.name is '围栏名称';
comment on column fence_geo_info.explain is '围栏描述';
comment on column fence_geo_info.city_code is '所属城市编码';
comment on column fence_geo_info.ad_code is '归属分区编码';
comment on column fence_geo_info.layer_code is '关联的图层编码;
comment on column fence_geo_info.region is '围栏坐标信息';
comment on column fence_geo_info.centre is '围栏中心点';
comment on column fence_geo_info.area is '围栏面积（单位：m2）';
comment on column fence_geo_info.custom_info is '自定义字段数据';
comment on column fence_geo_info.batch_id is '批量导入批次标识';
comment on column fence_geo_info.from_id is '来源围栏ID';
```

```
    comment on column fence_geo_info.geo_json is '冗余的围栏 geojson 信息';
    comment on column fence_geo_info.geo_hash is '围栏覆盖的 geohash 列表';
    comment on column fence_geo_info.date_range is '有效期（开始时间、结束时间）';
    comment on column fence_geo_info.time_bucket is '一天内的有效时间段(用分钟表示),
例如：{[360,480],[600,840]}';
    comment on column fence_geo_info.state is '围栏状态。0-生效；1-已删除；2-失效';
    --添加表的索引信息
    create index idx_fence_geo_centre on fence_geo_info using gist(centre);
    comment on index idx_fence_geo_centre is '围栏中心点字段索引';
    create index idx_fence_geo_city_code on fence_geo_info using btree(city_code
collate pg_catalog."default");
    comment on index idx_fence_geo_city_code is '围栏城市编码字段索引';
    create index idx_fence_geo_region on fence_geo_info using gist(region);
    comment on index idx_fence_geo_region is '围栏地理范围字段索引';
```

在电子围栏信息表的定义中，“region”字段使用“geometry”类型来定义围栏的区域信息，其中的数字“4326”表示使用的是大地坐标系（经纬度）；而“centre”字段则使用“point”类型来定义围栏的中心点。

4.3 步骤 1：构建 Spring Cloud 微服务工程代码

参考 4.2 节系统设计的内容，本节将搭建“PostgreSQL + PostGIS”数据库环境，并构建 Spring Cloud 微服务代码工程，以及集成所需要的第三方依赖。

4.3.1 搭建“PostgreSQL + PostGIS”数据库环境

本实例使用“PostgreSQL + PostGIS”数据库来存储和计算电子围栏的位置信息。接下来将在本地 Docker 下搭建“PostgreSQL + PostGIS”数据库环境，步骤如下。

1. 获取“PostgreSQL + PostGIS”的 Docker 镜像

向 Docker 环境中拉取“PostgresSQL 10.0 + PostGIS 2.4”版本的 Docker 镜像。命令如下：

```
docker pull kartoza/postgis:10.0-2.4
```

2. 安装“PostgreSQL + PostGIS”数据库

通过 Docker 命令运行“1.”小标题中获取的 Docker 镜像，以安装“PostgreSQL + PostGIS”数据库。命令如下：

```
docker run --name postgres1 -e POSTGRES_USER=gis -e POSTGRES_PASSWORD=123456
-p 54321:5432 -d kartoza/postgis:10.0-2.4
```

3. 验证“PostgreSQL + PostGIS”的安装结果

利用“docker ps”命令查看运行效果。可以看到，“PostgreSQL + PostGIS”镜像已经成功运行，如图 4-2 所示。

```
qiaodeMacBook-Pro-2:springcloud-action qiaojiang$ docker ps
CONTAINER ID   IMAGE                     COMMAND                  CREATED       STATUS       PORTS                      NAMES
5febc7d159c1   kartoza/postgis:10.0-2.4  "/bin/sh -c /docker-…"   5 hours ago   Up 5 hours   0.0.0.0:54321->5432/tcp    postgres1
```

图 4-2

利用“Navicat”数据库客户端工具连接运行的“PostGreSQLl + PostGIS”数据库，如图 4-3 所示。

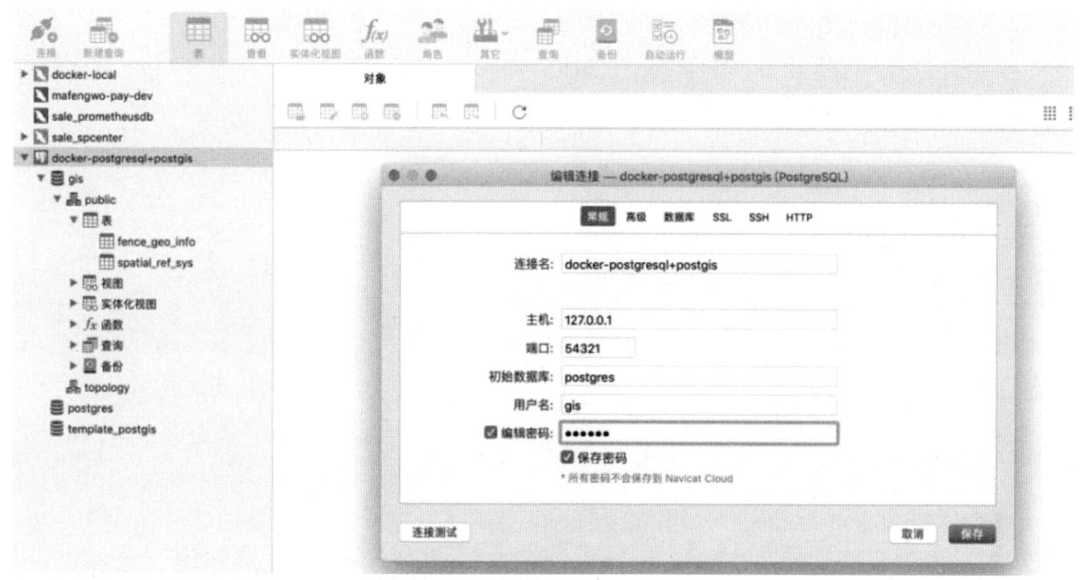

图 4-3

如果连接正常，则在“PostgreSQL + PostGIS”数据库中创建了一个名为“gis”的数据库。执行 4.2.2 节中所定义的建表语句，即可完成电子围栏微服务数据库的初始化。

4.3.2 创建 Spring Cloud 微服务工程

接下来创建 Spring Cloud 微服务代码工程。

1. 创建一个基本的 Maven 工程

利用 2.3.1 节介绍的方法创建一个 Maven 工程，其结构如图 4-4 所示。

```
▼ chapter04-efence [efence] ~/dev-tools/workspace/springcl
  ▼ src
    ▼ main
      ▶ java
      ▶ resources
      test
    efence.iml 2019/12/23, 6:32 下午, 10.78 kB
    pom.xml 2019/12/23, 6:32 下午, 2.45 kB
```

图 4-4

2. 引入 Spring Cloud 依赖，将 Maven 工程改造为微服务项目

（1）引入 Spring Cloud 微服务的核心依赖。

这里可以参考 2.5.2 节中的具体步骤。

（2）在工程代码的 resources 目录下，新建一个基础性配置文件——bootstrap.yml。

代码如下：

```
spring:
  application:
    name: efence
  profiles:
    active: debug
  cloud:
    consul:
      discovery:
        preferIpAddress: true
        instance-id:
${spring.application.name}:${spring.cloud.client.ipAddress}:${spring.applica
tion.instance_id:${server.port}}:@project.version@
        healthCheckPath: /actuator/health
server:
  port: 9090
```

（3）配置 Maven 资源文件的加载支持。

具体参考 2.5.2 节。

（4）创建微服务的入口程序类。

代码如下：

```
package com.wudimanong.efence;
import org.mybatis.spring.annotation.MapperScan;
...
```

```
import org.springframework.cloud.client.discovery.EnableDiscoveryClient;
@EnableDiscoveryClient
@SpringBootApplication
public class FenceApplication {
    public static void main(String[] args) {
        SpringApplication.run(FenceApplication.class, args);
    }
}
```

至此，Spring Cloud 车辆电子围栏微服务工程就构建出来了。

4.3.3 将 Spring Cloud 微服务注入 Consul

参考 2.5.1 节、2.5.3 节的内容，将“efence”微服务注入服务注册中心 Consul 中。然后运行所构建的“efence”微服务工程，可以看到该服务已经注册到 Consul 中了，如图 4-5 所示。

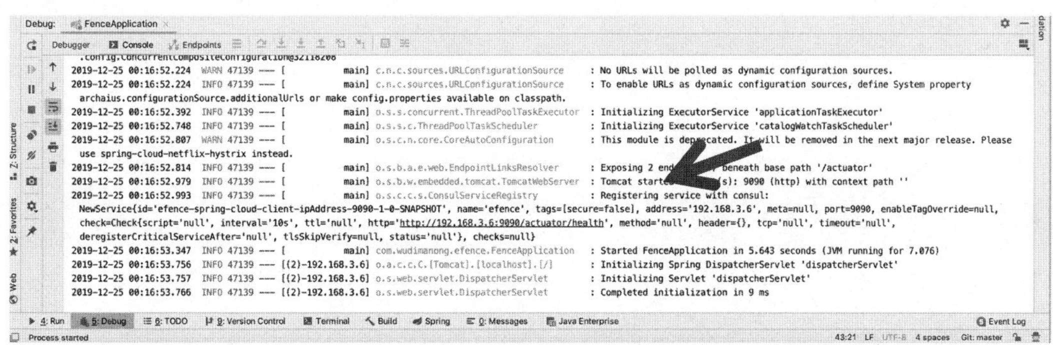

图 4-5

打开 Consul 控制台，“efence”微服务被注册到 Consul 中的效果如图 4-6 所示。

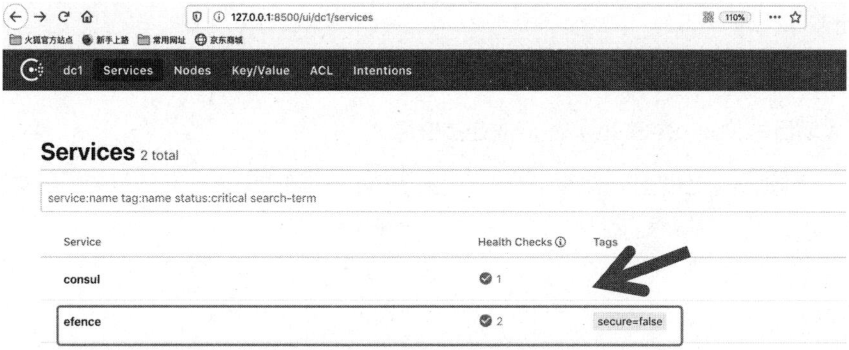

图 4-6

接下来将从“efence”微服务的业务功能层面集成所需要的其他组件。

4.3.4 集成 MyBatis，以访问 PostgreSQL 数据库

在车辆电子围栏微服务中，将基于 MyBatis 来访问 PostgreSQL 数据库。集成 MyBatis 的具体步骤如下。

1. 引入 MyBatis 框架的依赖，以及 PostgreSQL 的驱动程序

（1）在项目的 pom.xml 文件中，引入 MyBatis 框架的依赖：

```
<!--引入 MyBatis 的依赖-->
<dependency>
    <groupId>org.mybatis.spring.boot</groupId>
    <artifactId>mybatis-spring-boot-starter</artifactId>
    <version>2.0.1</version>
</dependency>
```

（2）引入数据库连接池，以及 PostgreSQL 的驱动程序。

在数据库连接池的选择上，这里选择 Druid 连接池；而在 JDBC 驱动程序的选择上，因为本实例使用的是 PostgreSQL 数据库，所以引入 PostgreSQL 的 JDBC 驱动程序。代码如下：

```
<!--引入 Druid 连接池的依赖-->
<dependency>
    <groupId>com.alibaba</groupId>
    <artifactId>druid</artifactId>
    <version>1.0.28</version>
</dependency>
<!—引入 PostgreSQL 的 JDBC 驱动程序-->
<dependency>
    <groupId>org.postgresql</groupId>
    <artifactId>postgresql</artifactId>
    <version>42.2.9</version>
</dependency>
```

2. 配置项目的 PostgreSQL 数据库的连接信息

在项目中创建一个新的配置文件——application.yml，用于配置与业务相关的配置信息。在其中添加 PostgreSQL 数据库的连接信息，具体代码如下：

```
spring:
  datasource:
    url: jdbc:postgresql://127.0.0.1:54321/gis
    username: gis
    password: 123456
    type: com.alibaba.druid.pool.DruidDataSource
```

```
    driver-class-name: org.postgresql.Driver
    separator: //
```

至此，完成了在 Spring Cloud 微服务中集成 MyBatis。

4.3.5 通过 MyBatis-Plus 简化 MyBatis 的操作

MyBatis 是 Java 开发领域目前普遍使用的数据库操作框架，常说的“SSM 组合”中的“M”指的就是 MyBatis。MyBatis 受欢迎的原因是，它可以灵活地定义 SQL 来操作数据库。

但在实际的编码过程中，对于一些简单的单表操作，如果在使用 MyBatis 时也创建 XML 文件来定义 SQL 映射，则不仅会增加开发工作量，也会造成代码冗余。所以，现在出现了一些像 MyBatis-Plus 这样的 MyBatis 增强工具，用于简化数据库开发。

在本实例的开发过程中，会通过 MyBatis-Plus 来简化数据库开发、提高效率。

1. Spring Boot 集成 MyBatis-Plus 框架

在微服务项目的 pom.xml 文件中，引入 MyBatis-Plus 的依赖。代码如下：

```
<!--MyBatis-Plus 的依赖-->
<dependency>
    <groupId>com.baomidou</groupId>
    <artifactId>mybatis-plus-boot-starter</artifactId>
    <version>3.3.0</version>
</dependency>
```

如果引入了 MyBatis-Plus 的依赖，则会自动引入 MyBatis 的依赖。为避免版本冲突，可以在 pom.xml 中删除前面引入的 MyBatis 依赖代码：

```
<!--引入 MyBatis 的依赖-->
<dependency>
    <groupId>org.mybatis.spring.boot</groupId>
    <artifactId>mybatis-spring-boot-starter</artifactId>
    <version>2.0.1</version>
</dependency>
```

2. 设置 MyBatis-Plus 的相关信息

在项目配置文件 application.yml 中，设置 MyBatis-Plus 的相关信息。代码如下：

```
#MyBatis-Plus 的集成配置
mybatis-plus:
  # MyBatis XML 映射文件的存放路径
  mapper-locations: classpath:mybatis/*.xml
  # PO 实体类扫描包的路径，在多个包路径之间用逗号分隔
```

```
    typeAliasesPackage: com.wudimanong.efence.dao.model
    global-config:
      db-config:
        #主键类型
        id-type: auto
        #字段策略
        field-strategy: not_empty
        #设置在将数据库字段映射为 Java 属性时，是否自动进行“驼峰”和“下画线”之间的转换
        column-underline: true
        #逻辑删除配置
        logic-delete-value: 0
        logic-not-delete-value: 1
        db-type: postgresql
      refresh: false
    configuration:
      #开启此配置后，会自动将“下画线格式的表字段”转换为“以驼峰命名的属性名称”
      map-underscore-to-camel-case: true
      cache-enabled: false
      #日志配置
      log-impl: org.apache.ibatis.logging.stdout.StdOutImpl
```

在以上配置中，指定了 MyBatis XML 映射文件的位置，以及数据库实体对象的包路径。其他的配置则是 MyBatis-Plus 提供的一些额外功能：支持逻辑删除，自动将“下画线格式的表字段”转换为“以驼峰命名的属性名称”等。

3. 验证 MyBatis-Plus 的集成效果

启动工程项目，集成 MyBatis-Plus 的效果如图 4-7 所示。

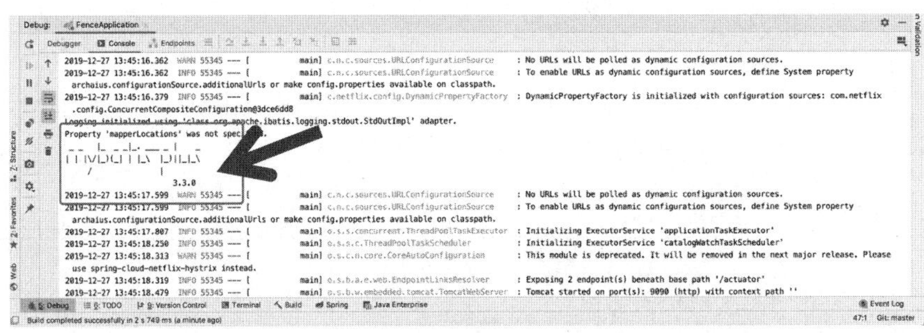

图 4-7

MyBatis-Plus 并没有改变 MyBatis 框架的任何特性及功能，只是基于 MyBatis 做了一些通用功能的抽象，从而简化开发、提高效率。所以，在使用 MyBatis-Plus 时，还是可以通过 XML 映射文件来处理一些复杂的数据库操作逻辑。

接下来将实现车辆电子围栏微服务的业务逻辑。

4.4 步骤 2：实现微服务的业务逻辑

参考 4.2 节的系统设计，以及 4.3 节搭建的微服务，接下来实现电子围栏微服务的业务逻辑。

4.4.1 定义服务接口层（Controller 层）

Controller 层是微服务面向外部调用的入口。它接收数据请求，将其转换为 Java 对象，并校验数据的合法性，在完成逻辑处理后对返回的数据进行统一的封装处理。

1. 接口数据的格式约定

关于接口请求方式及协议，按照实际项目中的普遍约定：对于无数据变更的查询类接口，采用“form 表单格式 + Get 请求方式”进行提交；对于存在数据变更的事务型接口，统一以“JSON 格式 + Post 请求方式”进行提交；所有的返回数据统一为 JSON 格式。

（1）约定返回报文数据格式对象。

在定义业务接口之前，需要先约定返回数据的格式对象。代码如下：

```
package com.wudimanong.efence.entity;
import com.fasterxml.jackson.annotation.JsonInclude;
...
import lombok.NoArgsConstructor;
@NoArgsConstructor
@AllArgsConstructor
@Builder
@Data
@JsonPropertyOrder({"code", "message", "data"})
public class ResponseResult<T> implements Serializable {
    private static final long serialVersionUID = 1L;
    /**
     * 返回的数据对象
     */
    @JsonInclude(JsonInclude.Include.NON_NULL)
    private T data;
    /**
     * 返回的编码
     */
    private Integer code;
    /**
     * 返回的描述信息
```

```
     */
    private String message;
    /**
     * 返回成功的响应码
     * @return 响应结果
     */
    public static ResponseResult<String> OK() {
        return packageObject("", GlobalCodeEnum.GL_SUCC_0000);
    }
    /**
     * 返回成功的响应数据
     * @param data 返回的数据
     * @param <T>  返回的数据类型
     * @return 响应结果
     */
    public static <T> ResponseResult<T> OK(T data) {
        return packageObject(data, GlobalCodeEnum.GL_SUCC_0000);
    }
    /**
     * 对返回的消息进行包装
     * @param data           返回的数据
     * @param globalCodeEnum 自定义的返回码枚举类型
     * @param <T>            返回的数据类型
     * @return 响应结果
     */
    public static <T> ResponseResult<T> packageObject(T data, GlobalCodeEnum
globalCodeEnum) {
        ResponseResult<T> responseResult = new ResponseResult<>();
        responseResult.setCode(globalCodeEnum.getCode());
        responseResult.setMessage(globalCodeEnum.getDesc());
        responseResult.setData(data);
        return responseResult;
  }
    /**
     * 在系统发生异常时的返回
     * @param <T> 返回的数据类型
     * @return 响应结果
     */
    public static <T> ResponseResult<T> systemException() {
        return packageObject(null, GlobalCodeEnum.GL_FAIL_9999);
  }
    /**
     * 在系统发生可感知异常时的返回
     * @param globalCodeEnum
     * @param <T>
```

```
     * @return
     */
    public static <T> ResponseResult<T> systemException(GlobalCodeEnum
globalCodeEnum) {
        return packageObject(null, globalCodeEnum);
    }
}
```

在上述代码中，定义了统一的响应数据包装类，并定义了处理成功及处理失败的返回方法。在后面定义业务接口层时，所有的接口返回数据都会通过此类进行包装，从而实现格式的统一。

（2）定义全局响应码枚举类。

在步骤（1）中涉及定义全局响应码的枚举类，其代码如下：

```
package com.wudimanong.efence.entity;
public enum GlobalCodeEnum {
    /**
     * 定义全局返回码
     */
    GL_SUCC_0000(0, "成功"),
    GL_FAIL_9996(996, "不支持的 HttpMethod"),
    GL_FAIL_9997(997, "HTTP 错误"),
    GL_FAIL_9998(998, "参数错误"),
    GL_FAIL_9999(999, "系统异常");
    /**
     * 编码
     */
    private Integer code;
    /**
     * 描述
     */
    private String desc;
    GlobalCodeEnum(Integer code, String desc) {
        this.code = code;
        this.desc = desc;
    }
    /**
     * 根据编码获取枚举类型
     *
     * @param code 编码
     * @return
     */
    public static GlobalCodeEnum getByCode(String code) {
        //判断编码是否为空
        if (code == null) {
```

```
            return null;
        }
        //循环处理
        GlobalCodeEnum[] values = GlobalCodeEnum.values();
        for (GlobalCodeEnum value : values) {
            if (value.getCode().equals(code)) {
                return value;
            }
        }
        return null;
    }
    public Integer getCode() {
        return code;
    }
    public String getDesc() {
        return desc;
    }
}
```

此枚举类为全局响应码定义类。它与 ResponseResult 全局响应数据包装类配合使用，可以实现接口响应码的统一定义。

2. 定义“新增图层”接口及“查询图层”接口

电子围栏的图层主要用于约定电子围栏的类型，便于业务管理，满足运营需求。在本实例中，只定义“新增图层”及“查询图层”这两个基本的接口。

（1）定义“新增图层”接口及“查询图层”接口的 Controller 层。代码如下：

```
package com.wudimanong.efence.controller;
import com.wudimanong.efence.entity.ResponseResult;
...
import org.springframework.web.bind.annotation.RestController;
@Slf4j
@RestController
@RequestMapping("/fence/layer")
public class FenceGeoLayerController {
    @Autowired
    FenceGeoLayerService fenceGeoLayerServiceImpl;
    /**
     * 定义“新增图层”接口
     *
     * @param fenceGeoLayerSaveDTO
     * @return
     */
    @PostMapping("/save")
```

```
    public ResponseResult<FenceGeoLayerBO> save(@RequestBody @Validated
FenceGeoLayerSaveDTO fenceGeoLayerSaveDTO) {
        return
ResponseResult.OK(fenceGeoLayerServiceImpl.save(fenceGeoLayerSaveDTO));
    }
    /**
     * 定义“查询图层”接口
     *
     * @param code
     * @return
     */
    @GetMapping("/getSingle")
    public ResponseResult<FenceGeoLayerBO> getSingle(
            @RequestParam(value = "code") String code) {
        return
ResponseResult.OK(fenceGeoLayerServiceImpl.getSingle(code));
    }
}
```

在上述代码中，通过 Spring MVC 提供的注解，定义了“新增图层”接口（/save）及“查询图层”接口（/getSingle）这两个接口。

- 在“/save”接口中，通过@PostMapping 注解定义了接口的 Post 请求方式，通过@RequestBody 注解约定了请求参数格式为 JSON。
- 在“/getSingle”接口中，通过@GetMapping 注解定义了接口的 Get 请求方式，其请求提交方式为“form 表单”提交，并通过@RequestParam 注解定义了接口的请求参数。

（2）定义“新增图层”接口及“查询图层”接口的请求参数对象。

定义“新增图层”接口（/save）的请求参数对象。代码如下：

```
package com.wudimanong.efence.entity.dto;
import com.wudimanong.efence.validator.EnumValue;
...
import lombok.Data;
@Data
public class FenceGeoLayerSaveDTO implements Serializable {
    /**
     * 图层的编码
     */
    @NotNull(message = "图层编码不能为空")
    private String code;
    /**
     * 图层的名称
     */
    @NotNull(message = "图层编码不能为空")
```

```
    private String name;
    /**
     * 电子围栏图层的分类。0-未知分类; 1-干预; 2-调度; 3-停车围栏; 4-运营范围; 5-技术
     */
    @EnumValue(intValues = {0, 1, 2, 3, 4, 5}, message = "围栏类型输入有误")
    private Integer businessType;
    /**
     * 归属城市编码
     */
    @NotNull(message = "归属城市编码不能为空")
    private Integer cityCode;
    /**
     * 地理范围。0-全球; 1-国内; 2-海外
     */
    @EnumValue(intValues = {0, 1, 2}, message = "地理范围输入有误")
    private Integer regionType;
    /**
     * 详细说明
     */
    private String explain;
    /**
     * 数据负责人，格式为“名字 + 邮箱”
     */
    private String owner;
}
```

在该对象中，针对具体的属性值校验，利用 Spring 框架的 Validation 机制提供的注解（如@NotNull）来对请求参数进行“是否必填”“取值范围”等数据合法性校验。

对于 Spring 框架的 Validation 机制不支持的一些数据校验方式，可以通过自定义注解来实现。例如，针对数据取值范围的校验，可以通过自定义@EnumValue 注解来处理，代码如下：

```
package com.wudimanong.efence.validator;
import static java.lang.annotation.ElementType.ANNOTATION_TYPE;
...
import javax.validation.Payload;
@Target({METHOD, FIELD, ANNOTATION_TYPE, CONSTRUCTOR, PARAMETER})
@Retention(RUNTIME)
@Documented
@Constraint(validatedBy = {EnumValueValidator.class})
public @interface EnumValue {
    //默认的错误消息
    String message() default "必须为指定值";
    String[] strValues() default {};
    int[] intValues() default {};
    Class<?>[] groups() default {};
```

```
        Class<? extends Payload>[] payload() default {};
        //在指定多个值时使用
        @Target({FIELD, METHOD, PARAMETER, ANNOTATION_TYPE})
        @Retention(RUNTIME)
        @Documented
        @interface List {
            EnumValue[] value();
        }
        class EnumValueValidator implements ConstraintValidator<EnumValue,
Object> {
            private String[] strValues;
            private int[] intValues;
            @Override
            public void initialize(EnumValue constraintAnnotation) {
                strValues = constraintAnnotation.strValues();
                intValues = constraintAnnotation.intValues();
            }
            @Override
            public boolean isValid(Object value, ConstraintValidatorContext
context) {
                if (value instanceof String) {
                    for (String s : strValues) {
                        if (s.equals(value)) {
                            return true;
                        }
                    }
                } else if (value instanceof Integer) {
                    for (Integer s : intValues) {
                        if (s == value) {
                            return true;
                        }
                    }
                }
                return false;
            }
        }
    }
```

“查询图层”接口（/getSingle）的请求参数比较简单，具体参考步骤（1）中接口参数的定义。

（3）定义“新增图层”接口及“查询图层”接口的返回参数对象。

“新增图层”接口及“查询图层”接口的返回参数格式是一样的，这里定义同一个数据对象。代码如下：

```
    package com.wudimanong.efence.entity.bo;
```

```
import com.fasterxml.jackson.annotation.JsonFormat;
import java.sql.Timestamp;
import lombok.Data;
@Data
public class FenceGeoLayerBO {
    private Integer id;
    private String code;
    private String name;
    private String businessType;
    private String explain;
    private Integer regionalType;
    private String owner;
    /**
     * 格式化日期显示
     */
    @JsonFormat(pattern = "yyyy-MM-dd HH:mm:ss", timezone = "GMT+8")
    private Timestamp createTime;
    @JsonFormat(pattern = "yyyy-MM-dd HH:mm:ss", timezone = "GMT+8")
    private Timestamp updateTime;
}
```

该接口返回数据对象由 ResponseResult 进行统一的数据报文格式包装。

3. 定义“批量导入电子围栏数据”接口

“批量导入电子围栏数据”接口是电子围栏微服务的核心功能，它提供建立符合运营要求的电子围栏数据，以及对电子围栏数据进行持续管理和维护的功能。在实际应用中，运营人员可通过此接口批量导入电子围栏数据。

（1）定义“批量导入电子围栏数据”接口的 Controller 层。代码如下：

```
package com.wudimanong.efence.controller;
import com.wudimanong.efence.entity.ResponseResult;
…
import org.springframework.web.bind.annotation.RestController;
@Slf4j
@RestController
@RequestMapping("/fence")
public class FenceGeoController {
    @Autowired
    FenceGeoService fenceGeoServiceImpl;
    /**
     * 批量导入电子围栏数据
     *
     * @param batchImportGeoFenceDTO
     * @return
```

```
     */
    @PostMapping("/batchImportGeoFence")
    public ResponseResult<List<BatchImportGeoFenceBO>>
batchImportGeoFence(
            @RequestBody @Validated BatchImportGeoFenceDTO
batchImportGeoFenceDTO) {
        return
ResponseResult.OK(fenceGeoServiceImpl.batchImportGeoFence(batchImportGeoFenc
eDTO));
    }
}
```

（2）定义“批量导入电子围栏数据”接口的请求参数对象。代码如下：

```
package com.wudimanong.efence.entity.dto;
import com.wudimanong.efence.entity.bo.GeoFenceBO;
...
import lombok.Data;
@Data
public class BatchImportGeoFenceDTO implements Serializable {
    /**
     * 城市编码
     */
    private String cityCode;
    /**
     * 管理归属分区编码
     */
    private String adCode;
    /**
     * 图层编码，关联图层信息
     */
    private String layerCode;
    /**
     * 要导入的电子围栏数据列表
     */
    @NotNull(message = "导入围栏数据不能为空")
    private List<GeoFenceBO> fences;
    /**
     * 批次导入 ID
     */
    private Integer batchId;
}
```

在上面的请求参数对象中，具体的围栏信息为 GeoFenceBO 类的 List 集合。GeoFenceBO 类的代码如下：

```
package com.wudimanong.efence.entity.bo;
```

```
import java.io.Serializable;
import javax.validation.constraints.NotNull;
import lombok.Data;
@Data
public class GeoFenceBO implements Serializable {
    /**
     * 电子围栏名称
     */
    @NotNull(message = "电子围栏名称不能为空")
    private String name;
    /**
     * 电子围栏描述
     */
    private String explain;
    /**
     * 城市编码
     */
    @NotNull(message = "归属城市编码不能为空")
    private String cityCode;
    /**
     * 管理归属区域编码
     */
    private String adCode;
    /**
     * 归属图层编码
     */
    @NotNull(message = "所属图层编码不能为空")
    private String layerCode;
    /**
     * 电子围栏的空间信息表示（尽量是完全合法的 GeoJson）
     */
    @NotNull(message = "围栏位置信息不能为空")
    private String regionGeoJson;
    /**
     * 自定义 JSON（自定义围栏业务信息）
     */
    private String customInfo;
    /**
     * 电子围栏生效日期区间
     */
    private String dateRange;
    /**
     * 电子围栏生效时间区间
     */
    private String timeBucket;
```

```
    /**
     * 状态。0-生效；1-已删除；2-失效（优先级低）
     */
    private Integer state;
}
```

（3）定义“批量导入电子围栏数据”接口的返回参数对象。

“批量导入电子围栏数据”接口的返回参数为 BatchImportGeoFenceBO 类的 List 集合，主要用于记录导入失败的围栏列表。代码如下：

```
package com.wudimanong.efence.entity.bo;
import lombok.Data;
@Data
public class BatchImportGeoFenceBO {
    /**
     * 在导入失败时，标识的电子围栏数据的索引位置
     */
    private Integer index;
    /**
     * 导入失败原因
     */
    private String message;
}
```

4. 定义“查询电子围栏数据”接口

“查询围栏电子围栏数据”接口，主要提供根据电子围栏 ID 查询电子围栏详细数据的功能。

（1）定义“查询电子围栏数据”接口的 Controller 层。

在“3.”小标题中创建的 FenceGeoController 类中，增加“查询电子围栏数据”的接口。代码如下：

```
/**
 * 根据电子围栏 ID 查询电子围栏数据
 *
 * @param fenceId
 * @return
 */
@GetMapping("/getGeofenceById")
public ResponseResult<GeoFenceBO> getGeofenceById(@RequestParam(value =
"fenceId") Integer fenceId) {
    return ResponseResult.OK(fenceGeoServiceImpl.getGeofenceById(fenceId));
}
```

（2）定义“查询电子围栏数据”接口的请求参数对象。

“查询电子围栏数据”接口的请求参数对象比较简单，具体参考步骤（1）中接口方法的参数对象定义。

（3）定义“查询电子围栏数据”接口的返回参数对象。

“查询电子围栏数据”接口的返回参数对象为前面“3.”小标题中定义的 GeoFenceBO 类。

5. 定义“判断坐标点是否在指定的电子围栏中”接口

在电子围栏微服务中，一个核心逻辑是：根据用户停车时上报的位置坐标点来判断用户是否将车辆停在指定的电子围栏中，如果未按照要求停车，或将车辆停在了明确划定的禁停区中，则需要对用户违停行为进行一定的处罚。

“判断坐标点是否在指定的电子围栏中”接口就提供了这样的功能。

（1）定义“判断坐标点是否在指定的电子围栏中”接口的 Controller 层。

在前面“3.”小标题中创建的 FenceGeoController 类中，增加“判断坐标点是否在指定的电子围栏中”接口。代码如下：

```
    /**
     * 判断坐标点是否在指定的电子围栏中
     *
     * @return
     */
    @GetMapping("/isContainPoint")
    public ResponseResult<ContainPointBO> isContainPoint(ContainPointDTO
containPointDTO) {
        return
ResponseResult.OK(fenceGeoServiceImpl.isContainPoint(containPointDTO));
    }
```

（2）定义“判断坐标点是否在指定的电子围栏中”接口的请求参数对象。代码如下：

```
    package com.wudimanong.efence.entity.dto;
    import java.io.Serializable;
    import lombok.Data;
    @Data
    public class ContainPointDTO implements Serializable {
        /**
         * 经度
         */
        private Double lon;
        /**
         * 纬度
         */
        private Double lat;
        /**
```

```
     * 电子围栏 ID
     */
    private Integer fenceId;
}
```

（3）定义“判断坐标点是否在指定的电子围栏中”接口的返回参数对象。代码如下：

```
package com.wudimanong.efence.entity.bo;
import java.io.Serializable;
import lombok.Builder;
import lombok.Data;
@Data
@Builder
public class ContainPointBO implements Serializable {
    private Boolean result;
}
```

4.4.2 开发业务层（Service 层）代码

业务层（Service 层）主要负责接收 Controller 层传递来的请求数据对象，完成正常及异常的业务逻辑处理逻辑。在业务逻辑的处理过程中，会涉及操作数据库及第三方组件（如 Redis 等）。

对于业务逻辑过于复杂的业务层（Service 层），可以对代码结构进行职责拆分，运用软件设计原则及设计模式让程序更易于扩展和维护。

1. 定义业务异常处理机制

在开发 Service 层代码时，除处理正常逻辑外，还需要处理异常逻辑：一旦数据或条件逻辑不满足程序正常运行的要求，则中断逻辑的处理，并透过 Controller 层向服务调用方返回异常信息。

在这种情况下，业务层（Service 层）需要封装待返回的业务异常信息，而 Controller 层也需要将其转换为接口的返回数据。为了减少业务层（Service 层）及 Controller 层的代码量，在实践中可以通过 Spring 提供的全局异常处理机制对业务异常进行统一的处理。步骤如下。

（1）定义通用的业务层（Service 层）业务异常类。代码如下：

```
package com.wudimanong.efence.exception;
public class ServiceException extends RuntimeException {
    private final Integer code;
    public ServiceException(Integer code, String message) {
        super(message);
        this.code = code;
    }
    public ServiceException(Integer code, String message, Throwable e) {
        super(message, e);
        this.code = code;
```

```
    }
    public Integer getCode() {
        return code;
    }
}
```

该类是一个继承了运行时异常的业务层（Service 层）异常基类，其他的业务异常类可以通过继承它来实现。

（2）通过@ControllerAdvice 注解定义系统全局异常处理类。代码如下：

```
package com.wudimanong.efence.exception;
import com.wudimanong.efence.entity.GlobalCodeEnum;
...
import org.springframework.web.bind.annotation.ResponseBody;
@Slf4j
@ControllerAdvice
public class GlobalExceptionHandler {
    /**
     * 通用的业务异常处理方法
     */
    @ExceptionHandler(ServiceException.class)
    @ResponseBody
    public ResponseResult<?> processServiceException(
            HttpServletResponse response, ServiceException e) {
        response.setStatus(HttpStatus.OK.value());
        response.setContentType("application/json;charset=UTF-8");
        ResponseResult result = new ResponseResult();
        result.setCode(e.getCode());
        result.setMessage(e.getMessage());
        log.error(e.toString() + "_" + e.getMessage(), e);
        return result;
    }
    /**
     *参数校验异常的处理方法 1
     *
     * @param response
     * @param e
     * @return
     */
    @ExceptionHandler(MethodArgumentNotValidException.class)
    @ResponseBody
    public ResponseResult<?> processValidException(HttpServletResponse
response, MethodArgumentNotValidException e) {
        response.setStatus(HttpStatus.INTERNAL_SERVER_ERROR.value());
        List<String> errorStringList = e.getBindingResult().getAllErrors()
```

```
                .stream().map(ObjectError::getDefaultMessage).collect(Colle
ctors.toList());
            String errorMessage = String.join("; ", errorStringList);
            response.setContentType("application/json;charset=UTF-8");
            log.error(e.toString() + "_" + e.getMessage(), e);
            return
ResponseResult.systemException(GlobalCodeEnum.GL_FAIL_9998);
        }
        /**
         * 参数校验异常的处理方法 2
         *
         * @param response
         * @param e
         * @return
         */
        @ExceptionHandler(BindException.class)
        @ResponseBody
        public ResponseResult<?> processValidException(HttpServletResponse
response, BindException e) {
            response.setStatus(HttpStatus.INTERNAL_SERVER_ERROR.value());
            List<String> errorStringList = e.getBindingResult().getAllErrors()
                .stream().map(ObjectError::getDefaultMessage).collect(Colle
ctors.toList());
            String errorMessage = String.join("; ", errorStringList);
            response.setContentType("application/json;charset=UTF-8");
            log.error(e.toString() + "_" + e.getMessage(), e);
            return
ResponseResult.systemException(GlobalCodeEnum.GL_FAIL_9998);
        }
        /**
         * 参数校验异常的处理方法 3
         *
         * @param response
         * @param e
         * @return
         */
        @ExceptionHandler(HttpRequestMethodNotSupportedException.class)
        @ResponseBody
        public ResponseResult<?> processValidException(HttpServletResponse
response,
              HttpRequestMethodNotSupportedException e) {
            response.setStatus(HttpStatus.INTERNAL_SERVER_ERROR.value());
            String[] supportedMethods = e.getSupportedMethods();
            String errorMessage = "此接口不支持" + e.getMethod();
            if (!ArrayUtils.isEmpty(supportedMethods)) {
```

```
            errorMessage += "（仅支持" + String.join(",", supportedMethods) +
"）";
        }
        response.setContentType("application/json;charset=UTF-8");
        log.error(e.toString() + "_" + e.getMessage(), e);
        return
ResponseResult.systemException(GlobalCodeEnum.GL_FAIL_9996);
    }
    /**
     * 未知系统异常的处理方法
     */
    @ExceptionHandler(Exception.class)
    @ResponseBody
    public ResponseResult<?> processDefaultException(HttpServletResponse
response, Exception e) {
        response.setStatus(HttpStatus.INTERNAL_SERVER_ERROR.value());
        response.setContentType("application/json;charset=UTF-8");
        log.error(e.toString() + "_" + e.getMessage(), e);
        return ResponseResult.systemException();
    }
}
```

在该全局异常处理类中，定义了通用业务异常的处理方法。这样，业务层（Service 层）在处理业务异常时，可以通过直接抛出业务层（Service 层）异常来实现 Controller 层调用异常的返回处理。业务层（Service 层）只需要在抛出业务异常时，设置相应的业务异常码及异常数据即可；Controller 层也无须进行额外的“try-catch”操作，由系统全局异常类处理即可。

此外，在该全局异常处理类中还定义了参数校验及未知系统异常的处理方法。

（3）定义业务层（Service 层）业务异常码的枚举类。

在具体的业务层（Service 层）异常处理中，可以按照业务类型来定义业务异常码及信息。定义枚举类的代码如下：

```
package com.wudimanong.efence.entity;
public enum BusinessCodeEnum {
    /**
     * 定义图层相关的异常码（例如以 1000 开头，可根据业务扩展）
     */
    BUSI_LAYER_FAIL_1000(1000, "图层信息已存在"),
    /**
     * 定义电子围栏操作相关的异常码（例如以 2000 开头，根据业务扩展)
     */
    BUSI_FENCE_FAIL_1000(2000, "围栏信息已存在");
    /**
     * 编码
```

```
     */
    private Integer code;
    /**
     * 描述
     */
    private String desc;
    BusinessCodeEnum(Integer code, String desc) {
        this.code = code;
        this.desc = desc;
    }
    /**
     * 根据编码获取枚举类型
     *
     * @param code 编码
     * @return
     */
    public static BusinessCodeEnum getByCode(String code) {
        //判断编码是否为空
        if (code == null) {
            return null;
        }
        //循环处理
        BusinessCodeEnum[] values = BusinessCodeEnum.values();
        for (BusinessCodeEnum value : values) {
            if (value.getCode().equals(code)) {
                return value;
            }
        }
        return null;
    }
    public Integer getCode() {
        return code;
    }
    public String getDesc() {
        return desc;
    }
}
```

此枚举类可根据具体业务情况进行扩展，例如将图层相关的业务异常码约定为 1000～1999，将电子围栏操作相关的业务异常码约定为 2000～2999。

2. 引入 MapStruct 实体映射工具

在系统分层结构中，在不同的分层之间需要进行大量的数据交互：在开发业务层（Service 层）代码的过程中，需要将“业务层（Service 层）的数据对象”转换为“持久层（Dao 层）的数据对象”，以及将“持久层（Dao 层）的数据对象”转换为“业务层（Service 层）输出的数据对象”。为了简化这部分代码逻辑，需要引入 MapStruct 实体映射工具。

（1）引入 MapStruct 的依赖。

在项目的 pom.xml 文件中引入 MapStruct 的依赖，代码如下：

```
<!--引入 MapStruct 的依赖-->
 <dependency>
     <groupId>org.mapstruct</groupId>
     <artifactId>mapstruct-jdk8</artifactId>
     <version>1.3.1.Final</version>
 </dependency>
 <dependency>
     <groupId>org.mapstruct</groupId>
     <artifactId>mapstruct-processor</artifactId>
     <version>1.3.1.Final</version>
 </dependency>
```

（2）配置 Maven 编译插件

此外，还需要在 pom.xml 文件的<build>标签中加入 Maven 编译插件，代码如下：

```
<!--提供给 MapStruct 使用的 Maven 编译插件 -->
<plugin>
    <groupId>org.apache.maven.plugins</groupId>
    <artifactId>maven-compiler-plugin</artifactId>
</plugin>
```

至此，完成了引入 MapStruct 实体映射工具的步骤。在接下来开发业务层（Service 层）的代码中将使用到该工具，以减少由于软件分层而增加的复制数据的代码。

3. 开发“新增图层”接口及“查询图层”接口的业务层（Service 层）代码

接下来开发“新增图层”接口及“查询图层”接口的业务层（Service 层）代码。

（1）定义“新增图层”接口及“查询图层”接口的业务层（Service 层）方法。代码如下：

```
package com.wudimanong.efence.service;
import com.wudimanong.efence.entity.bo.FenceGeoLayerBO;
import com.wudimanong.efence.entity.dto.FenceGeoLayerSaveDTO;
public interface FenceGeoLayerService {
    /**
     * 定义“新增图层”接口的业务层（Service 层）方法
     *
     * @param fenceGeoLayerSaveDTO
     * @return
     */
```

```
    FenceGeoLayerBO save(FenceGeoLayerSaveDTO fenceGeoLayerSaveDTO);
    /**
     * 定义“查询图层”接口的业务层（Service 层）方法
     *
     * @param code
     * @return
     */
    FenceGeoLayerBO getSingle(String code);
}
```

（2）开发“新增图层”接口及“查询图层”接口的业务层（Service 层）方法的代码。

“新增图层”接口及“查询图层”接口的业务层（Service 层）方法的实现类代码如下：

```
package com.wudimanong.efence.service.impl;
import com.wudimanong.efence.convert.FenceGeoLayerConvert;
...
import org.springframework.stereotype.Service;
@Slf4j
@Service
public class FenceGeoLayerServiceImpl implements FenceGeoLayerService {
    /**
     * 注入持久层（Dao 层）操作的依赖对象
     */
    @Autowired
    FenceGeoLayerDao fenceGeoLayerDao;
    /**
     * 实现“新增图层”接口的业务层（Service 层）方法
     *
     * @param fenceGeoLayerSaveDTO
     * @return
     */
    @Override
    public FenceGeoLayerBO save(FenceGeoLayerSaveDTO fenceGeoLayerSaveDTO) {
        //根据 code 判断电子围栏信息是否重复
        Map map = new HashMap<>();
        map.put("code", fenceGeoLayerSaveDTO.getCode());
        List<FenceGeoLayerPO> layerPOList =
fenceGeoLayerDao.selectByMap(map);
        if (layerPOList != null && layerPOList.size() > 0) {
            throw new FenceGeoLayerServiceException
(BusinessCodeEnum.BUSI_LAYER_FAIL_1000.getCode(),
                    BusinessCodeEnum.BUSI_LAYER_FAIL_1000.getDesc(),
fenceGeoLayerSaveDTO);
        }
```

```
            //将“业务层(Service 层)输出的数据对象”转换为“持久层(Dao 层)数据对象”,
这里使用 MapStruct 减少数据转换的代码量
            FenceGeoLayerPO fenceGeoLayerPO = FenceGeoLayerConvert.INSTANCE
                    .convertFenceGeoLayerPO(fenceGeoLayerSaveDTO);
            fenceGeoLayerPO.setCreateTime(new
Timestamp(System.currentTimeMillis()));
            fenceGeoLayerPO.setUpdateTime(new
Timestamp(System.currentTimeMillis()));
            //完成 MyBatis 及 MyBatis-Plus 支持的数据库 Insert 操作
            fenceGeoLayerDao.insert(fenceGeoLayerPO);
            //将“持久层(Dao 层)数据对象”转换为“业务层(Service 层)输出的数据对象”
            FenceGeoLayerBO fenceGeoLayerBO = FenceGeoLayerConvert.INSTANCE
                    .convertFenceGeoLayerBO(fenceGeoLayerPO);
            fenceGeoLayerBO.setRegionalType
(fenceGeoLayerSaveDTO.getRegionType());
            return fenceGeoLayerBO;
        }
        /**
         * 实现“查询图层”接口的业务层(Service 层)方法
         *
         * @param code
         * @return
         */
        @Override
        public FenceGeoLayerBO getSingle(String code) {
            Map map = new HashMap<>();
            map.put("code", code);
            List<FenceGeoLayerPO> layerPOList =
fenceGeoLayerDao.selectByMap(map);
            //将数据库对象转换为“业务层(Service 层)输出的 BO 对象”
            FenceGeoLayerBO fenceGeoLayerBO = FenceGeoLayerConvert.INSTANCE
                    .convertFenceGeoLayerBO(layerPOList.get(0));
            return fenceGeoLayerBO;
        }
    }
```

(3)编写 MapStruct 数据转化类。

在步骤(2)的代码中，分别实现了“新增图层”接口及“查询图层”接口的业务层(Service 层)方法。在实现的具体逻辑中，数据对象的复制使用了“2.”小标题中引入的 MapStruct 工具。涉及的代码片段如下：

```
    //将“业务层(Service 层)输出的数据对象”转换为“持久层(Dao 层)数据对象”,这里使用
MapStruct 减少数据转换的代码量
    FenceGeoLayerPO fenceGeoLayerPO = FenceGeoLayerConvert.INSTANCE
            .convertFenceGeoLayerPO(fenceGeoLayerSaveDTO);
```

```
...

//将数据库对象转换为业务层（Service 层）输出的 BO 对象
FenceGeoLayerBO fenceGeoLayerBO = FenceGeoLayerConvert.INSTANCE
        .convertFenceGeoLayerBO(fenceGeoLayerPO);
...
```

在上述代码片段中，涉及的具体映射转换类的代码如下：

```
package com.wudimanong.efence.convert;
import com.wudimanong.efence.dao.model.FenceGeoLayerPO;
...
import org.mapstruct.factory.Mappers;
@org.mapstruct.Mapper
public interface FenceGeoLayerConvert {
    FenceGeoLayerConvert INSTANCE =
Mappers.getMapper(FenceGeoLayerConvert.class);
    /**
     * 从“新增图层”接口的业务层（Service 层）输出的数据对象到持久层（Dao 层）数据对象
的转换方法
     *
     * @param salesCouponChannelsDTO
     * @return
     */
    @Mappings({
            @Mapping(source = "businessType", target = "type"),
            @Mapping(source = "owner", target = "createUser")
    })
    FenceGeoLayerPO convertFenceGeoLayerPO(FenceGeoLayerSaveDTO
salesCouponChannelsDTO);
    /**
     *从“新增图层”接口的持久层（Dao 层）数据对象到业务层（Service 层）输出的数据对象
的转换方法
     *
     * @param fenceGeoLayerPO
     * @return
     */
    @Mappings({
            @Mapping(source = "type", target = "businessType"),
            @Mapping(source = "createUser", target = "owner")
    })
    FenceGeoLayerBO convertFenceGeoLayerBO(FenceGeoLayerPO
fenceGeoLayerPO);
}
```

在上述数据映射接口方法中，如果“源对象”与“目标对象”的属性名称一致，则不需要通过

@Mapping 注解进行额外的字段映射，否则就需要通过@Mapping 注解进行字段映射。

在业务层（Service 层）实现逻辑中，涉及的数据库操作组件将在 4.4.3 节中实现。

4. 开发“批量导入电子围栏数据”接口的业务层（Service 层）代码

接下来开发“批量导入电子围栏数据”接口的业务层（Service 层）代码。

（1）定义业务层（Service 层）接口类 FenceGeoService。代码如下：

```
package com.wudimanong.efence.service;
import com.wudimanong.efence.entity.bo.BatchImportGeoFenceBO;
...
import java.util.List;
public interface FenceGeoService {
    /**
     * 定义“批量导入电子围栏数据”接口的业务层（Service 层）方法
     *
     * @param batchImportGeoFenceDTO
     * @return
     */
    List<BatchImportGeoFenceBO>
batchImportGeoFence(BatchImportGeoFenceDTO batchImportGeoFenceDTO);
}
```

（2）实现业务层（Service 层）接口类的方法。代码如下：

```
package com.wudimanong.efence.service.impl;
import com.wudimanong.efence.convert.FenceGeoConvert;
...
import org.springframework.transaction.annotation.Transactional;
@Slf4j
@Service
public class FenceGeoServiceImpl implements FenceGeoService {
    /**
     * 注入持久层（Dao 层）组件
     */
    @Autowired
    FenceGeoDao fenceGeoDao;
    /**
     * 实现“批量导入电子围栏数据”接口的业务层（Service 层）方法
     *
     * @param batchImportGeoFenceDTO
     * @return
     */
    @Transactional(rollbackFor = Exception.class)
    @Override
```

```
    public List<BatchImportGeoFenceBO>
batchImportGeoFence(BatchImportGeoFenceDTO batchImportGeoFenceDTO) {
        //对批量导入的围栏数据，进行数据校验及过滤（将校验方法单独拆分）
        Map<String, List<GeoFenceBO>> validateFenceData =
validateFenceData(batchImportGeoFenceDTO.getFences());
        //获取合法的电子围栏数据请求列表
        List<GeoFenceBO> successFenceData =
validateFenceData.get("success");
        if (successFenceData != null && successFenceData.size() > 0) {
            //通过 MapStruct，将 BO 数据对象转换为 PO 持久层（Dao 层）数据对象
            List<FenceGeoInfoPO> successFenceDataPO =
convertSuccessFenceData(batchImportGeoFenceDTO, successFenceData);
            //向数据库中批量导入电子围栏数据
            fenceGeoDao.batchInsert(successFenceDataPO);
        }
        //将获取数据请求列表中的不合法的电子围栏数据识别出来
        List<GeoFenceBO> failedFenceData = validateFenceData.get("fail");
        //将导入不合法的电子围栏数据列表转换为输出数据对象
        List<BatchImportGeoFenceBO> batchImportGeoFenceBOList =
convertFailFenceData(batchImportGeoFenceDTO,
                failedFenceData);
        return batchImportGeoFenceBOList;
    }
    /**
     * 电子围栏数据的合法性校验方法
     *
     * @param fenceBOList
     * @return
     */
    private Map<String, List<GeoFenceBO>>
validateFenceData(List<GeoFenceBO> fenceBOList) {
        //校验结果数据
        Map<String, List<GeoFenceBO>> validateResult = new HashMap<>();
        //合法数据
        List<GeoFenceBO> successGeoFenceBO = new ArrayList<>();
        //非法数据
        List<GeoFenceBO> failGeoFenceBO = new ArrayList<>();
        for (GeoFenceBO geoFenceBO : fenceBOList) {
            //根据实际的业务场景对数据进行过滤。这里不进行具体实现，直接返回原始的输入
数据
        }
        validateResult.put("success", fenceBOList);
        return validateResult;
    }
    /**
```

```
         * 将合法的电子围栏数据转换为 PostgreSQL 数据库持久层（Dao 层）对象的方法
         *
         * @param successFenceBOList
         * @return
         */
        private List<FenceGeoInfoPO>
convertSuccessFenceData(BatchImportGeoFenceDTO batchImportGeoFenceDTO,
                List<GeoFenceBO> successFenceBOList) {
            List<FenceGeoInfoPO> fenceGeoInfoPOList = new ArrayList<>();
            for (GeoFenceBO geoFenceBO : successFenceBOList) {
                FenceGeoInfoPO fenceGeoInfoPO =
FenceGeoConvert.INSTANCE.convertFenceGeoPO(geoFenceBO);
                //在将 GeoJson 转换为 Polygon 数据类型后进行数据设置
                Polygon regionPolygon =
GeoJsonUtil.convertPointArrayJsonToPolygon(geoFenceBO.getRegionGeoJson());
                fenceGeoInfoPO.setRegion(regionPolygon);
                //通过 PostGis 提供的 ST_Centroid(*)函数或 geoTools 工具包提供的方法来计
算电子围栏的中心点
                //通过 PostGis 提供的 ST_Area(*)函数或 geoTools 工具包提供的方法来计算电
子围栏的面积
                //设置导入批次 ID

fenceGeoInfoPO.setBatchId(batchImportGeoFenceDTO.getBatchId());
                fenceGeoInfoPOList.add(fenceGeoInfoPO);
            }
            return fenceGeoInfoPOList;
        }
        /**
         * 将不合法的电子围栏数据转换为错误结果对象列表的方法
         *
         * @param batchImportGeoFenceDTO
         * @param failFenceBOList
         * @return
         */
        private List<BatchImportGeoFenceBO>
convertFailFenceData(BatchImportGeoFenceDTO batchImportGeoFenceDTO,
                List<GeoFenceBO> failFenceBOList) {
            List<BatchImportGeoFenceBO> list = new ArrayList<>();
            //将不合法的数据转换为接口输出格式，这里仅作示范不进行具体实现
            return list;
        }
    }
```

上述代码实现了“批量导入电子围栏数据”接口的业务层（Service 层）方法，为了避免单个业务方法的代码量过大，对数据校验、转换等逻辑进行了单独的方法拆分。

（3）编写基于 MapStruct 的数据对象转化类。代码如下：

在步骤（2）中涉及的数据对象转换逻辑是通过定义 MapStruct 转换接口来实现的。代码如下：

```
package com.wudimanong.efence.convert;
import com.wudimanong.efence.dao.model.FenceGeoInfoPO;
...
import org.mapstruct.factory.Mappers;
@org.mapstruct.Mapper
public interface FenceGeoConvert {
    FenceGeoConvert INSTANCE = Mappers.getMapper(FenceGeoConvert.class);
    /**
     * 将“批量导入电子围栏数据”接口的“业务层（Service 层）输出的数据对象”转换为“持久层（Dao 层）的数据对象”
     * @param geoFenceBO
     * @return
     */
    @Mappings({})
    FenceGeoInfoPO convertFenceGeoPO(GeoFenceBO geoFenceBO);
    /**
     * 将“批量导入电子围栏数据”接口的“持久层(Dao 层)的数据对象”转换为“业务层(Service 层）输出的数据对象”
     * @param fenceGeoInfoPO
     * @return
     */
    @Mappings({})
    GeoFenceBO convertFenceGeoBO(FenceGeoInfoPO fenceGeoInfoPO);
}
```

（4）编写 GeoJsonUtil 工具类。

在步骤（2）的实现方法中，在将电子围栏位置 GeoJson 数据转换为“持久层（Dao 层）数据对象”FenceGeoInfoPO 时，使用了“Polygon”数据类型，涉及的 GeoJsonUtil 工具类的代码如下：

```
package com.wudimanong.efence.utils;
import com.alibaba.fastjson.JSON;
...
import org.postgis.Polygon;
public class GeoJsonUtil {
    /**
     *将 GeoJson 字符串数据转换为 PostGis 的 Polygon 数据类型
     *
     * @param geoJson
     * @return
     */
```

```
    public static Polygon convertPointArrayJsonToPolygon(String geoJson) {
        //将 GeoJson 对象转换为 Map 对象
        Map<String, Object> mapJson = JSON.parseObject(geoJson,
HashMap.class);
        //获取 GeoJson 中的坐标点列表数据
        String pointsJson = mapJson.get("coordinates").toString();
        List<List<List<BigDecimal>>> fencePoints =
JSON.parseObject(pointsJson, ArrayList.class);
        List<Point> pointList = new ArrayList<>();
        for (List<List<BigDecimal>> lists : fencePoints) {
            for (List<BigDecimal> pointValue : lists) {
                Double x = pointValue.get(0).doubleValue();
                Double y = pointValue.get(1).doubleValue();
                Point point = new Point(x, y);
                pointList.add(point);
            }
        }
        Point[] points = new Point[pointList.size()];
        pointList.toArray(points);
        LinearRing linearRing = new LinearRing(points);
        Polygon polygon = new Polygon(new LinearRing[]{linearRing});
        polygon.setSrid(4326);
        return polygon;
    }
    /**
     * 将“经纬度坐标”转换为 PostgreSQL 的 Point 数据类型
     *
     * @param lng
     * @param lat
     * @return
     */
    public static String getPointByLngAndLat(double lng, double lat) {
        Point point = new Point(lng, lat);
        PGgeometry pGgeometry = new PGgeometry();
        pGgeometry.setGeometry(point);
        String wktPoint = pGgeometry.getValue();
        return wktPoint;
    }

    /**
     * int 数组转换
     *
     * @param arr
     * @return
     */
```

```
    public static int NumberOf1(int[] arr) {
        int len = arr.length;
        int res = -1;
        if (len > 1) {
            res = arr[0];
            for (int i = 1; i < len; i++) {
                res = res ^ arr[i];
                System.out.println("-->" + res);
            }
        }
        return res;
    }
}
```

上述代码使用了 PostGIS 数据库驱动程序中的一些数据类型，将在 4.4.2 节持久层（Dao 层）的实现中进行具体介绍。

5. 开发“查询电子围栏数据”接口的业务层（Service 层）代码

接下来开发“查询电子围栏数据”接口的业务层（Service 层）代码，步骤如下。

（1）定义业务层（Service 层）方法。

在“4.”小标题下步骤（1）中定义的业务层（Service 层）接口 FenceGeoService 类中，增加“查询电子围栏数据”接口的业务层（Service 层）方法。代码如下：

```
/**
 * 定义“查询电子围栏数据”接口的业务层（Service 层）方法
 *
 * @param fenceId
 * @return
 */
GeoFenceBO getGeofenceById(Integer fenceId);
```

（2）实现业务层（Service 层）方法。

在“4.”小标题下步骤（2）中定义的业务层（Service 层）实现 FenceGeoServiceImpl 类中，增加“查询电子围栏数据”接口的业务层（Service 层）方法的实现。代码如下：

```
/**
 * 实现“查询电子围栏数据”接口的业务层（Service 层）方法
 *
 * @param fenceId
 * @return
 */
@Override
public GeoFenceBO getGeofenceById(Integer fenceId) {
```

```
        FenceGeoInfoPO fenceGeoInfoPO = fenceGeoDao.selectById(fenceId);
        GeoFenceBO geoFenceBO = null;
        if (fenceGeoInfoPO != null) {
            geoFenceBO = FenceGeoConvert.INSTANCE.convertFenceGeoBO
(fenceGeoInfoPO);
            geoFenceBO.setRegionGeoJson
(fenceGeoInfoPO.getRegion().toString());
        }
        return geoFenceBO;
    }
```

上述代码中涉及的数据对象转换方法，请参考“4.”小标题下步骤（3）中的转换代码。

6. 开发“判断坐标点是否在指定的电子围栏中”接口的业务层（Service 层）代码

（1）定义业务层（Service 层）方法。

在“4.”小标题下步骤（1）中定义的业务层（Service 层）接口 FenceGeoService 类中，增加“判断坐标点是否在指定的电子围栏中”接口的业务层（Service 层）方法。代码如下：

```
    /**
     * “判断坐标点是否在指定的电子围栏中”接口的业务层（Service 层）方法
     *
     * @param containPointDTO
     * @return
     */
    ContainPointBO isContainPoint(ContainPointDTO containPointDTO);
```

（2）实现业务层（Service 层）方法。

在“4.”小标题下步骤（2）中定义的业务层（Service 层）实现 FenceGeoServiceImpl 类中，增加“判断坐标点是否在指定的电子围栏中”接口的业务层（Service 层）方法的实现，具体代码如下：

```
    /**
     * “判断坐标点是否在指定的电子围栏中”接口的业务层（Service 层）方法
     *
     * @param containPointDTO
     * @return
     */
    @Override
    public ContainPointBO isContainPoint(ContainPointDTO containPointDTO) {
        String result = fenceGeoDao
                .isContainPoint(containPointDTO.getLon(),
containPointDTO.getLat(), containPointDTO.getFenceId());
        ContainPointBO containPointBO;
        if ("f".equals(result)) {
```

```
            containPointBO = ContainPointBO.builder().result(new
Boolean(false)).build();
        } else {
            containPointBO = ContainPointBO.builder().result(new
Boolean(true)).build();
        }
        return containPointBO;
    }
```

在上述的方法实现中涉及的通过“PostgreSQL + PostGIS”数据库操作进行空间位置判断的逻辑，请参考 4.4.3 节的实现。

4.4.3 开发 MyBatis 持久层（Dao 层）组件

1. 添加 MyBatis 接口代码的包扫描路径

在本实例中，持久层（Dao 层）的实现是以 MyBatis 及 MyBatis-Plus 框架的功能为基础的。在微服务入口类中，添加 MyBatis 接口代码的包扫描路径，具体如下：

```
package com.wudimanong.efence;
import org.mybatis.spring.annotation.MapperScan;
import org.springframework.boot.SpringApplication;
import org.springframework.boot.autoconfigure.SpringBootApplication;
import org.springframework.cloud.client.discovery.EnableDiscoveryClient;
@EnableDiscoveryClient
@SpringBootApplication
@MapperScan("com.wudimanong.efence.dao.mapper")
public class FenceApplication {
    public static void main(String[] args) {
        SpringApplication.run(FenceApplication.class, args);
    }
}
```

在上述代码中，通过@MapperScan 注解配置了 MyBatis 持久层（Dao 层）接口代码的包扫描路径。

2. 实现“新增图层”接口及“查询图层”接口的持久层（Dao 层）

“新增图层”接口及“查询图层”接口涉及的数据库操作主要是针对电子围栏图层信息表（fence_geo_layer）的。

（1）定义电子围栏图层信息表的数据库实体类。代码如下：

```
package com.wudimanong.efence.dao.model;
import com.baomidou.mybatisplus.annotation.KeySequence;
...
import lombok.Data;
```

```
@Data
@TableName("fence_geo_layer")
@KeySequence(value = "fence_geo_layer_id_seq")
public class FenceGeoLayerPO implements Serializable {
    private Integer id;
    private String code;
    private String name;
    private String explain;
    private String checkCity;
    private String cityCode;
    private Integer state;
    private Integer type;
    private Timestamp createTime;
    private Timestamp updateTime;
    private String createUser;
    private String updateUser;
}
```

由于要使用 MyBatis-Plus 工具来简化 MyBatis 操作，所以在定义数据库实体类时需要通过@TableName 注解指定具体的表名。

此外，本实例使用的数据库是 PostgreSQL，所以在 ID 主键生成策略上需要使用序列的方式（可以通过@KeySequence 注解指定具体的数据库序列）。

（2）创建 MyBatis-Plus 支持 PostgreSQL 主键生成器的配置类。

在定义步骤（1）中的数据库实体对象时，虽然通过@KeySequence 注解指定了序列，但还需要配置 MyBatis-Plus 对 PostgreSQL 主键生成器的支持。具体的配置类代码如下：

```
package com.wudimanong.efence.config;
import com.baomidou.mybatisplus.extension.incrementer.
PostgreKeyGenerator;
import org.springframework.context.annotation.Bean;
import org.springframework.context.annotation.Configuration;
@Configuration
public class MybatisPlusConfiguration {
    /**
     * 配置 MyBatis-Plus 对 PostgreSQL 主键生成器的支持
     *
     * @return
     */
    @Bean
    public PostgreKeyGenerator createPostgreKeyGenerator() {
        return new
com.baomidou.mybatisplus.extension.incrementer.PostgreKeyGenerator();
    }
}
```

（3）定义“电子围栏图层信息表”的持久层（Dao 层）组件。

定义“新增图层”接口及“查询图层”接口的持久层（Dao 层）组件。代码如下：

```
package com.wudimanong.efence.dao.mapper;

import com.baomidou.mybatisplus.core.mapper.BaseMapper;
import com.wudimanong.efence.dao.model.FenceGeoLayerPO;
import org.springframework.stereotype.Repository;

@Repository
public interface FenceGeoLayerDao extends BaseMapper<FenceGeoLayerPO> {
}
```

在上述持久层（Dao）接口中，并没有定义 MyBatis 的 XML 映射文件，这是因为使用 MyBatis-Plus 对 MyBatis 的操作进行了简化：通过继承 BaseMapper 泛型接口就能实现对数据库单表操作的封装。

3. 配置“MyBatis + MyBatis-Plus”对 PostGIS 空间字段类型的支持

在实现针对“电子围栏信息表”操作的持久层（Dao 层）组件之前，需要先引入 PostGIS 的驱动程序，以及配置“MyBatis + MyBatis-Plus”对 PostGIS 中特定空间数据类型映射的支持。步骤如下。

（1）引入 PostGIS 的驱动程序。

在本实例中，很多关于空间计算的功能使用了 PostGIS 数据库，所以需要在项目的 pom.xml 文件中引入 PostGIS 的驱动程序。代码如下：

```
<!--PostGIS 的驱动程序-->
<dependency>
    <groupId>org.postgis</groupId>
    <artifactId>postgis-jdbc</artifactId>
    <version>1.3.3</version>
</dependency>
```

（2）配置 MyBatis 对 PostGIS 中特定空间数据类型映射的支持。

在电子围栏信息表中，围栏范围等信息是通过 PostGIS 所支持的 Polygon、Point 等类型来存储的。

而在本实例中，数据库的操作是通过 MyBatis 框架来完成的，而默认情况下 MyBatis 不支持对 Polygon 及 Point 类型的 SQL 映射，所以需要引入额外的依赖支持。代码如下：

```
<!--MyBatis 无法识别 PostGIS 的 Geometry 等数据类型的适配-->
<dependency>
    <groupId>com.eyougo</groupId>
```

```
    <artifactId>mybatis-typehandlers-postgis</artifactId>
    <version>1.0</version>
    <exclusions>
        <!--排除 MyBatis 冲突依赖-->
        <exclusion>
            <groupId>org.mybatis</groupId>
            <artifactId>mybatis</artifactId>
        </exclusion>
    </exclusions>
</dependency>
```

（3）配置 MyBatis-Plus 对 PostGIS 中特定空间数据类型映射的支持。

引入步骤（2）中的依赖，实际上是引入了一些支持特定数据处理类型的 Handler。所以，在 MyBatis-Plus 中，需要配置这些处理 Handler 的包路径。在配置文件 application.yml 中加入如下配置：

```
#MyBatis-Plus 集成配置
mybatis-plus:
  ...
  #MyBatis-Plus 处理 PostGIS 数据类型映射的配置
  typeHandlersPackage: com.eyougo.mybatis.postgis.type
  ...
```

（4）引入支持 GIS 计算的 Java 依赖。

关于 GIS 计算，PostGIS 数据库提供了很多可以直接使用的函数。但如果想在 Java 代码中直接进行 GIS 计算，则需要使用一些开源工具，需要引入如下依赖：

```
<!—引入 Java  Geo 计算工具依赖，以实现在 Java 代码中进行 GIS 计算-->
<dependency>
     <groupId>org.geotools</groupId>
     <artifactId>gt-geojson</artifactId>
     <version>9.3</version>
 </dependency>
<dependency>
    <groupId>com.github.davidmoten</groupId>
    <artifactId>geo</artifactId>
    <version>0.7.6</version>
</dependency>
<!--引入 JTS 包，以支持拓扑服务中的常用算法-->
<dependency>
    <groupId>com.vividsolutions</groupId>
    <artifactId>jts</artifactId>
    <version>1.13</version>
</dependency>
```

```
<dependency>
    <groupId>com.vividsolutions</groupId>
    <artifactId>jts-io</artifactId>
    <version>1.14.0</version>
</dependency>
```

本实例代码中，空间计算的功能尚未完全实现，引入以上依赖仅仅在抛砖引玉，有这方面需要的读者可自行扩展。

4. 开发“批量导入电子围栏数据”等接口的持久层（Dao 层）代码

“批量导入电子围栏数据”接口、“查询电子围栏数据”接口及“判断坐标点是否在指定的电子围栏中”接口操作的都是电子围栏信息表（fence_geo_info）。

接下来开发操作电子围栏信息表的持久层（Dao 层）代码，步骤如下。

（1）定义业务层（Service 层）所依赖的持久层（Dao 层）接口。代码如下：

```
package com.wudimanong.efence.dao.mapper;
import com.baomidou.mybatisplus.core.mapper.BaseMapper;
...
import org.springframework.stereotype.Repository;
@Repository
public interface FenceGeoDao extends BaseMapper<FenceGeoInfoPO> {
    /**
     * 批量入库的方法
     *
     * @param fenceGeoInfoPOList
     * @return
     */
    int batchInsert(List<FenceGeoInfoPO> fenceGeoInfoPOList);
    /**
     * PostGIS 函数判断坐标点是否在指定的电子围栏中的持久层（Dao 层）方法
     *
     * @param lon
     * @param lat
     * @param fenceId
     * @return
     */
    String isContainPoint(@Param("lon") Double lon, @Param("lat") Double lat,
@Param("fenceId") Integer fenceId);
}
```

（2）编写 MyBatis 批量入库 XML 映射文件。

在步骤（1）的代码中，由于 MyBatis-Plus 封装了很多基础的数据库单表操作，所以简单的数据库单表操作都不需要定义额外的 SQL 映射。

但是批量数据插入及利用 PostGIS 的函数进行空间计算的操作，是 MyBatis-Plus 无法完成的，所以还需要定义 MyBatis 映射文件来实现。MyBatis 映射文件（FenceGeoDao.xml）的代码如下：

```
    <?xml version="1.0" encoding="UTF-8" ?>
    <!DOCTYPE mapper PUBLIC "-//mybatis.org//DTD Mapper 3.0//EN"
"http://mybatis.org/dtd/mybatis-3-mapper.dtd" >
    <mapper namespace="com.wudimanong.efence.dao.mapper.FenceGeoDao">
        <resultMap id="BaseResultMap"
type="com.wudimanong.efence.dao.model.FenceGeoInfoPO">
            <id column="id" property="id" jdbcType="INTEGER"/>
            <result column="name" property="name" jdbcType="VARCHAR"/>
            <result column="explain" property="explain" jdbcType="VARCHAR"/>
            <result column="city_code" property="cityCode" jdbcType="VARCHAR"/>
            <result column="ad_code" property="adCode" jdbcType="VARCHAR"/>
            <result column="layer_code" property="layerCode"
jdbcType="VARCHAR"/>
            <result column="region" property="region" jdbcType="OTHER"/>
            <result column="centre" property="centre" jdbcType="OTHER"/>
            <result column="area" property="area" jdbcType="DOUBLE"/>
            <result column="custom_info" property="customInfo"
jdbcType="VARCHAR"/>
            <result column="batch_id" property="batchId" jdbcType="INTEGER"/>
            <result column="from_id" property="fromId" jdbcType="INTEGER"/>
            <result column="geo_json" property="geoJson" jdbcType="VARCHAR"/>
            <result column="geo_hash" property="geoHash" jdbcType="OTHER"

typeHandler="com.wudimanong.efence.config.ArrayType2Handler"/>
            <result column="date_range" property="dateRange"
jdbcType="VARCHAR"/>
            <result column="time_bucket" property="timeBucket"
jdbcType="VARCHAR"/>
            <result column="state" property="state" jdbcType="INTEGER"/>
            <result column="update_time" property="updateTime"
jdbcType="TIMESTAMP"/>
            <result column="create_time" property="createTime"
jdbcType="TIMESTAMP"/>
            <result column="update_user" property="updateUser"
jdbcType="VARCHAR"/>
```

```
        <result column="create_user" property="createUser"
jdbcType="VARCHAR"/>
    </resultMap>
    <!--批量插入方法，数组类型需要单独指定自定义的“typeHandler”-->
    <insert id="batchInsert">
        INSERT INTO

fence_geo_info(name,explain,city_code,ad_code,layer_code,region,centre,area,
custom_info,batch_id,from_id,geo_json,geo_hash,date_range,time_bucket,state,
update_time,create_time,update_user,create_user)
        VALUES
        <foreach collection="list" item="item" separator=",">

(#{item.name},#{item.explain},#{item.cityCode},#{item.adCode},#{item.layerCo
de},#{item.region},#{item.centre},#{item.area},#{item.customInfo},#{item.bat
chId},#{item.fromId},#{item.geoJson},#{item.geoHash,typeHandler=com.wudimano
ng.efence.config.ArrayType2Handler},#{item.dateRange},#{item.timeBucket},#{i
tem.state},#{item.updateTime},#{item.createTime},#{item.updateUser},#{item.c
reateUser})
        </foreach>
    </insert>

    <!--利用 PostGIS 提供的函数来判断坐标点是否在指定的电子围栏中-->
    <select id="isContainPoint" resultType="java.lang.String">
        select ST_Within(ST_SetSRID(ST_MakePoint(#{lon},#{lat}), 4326),
region) from fence_geo_info where id=#{fenceId}
    </select>
</mapper>
```

（3）编写针对自定义“geo_hash”类型字段的 MyBatis 处理类。

由于 PostgreSQL 数据库支持数组类型，所以在电子围栏信息表的设计中，“geo_hash”类型字段是直接通过字符数组进行定义的。

但是，MyBatis 并不支持数组类型的 SQL 映射，所以对步骤（2）中的数组类型字段进行映射，需要使用自定义的“typeHandler”。代码如下：

```
package com.wudimanong.efence.config;
import java.sql.Array;
...
import org.apache.ibatis.type.TypeException;
@MappedJdbcTypes(JdbcType.ARRAY)
@MappedTypes(String[].class)
public class ArrayType2Handler extends BaseTypeHandler<Object[]> {
    private static final String TYPE_NAME_VARCHAR = "varchar";
    private static final String TYPE_NAME_INTEGER = "integer";
```

```
    private static final String TYPE_NAME_BOOLEAN = "boolean";
    private static final String TYPE_NAME_NUMERIC = "numeric";
    @Override
    public void setNonNullParameter(PreparedStatement ps, int i, Object[]
parameter, JdbcType jdbcType)
            throws SQLException {
        String typeName = null;
        if (parameter instanceof Integer[]) {
            typeName = TYPE_NAME_INTEGER;
        } else if (parameter instanceof String[]) {
            typeName = TYPE_NAME_VARCHAR;
        } else if (parameter instanceof Boolean[]) {
            typeName = TYPE_NAME_BOOLEAN;
        } else if (parameter instanceof Double[]) {
            typeName = TYPE_NAME_NUMERIC;
        }
        if (typeName == null) {
            throw new TypeException(
                    "ArrayType2Handler parameter typeName error, your type is
" + parameter.getClass().getName());
        }
        // 下面这 3 行是关键的代码，创建 Array，然后执行 ps.setArray(i, array)
        Connection conn = ps.getConnection();
        Array array = conn.createArrayOf(typeName, parameter);
        ps.setArray(i, array);
    }
    @Override
    public Object[] getNullableResult(ResultSet resultSet, String s) throws
SQLException {
        return getArray(resultSet.getArray(s));
    }
    @Override
    public Object[] getNullableResult(ResultSet resultSet, int i) throws
SQLException {
        return getArray(resultSet.getArray(i));
    }
    @Override
    public Object[] getNullableResult(CallableStatement callableStatement,
int i) throws SQLException {
        return getArray(callableStatement.getArray(i));
    }
    private Object[] getArray(Array array) {
        if (array == null) {
            return null;
        }
```

```
        try {
            return (Object[]) array.getArray();
        } catch (Exception e) {
        }
        return null;
    }
}
```

至此，完成了“电子围栏微服务”全部业务逻辑的实现。

4.5 步骤 3：演示电子围栏微服务的简单操作

经过前面步骤的操作完成了一个基于 Spring Cloud 微服务体系的电子围栏微服务的雏形。在正常情况下，为了方便使用，一般会开发一个功能完善的操作界面以简化电子围栏微服务的操作。

在本实例中，由于篇幅的原因不太可能实现这样的产品，接下来通过接口调用的方式，来简单演示电子围栏微服务的操作。

4.5.1 通过地图工具，定义电子围栏的 GeoJson 信息

在实际运营操作中，可以通过一些地图工具来划定电子围栏的坐标范围，并将这些信息以 GeoJson 数据形式批量导入系统。

通过在线地图工具划定一个电子围栏的范围。这里通过在线地图工具网站在“北京朝阳区望京”附近通过拖拽的方式定义了一个多边形的电子围栏，并将生成的 GeoJson 信息（右边的代码）复制出来，如图 4-8 所示。

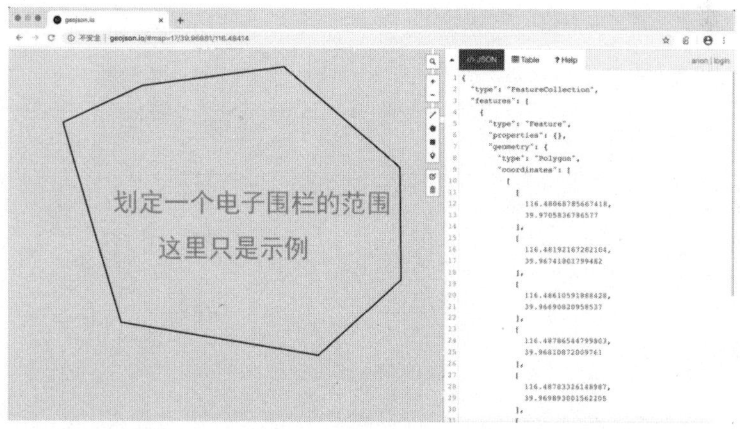

图 4-8

4.5.2 演示电子围栏微服务的简单操作

接下来，通过 Postman 接口工具调用电子围栏微服务的服务端接口，以演示简单的功能操作。

1. 调用“新增图层”接口，以创建电子围栏图层数据

调用“新增图层”接口，将电子围栏图层数据写入 PostgreSQL 数据库。“新增图层”接口的调用效果如图 4-9 所示。

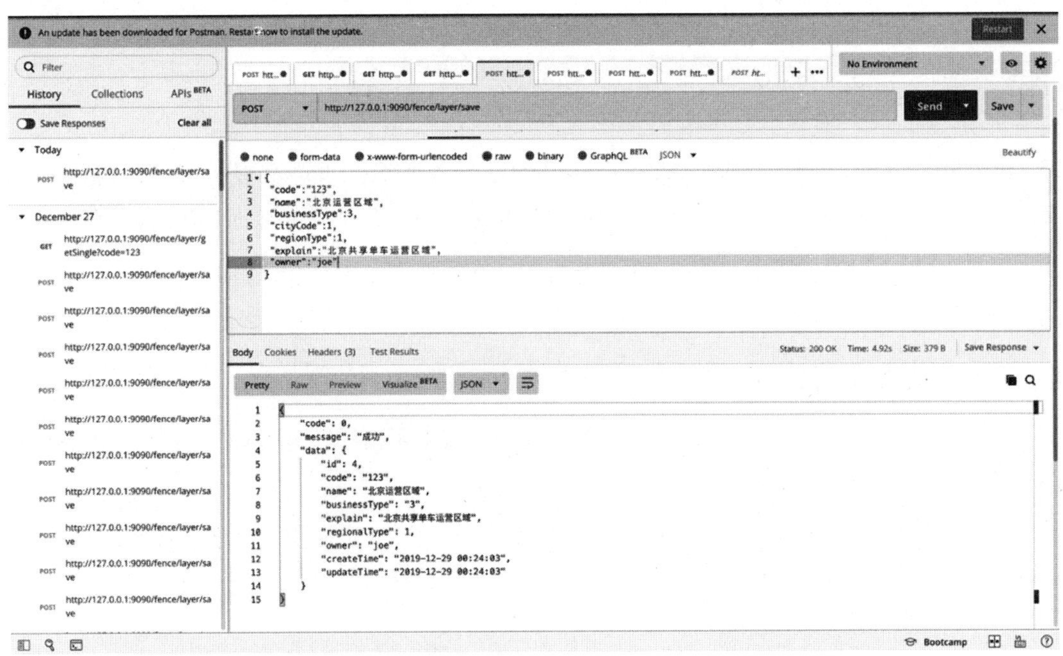

图 4-9

2. 调用“批量导入电子围栏数据”接口

（1）准备“批量导入电子围栏数据”接口的请求参数。代码如下：

```
{
 "cityCode":"1",
 "adCode":"2",
 "layerCode":"123",
 "fences":[
       {
         "name":"beijing-caoyan-wangjing",
         "explain":"望京片区运营范围",
         "cityCode":"1",
         "adCode":"2",
```

```
            "layerCode":"123",
            "regionGeoJson" :"{\"type\": \"Polygon\",
\"coordinates\":[[[116.48068785667418,39.9705836786577],[116.4819216728210 4,
39.96741801799482],[116.48610591888428,39.96690820958537],[116.4878654479980
3,39.96810872009761],[116.48783326148987,39.969893001562205],[116.4853549003
6011,39.97147168181156],[116.48236155509949,39.971167459808],[116.4806878566
7418,39.9705836786577]]]}",
           "batchId":1002
         }
       ]
    }
```

（2）调用“批量导入电子围栏数据”的接口，如图 4-10 所示。

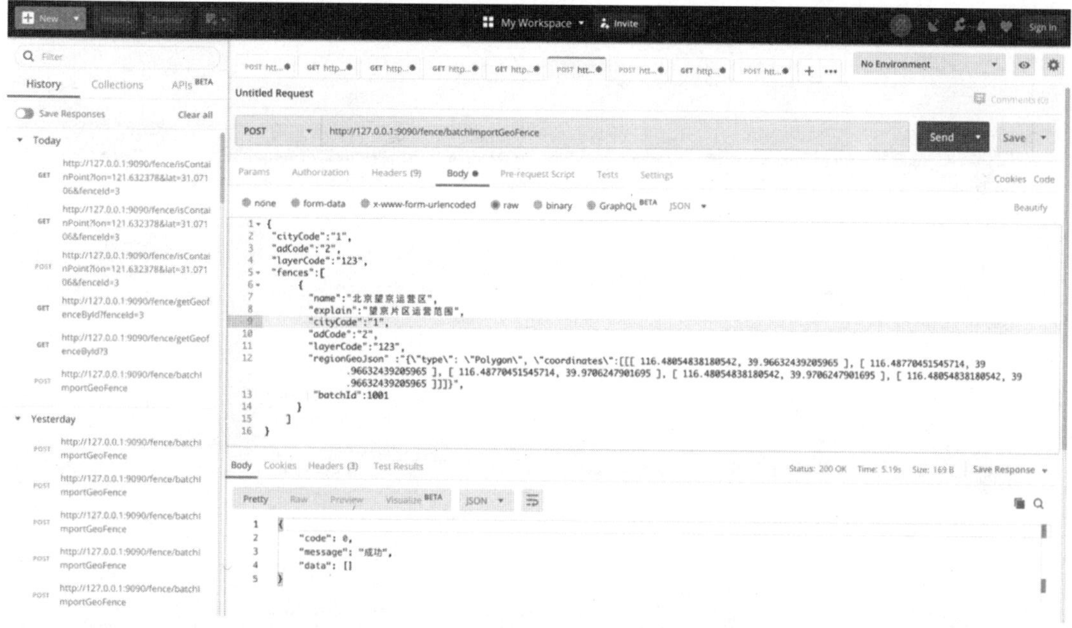

图 4-10

（3）查看电子围栏数据的入库效果。

查询数据库表，会发现围栏信息已经被导入 PostgreSQL 数据库表的电子围栏信息表（fence_geo_info）中，如图 4-11 所示。

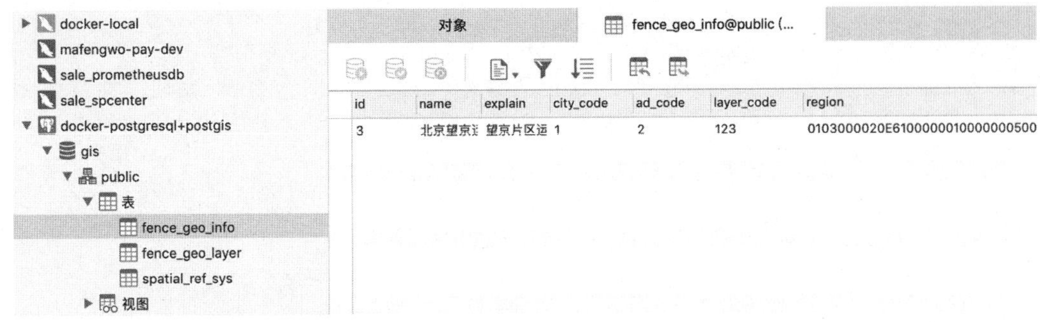

图 4-11

3. 调用“查询电子围栏数据”接口

如果接入了高德地图的 API，则可以将查询到的电子围栏的范围展示在地图上。

这里仅仅演示“查询电子围栏数据”接口功能，如图 4-12 所示。

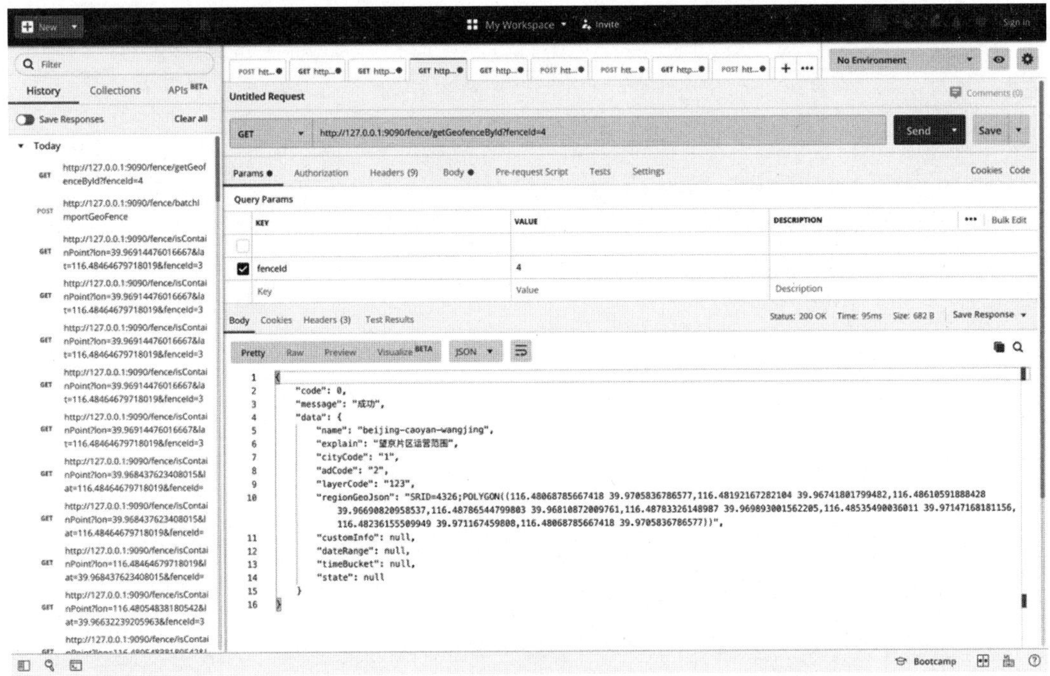

图 4-12

正常情况下，在通过围栏 ID 查询到电子围栏的具体 GeoJson 信息后，可以通过地图 API 来呈现电子围栏的地图范围。

4. 调用“判断坐标点是否在指定的电子围栏中”接口

在实际应用中，判断停车点是否越界是一个很频繁的操作。在本实例中，可以通过 PostGIS 空间数据库提供的函数来进行计算。接下来以接口调用的方式进行验证。

（1）判断坐标点是否在电子围栏外。

通过在线地图工具随机拾取一个电子围栏范围外的坐标点，如图 4-13 所示。

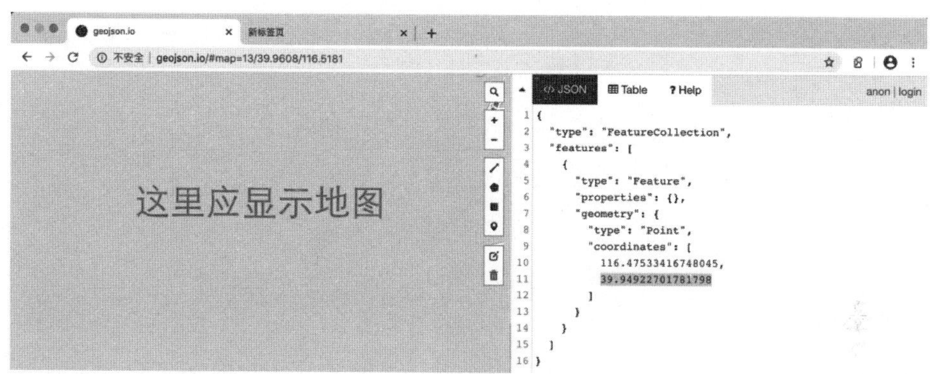

图 4-13

在获取经纬度信息后，调用“判断坐标点是否在指定的电子围栏中”接口，判断坐标点是否越界。请求链接及参数如下：

```
http://127.0.0.1:9090/fence/isContainPoint?lon=116.47533416748045&lat=39.94922701781798&fenceId=4
```

接口调用的返回结果如图 4-14 所示。

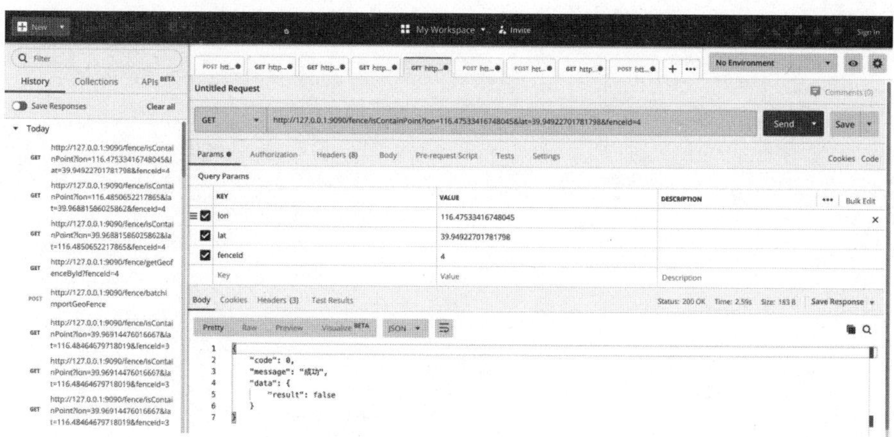

图 4-14

接口的返回值为“false”，这表明该坐标点处于指定电子围栏外。

（2）判断坐标点是否在电子围栏内。

随机拾取一个在电子围栏范围内的坐标点，如图 4-15 所示。

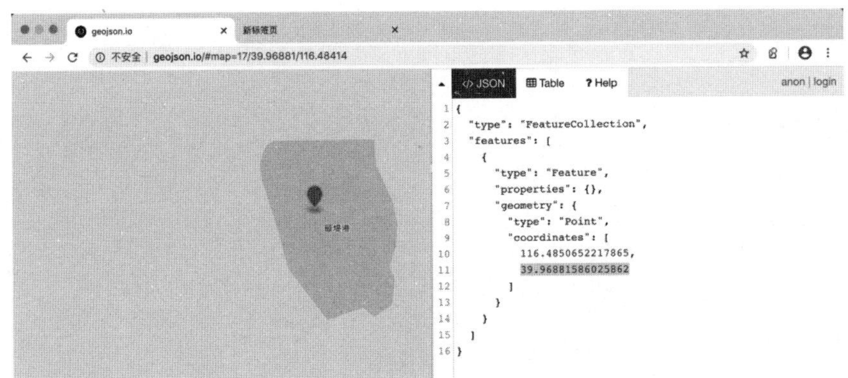

图 4-15

在获取经纬度信息后，调用“判断坐标点是否在指定的电子围栏中”接口，请求链接及参数如下：

```
http://127.0.0.1:9090/fence/isContainPoint?lon=116.4850652217865&lat=39.96881586025862&fenceId=4
```

接口调用的返回结果如图 4-16 所示。

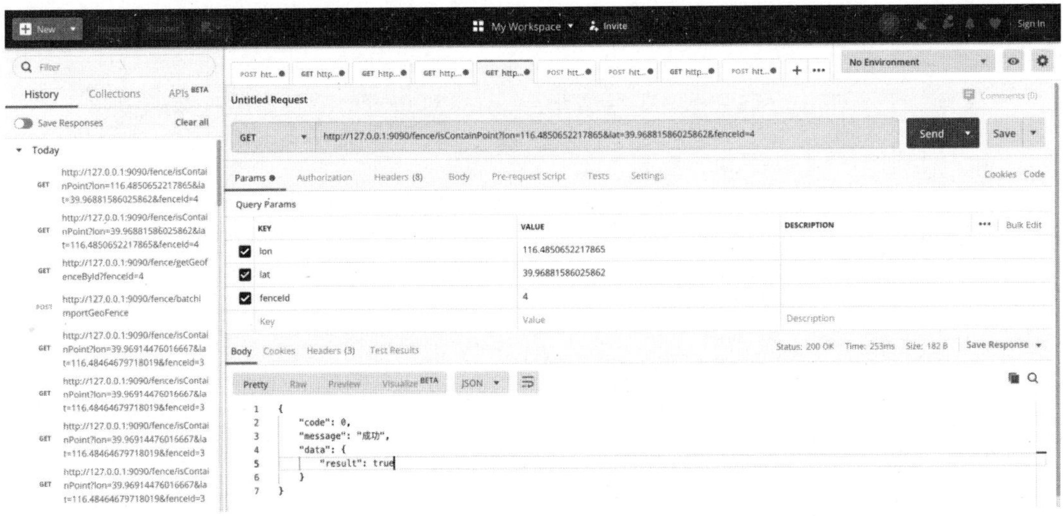

图 4-16

接口调用的返回值为“true”，表明该坐标点的确在电子围栏内。

通过上述操作可以发现，本实例所编写的电子围栏微服务已经具备基本的围栏数据定义及空间计算能力。读者可以在此基础上根据需要进一步丰富其功能，并结合地图开放接口做出更加酷炫的应用！

4.6 步骤 4：使用 Spring Cloud ConfigServer 配置中心

在日常的项目开发过程中，经常需要通过配置文件来配置系统参数：例如数据库的链接地址、Redis 及消息中间件（MQ）等的链接信息。

在前面的内容中，这些配置信息被配在“resources”本地目录的“*.yml”或“*.properties”文件中。但这样的方式会造成配置信息分散不好管理、多环境配置烦琐，以及数据库等敏感信息通过代码泄漏等问题。

在 Spring Cloud 微服务中提供了解决这些问题的配置中心——Spring Cloud ConfigServer。接下来将介绍如何构建及使用配置中心，以实现微服务配置信息的统一管理及多环境配置。

4.6.1 构建 Spring Cloud ConfigServer 配置中心微服务

参考 2.3.1、2.5.2 节中创建 Spring Boot 应用工程，以及对 Spring Boot 应用进行微服务改造的具体步骤，来构建 Spring Cloud ConfigServe 配置中心微服务。

1. 引入 Spring Cloud 微服务的核心依赖

这里可以参考 2.5.2 节中构建 Spring Cloud 微服务的步骤。但除引入核心依赖外，还需要在 pom.xml 文件中引入 Spring Cloud ConfigServer 的官方依赖包——spring-cloud-config-server。具体如下：

```
<dependency>
    <groupId>org.springframework.cloud</groupId>
    <artifactId>spring-cloud-config-server</artifactId>
</dependency>
```

2. 构建配置中心微服务

（1）建立 ConfigServer 的本地配置文件——bootstrap.yml。代码如下：

```
server:
  port: 8888
spring:
  profiles:
```

```
    active: debug
```

这里只是定义了服务端口。

（2）在 application.yml 配置文件中增加微服务“ConfigServer”的其他核心配置信息。代码如下：

```
spring:
  application:
    name: configserver
  cloud:
    config:
      server:
        #关闭微服务健康性检查
        health.enabled: false
        #配置 GitHub 仓库路径（可以换成自己的 GitLab 仓库地址）
        git:
          uri: https:// XXXXX/manongwudi/repos.git
          search-paths: 'common,{application}'
          #启动时克隆存储库
          clone-on-start: true
    #微服务发现的相关配置
    consul:
      discovery:
        prefer-ip-address: true
        tags: api
```

在以上配置中将 ConfigServer 与 Git 仓库集成，通过 Git 仓库来统一管理微服务的配置文件。在实践中，还可以通过内部流程来约定 Git 配置仓库的发布及管理流程。

（3）开发配置中心服务主类代码。代码如下：

```
package com.wudimanong.configserver;
import org.springframework.boot.SpringApplication;
import org.springframework.boot.autoconfigure.SpringBootApplication;
import org.springframework.cloud.config.server.EnableConfigServer;
@SpringBootApplication
@EnableConfigServer
public class ConfigServer {
    public static void main(String[] args) {
        SpringApplication.run(ConfigServer.class, args);
    }
}
```

通过@EnableConfigServer 注解开启配置中心服务。

（4）启动配置中心服务。运行效果如图 4-17 所示。

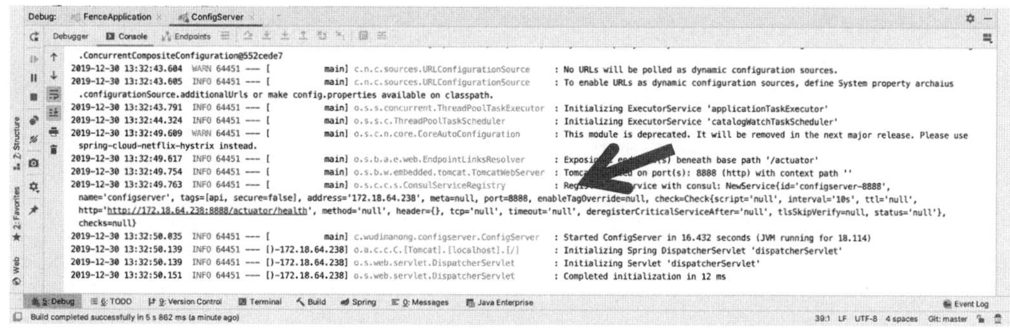

图 4-17

（5）查看 Consul 注册中心控制台，微服务“ConfigServer”的注册效果如图 4-18 所示。

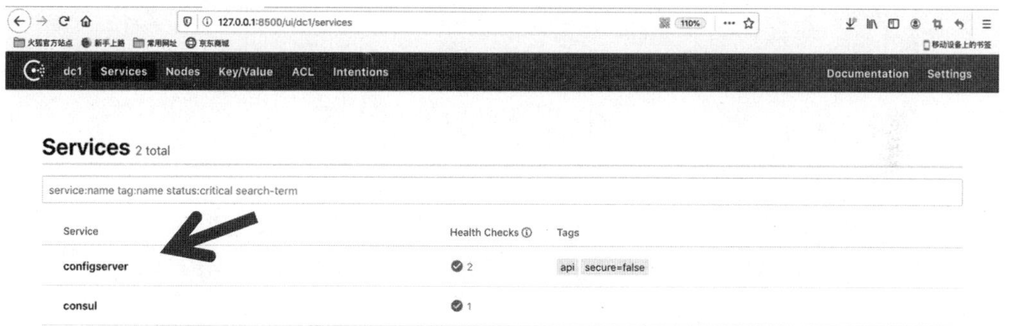

图 4-18

4.6.2 将微服务接入 Config 配置中心

经过 4.6.1 节的操作，“ConfigServer”作为一个独立的配置中心服务，已经可以被其他微服务使用了。下面以本实例的“efence”微服务为例，介绍如何将 Spring Cloud 微服务接入 Spring Cloud ConfigServer 配置中心。

1. 引入依赖并配置

（1）在微服务项目的 pom.xml 文件中引入 Spring Cloud Config 的依赖。代码如下：

```
<dependency>
    <groupId>org.springframework.cloud</groupId>
    <artifactId>spring-cloud-starter-config</artifactId>
 </dependency>
```

该依赖基于 Spring Boot 的自动配置原理，可以实现“开箱即用”。

（2）在项目 bootstrap.yml 配置文件中，配置 Config 服务的接入信息。代码如下：

```
spring:
  application:
    name: efence
  profiles:
    active: debug,dev
  cloud:
    consul:
      discovery:
        preferIpAddress: true
        instance-id:
${spring.application.name}:${spring.cloud.client.ipAddress}:${spring.applica
tion.instance_id:${server.port}}:@project.version@
        healthCheckPath: /actuator/health
    #开启 Spring Cloud Config 配置中心
    config:
      enabled: true
      uri: http://127.0.0.1:8888/
      label: master
server:
  port: 9090
```

在上述配置中，额外增加了 Config 相关的配置，并指定了配置中心服务的 URL。

2. 微服务从配置中心获取配置信息

如果将“efence”本地配置文件中的数据库配置删除，并将其数据库链接信息配置在 Git 仓库的“efence-dev.yml”文件中。启动“efence”微服务，就会发现微服务通过配置中心获取了相关配置，如图 4-19 所示。

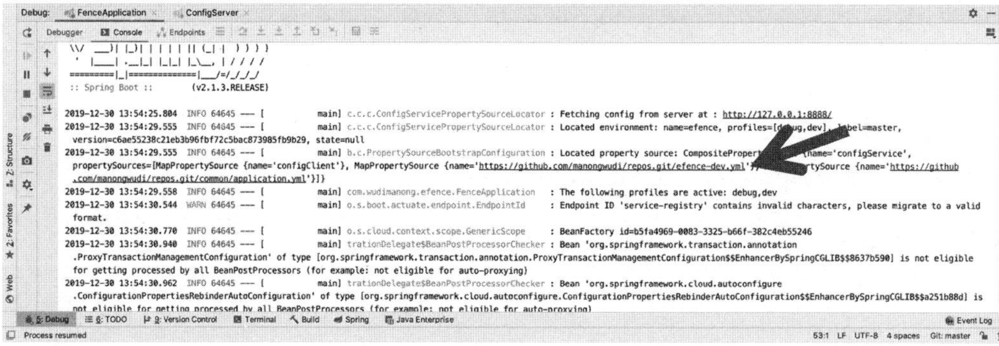

图 4-19

4.6.3 利用配置中心管理微服务的多环境配置

使用配置中心，可以统一管理多项目、多环境的配置文件。这里介绍一种比较通用的方式：

- 所有项目公用的配置（如 application.yml、application-*.yml 等通用配置）放在 common 文件夹中。
- 具体项目的个性配置文件（如 efence-*.yml），则需要单独定义存放位置。

在 4.6.1 节配置中心服务的本地配置中可以通过“search-paths: 'common,{application}'”占位符的方式来实现配置文件存放路径的匹配。

配置文件的具体存储仓库，在本实例中是通过 GitHub 公共仓库实现的。其中，配置文件的组织结构如下：

```
\- repos
+- common
|  +- application.yml
|  +- application-dev.yml
|  +- application-test.yml
|  +- application-production.yml
+- efence-dev.yml
+- efence-test.yml
+- efence-production.yml
+- user-dev.yml
+- user-test.yml
...
```

在以上文件结构中，如果“efence”微服务的启动环境为“dev”，则该微服务可以自动继承 application.yml 及 application-dev.yml 文件中的配置。

其他微服务的继承关系以此类推。利用这样的配置继承关系，就可以将全局公共配置或某个环境的公共配置进行抽象，从而简化配置项，以实现不同环境的配置管理。

访问 ConfigServer 的 URL，可以看到微服务获取的配置信息，如图 4-20 所示。

图 4-20

ConfigServer 访问配置信息的 URL 规则为：{url}/{label}/{application}-{profiles}。

4.7 本章小结

本实例实现了一个基于车辆电子围栏微服务，介绍了空间数据库组合“PostgreSQL + PostGIS”及其在 Java 应用中的使用方式。

为了简化编程，在实例中引入了 MyBatis-Plus 组件来简化数据库操作；引入了 MapStruct 来减少对象数据转换的代码量。

在本章的最后，还重点介绍了配置中心组件 Spring Cloud ConfigServer。它是 Spring Cloud 微服务体系中比较重要的组件，它可以将微服务的配置信息进行集中管理并支持多环境配置。

第 5 章

【实例】电子钱包系统

——用“Feign + Ribbon + Hystrix + Vue.js + Docker”实现微服务的“负载调用 + 熔断降级 + 部署”

很多互联网应用都有“电子钱包”的概念。例如，使用共享单车时，可以先将钱充值到电子钱包中，然后每次骑行后从电子钱包中扣费。这就是一种典型的电子钱包使用场景。

本实例将实现一个基于 Spring Cloud 微服务体系的电子钱包系统，该系统只有一个电子钱包微服务“wallet”。

因为电子钱包微服务的充值流程涉及对“支付系统”（参考第 6 章）的调用，所以，在本实例中还会介绍微服务体系下的服务发现、客户端服务负载调用及服务熔断降级等知识，这涉及 Feign、Ribbon、Hystrix 等 Spring Cloud 核心技术组件的运用。

此外，本实例的最后还会介绍 Spring Cloud 微服务场景下单元测试用例的编写方法，以及基于 Docker 的容器化部署方法。

通过本章，读者将学习到以下内容：

- 电子钱包微服务的设计流程。
- 用“Feign + Ribbon”实现微服务之间的客户端服务负载调用。
- 在 Spring Cloud 中集成 Hystrix 实现微服务的熔断降级。
- 利用 MyBatis-Plus 插件，简化 MyBatis 数据库操作。

- 基于 Vue.js 框架实现电子钱包微服务的充值界面。
- 基于 Docker 实现 Spring Cloud 微服务的容器化部署。

5.1 功能概述

电子钱包微服务是一种具备金融属性的虚拟账户系统。在互联网应用中，虚拟账户系统通过对接第三方支付公司或银行系统的接口，实现资金从“个人银行账户”到“互联网应用平台银行账户”，再到“电子钱包虚拟账户”的资金转移过程。这便是电子钱包账户充值的基本逻辑。电子钱包微服务会记录用户充值的资金，以及电子钱包的详细交易流水。

5.2 系统设计

在良好的应用架构设计中，电子钱包微服务一般是比较独立的余额账户系统：只提供余额账户开户、余额增加、余额减少，以及与之相关的流水记账等基本功能。而具体的余额充值、消费、退款等业务逻辑，则由独立拆分的“交易系统”去完成（本书不涉及）。

一般来说，“交易系统”与“支付系统”在完成电子钱包余额变动的交易逻辑后，会调用电子钱包微服务的接口以实现电子钱包账户余额的变动。

由于“交易系统”的复杂度非常高，且与业务强绑定，所以要实现完全统一的“交易系统”是一件比较困难的事。例如，在旅游业务中，购买机票、预订酒店这两个看似通用的场景，实际的交易流程却是很不一样的。

在设计电子钱包微服务的过程中也面临与上述类似的问题：如果只做一个单纯的“余额账户”系统，则在该系统前一定要有一个“交易系统”，来帮它处理余额充值、余额消费这样具有业务属性的交易逻辑。

在实际的系统设计实践中，是否需要将业务的交易逻辑拆分为独立的交易系统，需要从实际情况来考虑。例如，电子钱包系统中涉及的“余额充值”“余额消费”“余额退款”等交易流程，可预见的变化不会很大，所以暂时不必进行拆分。

综上所述，在本实例的实现过程中，会暂时将交易逻辑与账户逻辑耦合在同一个系统中，但在应用内代码的分层结构上会做相应的解耦。

5.2.1 系统流程设计

电子钱包微服务的前后端的交互流程如图 5-1 所示。

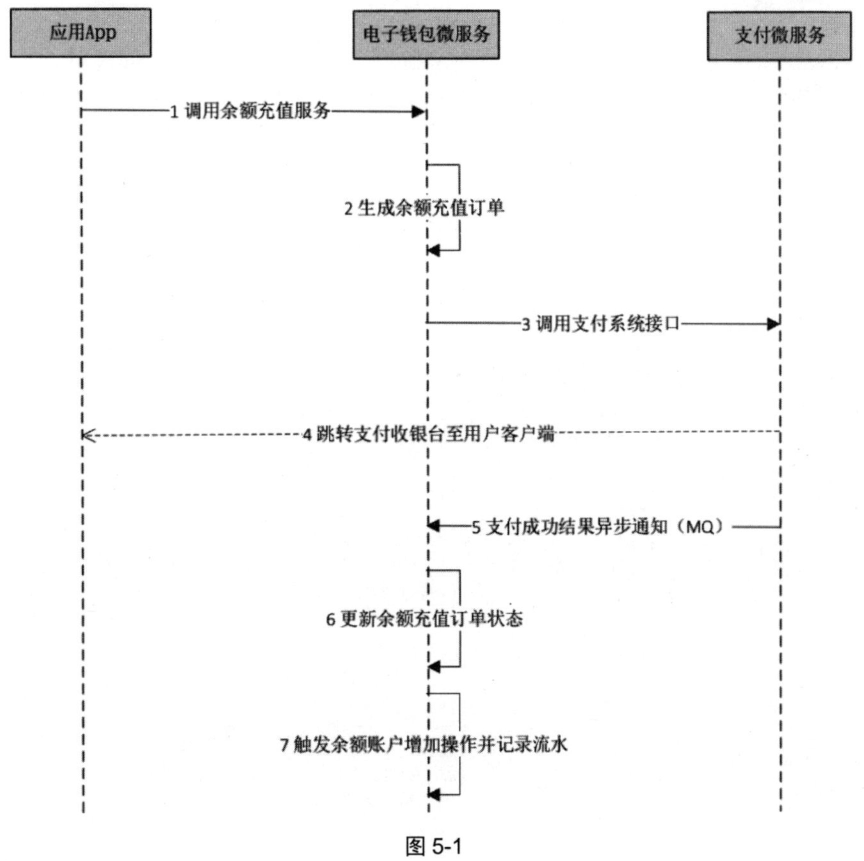

图 5-1

流程说明如下：

（1）应用 App 调用余额充值服务，电子钱包微服务先生成余额充值订单，然后调用“支付微服务”接口发起针对该笔充值交易的支付请求。

（2）之后的流程将脱离电子钱包微服务流程，由应用 App 与“支付微服务”直接交互完成支付行为。

（3）在用户支付成功后，支付系统以消息的方式将支付结果通知给电子钱包微服务。

（4）电子钱包微服务在收到支付系统的通知后，更新余额充值订单状态，并操作账户增加余额的逻辑。

（5）余额消费及余额退款等流程，可以直接通过账户逻辑来完成，无须与支付系统交互，但要确保“账户余额的变动”与“流水明细记录的数据”在事务上保持一致。

5.2.2 系统结构设计

在系统实现上，采用经典的 MVC 分层结构：通过“Feign + Ribbon”实现对“支付微服务”的调用；对于面向微服务体系外的应用访问，则通过 3.5 节介绍过的服务网关进行隔离。

电子钱包微服务的结构如图 5-2 所示。

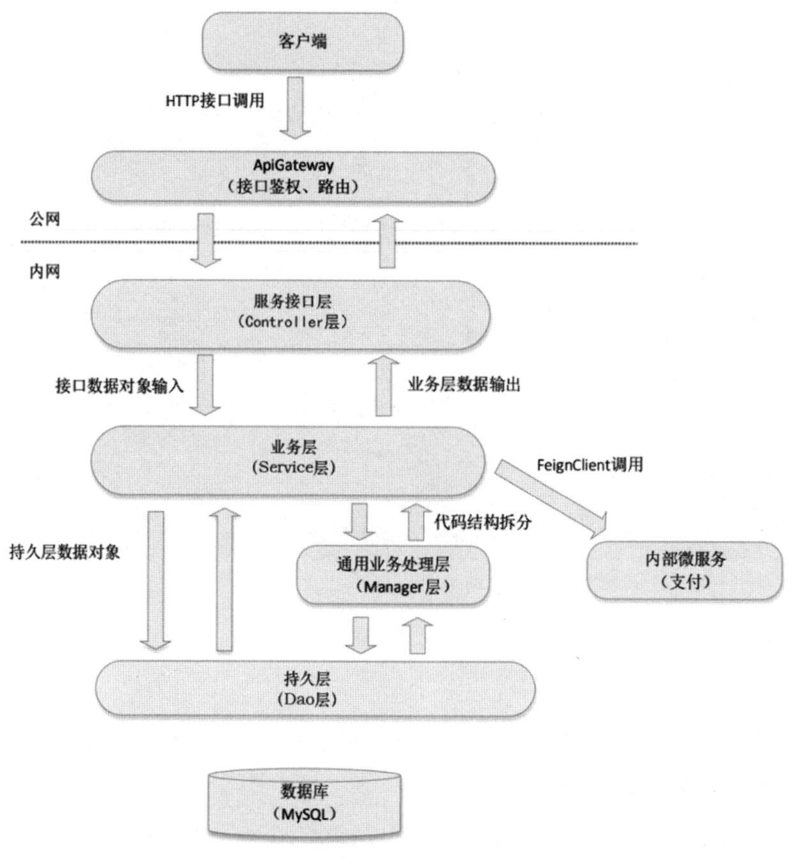

图 5-2

上述软件结构主要分为以下 3 层。

- 服务接口层（Controller 层）：定义服务接口。
- 业务层（Service 层）：处理业务逻辑并通过 Dao 层完成数据库相关操作。

- 持久层（Dao 层）：提供对数据库操作的接口封装。

在本实例中，在业务层(Service 层)与持久层(Dao 层)之间单独拆分出了一个 Manager 层，这主要是用于拆分某些复杂的业务层逻辑，避免业务层代码过于臃肿。

5.2.3 数据库设计

电子钱包微服务采用 MySQL 数据库，在本实例的实现中主要涉及“余额交易订单表”“余额账户信息表”及“余额账户流水记录表”。

- 代码 以下表在本书配置资源的“chapter05-wallet/src/main/resources/db.migration”目录下。

具体的表结构设计如下。

1. 余额交易订单表

“余额交易订单表”的作用是处理余额充值、余额退款等涉及支付流程的订单状态逻辑。具体的 SQL 代码如下：

```
create table user_balance_order
(
  id          bigint not null primary key auto_increment,
  order_id    varchar(50) comment '订单号 ID',
  user_id     varchar(60) comment '用户 ID',
  amount      bigint comment '交易金额',
  trade_type  varchar(20) comment '交易类型。charge-余额充值; refund-余额退款',
  currency    varchar(10) comment '币种',
  trade_no    varchar(32) comment '支付渠道流水号',
  status      varchar(2) comment '支付状态。0-待支付; 1-支付中; 2-支付成功; 3-支付失败',
  is_renew    int default 0 comment '是否自动续费充值。0-不自动续费; 1-自动续费',
  trade_time  timestamp default current_timestamp comment '交易时间',
  update_time timestamp null default current_timestamp on update current_timestamp comment '最后一次更新时间',
  create_time timestamp null default current_timestamp comment '创建时间'
);
alter table user_balance_order comment '余额交易订单表';
#添加索引信息
alter table user_balance_order add index unique_idx_order_id(order_id);
alter table user_balance_order add index idx_user_id(user_id);
alter table user_balance_order add index idx_trade_time(trade_time);
```

2. 余额账户信息表

余额账户信息表是电子钱包微服务的核心数据库表，主要用于记录用户电子钱包账户的基本信息及余额。具体的 SQL 代码如下：

```
create table user_balance
(
  id          bigint not null primary key auto_increment,
  user_id      varchar(60) comment '用户 ID',
  acc_no       varchar(60) comment '余额系统生成的账户唯一标识',
  acc_type     varchar(2) comment '余额账户类型。0-现金；1-赠送金',
  currency     varchar(10) comment '账户币种',
  balance      bigint comment '账户余额，以分为单位',
  update_time timestamp NULL DEFAULT CURRENT_TIMESTAMP ON UPDATE
CURRENT_TIMESTAMP COMMENT '最后一次更新时间',
  create_time timestamp NULL DEFAULT CURRENT_TIMESTAMP COMMENT '创建时间');
alter table user_balance comment '余额账户信息表';
#添加相应索引
alter table user_balance add index idx_ub_user_id(user_id);
alter table user_balance add index idx_ub_acc_no(acc_no);
```

3. 余额账户流水记录表

“余额账户流水记录表”用于记录余额账户的变动情况，是实现账户明细查询和余额对账核算的重要依据。具体的 SQL 代码如下：

```
create table user_balance_flow
(
  id            bigint not null primary key auto_increment,
  user_id        varchar(60) comment '业务方用户 ID',
  flow_no        varchar(64) comment '账户流水号，与业务方发起的流水号映射',
  acc_no         varchar(60) comment '账户唯一标识',
  busi_type      varchar(10) comment '余额流水业务类型。0-订单结费；1-购买月卡',
  amount         bigint comment '变动金额，以分为单位，区分正负，如+10、-10',
  currency       varchar(10) comment '币种',
  begin_balance bigint comment '变动前余额',
  end_balance    bigint comment '变动后余额',
  fund_direct    varchar(2) comment '借贷方向。00-借方；01-贷方',
  update_time timestamp null default current_timestamp on update
current_timestamp comment '最后一次更新时间',
  create_time timestamp null default current_timestamp comment '创建时间');
alter table user_balance_flow comment '余额账户流水记录表';
#创建相关索引信息
alter table user_balance_flow add index idx_ubf_user_id(user_id);
alter table user_balance_flow add index idx_ubf_acc_no(acc_no);
alter table user_balance_flow add index idx_ubf_flow_no(flow_no);
```

```
alter table user_balance_flow add index idx_ubf_create_time(create_time);
```

5.3 步骤 1：构建 Spring Cloud 微服务工程代码

接下来搭建电子钱包微服务所需要的 Spring Cloud 微服务工程，并集成所需要的第三方组件。

5.3.1 创建 Spring Cloud 微服务工程

1. 创建一个基本的 Maven 工程

利用 2.3.1 节介绍的方法创建一个 Maven 工程，完成后的工程代码结构如图 5-3 所示。

图 5-3

2. 引入 Spring Cloud 依赖，将其改造为微服务项目

（1）引入 Spring Cloud 微服务的核心依赖。

这里可以参考 2.5.2 节中的具体步骤。

（2）在工程代码的 resources 目录下新建一个基础性配置文件——bootstrap.yml。配置文件中的代码如下：

```
spring:
  application:
    name: wallet
  profiles:
    active: debug
  cloud:
    consul:
      discovery:
        preferIpAddress: true
        instance-id:
${spring.application.name}:${spring.cloud.client.ipAddress}:${spring.applica
tion.instance_id:${server.port}}:@project.version@
        healthCheckPath: /actuator/health
server:
  port: 9090
```

（3）在 2.5.2 节提到过，Spring Boot 并不会默认加载 bootstrap.yml 这个文件，所以需要在 pom.xml 中添加 Maven 资源相关的配置，具体参考 2.5.2 节内容。

（4）创建 Wallet 电子钱包微服务的入口程序类。代码如下：

```
package com.wudimanong.wallet;
import org.springframework.boot.SpringApplication;
import org.springframework.boot.autoconfigure.SpringBootApplication;
import org.springframework.cloud.client.discovery.EnableDiscoveryClient;
@EnableDiscoveryClient
@SpringBootApplication
public class WalletApplication {
    public static void main(String[] args) {
        SpringApplication.run(WalletApplication.class, args);
    }
}
```

至此，电子钱包微服务所需的 Spring Cloud 微服务工程就构建出来了。

5.3.2 将 Spring Cloud 微服务注入 Consul

参考 2.5.1 节、2.5.3 节的内容，将“wallet”微服务注入服务注册中心 Consul 中。然后运行所构建的“wallet”微服务工程，可以看到该服务已经注册到 Consul 中了，如图 5-4 所示。

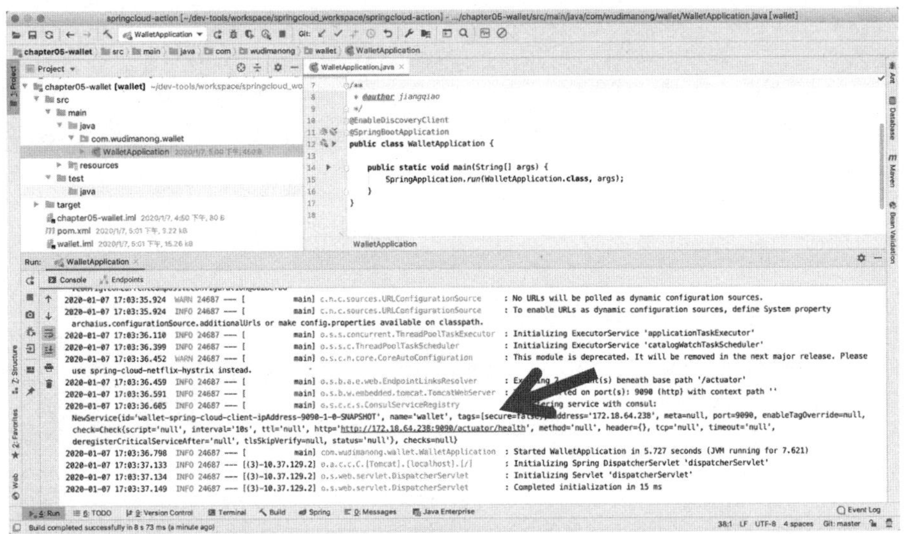

图 5-4

打开 Consul 控制台，“wallet”微服务被注册到 Consul 中的效果如图 5-5 所示。

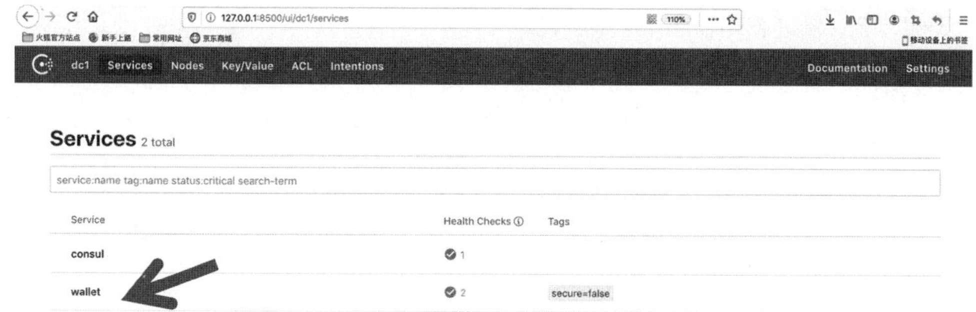

图 5-5

至此，从技术层面完成了 Spring Cloud 微服务的搭建过程。接下来继续集成开发“wallet”电子钱包微服务所依赖的其他组件。

5.3.3 集成 MyBatis，以访问 MySQL 数据库

在本实例中，使用 MyBatis 这个持久层（Dao 层）框架来操作数据库。

1. 引入 MyBatis 框架依赖，以及 MySQL 数据库驱动程序程序

具体步骤参考 2.3.3 节内容。

2. 配置项目数据库连接信息

在项目中创建一个新的配置文件 application.yml，在其中添加 MySQL 数据库的连接信息如下：

```
spring:
  datasource:
    url: jdbc:mysql://127.0.0.1:3306/wallet
    username: root
    password: 123456
    type: com.alibaba.druid.pool.DruidDataSource
    driver-class-name: com.mysql.jdbc.Driver
    separator: //
```

上述配置中涉及的数据库信息，可以参考 1.3.3 节通过 Docker 部署本地 MySQL 的步骤——创建一个名为“wallet”的数据库，并执行 5.2.3 节中所定义的 SQL 脚本。

5.3.4 通过 MyBatis-Plus 简化 MyBatis 的操作

1. Spring Boot 集成 MyBatis-Plus 框架

具体的集成和配置方法可参考 4.3.5 节内容。

2. 启动项目，验证 MyBatisPlus 集成效果

启动“wallet”电子钱包微服务工程，成功集成 MyBatis-Plus 框架的运行效果如图 5-6 所示。

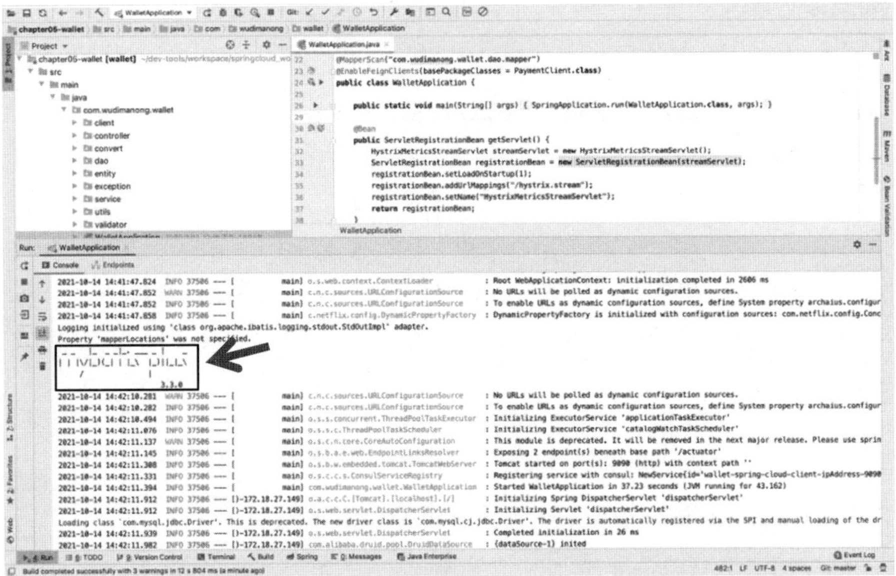

图 5-6

接下来将实现电子钱包微服务的业务逻辑。

5.4 步骤 2：实现微服务的业务逻辑

接下来实现电子钱包微服务的业务逻辑。

5.4.1 定义服务接口层（Controller 层）

定义电子钱包微服务实例所涉及的服务接口层（Controller 层）。

1. 约定接口数据格式

在定义 Controller 层接口之前，需要先约定接口的请求方式，以及统一的报文格式。具体的规范可以参考 4.1.1 节中“1.”小标题的内容。其中涉及的统一报文格式类（ResponseResult）及全局响应码枚举类（GlobalCodeEnum），可以复制 4.1.1 节“1.”小标题中的代码。

2. 定义“电子钱包开户”接口及“电子钱包查询”接口

“电子钱包开户”接口及“电子钱包查询”接口，主要用于管理电子钱包的基本信息。

（1）定义“电子钱包开户”接口及“电子钱包查询”接口的 Controller 层。代码如下：

```
package com.wudimanong.wallet.controller;
import com.wudimanong.wallet.entity.ResponseResult;
...
import org.springframework.web.bind.annotation.RequestMapping;
import org.springframework.web.bind.annotation.RestController;
@Slf4j
@RestController
@RequestMapping("/account")
public class UserAccountController {
    /**
     * 注入电子钱包业务层（Service 层）接口
     */
    @Autowired
    UserAccountService userAccountServiceImpl;
    /**
     * “电子钱包开户”接口
     *
     * @param accountOpenDTO
     * @return
     */
    @PostMapping("/openAcc")
    public ResponseResult<AccountOpenBO> openAcc(
            @RequestBody @Validated AccountOpenDTO accountOpenDTO) {
        return
ResponseResult.OK(userAccountServiceImpl.openAcc(accountOpenDTO));
    }
    /**
     * “电子钱包查询”接口
     *
     * @param accountQeuryDTO
     * @return
     */
    @GetMapping("/queryAcc")
    public ResponseResult<List<AccountBO>> queryAcc(@Validated
AccountQeuryDTO accountQeuryDTO) {
        return
ResponseResult.OK(userAccountServiceImpl.queryAcc(accountQeuryDTO));
    }
}
```

在上述代码中，通过 Spring MVC 提供的注解，定义了“电子钱包开户”接口（/openAcc）及“电子钱包查询”接口（/queryAcc）。

- “电子钱包开户”接口，通过@PostMapping 注解定义了接口的 Post 请求方式，通过

@RequestBody 注解约定了请求参数格式为 JSON。

- “电子钱包查询”接口，通过@GetMapping 注解定义了接口的 Get 请求方式，并约定了以“form”表单格式提交请求参数。

在上述接口定义中，对于接口参数的合法性，使用了 Spring 框架提供的@Validated 注解对 DTO 对象中的属性值进行校验。

（2）定义“电子钱包开户”接口及“电子钱包查询”接口的请求参数对象。

定义“电子钱包开户”接口（/openAcc）的请求参数对象。代码如下：

```
package com.wudimanong.wallet.entity.dto;
import com.wudimanong.wallet.validator.EnumValue;
...
import lombok.Data;
@Data
public class AccountOpenDTO implements Serializable {
    /**
     * 用户 ID
     */
    @NotNull(message = "用户 ID 不能为空")
    private Long userId;
    /**
     * 账户类型。0-现金账户；1-赠送金账户
     */
    @EnumValue(intValues = {0, 1}, message = "账户类型输入有误")
    private Integer accType;
    /**
     * 账户币种(仅支持人民币、美元两种账户)
     */
    @EnumValue(strValues = {"CNY", "USD"})
    private String currency;
}
```

定义“电子钱包查询”接口（/queryAcc）的请求参数对象。代码如下：

```
package com.wudimanong.wallet.entity.dto;
import javax.validation.constraints.NotNull;
import lombok.Data;
@Data
public class AccountQeuryDTO {
    /**
     * 用户 ID
     */
    @NotNull(message = "用户 ID 不能为空")
    private Long userId;
```

```
    /**
     * 账户类型。0-现金账户；1-赠送金账户
     */
    private Integer accType;
    /**
     * 账户币种(仅支持人民币、美元两种账户)
     */
    private String currency;
}
```

在上述 DTO 对象中，涉及的自定义参数校验注解@EnumValue 的定义可以参考 4.4.1 节“2.”小标题中的定义。

（3）定义“电子钱包开户”接口及“电子钱包查询”接口的返回参数对象。

定义“电子钱包开户”接口（/openAcc）的返回参数对象。代码如下：

```
package com.wudimanong.wallet.entity.bo;
import lombok.Data;
@Data
public class AccountOpenBO {
    /**
     * 用户 ID
     */
    private Long userId;
    /**
     * 钱包系统生成唯一账户编号
     */
    private String accNo;
    /**
     * 账户类型
     */
    private String accType;
    /**
     * 币种
     */
    private String currency;
}
```

在上述代码中，定义了电子钱包开户成功后返回的信息（包括生成的账户编号等）。

定义“电子钱包查询”接口（/queryAcc）的返回参数对象。代码如下：

```
package com.wudimanong.wallet.entity.bo;
import com.fasterxml.jackson.annotation.JsonFormat;
...
import lombok.Data;
```

```
@Data
public class AccountBO implements Serializable {
    /**
     * 主键 ID
     */
    private Integer id;
    /**
     * 用户编号
     */
    private Long userId;
    /**
     * 账户编码
     */
    private String accNo;
    /**
     * 账户类型
     */
    private String accType;
    /**
     * 币种
     */
    private String currency;
    /**
     * 账户余额
     */
    private Integer balance;
    /**
     * 格式化日期显示
     */
    @JsonFormat(pattern = "yyyy-MM-dd HH:mm:ss", timezone = "GMT+8")
    private Timestamp createTime;
    /**
     * 格式化日期显示
     */
    @JsonFormat(pattern = "yyyy-MM-dd HH:mm:ss", timezone = "GMT+8")
    private Timestamp updateTime;
}
```

“电子钱包开户”接口及“电子钱包查询”接口的返回参数对象，由 ResponseResult 统一数据报文格式进行包装。其中，“电子钱包查询”接口的返回参数设计，考虑到后续同一个用户可能存在不同类型账户，所以将其返回参数定义为“List<AccountBO>”集合的形式。

3. 定义“电子钱包充值”接口

接下来定义“电子钱包充值”接口。该接口的主要功能是：接收前端发起的钱包充值请求，并创建钱包充值订单，之后调用第 6 章的支付接口引导用户进行付款。

（1）定义“电子钱包充值”接口的 Controller 层。代码如下：

```
package com.wudimanong.wallet.controller;
import com.wudimanong.wallet.entity.ResponseResult;
...
import org.springframework.web.bind.annotation.RestController;
@Slf4j
@RestController
@RequestMapping("/account")
public class UserAccountTradeController {
    /**
     * 注入业务层（Service 层）依赖接口
     */
    @Autowired
    UserAccountTradeService userAccountTradeServiceImpl;
    /**
     * “电子钱包充值”接口
     *
     * @param accountChargeDTO
     * @return
     */
    @PostMapping("/chargeOrder")
    public ResponseResult<AccountChargeBO> chargeOrder(@RequestBody
@Validated AccountChargeDTO accountChargeDTO) {
        return ResponseResult.OK(userAccountTradeServiceImpl.chargeOrder
(accountChargeDTO));
    }
}
```

（2）定义“电子钱包充值”接口的请求参数对象。代码如下：

```
package com.wudimanong.wallet.entity.dto;
import com.wudimanong.wallet.validator.EnumValue;
import java.io.Serializable;
import javax.validation.constraints.NotNull;
import lombok.Data;
@Data
public class AccountChargeDTO implements Serializable {
    /**
     * 用户 ID
     */
    @NotNull(message = "用户 ID 不能为空")
    private Long userId;
    /**
     * 充值金额，以“分”为单位
```

```
     */
    private Integer amount;
    /**
     * 充值币种（仅支持人民币）
     */
    @EnumValue(strValues = {"CNY"})
    private String currency;
    /**
     * 支付类型。0-微信支付；1-支付宝支付
     */
    @EnumValue(intValues = {0, 1})
    private Integer paymentType;
    /**
     * 是否自动续费
     */
    @EnumValue(intValues = {0, 1})
    private Integer isRenew;
}
```

（3）定义“电子钱包充值”接口的返回参数对象。代码如下：

```
package com.wudimanong.wallet.entity.bo;
import java.io.Serializable;
import lombok.Data;
@Data
public class AccountChargeBO implements Serializable {
    /**
     * 用户 ID
     */
    private Long userId;
    /**
     * 充值金额，以“分”为单位
     */
    private Integer amount;
    /**
     * 充值币种
     */
    private String currency;
    /**
     * 充值业务订单号（由钱包系统生成）
     */
    private String orderId;
    /**
     * 支付系统唯一流水号（在调用支付系统后生成）
     */
    private String tradeNo;
```

```
    /**
     * 前端唤起支付收银台所需的额外信息
     */
    private String extraInfo;
}
```

4. 定义“电子钱包充值支付回调”接口

用户在发起钱包充值并完成支付后，用户电子钱包的余额并不会立刻增加，电子钱包微服务还需要等待支付系统的支付成功结果通知。

在接收到支付成功结果通知后，电子钱包微服务会先更新钱包充值订单的支付状态，然后以事务一致性的方式同步更新电子钱包的余额。

（1）定义“电子钱包充值支付回调”接口的 Controller 层。

在“3.”小标题中创建的 UserAccountTradeController 类中，增加“电子钱包充值支付回调”接口。代码如下：

```
/**
 * “电子钱包充值支付回调”接口
 *
 * @param payNotifyDTO
 * @return
 */
@PostMapping("/payNotify")
public ResponseResult<PayNotifyBO> receivePayNotify(@RequestBody
@Validated PayNotifyDTO payNotifyDTO) {
    return ResponseResult.OK(userAccountTradeServiceImpl.
receivePayNotify(payNotifyDTO));
}
```

（2）定义“电子钱包充值支付回调”接口的请求参数对象。代码如下：

```
package com.wudimanong.wallet.entity.dto;
import com.wudimanong.wallet.validator.EnumValue;
import java.io.Serializable;
import javax.validation.constraints.NotNull;
import lombok.Data;
@Data
public class PayNotifyDTO implements Serializable {
    /**
     * 商户支付订单号
     */
    @NotNull(message = "订单号不能为空")
    private String orderId;
    /**
```

```
     * 支付订单金额
     */
    private Integer amount;
    /**
     * 支付币种
     */
    private String currency;
    /**
     * 支付订单状态。0-待支付；1-支付中；2-支付成功；3-支付失败
     */
    @EnumValue(intValues = {2, 3}, message = "只接收支付状态为成功/失败的通知")
    private Integer payStatus;
}
```

（3）定义“电子钱包充值支付回调”接口的返回参数对象。代码如下：

```
package com.wudimanong.wallet.entity.bo;
import lombok.Builder;
import lombok.Data;
@Data
@Builder
public class PayNotifyBO {
    /**
     * 接收处理状态。success-成功；fail-失败
     */
    private String result;
}
```

该返回参数只有一个“result”属性：success 表示处理成功，fail 表示处理失败。如果处理失败，则支付系统（参考第 6 章）会对支付结果进行一定频率的重复回调。

5.4.2 开发业务层（Service 层）的代码

接下来开发电子钱包微服务业务层（Service 层）的代码。

1. 定义业务异常处理机制

关于业务层（Service 层）的业务异常处理机制，可以参考 4.4.2 节中“1.”小标题中的内容。需要专门定义的业务层（Service 层）异常码的枚举类，代码如下：

```
package com.wudimanong.wallet.entity;
public enum BusinessCodeEnum {
    /**
     * 电子钱包信息管理返回码的定义（以 1000 开头，根据业务扩展）
     */
    BUSI_ACCOUNT_FAIL_1000(1000, "该用户已开通该类型电子账户"),
    /**
```

```
     * 电子钱包交易相关返回码的定义（以2000开头，根据业务扩展）
     */
    BUSI_CHARGE_FAIL_2000(2000, "充值失败");
    /**
     * 编码
     */
    private Integer code;
    /**
     * 描述
     */
    private String desc;
    BusinessCodeEnum(Integer code, String desc) {
        this.code = code;
        this.desc = desc;
    }
    /**
     * 根据编码获取枚举类型
     *
     * @param code 编码
     * @return
     */
    public static BusinessCodeEnum getByCode(String code) {
        //判断编码是否为空
        if (code == null) {
            return null;
        }
        //循环处理
        BusinessCodeEnum[] values = BusinessCodeEnum.values();
        for (BusinessCodeEnum value : values) {
            if (value.getCode().equals(code)) {
                return value;
            }
        }
        return null;
    }
    public Integer getCode() {
        return code;
    }
    public String getDesc() {
        return desc;
    }
}
```

可在具体的业务层（Service 层）逻辑处理中，根据业务扩展枚举类。例如，将电子钱包信息管理相关的业务异常码值的范围约定为 1000～1999，将电子钱包充值交易操作相关的业务异常码约定为 2000 ~ 2999 等。

2. 引入 MapStruct 实体映射工具

具体参考 4.4.2 节中“2.”小标题中的内容。

3. 开发“电子钱包开户”接口及“电子钱包查询”接口的业务层（Service 层）代码

接下来开发“电子钱包开户”接口及“电子钱包查询”接口的业务层（Service 层）代码。

（1）定义“电子钱包开户”接口及“电子钱包查询”接口的业务层（Service 层）方法。代码如下：

```
package com.wudimanong.wallet.service;
import com.wudimanong.wallet.entity.bo.AccountBO;
import com.wudimanong.wallet.entity.bo.AccountOpenBO;
import com.wudimanong.wallet.entity.dto.AccountOpenDTO;
import com.wudimanong.wallet.entity.dto.AccountQeuryDTO;
import java.util.List;
public interface UserAccountService {
    /**
     * 定义“电子钱包开户”接口的业务层（Service 层）方法
     *
     * @param accountOpenDTO
     * @return
     */
    AccountOpenBO openAcc(AccountOpenDTO accountOpenDTO);
    /**
     * 定义“电子钱包查询”接口的业务层（Service 层）方法
     *
     * @param accountQeuryDTO
     * @return
     */
    List<AccountBO> queryAcc(AccountQeuryDTO accountQeuryDTO);
}
```

（2）开发“电子钱包开户”接口及“电子钱包查询”接口的业务层（Service 层）实现类的代码。

“电子钱包开户”接口及“电子钱包查询”接口的业务层（Service 层）实现类的代码如下：

```
package com.wudimanong.wallet.service.impl;
import com.wudimanong.wallet.convert.UserBalanceConvert;
...
import org.springframework.stereotype.Service;
@Service
public class UserAccountServiceImpl implements UserAccountService {
```

```
    /**
     * 注入持久层（Dao 层）接口依赖
     */
    @Autowired
    UserBalanceDao userBalanceDao;
    @Override
    public AccountOpenBO openAcc(AccountOpenDTO accountOpenDTO) {
        //判断在同一个用户 ID 下是否存在同一种类型的账户
        Map paramMap = new HashMap<>();
        paramMap.put("user_id", accountOpenDTO.getUserId());
        paramMap.put("acc_type", accountOpenDTO.getAccType());
        List<UserBalancePO> userBalancePOList =
userBalanceDao.selectByMap(paramMap);
        if (userBalancePOList != null && userBalancePOList.size() > 0) {
            throw new ServiceException
(BusinessCodeEnum.BUSI_ACCOUNT_FAIL_1000.getCode(),
                    BusinessCodeEnum.BUSI_ACCOUNT_FAIL_1000.getDesc());
        }
        //将"业务层输出的数据对象"转换为"持久层（Dao 层）数据对象"
        UserBalancePO userBalancePO =
UserBalanceConvert.INSTANCE.convertUserBalancePO(accountOpenDTO);
        //生成电子钱包账户编号
        String accountNo = getAccountNo();
        userBalancePO.setAccNo(accountNo);
        //设置电子钱包账户的初始余额
        userBalancePO.setBalance(0);
        //设置时间值
        userBalancePO.setCreateTime(new
Timestamp(System.currentTimeMillis()));
        userBalancePO.setUpdateTime(new
Timestamp(System.currentTimeMillis()));
        //通过持久层（Dao 层）组件将电子钱包账户信息写入数据库
        userBalanceDao.insert(userBalancePO);
        //封装返回的"业务层（Service 层）输出的数据对象"
        AccountOpenBO accountOpenBO =
UserBalanceConvert.INSTANCE.convertAccountOpenBO(userBalancePO);
        return accountOpenBO;
    }
    @Override
    public List<AccountBO> queryAcc(AccountQeuryDTO accountQeuryDTO) {
        //组合电子钱包账户查询条件
        Map paramMap = new HashMap<>();
        paramMap.put("user_id", accountQeuryDTO.getUserId());
        if (accountQeuryDTO.getAccType() != null) {
            paramMap.put("acc_type", accountQeuryDTO.getAccType());
```

```
        }
        if (accountQeuryDTO.getCurrency() != null
&& !"".equals(accountQeuryDTO.getCurrency())) {
            paramMap.put("currency", accountQeuryDTO.getCurrency());
        }
        List<AccountBO> accountBOList = new ArrayList<>();
        List<UserBalancePO> userBalancePOList =
userBalanceDao.selectByMap(paramMap);
        if (userBalancePOList != null && userBalancePOList.size() > 0) {
            //完成电子钱包账户“持久层（Dao 层）数据对象”到“业务层（Service 层）输出
的数据对象”的转换
            accountBOList =
UserBalanceConvert.INSTANCE.convertAccountBO(userBalancePOList);
        }
        return accountBOList;
    }
    /**
     * 生成电子钱包账户编号
     */
    private String getAccountNo() {
        //雪花算法 ID 生成器
        SnowFlakeIdGenerator idGenerator = new
SnowFlakeIdGenerator(IDutils.getWorkId(), 1);
        //以“日期 YYYYMMDD + 随机 ID”规则生成电子钱包账户编号
        return DateUtils.getStringByFormat(new Date(), DateUtils.sf1) +
idGenerator.nextId();
    }
}
```

（3）编写 MapStruct 数据转化类。

在步骤（2）的实现逻辑中，有关数据复制使用了“1.”小标题中引入的 MapStruct 工具。涉及的代码片段如下：

```
    //在电子钱包开户逻辑中使用 MapStruct 工具进行数据对象转换的代码片段
    //将“业务层（Service 层）输出的数据对象”转换为“持久层（Dao 层）数据对象”
    UserBalancePO userBalancePO =
UserBalanceConvert.INSTANCE.convertUserBalancePO(accountOpenDTO);
    ...
    //封装返回的“业务层（Service 层）输出数据对象”
    AccountOpenBO accountOpenBO =
UserBalanceConvert.INSTANCE.convertAccountOpenBO(userBalancePO);
    ...
    //在电子钱包查询逻辑中使用 MapStruct 工具进行数据对象转换的代码片段
    //完成电子钱包账户“持久层（Dao 层）数据对象”到“业务层（Service 层）输出的数据对象”
的转换
```

```
    accountBOList =
UserBalanceConvert.INSTANCE.convertAccountBO(userBalancePOList);
    ...
```

在上述代码片段中，具体映射转换类的代码如下：

```
    package com.wudimanong.wallet.convert;
    import com.wudimanong.wallet.dao.model.UserBalancePO;
    ...
    import org.mapstruct.factory.Mappers;
    @org.mapstruct.Mapper
    public interface UserBalanceConvert {
        UserBalanceConvert INSTANCE =
Mappers.getMapper(UserBalanceConvert.class);
        /**
         *将电子钱包开户逻辑的“业务层(Service 层)输出数据对象”转换成“持久层（Dao 层）数
据对象”
         *
         * @param accountOpenDTO
         * @return
         */
        @Mappings({})
        UserBalancePO convertUserBalancePO(AccountOpenDTO accountOpenDTO);
        /**
         *将电子钱包开户逻辑的“持久层（Dao 层）数据对象”转换为“业务层（Service 层）输出
数据对象”
         *
         * @param userBalancePO
         * @return
         */
        @Mappings({})
        AccountOpenBO convertAccountOpenBO(UserBalancePO userBalancePO);
        /**
         *将电子钱包查询逻辑的“持久层（Dao 层）数据对象”转换为“业务层（Service 层）输出
数据对象”
         *
         * @param userBalancePOList
         * @return
         */
        @Mappings({})
        List<AccountBO> convertAccountBO(List<UserBalancePO>
userBalancePOList);
    }
```

（4）编写“雪花算法”ID 生成器。

在上述业务层（Service 层）的具体逻辑中，电子钱包开户还涉及电子钱包账户编号的生成。

在本实例中，电子钱包账户编号的生成规则为“日期 YYYYMMDD + 随机生成 ID”，为了确保随机 ID 生成的幂等性，使用了“雪花算法”。

实现“雪花算法”的工具类的代码如下：

```
    package com.wudimanong.wallet.utils;
    public class SnowFlakeIdGenerator {
        //初始时间截
        private static final long INITIAL_TIME_STAMP = 1546272000000L;
        //机器 ID 所占的位数
        private static final long WORKER_ID_BITS = 10L;
        //数据标识 ID 所占的位数
        private static final long DATACENTER_ID_BITS = 5L;
        //支持的最大机器 ID，结果是 31 （这个移位算法可以很快计算出几位二进制数所能表示的最
大十进制数）
        private static final long MAX_WORKER_ID = ~(-1L << WORKER_ID_BITS);
        //支持的最大数据标识 ID，结果是 31
        private static final long MAX_DATACENTER_ID = ~(-1L <<
DATACENTER_ID_BITS);
        //序列在 ID 中占的位数
        private final long SEQUENCE_BITS = 12L;
        //机器 ID 的偏移量（12）
        private final long WORKERID_OFFSET = SEQUENCE_BITS;
        //数据中心 ID 的偏移量（12+5）
        private final long DATACENTERID_OFFSET = SEQUENCE_BITS + SEQUENCE_BITS;
        //时间截的偏移量(5+5+12)
        private final long TIMESTAMP_OFFSET = SEQUENCE_BITS + WORKER_ID_BITS +
DATACENTER_ID_BITS;
        //生成序列的掩码，这里为 4095 （0b111111111111=0xfff=4095）
        private final long SEQUENCE_MASK = ~(-1L << SEQUENCE_BITS);
        //工作节点 ID（0～31）
        private long workerId;
        //数据中心 ID（0～31）
        private long datacenterId;
        //毫秒内序列（0～4095）
        private long sequence = 0L;
        //上一次生成 ID 的时间截
        private long lastTimestamp = -1L;
        /**
         * 构造函数
         *
         * @param workerId 工作 ID （0～31）
         * @param datacenterId 数据中心 ID （0～31）
         */
        public SnowFlakeIdGenerator(long workerId, long datacenterId) {
```

```
        if (workerId > MAX_WORKER_ID || workerId < 0) {
            throw new IllegalArgumentException(String.format("WorkerID 不能
大于 %d 或小于 0", MAX_WORKER_ID));
        }
        if (datacenterId > MAX_DATACENTER_ID || datacenterId < 0) {
            throw new IllegalArgumentException(String.format("DataCenterID
不能大于 %d 或小于 0", MAX_DATACENTER_ID));
        }
        this.workerId = workerId;
        this.datacenterId = datacenterId;
    }
    /**
     * 获得下一个 ID （用同步锁保证线程安全）
     *
     * @return SnowflakeId
     */
    public synchronized long nextId() {
        long timestamp = System.currentTimeMillis();
        //如果当前时间小于上一次 ID 生成的时间戳，则说明系统时钟存在问题，应抛出异常
        if (timestamp < lastTimestamp) {
            throw new RuntimeException("Clock moved backwards.  Refusing to
generate id for %d milliseconds! ");
        }
        //如果是同一时间生成的，则以 ms 为单位进行序列计算
        if (lastTimestamp == timestamp) {
            sequence = (sequence + 1) & SEQUENCE_MASK;
            //如果 sequence 等于 0，则说明毫秒内序列已经增长到最大值
            if (sequence == 0) {
                //阻塞到下一个毫秒，获得新的时间戳
                timestamp = tilNextMillis(lastTimestamp);
            }
        } else {//时间戳改变，毫秒内序列重置
            sequence = 0L;
        }
        //上一次生成 ID 的时间截
        lastTimestamp = timestamp;
        //通过移位或运算，将结果拼到一起组成 64 位的 ID
        return ((timestamp - INITIAL_TIME_STAMP) << TIMESTAMP_OFFSET) |
(datacenterId << DATACENTERID_OFFSET) | (
                workerId << WORKERID_OFFSET) | sequence;
    }
    /**
     * 阻塞到下一个毫秒，直到获得新的时间戳
     *
     * @param lastTimestamp 上一次生成 ID 的时间截
```

```
     * @return 当前时间戳
     */
    protected long tilNextMillis(long lastTimestamp) {
        long timestamp = System.currentTimeMillis();
        while (timestamp <= lastTimestamp) {
            timestamp = System.currentTimeMillis();
        }
        return timestamp;
    }
}
```

（5）编写日期工具类。

接下来，创建用于在电子钱包账户生成编号过程中处理日期字符串的工具类。代码如下：

```
package com.wudimanong.wallet.utils;
import java.text.SimpleDateFormat;
import java.util.Date;
import lombok.extern.slf4j.Slf4j;
/**
 * 日期时间工具类
 */
@Slf4j
public class DateUtils {
    //线程局部变量
    public static final ThreadLocal<SimpleDateFormat> sf1 = new
ThreadLocal<SimpleDateFormat>() {
        @Override
        public SimpleDateFormat initialValue() {
            return new SimpleDateFormat("yyyyMMdd");
        }
    };
    public static final ThreadLocal<SimpleDateFormat> sf2 = new
ThreadLocal<SimpleDateFormat>() {
        @Override
        public SimpleDateFormat initialValue() {
            return new SimpleDateFormat("yyyy-MM-dd");
        }
    };
    public static final ThreadLocal<SimpleDateFormat> sf3 = new
ThreadLocal<SimpleDateFormat>() {
        @Override
        public SimpleDateFormat initialValue() {
            return new SimpleDateFormat("yyyyMMddHHmmss");
        }
    };
```

```
        /**
         * 时间格式化方法
         *
         * @param date
         * @param fromat
         * @return
         */
        public static String getStringByFormat(Date date,
ThreadLocal<SimpleDateFormat> fromat) {
            return fromat.get().format(date);
        }
    }
```

（6）编写获取 WorkID 的工具类。

在使用“雪花算法”ID 生成器时，需要传递 WorkID。这里创建一个获取 WorkID 的简单工具类，代码如下：

```
    package com.wudimanong.wallet.utils;
    import java.net.InetAddress;
    import lombok.extern.slf4j.Slf4j;
    @Slf4j
    public class IDutils {
        /**
         * WorkID 的获取方式为：根据机器 IP 地址获取工作进程 ID。如果线上机器的 IP 地址的二进制表示的最后 10 位不重复，则建议使用此种方式。例如机器的 IP 地址为“192.168.1.108”，二进制表示为“11000000 10101000”，截取最后 10 位“01 01101100”，转为十进制数为 364，则设置 workerID 为 364
         */
        public static int getWorkId() {//性能待优化
            int workId = 1;
            try {
                //获取机器 IP 地址的二进制表示
                InetAddress address = InetAddress.getLocalHost();
                String sIP = address.getHostAddress();
                sIP = sIP.replaceAll("\t", "").trim();
                String[] arr = sIP.split("\\.");
                String rs = "";
                for (String str : arr) {
                    String s = Integer.toBinaryString(Integer.parseInt(str));
                    if (s.length() < 8) {
                        int diff = 8 - s.length();
                        for (int i = 0; i < diff; i++) {
                            s = "0" + s;
                        }
                    }
```

```
                    rs += s;
                }
                if (!"".equals(rs)) {
                    //截取 IP 地址二进制表示的后 10 位
                    String last10 = rs.substring(rs.length() - 10, rs.length());
                    workId = Integer.parseInt(last10, 2);
                }
            } catch (Exception e) {
                e.printStackTrace();
                log.error(e.getMessage(), e);
            }
            return workId;
        }
}
```

以上代码实现了“电子钱包开户”接口及“电子钱包查询”接口的业务层（Service 层）逻辑，所依赖的持久层（Dao 层）操作组件可以参考 5.4.3 节的内容。

4. 开发“电子钱包充值”接口的业务层（Service 层）代码

接下来开发“电子钱包充值”接口的业务层（Service 层）代码。

（1）定义业务层（Service 层）接口类 UserAccountTradeService。代码如下：

```
package com.wudimanong.wallet.service;
import com.wudimanong.wallet.entity.bo.AccountChargeBO;
import com.wudimanong.wallet.entity.dto.AccountChargeDTO;
public interface UserAccountTradeService {
    /**
     * 定义“电子钱包充值”接口的业务层（Service 层）方法
     *
     * @param accountChargeDTO
     * @return
     */
    AccountChargeBO chargeOrder(AccountChargeDTO accountChargeDTO);
}
```

（2）实现业务层（Service 层）接口类的方法。代码如下：

```
package com.wudimanong.wallet.service.impl;
import com.wudimanong.wallet.client.PaymentClient;
...
import org.springframework.stereotype.Service;
@Service
public class UserAccountTradeServiceImpl implements UserAccountTradeService {
    @Autowired
    PaymentClient paymentClient;
```

```
        /**
         * 注入 Dao 层接口依赖
         */
        @Autowired
        UserBalanceOrderDao userBalanceOrderDao;
        @Override
        public AccountChargeBO chargeOrder(AccountChargeDTO accountChargeDTO) {
            //生成电子钱包充值订单信息
            UserBalanceOrderPO userBalanceOrderPO =
createChargeOrder(accountChargeDTO);
            try {
                userBalanceOrderDao.insert(userBalanceOrderPO);
            } catch (Exception e) {
                //抛出 Dao 层异常
                throw new
DAOException(BusinessCodeEnum.BUSI_CHARGE_FAIL_2000.getCode(),
                        BusinessCodeEnum.BUSI_CHARGE_FAIL_2000.getDesc(), e);
            }
            //调用支付系统接口
            //构建支付请求参数
            UnifiedPayDTO unifiedPayDTO = buildUnifiedPayDTO(accountChargeDTO,
userBalanceOrderPO);
            ResponseResult<UnifiedPayBO> responseResult =
paymentClient.unifiedPay(unifiedPayDTO);
            if
(!responseResult.getCode().equals(GlobalCodeEnum.GL_SUCC_0000.getCode())) {
                //支付失败的业务异常返回
                throw new ServiceException(responseResult.getCode(),
responseResult.getMessage());
            }
            //获取支付返回数据
            UnifiedPayBO unifiedPayBO = responseResult.getData();
            //封装返回的电子钱包充值订单信息
            AccountChargeBO accountChargeBO = UserBalanceOrderConvert.INSTANCE
                    .convertUserBalanceOrderBO(unifiedPayBO);
            accountChargeBO.setUserId(accountChargeDTO.getUserId());
            return accountChargeBO;
        }
        /**
         * 生成电子钱包充值订单信息的私有方法
         *
         * @param accountChargeDTO
         * @return
         */
```

```
    private UserBalanceOrderPO createChargeOrder(AccountChargeDTO
accountChargeDTO) {
        UserBalanceOrderPO userBalanceOrderPO =
UserBalanceOrderConvert.INSTANCE
                .convertUserBalanceOrderPO(accountChargeDTO);
        //生成电子钱包充值订单流水号
        String orderId = getOrderId();
        userBalanceOrderPO.setOrderId(orderId);
        //设置交易类型为“充值”
        userBalanceOrderPO.setTradeType(TradeType.CHARGE.getCode());
        //设置支付状态为“待支付”
        userBalanceOrderPO.setStatus("0");
        //设置交易时间
        userBalanceOrderPO.setTradeTime(new Timestamp
(System.currentTimeMillis()));
        //设置订单创建时间
        userBalanceOrderPO.setCreateTime(new Timestamp
(System.currentTimeMillis()));
        //设置订单初始更新时间
        userBalanceOrderPO.setUpdateTime(new Timestamp
(System.currentTimeMillis()));
        return userBalanceOrderPO;
    }
    /**
     * 以特定的规则生成电子钱包充值订单流水号的私有方法
     *
     * @return
     */
    private String getOrderId() {
        //“雪花算法”ID 生成器
        SnowFlakeIdGenerator idGenerator = new
SnowFlakeIdGenerator(IDutils.getWorkId(), 1);
        //以“日期 yyyyMMddHHmmss + 随机生存 ID”规则生成充值订单号
        return DateUtils.getStringByFormat(new Date(), DateUtils.sf3) +
idGenerator.nextId();
    }
    /**
     * 构建支付系统请求参数对象的私有方法
     *
     * @param userBalanceOrderPO
     * @return
     */
    private UnifiedPayDTO buildUnifiedPayDTO(AccountChargeDTO
accountChargeDTO, UserBalanceOrderPO userBalanceOrderPO) {
        UnifiedPayDTO unifiedPayDTO = new UnifiedPayDTO();
```

```
        //支付系统为接入方分配的应用 ID
        unifiedPayDTO.setAppId("10001");
        //支付业务订单号
        unifiedPayDTO.setOrderId(userBalanceOrderPO.getOrderId());
        //充值交易类型——余额充值
        unifiedPayDTO.setTradeType("topup");
        //支付渠道
        unifiedPayDTO.setChannel(accountChargeDTO.getPaymentType());
        //具体的支付渠道方式，可根据接入的支付产品设定
        unifiedPayDTO.setPayType("ALI_PAY_H5");
        //支付金额
        unifiedPayDTO.setAmount(accountChargeDTO.getAmount());
        //支付币种
        unifiedPayDTO.setCurrency(accountChargeDTO.getCurrency());
        //商户用户标识
        unifiedPayDTO.setUserId(String.valueOf
(accountChargeDTO.getUserId()));
        //商品标题，在实际情况下根据所购买的商品来定义相关内容
        unifiedPayDTO.setSubject("xiaomi 10 pro");
        //商品详情
        unifiedPayDTO.setBody("xiaomi 10 pro testing");
        //支付回调通知地址，可根据实际情况填充
        unifiedPayDTO.setNotifyUrl("http://www.baidu.com");
        //支付结果同步返回 URL，一般为用户前端页面，可根据实际情况填充
        unifiedPayDTO.setReturnUrl("http://www.baidu.com");
        return unifiedPayDTO;
    }
}
```

以上代码为“电子钱包充值”接口的业务层（Service 层）的完整代码。具体的逻辑为：

首先，完成余额充值业务订单的生成及持久化。

然后，通过 Feign 客户端代码（参见 5.5 节内容）实现对支付系统的调用。

最后，判断支付微系统调用结果——如果成功返回，则通过 MapStruct 组件转换返回参数对象；如果失败，则通过业务层（Service 层）异常处理机制向系统上层抛出支付处理错误异常。

5. 开发“电子钱包充值支付回调”接口的业务层（Service 层）代码

接下来开发“电子钱包充值支付回调”接口的业务层（Service 层）代码。

（1）定义业务层（Service 层）方法。

在“4.”小标题下步骤（1）中定义的业务层（Service 层）实现 UserAccountTradeService 类中，增加“电子钱包充值支付回调”接口的业务层（Service 层）方法。代码如下：

```
    /**
     * 定义“电子钱包充值支付回调”接口的业务层（Service 层）方法
     *
     * @param payNotifyDTO
     * @return
     */
    PayNotifyBO receivePayNotify(PayNotifyDTO payNotifyDTO);
```

（2）实现业务层（Service 层）方法。

在“4.”小标题下步骤（2）中定义的业务层（Service 层）实现 UserAccountTradeServiceImpl 类中，增加“电子钱包充值支付回调”接口业务层（Service 层）方法的实现。代码如下：

```
    /**
     * 电子钱包余额服务层接口依赖
     */
    @Autowired
    UserBalanceService userBalanceServiceImpl;
    /**
     * “电子钱包充值支付回调”接口的业务层（Service 层）方法的实现
     *
     * @param payNotifyDTO
     * @return
     */
    @Override
    public PayNotifyBO receivePayNotify(PayNotifyDTO payNotifyDTO) {
        //判断电子钱包充值订单支付状态是否为成功
        Map parmMap = new HashMap<>();
        parmMap.put("order_id", payNotifyDTO.getOrderId());
        List<UserBalanceOrderPO> userBalanceOrderPOList =
userBalanceOrderDao.selectByMap(parmMap);
        //如果电子钱包充值订单不存在，则返回失败结果
        if (userBalanceOrderPOList == null && userBalanceOrderPOList.size() <=
0) {
            return PayNotifyBO.builder().result("fail").build();
        }
        UserBalanceOrderPO userBalanceOrderPO = userBalanceOrderPOList.get(0);
        //判断电子钱包充值订单的支付状态，如果已经为成功状态，则说明已处理，返回成功结果
        if ("2".equals(userBalanceOrderPO.getStatus())) {
            return PayNotifyBO.builder().result("success").build();
        }
        //更新电子钱包充值订单支付状态为成功

userBalanceOrderPO.setStatus(String.valueOf(payNotifyDTO.getPayStatus()));
        //设置订单更新时间
```

```
        userBalanceOrderPO.setUpdateTime(new 
Timestamp(System.currentTimeMillis()));
        //更新状态
        userBalanceOrderDao.updateById(userBalanceOrderPO);
        //如果是支付成功回调通知，则完成电子钱包账户余额的增加
        if (payNotifyDTO.getPayStatus() == 2) {
            AddBalanceBO addBalanceBO = 
AddBalanceBO.builder().userId(userBalanceOrderPO.getUserId())
                    .amount(userBalanceOrderPO.getAmount()).busiType("charge").
accType("0")
                    .currency(userBalanceOrderPO.getCurrency()).build();
            //调用电子钱包余额业务层（Service 层）方法增加余额
            userBalanceServiceImpl.addBalance(addBalanceBO);
        }
        return PayNotifyBO.builder().result("success").build();
    }
```

该实现方法主要完成对电子钱包充值订单的状态更新，如果是支付成功回调通知，则完成电子钱包账户余额的增加，以及钱包账户变动流水的记录。

（3）定义“电子钱包余额服务层”的业务层（Service 层）方法。

从业务逻辑解耦的角度，将步骤（2）中所依赖的 UserBalanceService 接口（电子钱包余额服务层）进行单独的业务层（Service 层）拆分。

定义“电子钱包余额服务层”的业务层（Service 层）方法，代码如下：

```
    package com.wudimanong.wallet.service;
    import com.wudimanong.wallet.entity.bo.AddBalanceBO;
    public interface UserBalanceService {
        /**
         * 定义余额增加逻辑的业务层（Service 层）方法
         *
         * @param addBalanceBO
         * @return
         */
        boolean addBalance(AddBalanceBO addBalanceBO);
    }
```

在业务接口中定义了电子钱包余额增加的方法，该方法的入口参数 AddBalanceBO 类的代码如下：

```
    package com.wudimanong.wallet.entity.bo;
    import lombok.Builder;
    import lombok.Data;
    @Data
    @Builder
```

```
public class AddBalanceBO {
    /**
     * 用户 ID
     */
    private String userId;
    /**
     * 增加金额
     */
    private Integer amount;
    /**
     * 业务类型
     */
    private String busiType;
    /**
     * 账户类型
     */
    private String accType;
    /**
     * 币种
     */
    private String currency;
}
```

（4）开发“电子钱包余额服务层”的业务层（Service 层）实现类的代码。

开发“电子钱包余额服务层”业务接口实现类 UserBalanceServiceImpl 的代码。代码如下：

```
package com.wudimanong.wallet.service.impl;
import com.wudimanong.wallet.dao.mapper.UserBalanceDao;
...
import org.springframework.stereotype.Service;
import org.springframework.transaction.annotation.Transactional;
@Slf4j
@Service
public class UserBalanceServiceImpl implements UserBalanceService {
    @Autowired
    UserBalanceDao userBalanceDao;
    @Autowired
    UserBalanceFlowDao userBalanceFlowDao;
    @Transactional(rollbackFor = Exception.class)
    @Override
    public boolean addBalance(AddBalanceBO addBalanceBO) {
        //查询电子钱包账户余额
        Map param = new HashMap<>();
        param.put("user_id", addBalanceBO.getUserId());
        param.put("acc_type", addBalanceBO.getAccType());
```

```
        List<UserBalancePO> userBalancePOList = userBalanceDao.
selectByMap(param);
        UserBalancePO userBalancePO = userBalancePOList.get(0);
        userBalancePO.setBalance(userBalancePO.getBalance() +
addBalanceBO.getAmount());
        userBalancePO.setUpdateTime(new Timestamp
(System.currentTimeMillis()));
        userBalanceDao.updateById(userBalancePO);
        //生成电子钱包账户变动流水记录
        UserBalanceFlowPO userBalanceFlowPO = createUserBalanceFlow
(addBalanceBO, userBalancePO);
        //持久层（Dao 层）入库处理
        userBalanceFlowDao.insert(userBalanceFlowPO);
        return true;
    }
    /**
     * 生成电子钱包账户变动流水记录
     *
     * @param addBalanceBO
     * @param userBalancePO
     * @return
     */
    private UserBalanceFlowPO createUserBalanceFlow(AddBalanceBO
addBalanceBO, UserBalancePO userBalancePO) {
        UserBalanceFlowPO userBalanceFlowPO = new UserBalanceFlowPO();
        userBalanceFlowPO.setUserId(addBalanceBO.getUserId());
        //设置账户变动流水号
        userBalanceFlowPO.setFlowNo(getFlowId());
        //记录账户编号
        userBalanceFlowPO.setAccNo(userBalancePO.getAccNo());
        //记录业务类型
        userBalanceFlowPO.setBusiType(addBalanceBO.getBusiType());
        //记录变动金额
        userBalanceFlowPO.setAmount(addBalanceBO.getAmount());
        //币种
        userBalanceFlowPO.setCurrency(userBalancePO.getCurrency());
        //记录账户变动前的金额
        userBalanceFlowPO.setBeginBalance(userBalancePO.getBalance() -
addBalanceBO.getAmount());
        //记录账户变动后的金额
        userBalanceFlowPO.setEndBalance(userBalancePO.getBalance());
        //借贷方向，借方账户
        userBalanceFlowPO.setFundDirect("00");
        //设置创建时间
        userBalanceFlowPO.setCreateTime(new Timestamp
(System.currentTimeMillis()));
```

```
        //设置更新时间
        userBalanceFlowPO.setUpdateTime(new Timestamp
(System.currentTimeMillis()));
        return userBalanceFlowPO;
    }
    /**
     * 以特定的规则生成电子钱包账户变动流水号的私有方法
     * @return
     */
    private String getFlowId() {
        //"雪花算法"ID 生成器
        SnowFlakeIdGenerator idGenerator = new
SnowFlakeIdGenerator(IDutils.getWorkId(), 1);
        //以"日期 yyyyMMddHHmmss + 随机生成 ID 器"规则生成电子钱包余额变动流水号
        return DateUtils.getStringByFormat(new Date(), DateUtils.sf3) +
idGenerator.nextId();
    }
}
```

以上实现了电子钱包账户余额增加的逻辑。其中，通过@Transactional 事务注解实现了余额增加与账户变动流水记录数据库操作的事务一致性。

5.4.3 开发 MyBatis 持久层（Dao 层）组件

持久层（Dao 层）的实现，以 MyBatis 及 MyBatis-Plus 组合提供的功能为基础。需要先在程序主类中添加 MyBatis 接口代码的包扫描路径，代码如下：

```
package com.wudimanong.wallet;
import org.mybatis.spring.annotation.MapperScan;
...
import org.springframework.cloud.client.discovery.EnableDiscoveryClient;
@EnableDiscoveryClient
@SpringBootApplication
@MapperScan("com.wudimanong.wallet.dao.mapper")
public class WalletApplication {
    public static void main(String[] args) {
        SpringApplication.run(WalletApplication.class, args);
    }
}
```

在上述代码中，通过@MapperScan 注解定义 MyBatis 持久层（Dao 层）组件代码的包路径。

1. 实现"电子钱包开户"接口及"电子钱包查询"接口的持久层（Dao 层）

"电子钱包开户"接口及"电子钱包查询"接口涉及的数据库操作，主要以（user_balance）"余额账户信息表"为准。

（1）定义“余额账户信息表”的数据库实体类。代码如下：

```
package com.wudimanong.wallet.dao.model;
import com.baomidou.mybatisplus.annotation.TableName;
import java.sql.Timestamp;
import lombok.Data;
@Data
@TableName("user_balance")
public class UserBalancePO {
    /**
     * 主键 ID
     */
    private Integer id;
    /**
     * 用户编号
     */
    private Long userId;
    /**
     * 账户编号
     */
    private String accNo;
    /**
     * 账户类型
     */
    private String accType;
    /**
     * 币种
     */
    private String currency;
    /**
     * 账户余额
     */
    private Integer balance;
    /**
     * 创建时间
     */
    private Timestamp createTime;
    /**
     * 更新时间
     */
    private Timestamp updateTime;
}
```

由于要使用 MyBatis-Plus 工具来简化 MyBatis 的操作，所以，在数据库实体类的定义中需要通过@TableName 注解来指定具体的表名。

（2）定义“余额账户信息表”的持久层（Dao 层）接口。

定义“电子钱包开户”接口及“电子钱包查询”接口的业务层（Service 层）所依赖的持久层（Dao 层）接口。代码如下：

```
package com.wudimanong.wallet.dao.mapper;
import com.baomidou.mybatisplus.core.mapper.BaseMapper;
import com.wudimanong.wallet.dao.model.UserBalancePO;
import org.springframework.stereotype.Repository;
@Repository
public interface UserBalanceDao extends BaseMapper<UserBalancePO> {
}
```

至此，“电子钱包开户”接口及“电子钱包查询”接口所涉及的 Controller、Service 及 Dao 层的代码就编写完成了。

2. 开发“电子钱包充值”接口的持久层（Dao 层）代码

电子钱包充值持久层（Dao 层）主要涉及对“余额交易订单表”（user_balance_order）的操作。

（1）定义“余额交易订单表”的数据库实体类。代码如下：

```
package com.wudimanong.wallet.dao.model;
import com.baomidou.mybatisplus.annotation.TableName;
import java.sql.Timestamp;
import lombok.Data;
@Data
@TableName("user_balance_order")
public class UserBalanceOrderPO {
    /**
     * 主键 ID
     */
    private Integer id;
    /**
     * 充值订单 ID
     */
    private String orderId;
    /**
     * 用户 ID
     */
    private String userId;
    /**
     * 充值订单金额
     */
    private Integer amount;
```

```
    /**
     * 订单交易类型。charge-余额充值；refund-余额退款
     */
    private String tradeType;
    /**
     * 币种
     */
    private String currency;
    /**
     * 支付流水号
     */
    private String tradeNo;
    /**
     * 支付状态。0-待支付；1-支付中；2-支付成功；3-支付失败
     */
    private String status;
    /**
     * 是否自动续费充值。0-不自动续费；1-自动续费
     */
    private Integer isRenew;
    /**
     * 交易发生时间
     */
    private Timestamp tradeTime;
    /**
     * 创建时间
     */
    private Timestamp createTime;
    /**
     * 更新时间
     */
    private Timestamp updateTime;
}
```

（2）定义“余额交易订单表”的持久层（Dao 层）接口。

代码如下：

```
package com.wudimanong.wallet.dao.mapper;
import com.baomidou.mybatisplus.core.mapper.BaseMapper;
import com.wudimanong.wallet.dao.model.UserBalanceOrderPO;
import org.springframework.stereotype.Repository;
@Repository
public interface UserBalanceOrderDao extends BaseMapper<UserBalanceOrderPO> {
}
```

3. 开发“电子钱包充值支付回调”接口的持久层（Dao 层）代码

“电子钱包充值支付回调”接口的持久层（Dao 层）实现，涉及“余额交易订单表”（user_balance_order）、“余额账户信息表”（user_balance）和“余额账户流水记录表”（user_balance_flow）的操作。

“余额交易订单表”及“余额账户信息表”的持久层（Dao 层）接口，已经在“1.”小标题及“2.”小标题中实现了，因此这里只需要实现“余额账户流水记录表”（user_balance_flow）的持久层（Dao 层）。

（1）定义“余额账户流水记录表”的数据库实体类。代码如下：

```
package com.wudimanong.wallet.dao.model;
import com.baomidou.mybatisplus.annotation.TableName;
import java.sql.Timestamp;
import lombok.Data;
@Data
@TableName("user_balance_flow")
public class UserBalanceFlowPO {
    /**
     * 主键 ID
     */
    private Integer id;
    /**
     * 用户 ID
     */
    private String userId;
    /**
     * 账户变动流水号
     */
    private String flowNo;
    /**
     * 账户编号
     */
    private String accNo;
    /**
     * 业务类型
     */
    private String busiType;
    /**
     * 变动金额
     */
    private Integer amount;
    /**
     * 币种
```

```
     */
    private String currency;
    /**
     * 变动前的金额
     */
    private Integer beginBalance;
    /**
     * 变动后的金额
     */
    private Integer endBalance;
    /**
     * 借贷方向
     */
    private String fundDirect;
    /**
     * 更新时间
     */
    private Timestamp updateTime;
    /**
     * 创建时间
     */
    private Timestamp createTime;
}
```

（2）定义“余额账户流水记录表”的持久层（Dao 层）接口。

```
package com.wudimanong.wallet.dao.mapper;
import com.baomidou.mybatisplus.core.mapper.BaseMapper;
import com.wudimanong.wallet.dao.model.UserBalanceFlowPO;
import org.springframework.stereotype.Repository;
@Repository
public interface UserBalanceFlowDao extends BaseMapper<UserBalanceFlowPO> {
}
```

至此，完成了电子钱包微服务业务层（Service 层）功能的开发。

5.5 步骤 3：集成“Feign + Ribbon + Hystrix”实现微服务的“远程通信 + 负载调用 + 熔断降级”

在微服务技术体系中，除要实现服务注册、服务发现外，还要实现微服务之间的远程通信、负载均衡及熔断降级等功能，并为此提供可靠的技术解决方案。

本节的内容将演示 Spring Cloud 微服务中实现远程通信、负载均衡及熔断降级功能。

演示的内容为：在“电子钱包微服务”与“支付微服务”（参考第 6 章）之间通过“Feign + Ribbon + Hystrix”实现电子钱包充值支付功能。

5.5.1 集成微服务通信组件“Feign + Ribbon”

在 Spring Cloud 微服务技术中，服务之间最常用的通信方式是基于 Feign 的 HTTP 调用方式。Feign 组件的底层通过集成 Ribbon 客户端负载均衡组件，来实现对目标微服务实例的负载均衡调用。

在工程代码的 pom.xml 文件中，引入 Feign 依赖，代码如下：

```
<!--引入 Feign 依赖-->
<dependency>
    <groupId>org.springframework.cloud</groupId>
    <artifactId>spring-cloud-starter-openfeign</artifactId>
</dependency>
```

引入此依赖后，也会自动引入 Ribbon 依赖。

5.5.2 开发调用“支付微服务”的 FeignClient 客户端代码

接下来开发调用“支付微服务”的 FeignClient 客户端调用代码。

1. 开发“支付微服务”客户端代码

接下来基于@FeignClient 注解开发调用“支付微服务”的客户端代码。

（1）定义“统一支付”接口的 FeignClient 接口。

具体的服务方法及参数，与“支付微服务”所定义的接口一致。具体代码如下：

```
package com.wudimanong.wallet.client;
import com.wudimanong.wallet.client.bo.UnifiedPayBO;
...
import org.springframework.web.bind.annotation.RequestBody;
@FeignClient(value = "payment")
public interface PaymentClient {
    /**
     * “统一支付”接口
     *
     * @param unifiedPayDTO
     * @return
     */
    @PostMapping("/pay/unifiedPay")
```

```
    public ResponseResult<UnifiedPayBO> unifiedPay(@RequestBody @Validated
UnifiedPayDTO unifiedPayDTO);
  }
```

上述接口通过@FeignClient 注解定义了调用“支付微服务”的 FeignClient 客户端代码，注解中“value”属性的取值为“支付微服务”在注册中心 Consul 中的服务名称。

在“电子钱包微服务”调用“支付微服务”的过程中，“电子钱包微服务”的 Feign 客户端会通过 Ribbon 组件从注册中心获取“支付微服务”的实例地址列表，以此实现客户端负载调用。

（2）定义“统一支付”接口的请求参数对象。

“统一支付”接口的请求参数对象，与在“支付微服务”中定义的“统一支付”接口的请求参数对象完全一致（参考第 6 章内容）。具体代码如下：

```
package com.wudimanong.wallet.client.dto;
import com.wudimanong.wallet.validator.EnumValue;
...
import lombok.Data;
@Data
public class UnifiedPayDTO implements Serializable {
    /**
     * 接入方应用 ID
     */
    @NotNull(message = "应用 ID 不能为空")
    private String appId;
    /**
     * 接入方支付订单 ID，必须在接入方系统唯一（如电子钱包微服务）
     */
    @NotNull(message = "支付订单 ID 不能为空")
    private String orderId;
    /**
     * 交易类型。用于标识具体的业务类型，如 topup 表示钱包充值等，可以根据具体业务定义
     */
    @EnumValue(strValues = {"topup"})
    private String tradeType;
    /**
     * 支付渠道。0-微信支付，1-支付宝支付
     */
    @EnumValue(intValues = {0, 1})
    private Integer channel;
    /**
     * 支付产品定义，用于区分具体的渠道支付产品，具体可根据实际情况定义
     */
    private String payType;
    /**
```

```
     * 支付金额，以“分”为单位，数值必须大于 0
     */
    private Integer amount;
    /**
     * 支付币种，默认为 CNY
     */
    @EnumValue(strValues = {"CNY"})
    private String currency;
    /**
     * 商户系统唯一标识用户身份的 ID
     */
    @NotNull(message = "用户 ID 不能为空")
    private String userId;
    /**
     * 商品标题，一般支付渠道对此会有要求
     */
    @NotNull(message = "商品标题不能为空")
    private String subject;
    /**
     * 商品描述信息
     */
    private String body;
    /**
     * 支付扩展信息
     */
    private Object extraInfo;
    /**
     * 用于发送“异步支付结果通知”的服务端地址
     */
    @NotNull(message = "支付通知地址不能为空")
    private String notifyUrl;
    /**
     * 同步支付结果的跳转地址（支付成功后同步跳转回商户界面的 URL）
     */
    private String returnUrl;
}
```

（3）定义“统一支付”接口的返回参数对象。代码如下：

```
package com.wudimanong.wallet.client.bo;
import java.io.Serializable;
import lombok.Builder;
import lombok.Data;
@Data
@Builder
public class UnifiedPayBO implements Serializable {
```

```
    /**
     * 商户支付订单号
     */
    private String orderId;
    /**
     * 由第三方支付渠道生成的预支付订单号
     */
    private String tradeNo;
    /**
     * 支付订单的金额
     */
    private Integer amount;
    /**
     * 支付币种
     */
    private String currency;
    /**
     * 支付渠道的编码
     */
    private String channel;
    /**
     * 特殊支付场景所需要传递的额外支付信息
     */
    private String extraInfo;
    /**
     * 支付订单状态。0-待支付；1-支付中；2-支付成功；3-支付失败
     */
    private Integer payStatus;
}
```

2. 配置微服务客户端 FeignClient 的支持

在“1.”小标题中完成了“支付微服务”客户端代码编写。如果要让 FeignClient 客户端的类生效，还需要在“电子钱包微服务”的入口类中进行注解配置，代码如下：

```
package com.wudimanong.wallet;
import com.wudimanong.wallet.client.PaymentClient;
...
import org.springframework.cloud.openfeign.EnableFeignClients;
@EnableDiscoveryClient
@SpringBootApplication
@MapperScan("com.wudimanong.wallet.dao.mapper")
@EnableFeignClients(basePackageClasses = PaymentClient.class)
public class WalletApplication {
    public static void main(String[] args) {
        SpringApplication.run(WalletApplication.class, args);
```

```
        }
    }
```

可以看到，在微服务入口程序类中，通过@EnableFeignClients 注解开启了 FeignClient 功能，并指定了需要实例化的 PaymentClient 接口类。

如果此时将“电子钱包微服务”与“支付微服务”同时注册到服务注册中心 Consul 中，则可以实现“电子钱包微服务”对“支付微服务”的远程调用。

5.5.3 微服务熔断降级的概念

在微服务架构中，随着服务调用链路变长，为了防止出现级联“雪崩”，常用“熔断降级”作为服务保护的重要机制，它们是确保微服务架构稳定运行的关键手段。

在高并发情况下，如果链路中的某个服务出现不可用的情况，则可能会导致整个链路的网络调用出现大量的延时。在瞬间流量“洪峰”的冲击下，这些增加的延时很可能导致链路中所有微服务的可用线程资源被耗尽，从而造成服务“雪崩”。

所以，无论是“服务调用方”，还是“服务提供方”，都要从保证服务可用性的角度，提供相应的过载保护机制。“熔断降级”分为“熔断”和“降级”两层含义。

1. 熔断的概念

对于“服务调用方”来说，需要将所依赖服务的调用设置为可接受的超时时间，一旦发现依赖服务在一定的时间内出现多次调用超时或失败，则及时对该依赖服务进行“熔断”，即在一定的时间内，对需要调用该依赖服务的请求进行 fallback 处理，待依赖服务恢复后，再恢复将对其的请求调用。

2. 降级的概念

对于“服务提供方”来说，则要对微服务本身进行“限流保护”，即根据服务的整体负载能力设计相应的降级策略。例如，对一定时间内的流量进行限制——假设 1s 内服务最多只能处理 10 个请求，那么 1s 内的第 11 个请求就会被拒绝。

3. 实现熔断降级功能的技术组件

要实现微服务熔断降级，需要一定的技术组件来支持。在 Spring Cloud 微服务中，最著名的熔断降级组件是 Netflix 公司开源的 Hystrix 组件。

此外，阿里巴巴开源的 Sentinel 组件最近也比较流行，但它与 Hystrix 本质上都是基于客户端的熔断降级组件，对微服务本身有侵入。

于是又出现了以 Istio 为代表的 Service Mesh（服务网格）微服务架构。关于 Service Mesh 微服务架构的内容超出了本书讨论的范畴，感兴趣的朋友可以查阅相关资料，这里只介绍 Hystrix 的使用方式。

5.5.4 集成 Hystrix 实现微服务的熔断降级

接下来集成 Hystrix，以实现“电子钱包微服务”对“支付微服务”调用的熔断降级。

1. 集成 Hystrix 的依赖

（1）在项目工程的 pom.xml 文件中，引入 Hystrix 的依赖。代码如下：

```
<!--引入 Hystrix 的依赖-->
<dependency>
    <groupId>org.springframework.cloud</groupId>
    <artifactId>spring-cloud-starter-netflix-hystrix</artifactId>
</dependency>
```

在引入 Hystrix 的 starter 依赖后，就可以实现对 Hystrix 组件的“开箱即用”了。

（2）在微服务的入口程序类中，通过@EnableCircuitBreaker 注解启用 Hystrix 断路器功能。代码如下：

```
package com.wudimanong.wallet;
import com.wudimanong.wallet.client.PaymentClient;
...
import org.springframework.cloud.client.discovery.EnableDiscoveryClient;
import org.springframework.cloud.openfeign.EnableFeignClients;
@EnableCircuitBreaker
@EnableDiscoveryClient
@SpringBootApplication
@MapperScan("com.wudimanong.wallet.dao.mapper")
@EnableFeignClients(basePackageClasses = PaymentClient.class)
public class WalletApplication {
    public static void main(String[] args) {
        SpringApplication.run(WalletApplication.class, args);
    }
}
```

（3）在项目的 bootstrap.yml 配置文件中配置 Feign 对 Hystrix 的支持。代码如下：

```
#配置 Feign 对 Hystrix 的支持
feign:
  hystrix:
    enabled: true
```

开启 Hystrix 断路器后并不会立刻生效。Spring Cloud 微服务是通过 Feign 来通信的，而默认情况下 Feign 是禁用 Hystrix 的，所以，需要在 Feign 中开启对 Hystrix 的支持，这样 FeignClient 客户端在微服务之间进行调用时，才能在感知服务调用异常的情况下将错误指标信息反馈给 Hystrix。Hystrix 才能根据相关指标来开启/关闭断路器，从而实现对依赖服务调用的熔断降级。

2. 开发 FeignClient 微服务熔断降级代码

（1）在 FeignClient 微服务调用代码中指定熔断处理逻辑。

在 5.5.2 节中，通过@FeignClient 注解定义了调用“支付微服务”的代码，并通过其 value 属性值指定了目标微服务的名称。除此之外，还可以通过 fallback、fallbackFactory 属性值来指定对依赖服务的熔断处理逻辑，代码如下：

```
    package com.wudimanong.wallet.client;
    import com.wudimanong.wallet.client.bo.UnifiedPayBO;
    ...
    import org.springframework.web.bind.annotation.RequestBody;
    @FeignClient(value = "payment", configuration = PaymentConfiguration.class,
fallbackFactory = PaymentClientFallbackFactory.class)
    public interface PaymentClient {
        /**
         * 定义调用支付微服务“统一支付”接口的客户端
         *
         * @param unifiedPayDTO
         * @return
         */
        @PostMapping("/pay/unifiedPay")
        public ResponseResult<UnifiedPayBO> unifiedPay(@RequestBody @Validated
UnifiedPayDTO unifiedPayDTO);
    }
```

可以看到，在上述 FeignClient 接口中通过 fallbackFactory 属性值指定了对依赖服务的“熔断”逻辑处理类。

（2）开发 fallbackFactory 属性的熔断处理代码。

这里说明一下，相比较于 fallback 属性， fallbackFactory 属性指定的熔断处理逻辑可以更好地捕捉异常信息。所以，这里使用 fallbackFactory 属性来指定“支付微服务”的熔断处理类。代码如下：

```
    package com.wudimanong.wallet.client;
    import com.wudimanong.wallet.client.bo.UnifiedPayBO;
```

```
    ...
    import lombok.extern.slf4j.Slf4j;
    @Slf4j
    public class PaymentClientFallbackFactory implements
FallbackFactory<PaymentClient> {
        @Override
        public PaymentClient create(Throwable cause) {
            return new PaymentClient() {
                @Override
                public ResponseResult<UnifiedPayBO> unifiedPay(UnifiedPayDTO
unifiedPayDTO) {
                    log.info("支付服务调用降级逻辑处理...");
                    log.error(cause.getMessage());
                    return
ResponseResult.serviceException(BusinessCodeEnum.BUSI_PAY_FAIL_2001.getCode(
),
                            BusinessCodeEnum.BUSI_PAY_FAIL_2001.getDesc());
                }
            };
        }
    }
```

该 fallbackFactory 类通过实现 PaymentClient 接口，定义了对应远程服务接口的熔断代码。

（3）增加对依赖服务熔断降级处理的接口响应信息封装。

为了专门处理对依赖服务熔断降级的异常响应，在 5.4.1 节“1.”小标题中“统一报文格式类 ResponseResult”中增加如下方法：

```
    /**
     * 对依赖服务熔断降级结果响应信息的封装
     *
     * @param code
     * @param message
     * @param <T>
     * @return
     */
    public static <T> ResponseResult<T> serviceException(Integer code, String
message) {
        ResponseResult<T> responseResult = new ResponseResult<>();
        responseResult.setCode(code);
        responseResult.setMessage(message);
        return responseResult;
    }
```

此外，在业务异常码枚举类 BusinessCodeEnum 中增加一个提示“支付服务熔断故障”的提示码，代码片段如下：

```
BUSI_PAY_FAIL_2001(2001, "支付系统故障，请稍后重试");
```

（4）增加对“熔断降级”处理类的实例化配置。

在@FeignClient 注解中还有一个 configuration 属性，它用来实现对 fallbackFactory 属性所指定的熔断降级处理逻辑类的实例化配置，代码如下：

```
package com.wudimanong.wallet.client;
import org.springframework.context.annotation.Bean;
import org.springframework.context.annotation.Configuration;
@Configuration
public class PaymentConfiguration {
    @Bean
    PaymentClientFallbackFactory paymentClientFallbackFactory() {
        return new PaymentClientFallbackFactory();
    }
}
```

至此，完成了“电子钱包微服务”集成 Hystrix 组件实现微服务熔断降级的功能。5.5.5 节的内容将通过测试“电子钱包微服务”对“支付微服务”的调用，来演示微服务熔断降级的效果。

5.5.5 测试 Hystrix 熔断降级的生效情况

为了验证 Hystrix 熔断降级的生效情况，接下来模拟“支付微服务”（具体实现参考第 6 章）的调用场景。

1. 集成 HystrixDashboard 的依赖

（1）Hystrix 提供了简单的观测界面，在 pom.xml 文件中引入 HystrixDashboard 的依赖。代码如下：

```
<!--引入 HystrixDashboard 的依赖-->
<dependency>
    <groupId>org.springframework.cloud</groupId>
<artifactId>spring-cloud-starter-hystrix-dashboard</artifactId>
<version>1.4.7.RELEASE</version>
</dependency>
```

（2）在微服务应用入口类中加上@EnableHystrixDashboard 注解，以启用 HystrixDashboard。代码如下：

```
@EnableHystrixDashboard
@EnableCircuitBreaker
@EnableDiscoveryClient
@SpringBootApplication
```

```
@MapperScan("com.wudimanong.wallet.dao.mapper")
@EnableFeignClients(basePackageClasses = PaymentClient.class)
public class WalletApplication {
    public static void main(String[] args) {
        SpringApplication.run(WalletApplication.class, args);
    }
}
```

（3）配置 Spring Boot 2.0 中 Hystrix 的指标访问路径。

由于本实例采用的是 Spring Boot 2.0 以上的版本，该版本默认的 Hystrix 路径不是“/hystrix.stream”。为了正常访问 Hystrix 的指标，还需要在 WalletApplication 入口类中添加以下代码：

```
@Bean
public ServletRegistrationBean getServlet() {
    HystrixMetricsStreamServlet streamServlet = new
HystrixMetricsStreamServlet();
    ServletRegistrationBean registrationBean = new
ServletRegistrationBean(streamServlet);
    registrationBean.setLoadOnStartup(1);
    registrationBean.addUrlMappings("/hystrix.stream");
    registrationBean.setName("HystrixMetricsStreamServlet");
    return registrationBean;
}
```

2. 测试 Hystrix 的熔断降级的生效情况

（1）查看 Hystrix 的 Dashborad 界面。

启动“电子钱包微服务”，输入“http://{地址}:{端口}/hystrix”就可以看到 Hystrix 的 Dashboard 界面。

（2）查看具体的熔断器运行指标信息。

如果要看到熔断器的具体运行指标，则需要在步骤（1）所演示的界面输入框中输入具体的监控地址“http://{地址}:{端口}/hystrix.stream”，之后单击“Monitor Stream”按钮，如图 5-7 所示。效果如图 5-8 所示。

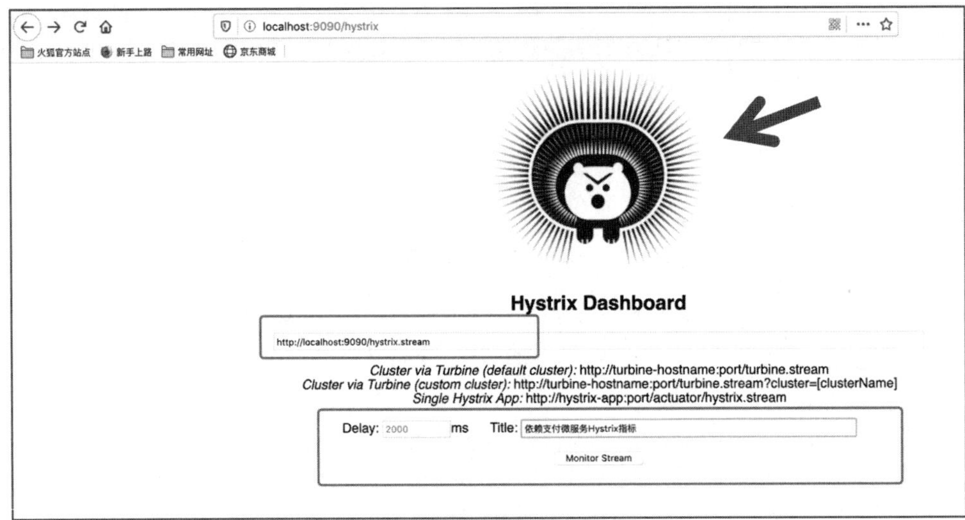

图 5-7

图 5-8

（3）查看熔断器自动打开的效果。

可以看到，此时该界面还没有任何指标。如果多次触发对“支付微服务”的调用，则可以看到如图 5-9 所示的情况。

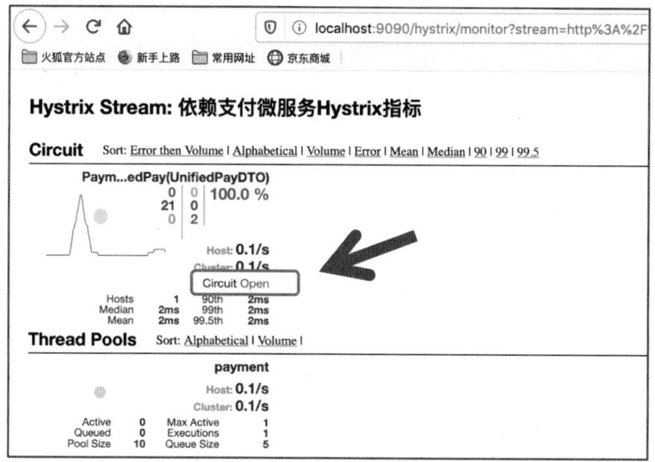

图 5-9

在上述操作中，通过 Postman 多次调用“电子钱包充值”接口，由于此时其依赖的“支付微服务”还不能正常访问，所以，Hystrix 会在收集到异常指标后打开熔断器。之后，针对“电子钱包充值”接口的请求将直接进入熔断逻辑，而不再对“支付微服务”发起网络调用。

（4）查看熔断器自动关闭的效果。

假设“支付微服务”能被正常访问，Hystrix 在经过一定的尝试性访问后发现依赖服务已经恢复，则会自动关闭熔断器，效果如图 5-10 所示。

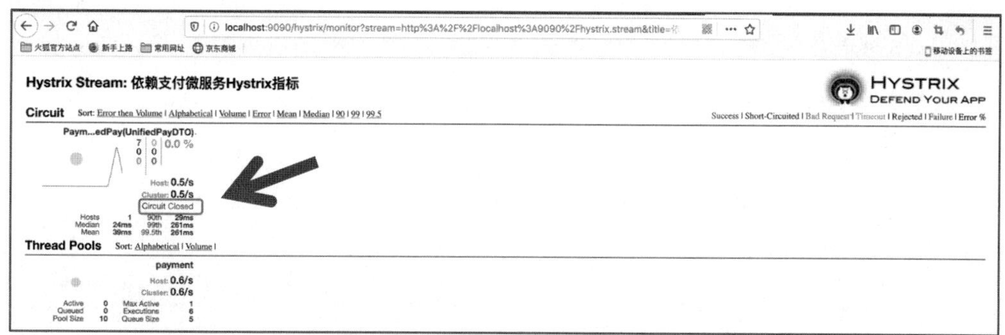

图 5-10

通过 Hystrix，可以快速实现微服务的熔断降级。对于严重依赖外部服务的系统来说，这是一种非常重要的可靠性保障机制。

5.6 步骤 4：基于 Vue.js 开发电子钱包微服务的充值界面

在前面的章节中，基本完成了电子钱包微服务的后端逻辑。但对于涉及用户支付的场景来说，电子钱包微服务还涉及部分前端处理逻辑——例如，在支付时需要通过浏览器跳转到“支付宝的支付界面”等。

本节基于 Vue.js 前端开发框架来编写电子钱包微服务的充值界面。

5.6.1 认识 Vue.js

Vue.js 是目前流行的前端开发框架，与之类似的前端框架还有 React、Angular 等。

> Vue.js 目前生态繁荣，有很多基于 Vue.js 的移动端、PC 端开源组件库可以使用，如 elementUI、Vant 等。

关于 Vue.js 的更多细节，可以参考更专业的书籍或资料。本节主要基于 Vue.js 及其 elmentUI 组件来实现电子钱包微服务的充值界面。

5.6.2 搭建 Node.js 环境

在实际的项目开发中，Vue.js 会依赖一些 Vue.js 插件及打包工具。目前，前端开发者一般会通过 Node.js 提供的 NPM 工具来实现 Vue.js 插件包的管理。

1. 安装 Node.js

下载对应操作系统环境的 Node.js 安装版本，并安装。

在安装完成后，可以通过命令查看 Node.js 及 NPM 的版本信息。命令如下：

```
$ node -v
v12.14.1
$ npm -v
6.13.4
```

如上述命令运行正常，则说明 Node.js 环境安装成功。

2. 安装淘宝镜像（cnpm）

由于官方的 NPM 工具安装速度比较慢，所以国内开发者可以使用淘宝镜像（cnpm）。安装命令如下：

```
$ sudo  npm install -g cnpm --registry=https://registry.npm.taobao.org
```

通过 cnpm 命令安装 vue-cli，命令如下：

```
$ sudo cnpm install -g vue-cli
```

5.6.3 创建电子钱包微服务的 Vue.js 前端工程

1. 配置电子钱包微服务的 Vue.js 前端工程

（1）通过 Vue.js 提供的脚手架初始化一个标准的 Vue.js 工程。命令如下：

```
$ vue init webpack chapter05-wallet-ui
```

（2）填写项目的基本信息。代码如下：

```
? Project name chapter05-wallet-ui
? Project description A Vue.js project
? Author wudimanong <wudimanong@wudimanong.com>
? Vue build standalone
? Install vue-router? Yes
? Use ESLint to lint your code? Yes
? Pick an ESLint preset Standard
? Set up unit tests Yes
? Pick a test runner jest
? Setup e2e tests with Nightwatch? Yes
? Should we run `npm install` for you after the project has been created?
(recommended) npm
  vue-cli · Generated "chapter05-wallet-ui".
# Installing project dependencies ...
# ========================
...
# Project initialization finished!
# ========================
To get started:
  cd chapter05-wallet-ui
  npm run dev
Documentation can be found at https://vuejs-templates.github.io/webpack
```

（3）通过 VsCode 工具打开 Vue.js 的项目结构，具体说明如下：

```
|--build    最终发布的代码的存放位置
|--config   配置路径、端口号等信息，刚开始学习时选择默认配置
|--node_modules npm  加载的项目依赖模块
|--src  这里是开发的主要目录，基本上要做的事情都在这里，其中包含几个目录及文件
    |--assets        放置一些图片，如 logo 等
    |--components    其中放置的是组件文件
    |--App.vue       项目入口文件
    |--main.js       项目的核心文件
    |--router         "访问路径"与"在 components 中定义的组件"的路由映射关系
```

```
|--static          静态资源目录，如图片、字体等
|--test            初始测试目录，可删除
   |--index.html      首页入口文件，可以添加一些meta信息或者统计之类的代码
|--package.json       项目配置文件
|--dist               编译打包后前端资源的输出目录
```

（4）进入项目根目录编译、运行服务。命令如下：

```
$ cnpm install
✔ Installed 58 packages
✔ Linked 0 latest versions
✔ Run 0 scripts
✔ All packages installed (used 34ms(network 31ms), speed 0B/s, json 0(0B), tarball 0B)
```

（5）运行 Vue.js 项目。命令如下：

```
$ npm run dev
> wallet@1.0.0 dev /Users/qiaojiang/dev-tools/workspace/workspace_vue/wallet
> webpack-dev-server --inline --progress --config build/webpack.dev.conf.js
13% building modules 27/29 modules 2 active ...pace/workspace_vue/wallet/src/App.vue{ parser: "babylon" } is deprecated; we now treat it as { parser: "babel" }.
 95% emitting
 DONE  Compiled successfully in 5165ms                          16:49:17
 I  Your application is running here: http://localhost:8080
```

如果出现如图 5-11 所示界面，则说明 Vue.js 项目初始化成功，可以进行前端界面的开发了。

图 5-11

2. 安装 ElementUI 组件

继续安装 ElementUI 组件，步骤如下。

（1）进入 Vue.js 项目根目录，安装 ElementUI 组件。命令如下：

```
$ cnpm i element-ui -S
```

（2）执行成功后，在项目的 package.json 文件中就会出现 ElementUI 组件的依赖。例如：

```
"dependencies": {
    "axios": "^0.19.2",
    "element-ui": "^2.13.0",
    "vue": "^2.5.2",
    "vue-router": "^3.0.1"
  },
```

（3）以完整引入的方式在 Vue.js 项目的“src/main.js”文件中添加 ElementUI。内容如下：

```
import Vue from 'vue'
import App from './App'
import router from './router'
// 引入 ElementUI 框架
import ElementUI from 'element-ui'
// 引入 ElementUI 框架的样式文件
import 'element-ui/lib/theme-chalk/index.css'
// 引入 Axios
// eslint-disable-next-line no-unused-vars
import axios from 'axios'
// Vue.js 使用 ElementUI
Vue.use(ElementUI)
Vue.config.productionTip = false
/* eslint-disable no-new */
new Vue({
  el: '#app',
  router,
  components: { App },
  template: '<App/>'
})
```

在上述代码中也引入了 Axios，这是因为 Vue.js 本身并不支持 Ajax 数据访问，所以要借助 Axios 来完成。其安装命令如下：

```
$ cnpm install axios -save
```

5.6.4 编写电子钱包微服务的前端功能

1. 编写电子钱包微服务的余额展示界面

接下来基于 Vue.js 编写电子钱包微服务的余额展示界面，用于显示电子钱包余额，并提供“充值”按钮，效果如图 5-12 所示。

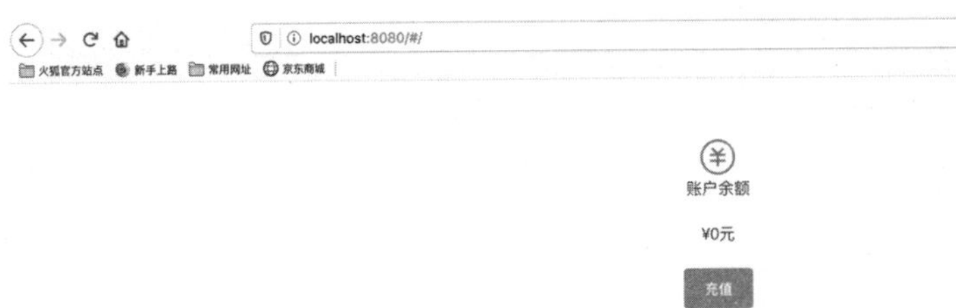

图 5-12

以上界面基于“Vue.js + ElementUI 组件”实现，通过访问“电子钱包微服务”的“电子钱包查询”接口来显示余额。具体编写步骤如下。

（1）在 Vue.js 工程的“src/components/”目录下，创建一个名为“QueryAcc.vue”的文件。代码如下：

```
<template>
  <div id="queryAcc">
    <!-- 由于Element-UI 官方支持的 ICON 图标比较少，所以我们自定义了一个货币图标-->
    <i class="el-icon-cny"/><br/>
    <div>
      <span>账户余额</span>
    </div><br/>
    <!--调用后端“余额查询”接口进行数据渲染-->
    <div>
      {{balance}}
    </div>
    <!--使用 Element-UI 组件添加“充值”按钮-->
    <br/>
    <el-row>
        <el-button type="info" @click="toCharge">充值</el-button>
    </el-row>
    <router-view/>
  </div>
</template>
<script>
// 引入 axios
// eslint-disable-next-line no-unused-vars
import axios from 'axios'
export default {
  name: 'App',
  // 定义页面数据
  data () {
    return {
      balance: ''
    }
```

```
    },
    // 在 Vue.js 的 created 生命周期中实现向后端微服务查询余额的功能
    created () {
      this.getBalance()
    },
    methods: {
      // 获取用户余额的方法
      getBalance: function () {
        // 调用“电子钱包查询”接口查询余额信息。这里的 userId 是在开户时所设置的，在真实
环境中是通过会话动态获取的
        axios.get('/api/account/queryAcc?userId=10001&accType=0').
then(response => {
          // 通过接口返回的数据为显示变量赋值
          this.balance = '¥' + response.data.data[0].balance / 100 + '元'
          console.log(response.data)
        }, response => {
          console.log('error')
        })
      },
      // 通过单击“充值”按钮跳转到钱包充值界面
      toCharge: function () {
        // 路由打开充值界面，这里以“重新打开新窗口”的方式进行页面跳转
        let routeData = this.$router.resolve({ path: '/charge', query: { userId:
10001 } })
        window.open(routeData.href, '_blank')
      }
    }
  }
  </script>
  <style>
  #queryAcc {
    font-family: 'Avenir', Helvetica, Arial, sans-serif;
    -webkit-font-smoothing: antialiased;
    -moz-osx-font-smoothing: grayscale;
    text-align: center;
    color: #2c3e50;
    margin-top: 60px;
  }
  .el-icon-cny{
      background: url(../../src/assets/cny.png) center no-repeat;
      background-size: cover;
  }
  .el-icon-cny:before{
      content: "替";
      font-size: 35px;
```

```
    visibility: hidden;
}
</style>
```

上述代码为 Vue.js 界面模板代码，其中，引入了 ElementUI 组件作为视图组件，并编写了相关的 JavaScript 函数来完成对后端接口的访问及按钮事件的响应，还定义了一个 ElementUI 组件来显示“¥”图标。

（2）修改“config/index.js”文件，设置跨域配置 proxyTable。

在步骤（1）的代码中，在查询电子钱包余额的 JS 方法中，并没有指定具体的服务端地址，代码如下：

```
axios.get('/api/account/queryAcc?userId=10001&accType=0'))
```

这是因为，对服务端地址的管理，统一在 Vue.js 工程的 config/index.js 文件中进行了配置。配置代码片段如下：

```
...
module.exports = {
  dev: {
    // 访问路径
    assetsSubDirectory: 'static',
    assetsPublicPath: '/',
    // 修改 config/index.js 文件，设置跨域配置 proxyTable
    proxyTable: {
      '/api': {
        target: 'http://localhost:9090/',
        changeOrigin: true,
        pathRewrite: {
          '^/api': '/'
        }
      }
    },
...
```

在上述配置中，通过“路径代理匹配”将以“/api 路径”开头的请求都匹配到 target 目标地址。

（3）修改“router/index.js”路由配置。

为了让 Vue.js 模板组件能够被正常访问，还需要进行路由映射配置，代码如下：

```
import Vue from 'vue'
import Router from 'vue-router'
import HelloWorld from '@/components/HelloWorld'
import Charge from '@/components/Charge'
import QueryAcc from '@/components/QueryAcc'
Vue.use(Router)
```

```
export default new Router({
  routes: [
    {
      path: '/',
      name: 'QueryAcc',
      component: QueryAcc
    },
    {
      path: '/charge',
      name: 'Charge',
      component: Charge
    }, {
      path: '/hello',
      name: 'HelloWorld',
      component: HelloWorld
    }]
})
```

可以看到，项目的根访问路径指向了 QueryAcc 组件。这样，直接访问 Vue.js 项目的根 URL，即可看到如图 5-12 所示的界面。

2. 编写“电子钱包充值”界面

在电子钱包余额界面中有一个“充值”按钮，单击该按钮即可进入“充值金额”及“支付方式”选择界面，如图 5-13 所示。

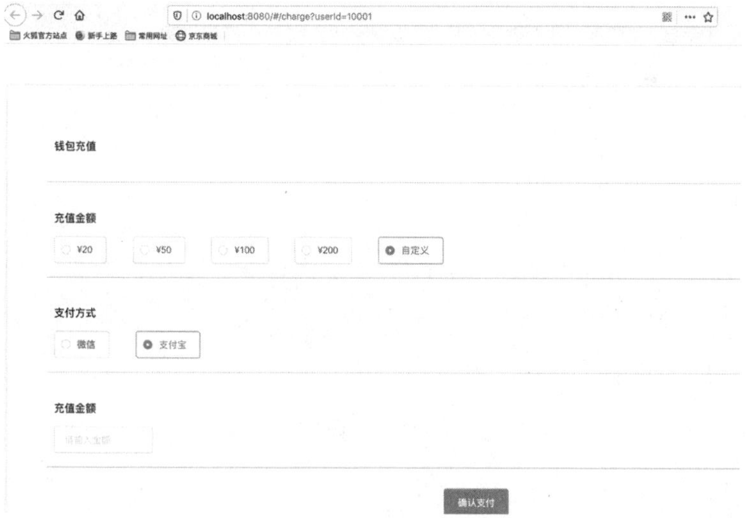

图 5-13

与“电子钱包余额”展示界面一样，该界面也是通过 Vue.js 模板编写的。步骤如下。

在“src/components”目录创建 Charge.vue 文件。

该代码文件在本书配置资源的“chapter05-wallet-ui/src/components”目录下。

上述代码文件提供了“充值金额”“支付方式的选择”和“数据校验逻辑”，并通过接入电子钱包微服务“电子钱包充值”接口来完成充值支付请求，并根据接口返回支付参数进行前端逻辑处理。

在本实例中，支付方式采用的是“支付宝网页支付”，因此，在电子钱包微服务的“电子钱包充值”接口返回支付宝的 form 表单数据后，浏览器会从用户界面跳转至支付宝支付界面。

5.6.5 测试“电子钱包充值”前后端交互流程

通过前面的步骤，完成了一个前后端分离的电子钱包微服务。接下来测试整体效果。

1. 创建电子钱包账户

在正常的流程中，在用户注册时要调用“电子钱包开户”接口。通过 Postman 工具调用“电子钱包开户”接口的效果如图 5-14 所示。

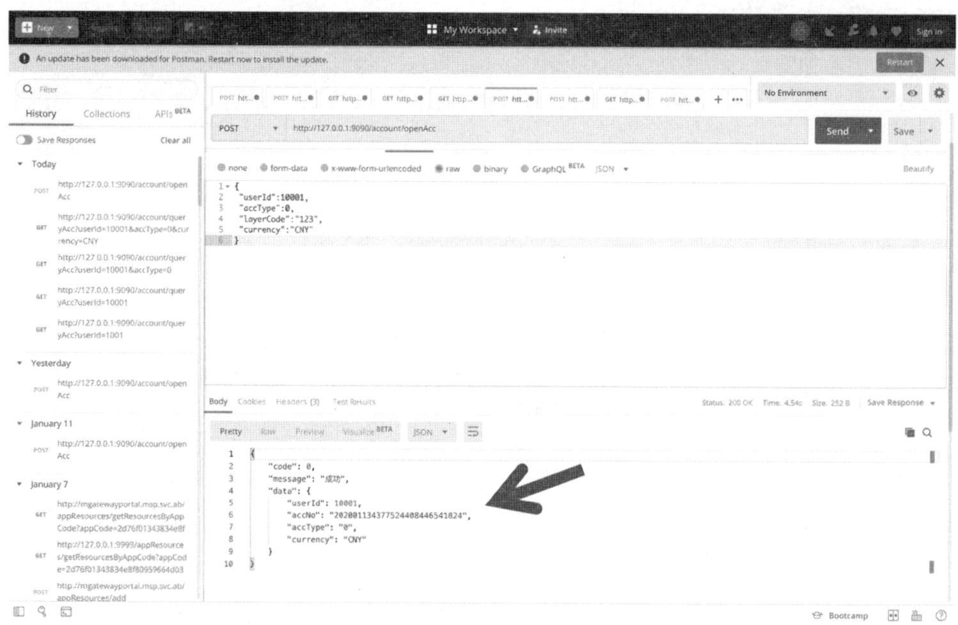

图 5-14

2. 单击"充值支付"按钮完成用户支付

在 5.6.4 节的界面中选择"支付宝支付方式"发起支付请求，此时会调用电子钱包微服务的"电子钱包充值"接口，并完成电子钱包充值订单的创建。

之后，会向"支付微服务"（参考第 6 章的实现）发起支付调用。如果调用正常，则支付系统会返回支付宝的"form 表单封装参数"。

最终用户浏览器会跳转到支付宝界面，如图 5-15 所示。

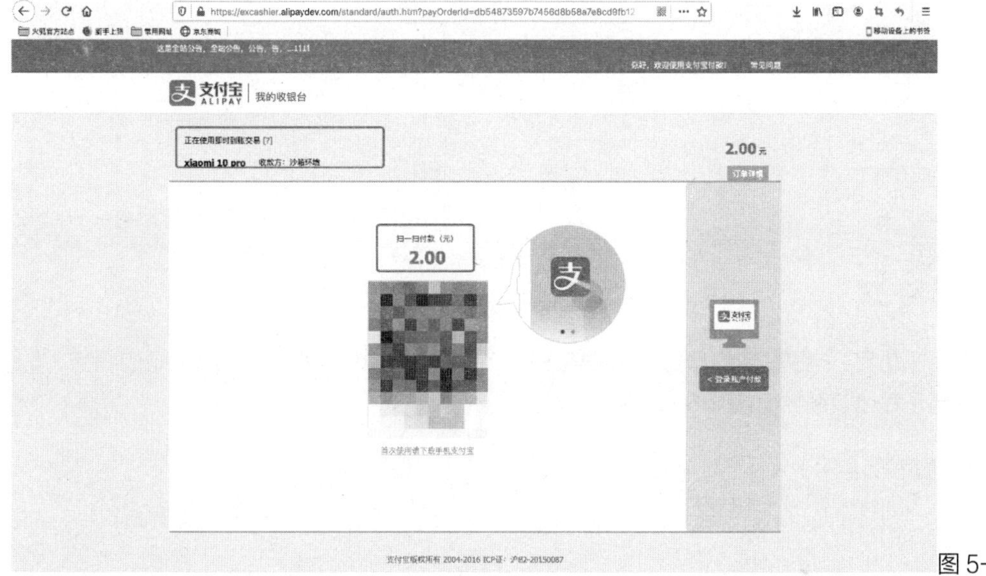

图 5-15

此时用户通过支付宝客户端扫码或者使用支付宝账号密码，即可完成付款动作。

3. 模拟"电子钱包充值支付回调"

在"2."小标题中支付步骤操作正常的情况下，在用户完成支付后，"支付宝系统"会向"支付微服务"发送支付结果通知回调，"支付微服务"在处理完自身逻辑后会向"电子钱包微服务"的"电子钱包充值支付回调"接口发起支付结果回调。

但是，上述流程的执行需要将"电子钱包微服务"及"支付微服务"进行完整的部署，并且"支付微服务"还需要具有可供外网访问的回调域名，模拟起来比较困难。这里假设用户已经完成支付，通过模拟调用"电子钱包微服务"的"电子钱包充值支付回调"接口来完成充值，如图 5-16 所示。

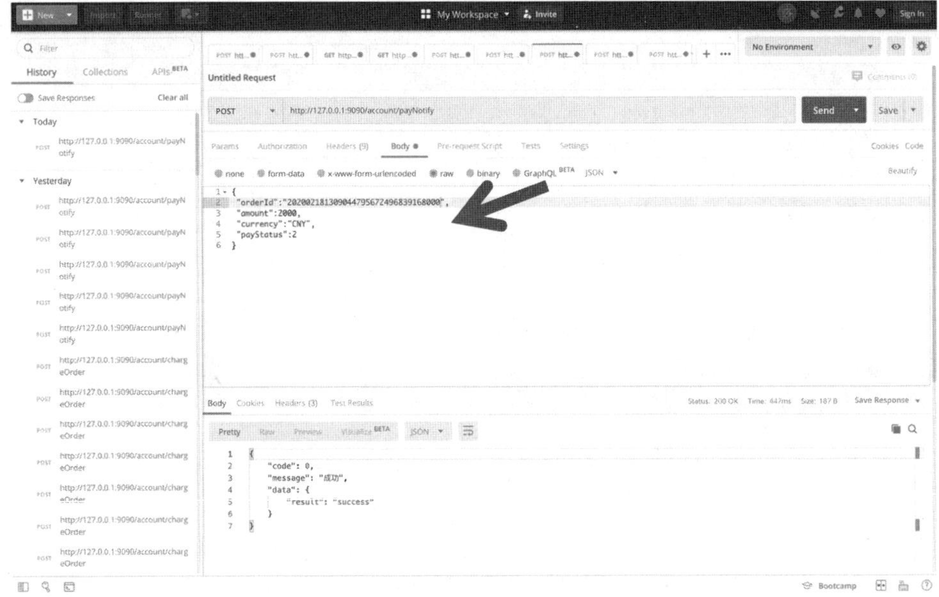

图 5-16

在调用成功后，电子钱包微服务会完成电子钱包余额的增加，以及账户流水记录的保存。此时，查看到的电子钱包余额界面如图 5-17 所示。

图 5-17

5.7 步骤 5：用 Docker 部署 Spring Cloud 微服务

随着服务数量及规模越来越大，如果微服务的部署及运维仍然采用传统方式，则会大大增加运维成本。因此，微服务模体系下的运维一定是 Devops（开发运维）模式：通过构建自动化运维发布平台，来打通产品、开发、测试及运维的协作流程，从而从整体上提高研发效率。

随着以 Docker 为代表的容器化技术的普及，现在大部分公司的 Devops 实践都会采用容器(如 Docker、K8s) 的方式来发布微服务，并通过容器的弹性伸缩能力来实现快速扩容和缩容，从而更快地响应业务、更好地利用资源。

目前，Devops 最流行的部署方案是基于 K8s 的集群方案。但是，K8s 本身对 Docker 容器技术也存在一定的依赖。所以，在接触 K8s 技术之前，先了解下基于 Docker 是如何实现 Spring Cloud 微服务的容器化部署的。

5.7.1 认识 Docker

Docker 是一个开源的应用容器引擎，也是目前最流行的应用部署方式。通过它，可以把应用及其依赖打包到一个可移植的镜像中，之后，可以利用 Docker 提供的部署机制将其发布至任何安装了 Docker 容器的系统中。

Docker 的核心概念如图 5-18 所示。

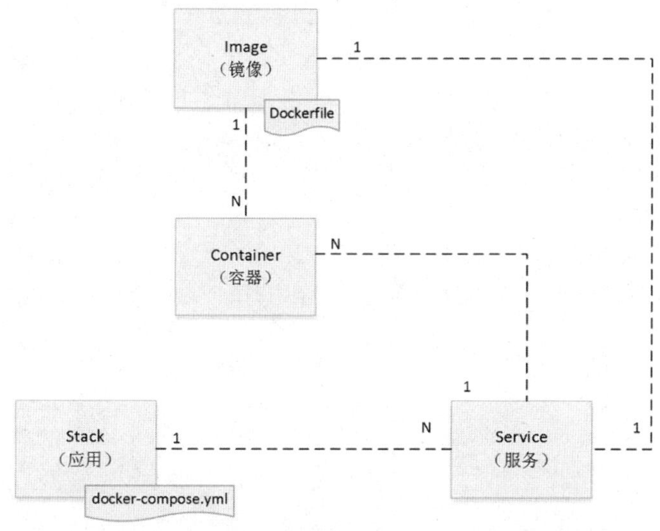

图 5-18

- Image (镜像) : 一个可执行文件，包含应用代码、依赖库、运行环境 (如 JRE 等) 、环境变量及配置等信息。通过镜像可以启动一个应用。镜像的构建过程通过 Dockefile 文件描述。
- Container (容器) : 使用 Image 启动的一个进程实例。它与镜像之间为“一对多”的关系，一个镜像可以启动多个容器实例。
- Service (服务) : 一组提供对外服务的 Container，这些 Container 使用同一个 Image 镜像，它与镜像为“一对一”的关系，与容器为“一对多”的关系。Service 由

docker-compose.yml 文件定义。

- Stack（应用）：一组 Service，相互协作对外提供服务，可以将其看作是一个完整的应用。在一些复杂的场景中，会将其拆分为多个 Stack（具体在 docker-compose.yml 文件中配置）。

5.7.2 利用 Dockerfile 文件构建微服务镜像

1. 创建 Dockerfile 文件

要让 Spring Cloud 微服务运行在 Docker 容器中，则需要先构建 Docker 镜像。构建过程需要使用 Dockerfile 文件来描述。

在项目“src/main/docker”目录下，创建 Dockerfile 文件，代码如下：

```
FROM java:8
VOLUME /tmp
RUN mkdir /app
ADD wallet-1.0-SNAPSHOT.jar /app/wallet.jar
ADD runboot.sh /app/
RUN bash -c 'touch /app/wallet.jar'
WORKDIR /app
RUN chmod a+x runboot.sh
EXPOSE 9090
CMD /app/runboot.sh
```

在上述 Dockerfile 文件中，定义了运行的 JDK 环境为 JDK 1.8、容器运行的目录为“/app”，并添加了所需的依赖（JAR 包）等信息，最后定义了执行命令的“/app/runboot.sh”脚本。

“runboot.sh”脚本的代码如下：

```
sleep 10
java -Djava.security.egd=file:/dev/./urandom -jar  /app/wallet.jar
```

至此，描述电子钱包微服务 Docker 镜像的 Dockerfile 文件就定义好了。

2. 在项目 pom.xml 文件中添加 Maven 插件依赖

为了在 Maven 项目中执行 Docker 镜像构建命令，还需要在项目 pom.xml 文件添加 Docker Maven 插件依赖，代码如下：

```
<!--Docker Maven 插件依赖-->
<plugin>
    <groupId>com.spotify</groupId>
    <artifactId>docker-maven-plugin</artifactId>
    <configuration>
        <imageName>${project.name}:${project.version}</imageName>
```

```
<dockerDirectory>${project.basedir}/src/main/docker</dockerDirectory>
            <skipDockerBuild>false</skipDockerBuild>
            <resources>
                <resource>
                    <directory>${project.build.directory}</directory>
                    <include>${project.build.finalName}.jar</include>
                </resource>
            </resources>
        </configuration>
    </plugin>
```

3. 构建微服务的 Docker 镜像

（1）通过 Maven 命令构建微服务的 Docker 镜像。命令如下：

```
mvn clean package docker:build
```

执行效果如图 5-19 所示。

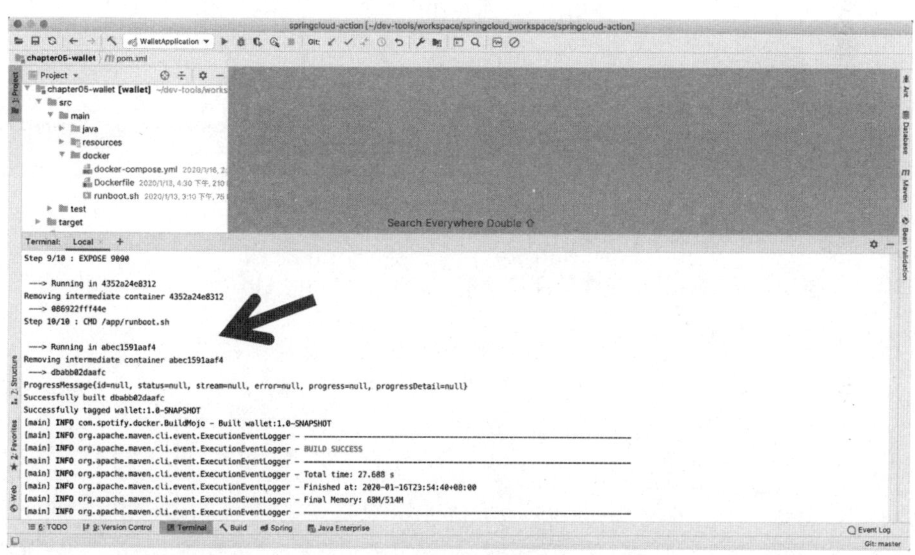

图 5-19

（2）通过 Docker 命令查看构建的镜像信息。命令如下：

```
docker images
```

可以看到构建的 Docker 镜像信息如图 5-20 所示。

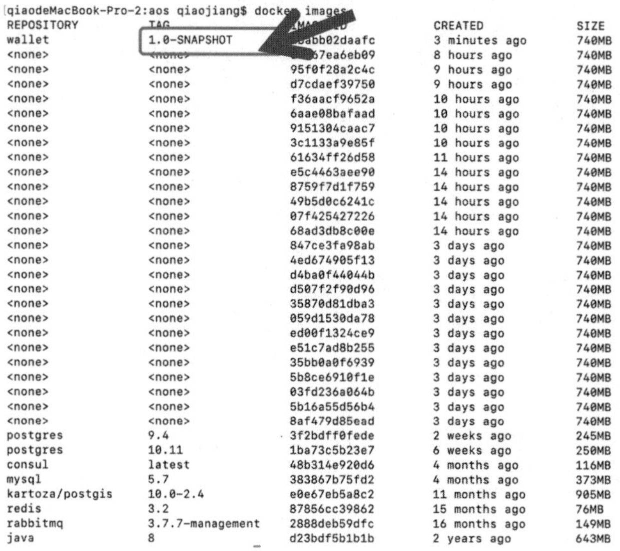

```
qiaodeMacBook-Pro-2:aos qiaojiang$ docke[illegible] images
REPOSITORY          TAG                   [illegible]          CREATED            SIZE
wallet              1.0-SNAPSHOT          [illegible]abb02daafc   3 minutes ago      740MB
<none>              [illegible]           [illegible]67ea6eb09    8 hours ago        740MB
<none>              <none>                95f0f28a2c4c         9 hours ago        740MB
<none>              <none>                d7cdaef39750         9 hours ago        740MB
<none>              <none>                f36aacf9652a         10 hours ago       740MB
<none>              <none>                6aae08bafaad         10 hours ago       740MB
<none>              <none>                9151304caac7         10 hours ago       740MB
<none>              <none>                3c1133a9e85f         10 hours ago       740MB
<none>              <none>                61634ff26d58         11 hours ago       740MB
<none>              <none>                e5c4463aee90         14 hours ago       740MB
<none>              <none>                8759f7d1f759         14 hours ago       740MB
<none>              <none>                49b5d0c6241c         14 hours ago       740MB
<none>              <none>                07f425427226         14 hours ago       740MB
<none>              <none>                68ad3db8c00e         14 hours ago       740MB
<none>              <none>                847ce3fa98ab         3 days ago         740MB
<none>              <none>                4ed674905f13         3 days ago         740MB
<none>              <none>                d4ba0f44044b         3 days ago         740MB
<none>              <none>                d507f2f90d96         3 days ago         740MB
<none>              <none>                35870d81dba3         3 days ago         740MB
<none>              <none>                059d1530da78         3 days ago         740MB
<none>              <none>                ed00f1324ce9         3 days ago         740MB
<none>              <none>                e51c7ad8b255         3 days ago         740MB
<none>              <none>                35bb0a0f6939         3 days ago         740MB
<none>              <none>                5b8ce6910f1e         3 days ago         740MB
<none>              <none>                03fd236a064b         3 days ago         740MB
<none>              <none>                5b16a55d56b4         3 days ago         740MB
<none>              <none>                8af479d85ead         3 days ago         740MB
postgres            9.4                   3f2bdff0fede         2 weeks ago        245MB
postgres            10.11                 1ba73c5b23e7         6 weeks ago        250MB
consul              latest                48b314e920d6         4 months ago       116MB
mysql               5.7                   383867b75fd2         4 months ago       373MB
kartoza/postgis     10.0-2.4              e0e67eb5a8c2         11 months ago      905MB
redis               3.2                   87856cc39862         15 months ago      76MB
rabbitmq            3.7.7-management      2888deb59dfc         16 months ago      149MB
java                8                     d23bdf5b1b1b         2 years ago        643MB
```

图 5-20

5.7.3 创建 docker-compose.yml 文件

有了 Docker 镜像，如何将镜像作为容器启动，以及在该镜像中启动哪些服务、它的资源限制及网络使用什么方式，这些都是在 docker-compose.yml 文件中定义的。

1. 创建 docker-compose.yml 文件

创建用于描述微服务 Docker 镜像应用信息的 docker-compose.yml 文件。代码如下：

```
version: '3.2'
services:
  wallet:
    image: wallet:1.0-SNAPSHOT
    hostname: wallet
    environment:
      - SPRING_PROFILES_ACTIVE=${SPRING_PROFILES_ACTIVE:-debug}
    ports:
      - "9090:9090"
    networks:
      - mynet
networks:
  mynet:
    external: true
```

在上述 docker-compose.yml 文件中，定义了一个“wallet”服务，并针对该服务描述了其所使用的 Docker 镜像、环境变量参数、容器端口映射及网络等信息。

在 services 属性下，可以定义具体的服务。Service（服务）与 stack（应用，如 wallet）的关系在 docker-compose.yml 文件中可以定义为“一对多”的关系。由于本例中没有定义多个 statck，所以这里并没有使用 service 来进行配置。

2. 配置 Docker 网络

在本实例中，电子钱包微服务所依赖的数据库、Consul 等服务，需要通过 Docker 端口映射的方式才能与电子钱包微服务连接。

如果没有设置特殊的网络，则 Docker 中的应用是无法直接与容器外的主机进行通信的。

这里 Docker 容器网络要与宿主机采用 Bridge 方式连接。通过自定义网络“mynet”，并在“mynet”中设置与宿主机处于同一个网段（具体 IP 地址段根据自己实际的实验环境而定）。在 Docker 中创建网络的命令如下：

```
docker network create -o parent=en0 --driver=bridge --subnet=172.18.64.2/24
--ip-range=172.18.64.239/24 --gateway=172.18.64.1 mynet
```

Spring Cloud 微服务在容器中运行时，默认会读取 application.yml 中的数据库等配置信息。在使用 Docker 之前，这些配置都是通过 127.0.0.1 这样的本地网络 IP 地址来进行访问的，但在容器中这是无法连通的，需要新建一个 application-test.yml 的配置文件，代码如下：

```
#数据库宿主机的地址
spring:
  datasource:
    url: jdbc:mysql://host.docker.internal:3306/wallet
    username: root
    password: 123456
  #注册中心的地址
  cloud:
    consul:
      host: host.docker.internal
      port: 8500
```

在以上代码中，重新定义了数据库的连接信息，以及 Consul 注册中心的地址。这里为了简化，使用 host.docker.internel 来表示宿主机的 IP 地址。

3. 让微服务程序读取到 application-test.yml 文件

如何才能让在 Docker 容器中运行的 Spring Cloud 微服务程序读取到 application-test.yml 文件呢?

可以通过 docker-compose.yml 文件中的系统环境变量来进行匹配，设置系统环境变量 "spring.profiles.active=test"，命令如下：

```
export SPRING_PROFILES_ACTIVE=test
```

该系统环境变量与 docker-compose.yml 文件中的 environment 属性值匹配，这样在启动容器时，就能知道 "spring.profiles.active" 的环境被设置为 "test" 了。

5.7.4 通过 Docker 容器化部署微服务

1. 启动微服务 Docker 镜像

（1）切换到项目的 "src/main/docker" 目录下，启动微服务的 Docker 镜像。命令如下：

```
$ docker-compose up -d
Recreating docker_wallet_1 ... done
```

如果没有错误，则表示成功启动了微服务镜像。

（2）查看启动的微服务容器。命令如下：

```
docker ps
```

微服务容器镜像的运行效果如图 5-21 所示。

```
qiaodeMacBook-Pro-2:aos qiaojiang$ docker ps
CONTAINER ID   IMAGE                COMMAND                  CREATED              STATUS              PORTS                                                                  NAMES
5e309fe8b13c   wallet:1.0-SNAPSHOT  "/bin/sh -c /app/run…"   About a minute ago   Up About a minute   0.0.0.0:9090->9090/tcp                                                 docker_wallet_1
17c8233636fa   consul               "docker-entrypoint.s…"   3 months ago         Up 9 hours          8300-8302/tcp, 8301-8302/udp, 8600/tcp, 8600/udp, 0.0.0.0:8500->8500/tcp   consul
6697e4109b66   mysql:5.7            "docker-entrypoint.s…"   3 months ago         Up 11 hours         0.0.0.0:3306->3306/tcp, 33060/tcp                                      dev-mysql
```

图 5-21

2. 查看微服务容器信息

（1）通过容器 ID 查看微服务容器的启动日志。命令如下：

```
docker logs -f 5e309fe8b13c
```

容器启动日志如图 5-22 所示。

```
qiaodeMacBook-Pro-2:aos qiaojiang$ docker logs -f 5e309fe8b13c
2020-01-16 16:29:09.193  INFO 8 --- [           main] trationDelegate$BeanPostProcessorChecker : Bean 'org.springframework.cloud.autoconfigure.ConfigurationPropertiesRebinderAutoConfiguration' of type [or
g.springframework.cloud.autoconfigure.ConfigurationPropertiesRebinderAutoConfiguration$$EnhancerBySpringCGLIB$$6fbe5f35] is not eligible for getting processed by all BeanPostProcessors (for example: not e
ligible for auto-proxying)

  .   ____          _            __ _ _
 /\\ / ___'_ __ _ _(_)_ __  __ _ \ \ \ \
( ( )\___ | '_ | '_| | '_ \/ _` | \ \ \ \
 \\/  ___)| |_)| | | | | || (_| |  ) ) ) )
  '  |____| .__|_| |_|_| |_\__, | / / / /
 =========|_|==============|___/=/_/_/_/
 :: Spring Boot ::        (v2.1.3.RELEASE)

2020-01-16 16:29:09.418  INFO 8 --- [           main] com.wudimanong.wallet.WalletApplication  : The following profiles are active: test
2020-01-16 16:29:11.132  WARN 8 --- [           main] o.s.boot.actuate.endpoint.EndpointId     : Endpoint ID 'service-registry' contains invalid characters, please migrate to a valid format.
2020-01-16 16:29:11.662  INFO 8 --- [           main] o.s.cloud.context.scope.GenericScope     : BeanFactory id=e4fa6d66-8b3e-3ac2-8f69-48bd73363e8f
2020-01-16 16:29:11.880  INFO 8 --- [           main] trationDelegate$BeanPostProcessorChecker : Bean 'org.springframework.transaction.annotation.ProxyTransactionManagementConfiguration' of type [org.spri
ngframework.transaction.annotation.ProxyTransactionManagementConfiguration$$EnhancerBySpringCGLIB$$53a45c38] is not eligible for getting processed by all BeanPostProcessors (for example: not eligible for
auto-proxying)
2020-01-16 16:29:11.930  INFO 8 --- [           main] trationDelegate$BeanPostProcessorChecker : Bean 'org.springframework.cloud.autoconfigure.ConfigurationPropertiesRebinderAutoConfiguration' of type [or
g.springframework.cloud.autoconfigure.ConfigurationPropertiesRebinderAutoConfiguration$$EnhancerBySpringCGLIB$$6fbe5f35] is not eligible for getting processed by all BeanPostProcessors (for example: not e
ligible for auto-proxying)
2020-01-16 16:29:12.728  INFO 8 --- [           main] o.s.b.w.embedded.tomcat.TomcatWebServer  : Tomcat initialized with port(s): 9090 (http)
2020-01-16 16:29:12.780  INFO 8 --- [           main] o.apache.catalina.core.StandardService   : Starting service [Tomcat]
2020-01-16 16:29:12.780  INFO 8 --- [           main] org.apache.catalina.core.StandardEngine  : Starting Servlet engine: [Apache Tomcat/9.0.16]
2020-01-16 16:29:12.803  INFO 8 --- [           main] o.a.catalina.core.AprLifecycleListener   : The APR based Apache Tomcat Native library which allows optimal performance in production environments was
not found on the java.library.path: [/usr/java/packages/lib/amd64:/usr/lib/x86_64-linux-gnu/jni:/lib/x86_64-linux-gnu:/usr/lib/x86_64-linux-gnu:/usr/lib/jni:/lib:/usr/lib]
2020-01-16 16:29:12.971  INFO 8 --- [           main] o.a.c.c.C.[Tomcat].[localhost].[/]       : Initializing Spring embedded WebApplicationContext
2020-01-16 16:29:12.971  INFO 8 --- [           main] o.s.web.context.ContextLoader            : Root WebApplicationContext: initialization completed in 3520 ms
2020-01-16 16:29:13.186  WARN 8 --- [           main] c.n.c.sources.URLConfigurationSource     : No URLs will be polled as dynamic configuration sources.
2020-01-16 16:29:13.187  INFO 8 --- [           main] c.n.c.sources.URLConfigurationSource     : To enable URLs as dynamic configuration sources, define System property archaius.configurationSource.additi
onalUrls or make config.properties available on classpath.
2020-01-16 16:29:13.224  INFO 8 --- [           main] c.netflix.config.DynamicPropertyFactory  : DynamicPropertyFactory is initialized with configuration sources: com.netflix.config.ConcurrentCompositeCon
figuration@3d9c13b5
Logging initialized using 'class org.apache.ibatis.logging.stdout.StdOutImpl' adapter.
Property 'mapperLocations' was not specified.
 _ _   |_  _ _|_. ___ _ |    _
| | |\/|_)(_| | |_\  |_)||_|_\
     /               |
                        3.3.0
2020-01-16 16:29:15.904  WARN 8 --- [           main] c.n.c.sources.URLConfigurationSource     : No URLs will be polled as dynamic configuration sources.
2020-01-16 16:29:15.905  INFO 8 --- [           main] c.n.c.sources.URLConfigurationSource     : To enable URLs as dynamic configuration sources, define System property archaius.configurationSource.additi
onalUrls or make config.properties available on classpath.
2020-01-16 16:29:16.305  INFO 8 --- [           main] o.s.s.concurrent.ThreadPoolTaskExecutor  : Initializing ExecutorService 'applicationTaskExecutor'
2020-01-16 16:29:17.143  INFO 8 --- [           main] o.s.s.c.ThreadPoolTaskScheduler          : Initializing ExecutorService 'catalogWatchTaskScheduler'
2020-01-16 16:29:17.253  WARN 8 --- [           main] o.s.c.n.core.CoreAutoConfiguration       : This module is deprecated. It will be removed in the next major release. Please use spring-cloud-netflix-hy
strix instead.
2020-01-16 16:29:17.266  INFO 8 --- [           main] o.s.b.a.e.web.EndpointLinksResolver      : Exposing 2 endpoint(s) beneath base path '/actuator'
2020-01-16 16:29:17.612  INFO 8 --- [           main] o.s.b.w.embedded.tomcat.TomcatWebServer  : Tomcat started on port(s): 9090 (http) with context path ''
2020-01-16 16:29:17.659  INFO 8 --- [           main] o.s.c.c.s.ConsulServiceRegistry          : Registering service with consul: NewService{id='wallet-spring-cloud-client-ipAddress-9090-1-0-SNAPSHOT', na
me='wallet', tags=[secure=false], address='172.18.64.2', meta=null, port=9090, enableTagOverride=null, check=Check{script='null', interval='10s', ttl='null', http='http://172.18.64.2:9090/actuator/health'
, method='null', header={}, tcp='null', timeout='null', deregisterCriticalServiceAfter='null', tlsSkipVerify=null, status='null'}, checks=null}
2020-01-16 16:29:17.809  INFO 8 --- [           main] com.wudimanong.wallet.WalletApplication  : Started WalletApplication in 11.348 seconds (JVM running for 12.512)
```

图 5-22

通过上述操作可以看到，电子钱包微服务在 Docker 容器中已经成功启动了。

（2）如果要查看 Docker 容器的更多信息，则需要进入容器终端。命令如下：

```
docker exec -it 5e309fe8b13c /bin/bash
```

之后就可以在容器内部进行一些操作，如测试网络等。在网络通畅的情况下，也可以直接通过访问容器“IP 地址 + 端口号”来进行微服务访问测试。

5.8 本章小结

本实例的综合性较强，部分逻辑的演示涉及“支付微服务”的内容，在学习时可以同步参考第 6 章的内容。

第6章 【实例】支付系统——用“Redis分布式锁 + Mockito”实现微服务场景下的“支付逻辑 + 代码测试”

第5章介绍了电子钱包系统如何通过接入支付系统来完成充值操作。在现实生活中类似的场景还有很多，诸如电商App在线购物、共享单车骑行付费、在线点外卖等，这些功能都离不开支付系统的支撑。不同发展程度的公司对于支付系统的需求也是不同的。

- 对于初创公司来说，由于其产品比较简单，涉及的支付方式也不多，在这种情况下支付系统可能会与业务系统耦合在一起。
- 对于有一定规模的公司来说，如果其产品形态多样，对支付渠道、资金流管理有更多要求，则需要将支付系统从业务系统中拆分出来作为独立的系统，以使支付系统可以提供平台化的能力。

本实例将利用Spring Cloud来构建相对独立的支付系统，并实现多渠道、多租户等平台能力，该系统只有一个支付微服务“payment”。在支付微服务中，订单防重也是一个比较重要的问题，所以本实例也会演示基于Redis的分布式锁机制。本章还会介绍Spring Cloud微服务场景下的单元测试代码的编写方法，这对于实际编程工作而言是非常有用的。

通过本章，读者将学习到以下内容：

- Spring Cloud微服务的构建及组件应用。
- 支付微服务的通用系统设计方法及运行流程。
- 支付宝渠道PC端/移动端支付方式的接入。

- 基于 Redis 的分布式锁机制的实现。
- Spring Cloud 微服务单元测试代码的编写。

6.1 功能概述

支付微服务，是一种通过连接第三方支付渠道与业务接入方，以实现收付款业务的中间系统。它可以通过简化的接入方式，帮助业务接入方避免多种支付方式接入带来的系统复杂度；也可以通过支持多租户，来满足接入方对不同支付渠道的支付需求。

举个例子：某公司有外卖和酒店两种业务。外卖业务要求支持支付宝和微信两种支付方式，但酒店业务除需要支持这两种支付方式外，还需要支持银行卡支付方式。这两种业务虽然属于同一集团公司，但是却是属于不同法律主体的子公司。所以从财务上，需要将这两种业务申请的支付宝、微信接入商户号分开，避免资金流混乱。

因此，支付微服务，不仅需要具备基本的多支付渠道接入能力，还需要具备多商户接入、多维度渠道路由功能，并通过参数化配置设计来实现渠道接入代码的复用。

例如，上述两种业务都需要接入支付宝、微信渠道，但是由于支付渠道商户信息不一致，所以支付所需要的渠道参数也不一样。通过渠道参数路由配置设计，可以使支付渠道接入代码被复用，从而支持多个接入方。

6.2 系统设计

支付微服务的拆分粒度，可以根据系统的复杂程度和公司的规模来设计。

例如，有些公司需要的支付渠道非常多样，除需要支持支付宝、微信外，还需要支持多种银行卡支付；除支持收款功能外，还需要支持付款、退款功能。对于这些情况，就需要将支付微服务拆分为多个子系统（例如付款服务、路由服务等），从而更好地满足支付资金安全及渠道管理的需要。

> 如果支付方式比较单一（如只有支付宝或微信），业务接入方也并没有那么多，则不进行过度的拆分设计反而会使系统运行得更好。

在本实例中，支付微服务不会被拆分为多个子服务，但是会在编码的过程中采用模块化设计，从而更好地满足未来系统扩展的需求。

6.2.1 支付流程设计

从系统流程设计上看，支付微服务的前后端交互流程如图 6-1 所示。

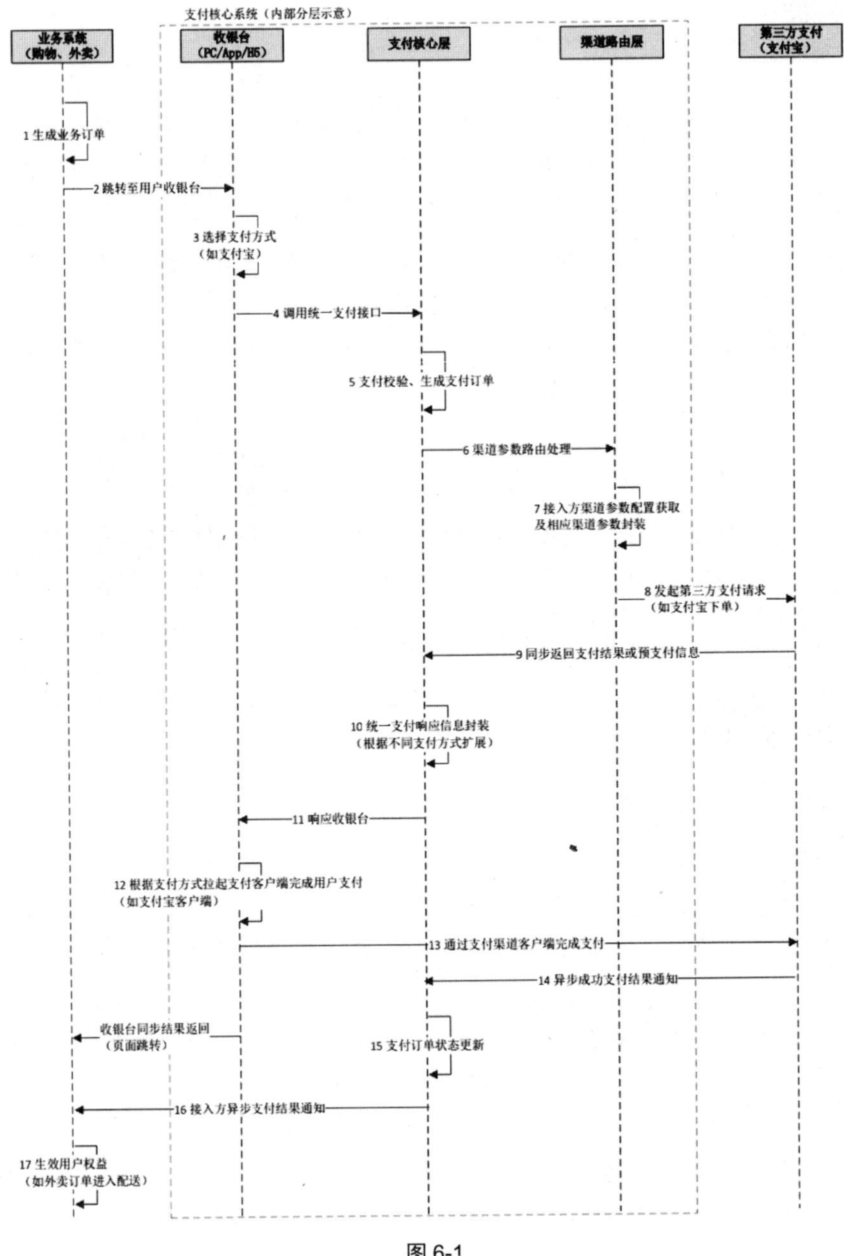

图 6-1

具体说明如下。

（1）业务系统（指外卖、电商等接入支付场景的系统）在完成自身业务订单逻辑后，会引导用户进入支付收银台界面。一般情况下，支付收银台可以由支付微服务统一提供，也可以由各业务系统应用端自行定制，具体方案根据实际需要来确定。

（2）在用户选择完支付方式后，收银台系统会将支付请求发送到支付微服务。

（3）支付微服务在完成支付订单逻辑后，会对支付渠道请求参数进行封装，并根据具体的支付方式和形态完成第三方支付请求。

在某些支付场景下，支付微服务并不能直接请求第三方支付，例如，需要前端浏览器跳转的电脑网页支付方式。在这种情况下，支付微服务会根据渠道支付要求完成请求参数的封装，并通过接口响应至收银台系统，由收银台系统完成用户浏览器跳转至第三方支付界面的逻辑。

（4）对于需要拉起用户支付客户端的支付方式（如支付宝/微信 App 支付等），支付微服务需要先进行“预支付”操作。

（5）在“预支付”操作完成后，支付微服务将渠道返回的预支付订单信息返回给支付收银台。之后，支付收银台根据支付场景拉起相应的支付客户端（如支付宝 App），从而完成客户端支付。

（6）在用户完成支付后，第三方支付渠道将支付结果通知给支付微服务，支付微服务在完成订单状态逻辑更新后会同步将支付结果通知到具体的业务系统，从而完成整个支付流程的闭环。

在用户完成支付动作后，由于整个支付过程涉及多个系统的前后端交互，所以，实际上支付结果是不能同步被支付系统感知的。因此在现有支付流程中，所有的支付结果都会以异步通知为准。而为了确保支付结果通知的可靠性，支付系统会对支付结果的通知方式进行一定的重复机制设计，以确保支付通知最大限度被成功接收。

上述逻辑基本涵盖了目前“电脑端”及“移动端”支付方式的后端逻辑。由于涉及前后端的协作，所以，在接口参数设计上需要考虑一定的扩展性。

如果还涉及银行卡之类的支付方式，则需要考虑信息安全、银行卡类型识别、银行卡多渠道路由等逻辑。但目前在互联网应用中直接使用银行卡支付的场景比较少，所以这里就不再讨论了。

6.2.2 系统结构设计

在系统实现上采用经典 MVC 分层模式，支付微服务的结构如图 6-2 所示。

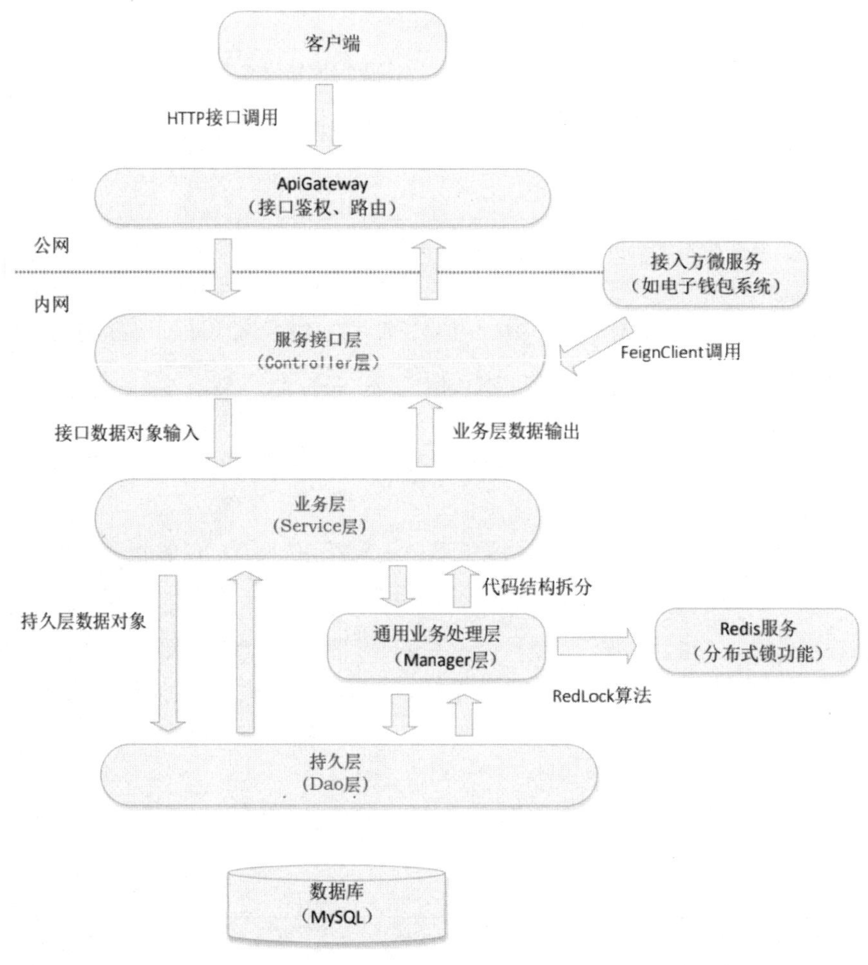

图 6-2

在上述系统结构中，主要分为 3 层：服务接口层（Controller 层）、业务层（Service 层）及持久层（Dao 层）。

- 服务接口层（Controller 层）：用于定义服务接口。
- 业务层（Service 层）：用于处理业务逻辑并通过 Dao 层完成数据库相关操作。
- 持久层（Dao 层）：用于封装对 MySQL 数据库操作的接口。

> 在本实例中，在业务层与持久层之间单独拆分出了一个 Manager 层，主要是用于业务层逻辑代码的模块化拆分，避免业务层代码过于臃肿。

关于 Manager 层的拆分，在本实例中，主要体现在使用工厂模式对多个支付渠道的对接代码进行解耦上。

6.2.3 数据库设计

根据系统的复杂程度及实际需要，在设计支付微服务数据库模型时，需要设计很多表。其中，比较核心的表有：①支付订单信息表；②支付通知信息表；③支付渠道路由配置表；④渠道参数信息表。

- 代码 以下表在本书配置资源的"chapter06-payment/src/main/resources/db.migration"目录下。

具体的表结构设计如下。

1. 支付订单信息表

支付订单信息表是支付微服务中最重要的数据表，是用户支付的凭据，也是后续进行资金清洁算及数据统计的关键数据。具体的 SQL 代码如下：

```
create table pay_order (
 id bigint not null primary key auto_increment,
 order_id varchar (50) comment '业务方订单号(业务方系统唯一)',
 trade_type varchar (30) comment '业务交易类型，例如 topup 表示钱包充值',
 amount bigint comment '交易金额，以分为单位',
 currency varchar (10) comment '币种',
 status varchar (2) comment '支付状态。0-待支付；1-支付中；2-支付成功；3-支付失败',
 channel varchar (10) comment '支付渠道编码。0-微信支付，1-支付宝支付',
 pay_type varchar (30) comment '渠道支付方式。ali_pay_pc-支付宝电脑网页支付；
ali_pay_app-支付宝移动应用支付',
 pay_id varchar (50) comment '支付平台自己生成的唯一订单流水号，用于与第三方渠道交互',
 trade_no varchar (32) comment '支付渠道流水号',
 user_id varchar (60) comment '业务方用户 ID',
 create_time timestamp null default current_timestamp comment '支付创建时间',
 update_time timestamp null default current_timestamp on update
current_timestamp comment '最后一次更新时间',
 remark varchar(128)  comment '订单备注信息'
);
alter table pay_order comment '支付订单信息表';
```

```
#添加索引信息
alter table pay_order add index unique_idx_pay_id ( pay_id );
alter table pay_order add index idx_order_id ( order_id );
alter table pay_order add index idx_create_time ( create_time );
```

2. 支付通知信息表

支付通知信息表，主要记录支付渠道支付结果通知报文信息，以便后续出现订单争议时反查系统的交互过程。

此外，该表还会承担“支付微服务”向“接入方业务系统”同步支付状态时的辅助逻辑：例如设置向业务方通知的次数、最近通知时间等，实现在向业务方通知失败的情况下重复通知的逻辑（如通知 5 次，持续通知 24 小时等逻辑）。具体的 SQL 代码如下：

```
create table pay_notify (
 id bigint not null primary key auto_increment,
 pay_id varchar (50) comment '支付平台订单流水号',
 channel varchar (10) comment '支付渠道编码。0-微信支付，1-支付宝支付',
 status varchar (2) comment '支付通知状态。1-支付中；2-支付成功；3-支付失败',
 fullinfo text comment '渠道通知原始报文信息',
 order_id varchar (50) comment '业务方订单流水号',
 verify varchar (2) comment '报文签名验证结果。0-验证成功；1-签名验证失败',
 merchant_id varchar (30) comment '支付渠道商户号，用于精准识别渠道参数',
 receive_status varchar (2) comment '接收处理状态。1-已接收；2-已处理；3-已同步至业务方',
 notify_count int comment '业务方通知次数',
 notify_time timestamp comment '业务方最近通知时间',
 update_time timestamp null default current_timestamp on update current_timestamp comment '最后一次更新时间',
 create_time timestamp null default current_timestamp comment '交易创建时间'
);
alter table pay_order comment '支付通知信息表，记录支付渠道通知报文及业务方通知状态信息';
#添加索引信息
alter table pay_notify add index unique_idx_pay_id ( pay_id );
alter table pay_notify add index idx_order_id ( order_id );
```

3. 支付渠道路由配置表

支付渠道路由配置表，主要用于配置“业务接入方”使用多个支付渠道时的路由信息。这样，同一个“业务接入方”可以根据应用标识及业务类型，相对灵活地选择合适的支付渠道。具体的 SQL 代码如下：

```
create table pay_channel_route_config (
 id bigint not null primary key auto_increment,
 app_id varchar (50) comment '业务接入方应用标识',
```

```
    trade_type varchar (10) comment '业务接入方业务类型',
    channel varchar (10) comment '支付渠道编码。0-微信支付，1-支付宝支付',
    pay_type varchar (30) comment '渠道具体支付方式',
    partner varchar (50) comment '具体支付渠道接入唯一账号标识',
    status varchar(2) NOT NULL DEFAULT '0' COMMENT '状态。 0-可用，1-不可用',
    update_time timestamp null default current_timestamp on update
current_timestamp comment '最后一次更新时间',
    create_time timestamp null default current_timestamp comment '创建时间'
   );
   alter table pay_channel_route_config comment '支付渠道路由配置表';
   #添加索引信息
   alter table pay_channel_route_config add index idx_app_id (app_id);
```

4. 渠道参数信息表

渠道参数信息表主要用于在支付路由成功后获取具体的支付渠道接入账号，以及所对应的系统参数信息（例如接口密钥，证书类型等）。

通过对渠道参数的配置化设计，使得同一份渠道接入代码可以灵活地支持该渠道下多个支付渠道商户的接入，从而使得支付微服务具备一定的 SASS 能力。具体的 SQL 代码如下：

```
   create table pay_channel_param (
    id bigint not null primary key auto_increment,
    partner varchar (50) comment '具体支付渠道接入唯一账号标识',
    sign_type varchar(10) NOT NULL COMMENT '签名加密方式，如：RSA、MD5、3DES',
    key_type varchar(30) COMMENT '证书类型。如：publickey-公钥；privatekey-私钥，
若签名加密方式为对称加密，则约定为私钥类型',
    key_context text COMMENT '证书文本内容',
    expire_time timestamp NOT NULL DEFAULT CURRENT_TIMESTAMP COMMENT '证书到期
时间',
    status varchar(2) NOT NULL DEFAULT '0' COMMENT '状态。0-可用；1-不可用',
    update_time timestamp null default current_timestamp on update
current_timestamp comment '最后一次更新时间',
    create_time timestamp null default current_timestamp comment '创建时间',
    remark varchar(128) COMMENT '证书用途描述'
   );
   alter table pay_channel_param comment '渠道参数信息表，存储支付渠道密钥、加密方
式等信息';
   #添加索引信息
   alter table pay_channel_param add index idx_partner(partner);
```

6.3 步骤 1：构建 Spring Cloud 微服务工程代码

接下来搭建支付微服务所需要的 Spring Cloud 微服务工程，并集成所需要的其他第三方组件。

6.3.1 创建 Spring Cloud 微服务工程

1. 创建一个基本的 Maven 工程

利用 2.3.1 节介绍的方法创建一个 Maven 工程，完成后的工程代码结构如图 6-3 所示。

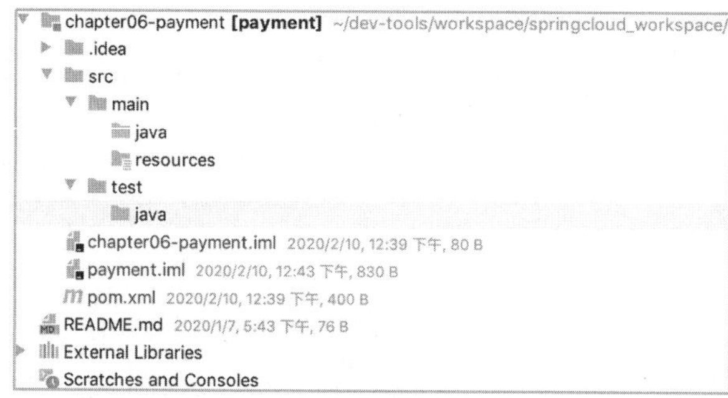

图 6-3

2. 引入 Spring Cloud 依赖，将其改造为微服务项目

（1）引入 Spring Cloud 微服务的核心依赖。

这里可以参考 2.5.2 节中的具体步骤。

（2）在工程代码的 resources 目录新建一个基础性配置文件——bootstrap.yml。配置文件中的代码如下：

```
spring:
  application:
    name: payment
  profiles:
    active: debug
  cloud:
    consul:
      discovery:
        preferIpAddress: true
        instance-id: ${spring.application.name}:${spring.cloud.client.
ipAddress}:${spring.application.instance_id:${server.port}}:@project.version@
        healthCheckPath: /actuator/health
```

```
server:
  port: 9091
```

（3）在 2.5.2 节提到过，Spring Boot 并不会默认加载 bootstrap.yml 这个文件，所以需要在 pom.xml 中添加 Maven 资源相关的配置，具体参考 2.5.2 节内容。

（4）创建“Payment”支付微服务的入口程序类。代码如下：

```
package com.wudimanong.payment;
import org.springframework.boot.SpringApplication;
import org.springframework.boot.autoconfigure.SpringBootApplication;
import org.springframework.cloud.client.discovery.EnableDiscoveryClient;
@EnableDiscoveryClient
@SpringBootApplication
public class PaymentApplication {
    public static void main(String[] args) {
        SpringApplication.run(PaymentApplication.class, args);
    }
}
```

至此，支付微服务所需的 Spring Cloud 微服务工程就构建出来了。

6.3.2 将 Spring Cloud 微服务注入 Consul

参考 2.5.1 节、2.5.3 节的内容，将“payment”微服务注入服务注册中心 Consul 中。然后运行所构建的“payment”微服务工程，可以看到该服务已经注册到 Consul 中了，如图 6-4 所示。

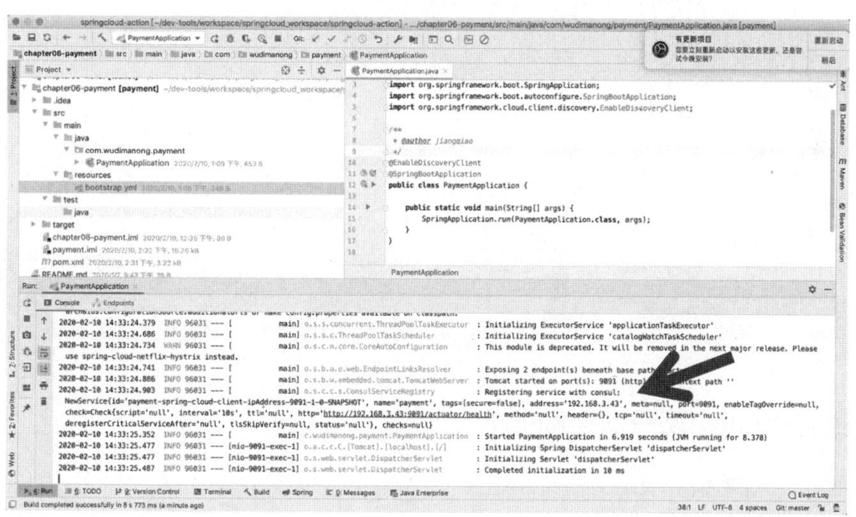

图 6-4

打开 Consul 控制台，“payment”微服务被注册到 Consul 中的效果如图 6-5 所示。

图 6-5

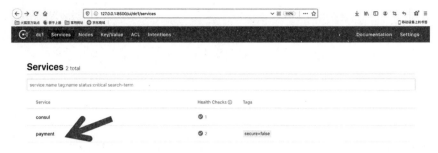

至此，就从技术层面完成了 Spring Cloud 微服务的搭建过程。接下来继续集成开发“payment”支付微服务所依赖的其他组件。

6.3.3 集成 MyBatis，以访问 MySQL 数据库

在本实例中，使用 MyBatis 这个持久层框架来操作数据库。

1. 引入 MyBatis 框架依赖，以及 MySQL 数据库驱动程序

具体步骤参考 2.3.3 节内容。

2. 配置项目数据库连接信息

在项目中创建一个新的配置文件 application.yml，添加 MySQL 数据库的连接信息如下：

```
spring:
  datasource:
    url: jdbc:mysql://127.0.0.1:3306/payment
    username: root
    password: 123456
    type: com.alibaba.druid.pool.DruidDataSource
    driver-class-name: com.mysql.jdbc.Driver
    separator: //
```

上述配置中涉及的数据库信息，可以参考 1.3.3 节通过 Docker 部署本地 MySQL 的步骤——创建一个名为 payment 的数据库，并执行 6.2.3 节中所定义的 SQL 脚本。

6.3.4 通过 MyBatis-Plus 简化 MyBatis 的操作

1. Spring Boot 集成 MyBatis-Plus 框架

具体的集成和配置方法可参考 4.3.5 节内容。

2. 启动项目，验证 MyBatis-Plus 集成效果

启动“payment”支付微服务工程，成功集成 MyBatis-Plus 框架的运行效果如图 6-6 所示。

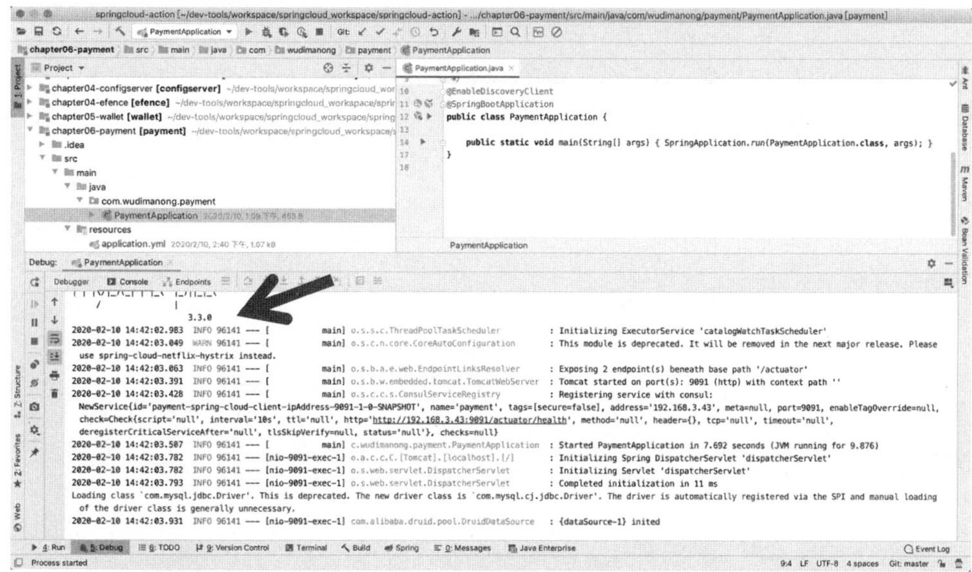

图 6-6

6.4 步骤 2：实现基于 Redis 的分布式锁

Redis 是一款性能优异的 Key-Value 存储系统。目前绝大多数互联网公司的应用都在使用它来提升服务的性能。除此之外，还有很多场景会使用 Redis 作为分布式锁来实现并发控制相关的逻辑，例如，本实例中的支付订单防重，就需要实用分布式锁功能——除依赖数据库判断外，还需要考虑高并发情况下的数据幂等问题。

本节将使用 Redis 来实现高性能的分布式锁，并在具体的支付订单防重逻辑中使用它。

6.4.1 配置 Redis 服务

在生产环境中，Redis 一般会以集群的方式对外提供可靠的服务。不过这里为了方便测试开发，只是基于 Docker 容器快速部署一个 Redis 服务。

1. 快速部署一个 Redis 服务

（1）使用 Docker 安装 Redis。具体命令如下：

```
    docker run -p 6379:6379 -v $PWD/data:/data  -d redis:3.2 redis-server
--requirepass "123456"  --appendonly yes
```

命令说明如下。

- -p 6379:6379：将容器的 6379 端口映射到主机的 6379 端口。
- -v $PWD/data:/data：将主机当前目录下的 data 目录挂载到容器的“/data”目录。
- redis-server --requirepass "123456" --appendonly yes：在容器执行“redis-server”命令，打开 Redis 持久化配置并设置访问密码为“123456”。

（2）通过“docker ps”命令查看到 Redis 服务的 Docker 容器信息，如图 6-7 所示。

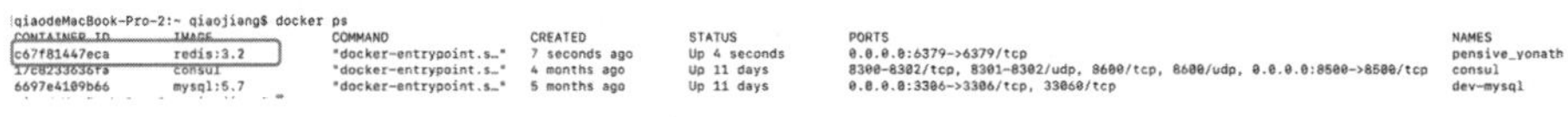

图 6-7

2. 访问 Docker 容器中的 Redis 服务

接下来，通过 Redis 客户端工具来访问 Docker 容器中的 Redis 服务。这里下载 Redis 官方的客户端工具，步骤如下。

（1）通过 Redis 官方下载链接下载 Redis 安装包。

下载一个官方的 Redis 发布版本，但不使用它的服务功能，只将其作为客户端测试使用。

（2）下载后将其解压缩至指定目录，然后使用“make”命令进行编译。

（3）编译成功后切换至“/src”目录下，通过“redis-cli”命令链接 Docker 容器中的 Redis 服务。具体命令如下：

```
./redis-cli -h 127.0.0.1 -p 6379 -a 123456
```

（4）连接上后为了测试 Redis 服务的可用性，可以通过“set/get”命令进行赋值和取值。命令如下：

```
127.0.0.1:6379> set a 123
OK
127.0.0.1:6379> get a
"123"
```

在上述命令中，通过“set”命令设置属性“a”的值，并通过“get”命令获取属性“a”的值。这说明基于 Docker 运行的 Redis 服务是可用的，能够满足开发的需求。

6.4.2 集成 Redis 客户端访问组件

接下来在 Spring Cloud 微服务中，通过集成 Redis 依赖来实现应用对 Redis 服务的访问和操作。

与在 Spring Boot 应用中集成 MyBatis 框架一样，Spring Boot 针对 Redis 服务的集成也提供了现成的 Starter 依赖，只需引入即可，代码如下：

```
<dependency>
    <groupId>org.springframework.boot</groupId>
    <artifactId>spring-boot-starter-data-redis</artifactId>
</dependency>
```

引入该依赖后，Spring Boot 应用就具备了访问和操作 Redis 的能力。

而 spring-boot-starter-data-redis 依赖的基本原理也是利用了 Spring Boot 框架提供的自动配置能力，具体细节大家可以阅读该依赖的源码。

在 Spring Boot 的应用配置文件中，配置对 Redis 服务的连接信息，配置如下：

```
spring:
  datasource:
    url: jdbc:mysql://127.0.0.1:3306/payment
    username: root
    password: 123456
    type: com.alibaba.druid.pool.DruidDataSource
    driver-class-name: com.mysql.jdbc.Driver
    separator: //
  #Redis 服务的地址
  redis:
    host: 127.0.0.1
    port: 6379
    password: 123456
```

完成后，Spring Boot 应用就可以通过 RedisTemplate 方便地操作和使用 Redis 服务了，具体的使用方法将在“支付微服务”的业务逻辑开发中演示。

6.4.3 理解 Redis 分布式锁的原理

Redis 服务本身并不提供分布式锁功能，但是作为全局 Key-Value 存储系统，客户端可以利用 Redis 提供的基本功能，并通过一定的算法设计来实现分布式锁功能。

网上有不少博客文章及代码库描述了如何使用 Redis 来实现分布式锁，但是许多实现相对比较简单，安全性也比较低。

在 Redis 的官方文档中，推荐了一种叫作“RedLock”的算法可以实现基于 Redis 集群的分布式锁功能。该算法已经有多种语言版本的 Redis 客户端实现库。其中，Java 领域最为知名的是 Redisson 库，它不仅实现了分布式锁，还实现了一套复杂的 Redis 分布式数据结构。

目前，在 Spring Boot 2.x 以上版本中使用 Redis 时，Redis 客户端库已经默认使用了 lettuce 库（一种比 Redisson Jedis 线程更安全、更轻量级的 Java Redis 客户端库）。

在实践中往往会选择基于 RedLock 算法自行实现分布式锁。本实例中将基于 RedLock 算法来实现 Redis 分布式锁。

先来看看 RedLock 算法的运行原理，如图 6-8 所示。

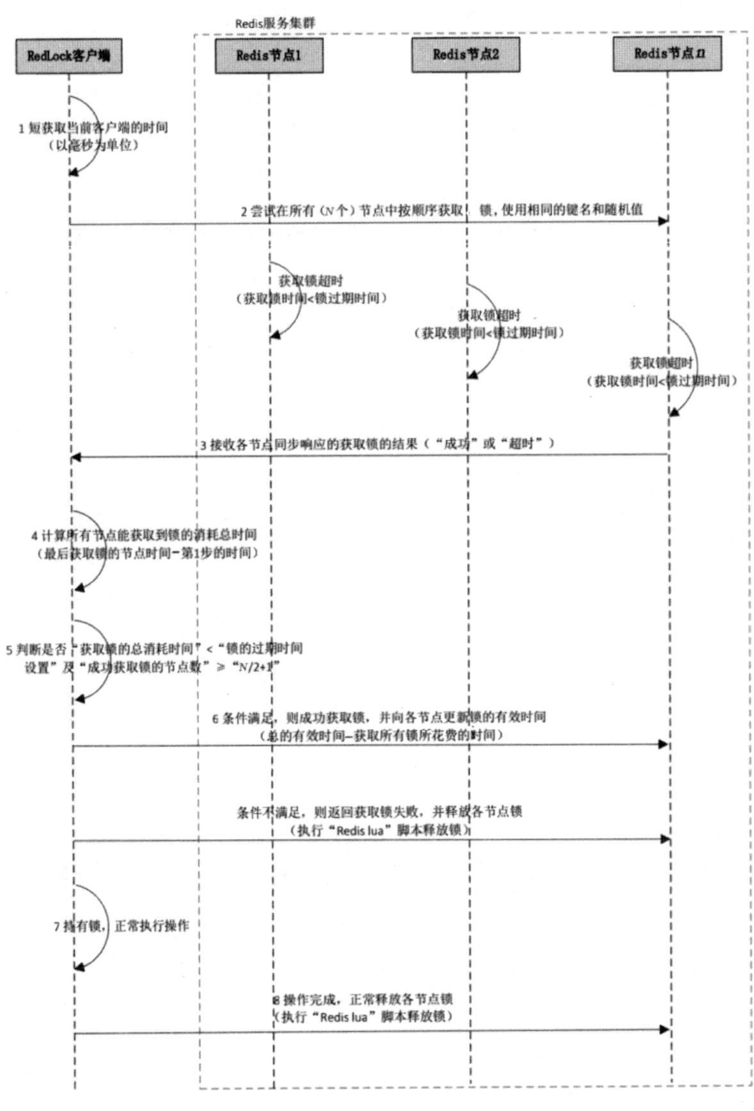

图 6-8

RedLock 算法是一种 Redis 集群环境下的分布式锁解决方案，可以有效地防止单节点故障问题。运行原理说明如下：

（1）RedLock 客户端获取系统的当前时间，以毫秒（ms）为单位。

（2）RedLock 客户端使用相同的键（Key）名和随机值，依次向 Redis 集群中的每个节点发起获取锁的请求。

在向每个 Redis 节点获取锁的过程中，客户端会以比锁过期时间小得多的时间来设定超时机制，例如，锁的整个超时时间为 10s，集群有 5 个节点，那么每个节点获取锁的超时时间可能会被限制在 5ms ~ 50ms 之间。这是为了防止在某个节点不可用的情况下，客户端等待时间过长从而造成阻塞。

（3）在收到 Redis 集群各节点获取锁结果的反馈后，RedLock 客户端会对锁的获取情况进行判断：如果获取各节点锁的总时间小于锁的超时设置，并且成功获取锁的节点数目大于“*N*/2+1”个（例如 5 个节点至少要有 3 个节点成功获取锁），则 RedLock 客户端认为成功获取分布式锁；否则认为分布式锁获取失败，并依次释放各个节点已获取的锁信息。

（4）在成功获取分布式锁后，就可以安全地执行需要锁保护的操作，并在完成后依次释放各节点所持有的锁信息。

实现 RedLock 算法的 Redis 客户端，基本上可以保证分布式锁在有效性及安全性方面的几个基本要求。

- 互斥：任何时刻只能有一个客户端获取锁。
- 无死锁：即使锁定资源的服务发生崩溃或者分区，仍然能释放锁。
- 容错性：只要多数 Redis 节点（一半以上）在使用，客户端就可以获取和释放锁。

6.4.4 实现 Redis 分布式锁的客户端代码

通过 6.4.3 节的学习，我们对实现 Redis 分布式锁的基本算法有了一定的了解。在实践中，可以依据 RedLock 算法自行实现分布式锁的客户端。

在 Spring Boot 项目中，也可以直接使用 Spring 框架所提供的基于 RedLock 算法的分布式锁的实现库。

1. 集成 Spring Integration 依赖

本实例所使用的 RedLock 客户端为 Spring Integration 依赖中的实现。集成步骤如下。

（1）在工程的 pom.xml 文件中，引入 Spring Integration 依赖。代码如下：

```
<!-- spring integration -->
<dependency>
    <groupId>org.springframework.boot</groupId>
    <artifactId>spring-boot-starter-integration</artifactId>
</dependency>
<!-- Spring Integration 与 Redis 结合，实现 Redis 分布式锁 -->
<dependency>
    <groupId>org.springframework.integration</groupId>
    <artifactId>spring-integration-redis</artifactId>
</dependency>
```

目前 Spring 所提供的分布式锁相关的代码被迁移到“Spring Integration”子项目中了，所以这里引入其相关依赖。

（2）编写 RedLock 分布式锁的配置类。代码如下：

```
package com.wudimanong.payment.config;
import org.springframework.context.annotation.Bean;
import org.springframework.context.annotation.Configuration;
import org.springframework.data.redis.connection.RedisConnectionFactory;
import org.springframework.integration.redis.util.RedisLockRegistry;
@Configuration
public class RedisLockConfiguration {
    @Bean
    public RedisLockRegistry redisLockRegistry(RedisConnectionFactory
redisConnectionFactory) {
        return new RedisLockRegistry(redisConnectionFactory, "payment");
    }
}
```

在以上配置中，代码加载的前提是，微服务工程已经集成并配置了对 Redis 服务的访问及连接信息（参考 6.4 节）。

2. 分布式锁的使用方式

在业务中使用分布式锁，代码片段如下：

```
/**
 * 引入 Redis 分布式锁的依赖
 */
@Autowired
private RedisLockRegistry redisLockRegistry;
```

```
    @Override
    public UnifiedPayBO unifiedPay(UnifiedPayDTO unifiedPayDTO) {
        ...
        //创建 Redis 分布式锁
        Lock lock = redisLockRegistry.obtain(redisLockPrefix +
unifiedPayDTO.getOrderId());
        try {
            //尝试获取锁
            boolean isLock = lock.tryLock(1, TimeUnit.SECONDS);
            if (isLock) {
                //执行业务逻辑
                ...
            }
        } catch (InterruptedException e) {
            e.printStackTrace();
        } finally {
            //释放分布式锁
            lock.unlock();
        }
        ...
    }
```

在上方代码中，通过注入 RedisLockRegistry 实例对象，实现了分布式锁的相关操作——用 obtain()方法创建锁、用 tryLock()方法获取锁、用 unlock()方法释放锁。关于 Redis 分布式锁的使用细节将在 6.5 节中演示。

6.5 步骤 3：实现微服务的业务逻辑

经过前面的步骤，完成了系统结构及数据库的设计，构建了开发“支付微服务”所需的工程代码，并集成了实现 Redis 的分布式锁的依赖。接下来将实现“支付微服务”的业务逻辑。

6.5.1 定义服务接口层（Controller 层）

接下来定义“支付微服务”所涉及的 Controller 层服务接口。

1. 约定接口数据格式

在定义正式服务接口的 Controller 层之前，需要先约定接口的请求方式，以及统一的报文格式。具体的规范可以参考 4.1.1 节中“1.”小标题中的内容。

其中涉及的统一报文格式类（ResponseResult）及全局响应码枚举类（GlobalCodeEnum），可以复制 4.1.1 节中“1.”小标题中的代码。

2. 定义“统一支付”接口

“统一支付”接口是“支付微服务”的核心接口，该接口用于接收和处理不同支付类型的支付请求——根据不同的支付渠道及方式，实现相应的支付流程及参数适配。

不同的支付渠道及方式，对支付流程及请求参数有不同的要求。但是，支付系统需要通过统一的支付接入，来降低不同支付渠道接入的复杂度。这也是目前很多互联网公司构建独立支付系统的重要原因。

（1）定义“统一支付”接口的 Controller 层。代码如下：

```
package com.wudimanong.payment.controller;
import com.wudimanong.payment.entity.ResponseResult;
...
import org.springframework.web.bind.annotation.RequestMapping;
import org.springframework.web.bind.annotation.RestController;
@Slf4j
@RestController
@RequestMapping("/pay")
public class PayController {
    /**
     * 业务层（Service 层）依赖接口
     */
    @Autowired
PayService payServiceImpl;
    /**
     * 定义“统一支付”接口
     */
    @PostMapping("/unifiedPay")
    public ResponseResult<UnifiedPayBO> unifiedPay(@RequestBody @Validated
UnifiedPayDTO unifiedPayDTO) {
        return
ResponseResult.OK(payServiceImpl.unifiedPay(unifiedPayDTO));
    }
}
```

在上述代码中，通过@PostMapping 注解定义了“统一支付”接口（/unifiedPay）。该接口的请求方式为“Post”，请求参数为 JSON 格式（通过@RequestBody 注解约定）。

（2）定义“统一支付”接口的请求参数对象。代码如下：

```
package com.wudimanong.payment.entity.dto;
import com.wudimanong.payment.validator.EnumValue;
import java.io.Serializable;
```

```
import javax.validation.constraints.NotNull;
import lombok.Data;
@Data
public class UnifiedPayDTO implements Serializable {
    /**
     * 接入方应用 ID
     */
    @NotNull(message = "应用 ID 不能为空")
    private String appId;
    /**
     * 接入方支付订单 ID，必须在接入方系统唯一（如电子钱包系统）
     */
    @NotNull(message = "支付订单 ID 不能为空")
    private String orderId;
    /**
     * 交易类型。用于标识具体的业务类型，如 topup 表示钱包充值等。
     */
    @EnumValue(strValues = {"topup"})
    private String tradeType;
    /**
     * 支付渠道。0-微信支付，1-支付宝支付
     */
    @EnumValue(intValues = {0, 1})
    private Integer channel;
    /**
     * 支付产品定义。用于区分具体的渠道支付产品，可根据实际情况定义
     */
    private String payType;
    /**
     * 支付金额，以“分”为单位，数值必须大于 0
     */
    private Integer amount;
    /**
     * 支付币种，默认为 CNY
     */
    @EnumValue(strValues = {"CNY"})
    private String currency;
    /**
     * 接入方系统唯一标识用户身份的 ID
     */
    @NotNull(message = "用户 ID 不能为空")
    private String userId;
    /**
     * 商品标题，支付所购买的商品标题
     */
```

```
    @NotNull(message = "商品标题不能为空")
    private String subject;
    /**
     * 商品描述信息
     */
    private String body;
    /**
     * 支付扩展信息，例如针对某些支付渠道的特殊请求参数的补充
     */
    private Object extraInfo;
    /**
     * 异步支付结果通知地址
     */
    @NotNull(message = "支付通知地址不能为空")
    private String notifyUrl;
    /**
     * 同步支付结果跳转地址（支付成功后同步跳转回接入方系统界面的 URL）
     */
    private String returnUrl;
}
```

上述代码定义了“统一支付”接口的请求参数类，属性涵盖了大部分支付场景所需的请求信息（如支付金额、币种、支付渠道及其方式等）。

此外，该参数对象中涉及的自定义参数校验注解@EnumValue 的定义可以参考 4.4.1 节中“2.”小标题中的具体说明。

（3）定义“统一支付”接口的返回参数对象。代码如下：

```
package com.wudimanong.payment.entity.bo;
import java.io.Serializable;
import lombok.AllArgsConstructor;
import lombok.Builder;
import lombok.Data;
import lombok.NoArgsConstructor;
@Data
@Builder
@NoArgsConstructor
@AllArgsConstructor
public class UnifiedPayBO implements Serializable {
    /**
     * 商户支付订单号
     */
    private String orderId;
    /**
     * 第三方支付渠道生成的预支付订单号
```

```
     */
    private String tradeNo;
    /**
     * 支付订单金额
     */
    private Integer amount;
    /**
     * 支付币种
     */
    private String currency;
    /**
     * 支付渠道编码
     */
    private String channel;
    /**
     * 特殊支付场景所需要传递的额外支付信息
     */
    private String extraInfo;
    /**
     * 支付订单状态。0-待支付；1-支付中；2-支付成功；3-支付失败
     */
    private Integer payStatus;
}
```

3. 定义“渠道支付结果通知”接口

在支付流程中，用户支付结果一般以后端服务的通知为准。所以，“支付微服务”除需要定义“统一支付”接口处理支付请求外，还需要根据第三方支付渠道接口的约定，开发接收支付渠道支付结果通知的接口，并以此完成支付订单状态的更新，以及向接入方业务系统同步用户支付结果。

不同支付渠道的支付结果通知接口各不相同，需要根据具体情况而定。这里以“支付宝异步支付结果通知”接口为例。

在用户使用支付宝完成支付后，支付宝会根据支付 API 中接入方传入的“notify_url”字段，以 POST 请求的方式，将支付结果通知到接入方系统（例如本实例的“支付微服务”）。

（1）定义“支付宝异步支付结果通知”接口的 Controller 层。代码如下：

```
package com.wudimanong.payment.controller;
import com.wudimanong.payment.entity.dto.AliPayReceiveDTO;
import com.wudimanong.payment.service.PayNotifyService;
...
import org.springframework.web.bind.annotation.RequestMapping;
import org.springframework.web.bind.annotation.RestController;
@Slf4j
@RestController
```

```
@RequestMapping("/notify")
public class PayNotifyController {
    @Autowired
    PayNotifyService payNotifyServiceImpl;
    /**
     * 定义“支付宝异步支付结果通知”接口
     */
    @PostMapping("/aliPayReceive")
    public String aliPayReceive(AliPayReceiveDTO aliPayReceiveDTO) {
        return payNotifyServiceImpl.aliPayReceive(aliPayReceiveDTO);
    }
}
```

上述接口需要根据“支付宝异步支付结果通知”接口的文档规范进行定义。

（2）定义“支付宝异步支付结果通知”接口的请求参数对象。代码如下：

```
package com.wudimanong.payment.entity.dto;
import lombok.Data;
@Data
public class AliPayReceiveDTO {
    /**
     * 通知时间（格式为“yyyy-MM-dd HH:mm:ss”）
     */
    private String notify_time;
    /**
     * 通知类型，例如：trade_status_sync
     */
    private String notify_type;
    /**
     * 通知校验 ID
     */
    private String notify_id;
    /**
     * 格式编码，如 UTF-8
     */
    private String charset;
    /**
     * 接口版本，固定为 1.0
     */
    private String version;
    /**
     * 签名类型，目前为 RSA2
     */
    private String sign_type;
    /**
```

```
 * 签名信息
 */
private String sign;
/**
 * 授权方的 app_id（本接口暂不开放第三方应用授权，所以 auth_app_id=app_id）
 */
private String auth_app_id;
/**
 * 支付宝交易号
 */
private String trade_no;
/**
 * 开发者的 app_id(支付宝分配给开发者的应用 ID)
 */
private String app_id;
/**
 * 商户订单号
 */
private String out_trade_no;
/**
 * 商户业务 ID，主要是在退款通知中返回退款申请的流水号
 */
private String out_biz_no;
/**
 * 买家支付宝用户号
 */
private String buyer_id;
/**
 * 卖家支付宝用户号
 */
private String seller_id;
/**
 * 交易状态，TRADE_SUCCESS 表示交易成功
 */
private String trade_status;
/**
 * 订单金额
 */
private Double total_amount;
/**
 * 实收金额
 */
private Double receipt_amount;
/**
 * 开票金额
```

```
     */
    private Double invoice_amount;
    /**
     * 用户在交易中支付的金额，单位为“元”，精确到小数点后 2 位
     */
    private Double buyer_pay_amount;
    /**
     * 使用集分宝支付的金额，单位为“元”，精确到小数点后 2 位
     */
    private Double point_amount;
    /**
     * 在退款通知中返回的总退款金额，单位为“元”，精确到小数点后 2 位
     */
    private Double refund_fee;
    /**
     * 订单标题
     */
    private String subject;
    /**
     * 商品描述
     */
    private String body;
    /**
     * 交易创建时间，格式为“yyyy-MM-dd HH:mm:ss”
     */
    private String gmt_create;
    /**
     * 交易付款时间，格式为“yyyy-MM-dd HH:mm:ss”
     */
    private String gmt_payment;
    /**
     * 交易退款时间，格式为“yyyy-MM-dd HH:mm:ss”
     */
    private String gmt_refund;
    /**
     * 交易结束时间，格式为“yyyy-MM-dd HH:mm:ss”
     */
    private String gmt_close;
    /**
     * 支付成功的各个渠道金额信息，格式为[{}]
     */
    private String fund_bill_list;
    /**
     * 优惠券信息，格式为[{}]
     */
```

```
    private String voucher_detail_list;
    /**
     * 公共回传参数，如果请求时传递了该参数，则在返给商户时会在异步通知时原样返回该参数
     */
    private String passback_params;
}
```

上述请求参数对象，完全以“支付宝异步支付结果通知”接口的定义为准。在实际开发中，遵循具体的支付渠道规范即可。

6.5.2 开发业务层（Service 层）代码

接下来开发“支付微服务”业务层（Service 层）代码。

1. 定义业务异常处理机制

关于业务层（Service 层）的业务异常处理机制，可以参考 4.4.2 节中“1.”小标题中的内容。

定义的业务层（Service 层）异常码的枚举类，具体代码如下：

```
package com.wudimanong.payment.entity;
public enum BusinessCodeEnum {
    /**
     * “支付微服务”内部错误逻辑返回码定义（以 1000 开头，根据业务扩展）
     */
    BUSI_PAY_FAIL_1000(1000, "支付已成功，请勿重复支付"),
    BUSI_PAY_FAIL_1001(1001, "支付请求处理中，请稍后重试"),
    /**
     * 支付渠道错误码封装（以 2000 开头，根据业务扩展）
     */
    BUSI_CHANNEL_FAIL_2000(2000, "支付宝报文组装错误");
    /**
     * 编码
     */
    private Integer code;
    /**
     * 描述
     */
    private String desc;
    BusinessCodeEnum(Integer code, String desc) {
        this.code = code;
        this.desc = desc;
    }
    /**
     * 根据编码获取枚举类型
     */
    public static BusinessCodeEnum getByCode(String code) {
```

```
        //判断是否为空
        if (code == null) {
            return null;
        }
        //循环处理
        BusinessCodeEnum[] values = BusinessCodeEnum.values();
        for (BusinessCodeEnum value : values) {
            if (value.getCode().equals(code)) {
                return value;
            }
        }
        return null;
    }
    public Integer getCode() {
        return code;
    }
    public String getDesc() {
        return desc;
    }
}
```

此枚举类可根据具体的业务层（Service 层）逻辑进行扩展，例如，将“支付微服务”内部的业务异常码的取值范围约定为 1000～1999，而将与具体支付渠道相关的业务异常码约定为 2000～2999。

2. 引入 MapStruct 实体映射工具

具体参考 4.4.2 节中“2.”小标题中的内容。

3. 开发“统一支付”接口的业务层（Service 层）代码

接下来开发“统一支付”接口的业务层（Service 层）代码。

（1）定义业务层（Service 层）接口类 PayService。代码如下：

```
package com.wudimanong.payment.service;
import com.wudimanong.payment.entity.bo.UnifiedPayBO;
import com.wudimanong.payment.entity.dto.UnifiedPayDTO;
public interface PayService {
    /**
     * 定义“统一支付”接口的业务层（Service 层）方法
     *
     * @param unifiedPayDTO
     * @return
     */
    UnifiedPayBO unifiedPay(UnifiedPayDTO unifiedPayDTO);
}
```

（2）实现业务层（Service 层）接口类的方法。代码如下：

```
package com.wudimanong.payment.service.impl;
import com.wudimanong.payment.convert.UnifiedPayConvert;
...
import org.springframework.integration.redis.util.RedisLockRegistry;
import org.springframework.stereotype.Service;
@Slf4j
@Service
public class PayServiceImpl implements PayService {
    /**
     * 定义分布锁 Redis 锁的前缀
     */
    public final String redisLockPrefix = "pay-order&";
    /**
     * 引入 Redis 分布式锁的依赖
     */
    @Autowired
    private RedisLockRegistry redisLockRegistry;
    /**
     * 支付订单持久层（Dao 层）接口的依赖
     */
    @Autowired
    PayOrderDao payOrderDao;
    /**
     * 支付渠道处理工厂类的依赖
     */
    @Autowired
    PayChannelServiceFactory payChannelServiceFactory;
    @Override
    public UnifiedPayBO unifiedPay(UnifiedPayDTO unifiedPayDTO) {
        //返回数据对象
        UnifiedPayBO unifiedPayBO = null;
        //创建 Redis 分布式锁
        //支付防并发安全逻辑——通过“前缀 + 接入方业务订单号”获取 Redis 分布式锁（同
一笔订单，同一时刻只允许一个线程处理）
        Lock lock = redisLockRegistry.obtain(redisLockPrefix +
unifiedPayDTO.getOrderId());
        //持有锁，等待时间为 1s
        boolean isLock = false;
        try {
            isLock = lock.tryLock(1, TimeUnit.SECONDS);
        } catch (InterruptedException e) {
            e.printStackTrace();
        }
```

```
        if (isLock) {
            //数据库级别订单状态防重判断
            boolean isRepeatPayOrder = isSuccessPayOrder(unifiedPayDTO);
            if (isRepeatPayOrder) {
                throw new
ServiceException(BusinessCodeEnum.BUSI_PAY_FAIL_1000.getCode(),
                        BusinessCodeEnum.BUSI_PAY_FAIL_1000.getDesc());
            }
            //支付订单入库
            String payId = this.payOrderSave(unifiedPayDTO);
            //获取具体的支付渠道服务类的实例
            PayChannelService payChannelService = payChannelServiceFactory
                    .createPayChannelService(unifiedPayDTO.getChannel());
            //调用渠道支付方法设置支付平台订单流水号
            unifiedPayDTO.setOrderId(payId);
            unifiedPayBO = payChannelService.pay(unifiedPayDTO);
            //释放分布式锁
            lock.unlock();
        } else {
            //如果持有锁超时，则说明请求正在被处理，提示用户稍后重试
            throw new
ServiceException(BusinessCodeEnum.BUSI_PAY_FAIL_1001.getCode(),
                    BusinessCodeEnum.BUSI_PAY_FAIL_1001.getDesc());
        }
        return unifiedPayBO;
    }
    /**
     * 从数据库级别判断是否为成功支付订单的私有方法
     *
     * @param unifiedPayDTO
     * @return
     */
    private boolean isSuccessPayOrder(UnifiedPayDTO unifiedPayDTO) {
        Map<String, Object> parm = new HashMap<>();
        parm.put("order_id", unifiedPayDTO.getOrderId());
        List<PayOrderPO> payOrderPOList = payOrderDao.selectByMap(parm);
        if (payOrderPOList != null && payOrderPOList.size() > 0) {
            //判断在支付订单中是否存在支付状态为“成功”的订单，若存在，则不处理新的支
付请求
            List<PayOrderPO> successPayOrderList = payOrderPOList.stream()
                    .filter(o ->
"2".equals(o.getStatus())).collect(Collectors.toList());
            if (successPayOrderList != null && successPayOrderList.size() >
0) {
                return true;
```

```
                }
            }
            return false;
        }
        /**
         * 支付订单入库方法
         *
         * @param unifiedPayDTO
         * @return
         */
        private String payOrderSave(UnifiedPayDTO unifiedPayDTO) {
            //用 MapStruct 工具进行实体对象类型转换
            PayOrderPO payOrderPO =
UnifiedPayConvert.INSTANCE.convertPayOrderPO(unifiedPayDTO);
            //设置支付状态为"待支付"
            payOrderPO.setStatus("0");
            //生成支付平台流水号
            String payId = createPayId();
            payOrderPO.setPayId(payId);
            //订单创建时间
            payOrderPO.setCreateTime(new
Timestamp(System.currentTimeMillis()));
            //订单更新时间
            payOrderPO.setUpdateTime(new
Timestamp(System.currentTimeMillis()));
            //订单入库操作
            payOrderDao.insert(payOrderPO);
            return payOrderPO.getPayId();
        }
        /**
         * 生成支付平台订单号
         *
         * @return
         */
        private String createPayId() {
            //获取 10000～99999 的随机数
            Integer random = new Random().nextInt(99999) % (99999 - 10000 + 1) +
10000;
            //时间戳 + 随机数
            String payId = DateUtils.getStringByFormat(new Date(), DateUtils.sf3)
+ String.valueOf(random);
            return payId;
        }
    }
```

上述实现类的代码包含了完整的支付处理逻辑。主要逻辑为：

①使用 Redis 分布式锁来防止用户产生并发的支付请求。

②通过数据库查询来判断支付订单的处理状态，避免重复支付（为了减少主方法逻辑的代码量，这部分逻辑被单独拆分至 isSuccessPayOrder()私有方法中）。

③在发送第三方支付请求前，“支付微服务”会生成支付订单流水。这部分代码被拆分至 payOrderSave()私有方法中。

（3）编写 MapStruct 数据转化类。

在步骤（2）中的 payOrderSave()方法中，生成“持久层（Dao 层）数据对象”使用 MapStruct 工具来减少代码量，涉及的代码片段如下：

```
//使用 MapStruct 工具进行实体数据对象转换
PayOrderPO payOrderPO =
UnifiedPayConvert.INSTANCE.convertPayOrderPO(unifiedPayDTO);
...
```

在上述代码片段中，具体映射转换类的代码如下：

```
package com.wudimanong.payment.convert;
import com.wudimanong.payment.dao.model.PayOrderPO;
...
import org.mapstruct.factory.Mappers;
@org.mapstruct.Mapper
public interface UnifiedPayConvert {
    UnifiedPayConvert INSTANCE =
Mappers.getMapper(UnifiedPayConvert.class);
    /**
     * 生成支付订单“业务层（Service 层）输出数据对象”的转换方法
     *
     * @param unifiedPayDTO
     * @return
     */
    @Mappings({
            @Mapping(target = "extraInfo", ignore = true)
    })
    UnifiedPayBO convertUnifiedPayBO(UnifiedPayDTO unifiedPayDTO);
    /**
     * “支付请求参数对象”到“支付订单持久层（Dao 层）实体类对象” 的转换方法
     *
     * @param unifiedPayDTO
     * @return
     */
    @Mappings({})
```

```
    PayOrderPO convertPayOrderPO(UnifiedPayDTO unifiedPayDTO);
}
```

（4）编写生成支付订单号所需要的日期工具类。

在步骤（2）的逻辑中还需要生成“支付微服务”订单流水。这是在私有方法 createPayId()中通过“日期时间戳 + 随机数”来实现的，其中会依赖一个 DateUtils 日期工具类，具体代码如下：

```
package com.wudimanong.payment.utils;
import java.text.SimpleDateFormat;
import java.util.Date;
import lombok.extern.slf4j.Slf4j;
@Slf4j
public class DateUtils {
    //线程局部变量
    public static final ThreadLocal<SimpleDateFormat> sf1 = new
ThreadLocal<SimpleDateFormat>() {
        @Override
        public SimpleDateFormat initialValue() {
            return new SimpleDateFormat("yyyyMMdd");
        }
    };
    public static final ThreadLocal<SimpleDateFormat> sf2 = new
ThreadLocal<SimpleDateFormat>() {
        @Override
        public SimpleDateFormat initialValue() {
            return new SimpleDateFormat("yyyy-MM-dd");
        }
    };
    public static final ThreadLocal<SimpleDateFormat> sf3 = new
ThreadLocal<SimpleDateFormat>() {
        @Override
        public SimpleDateFormat initialValue() {
            return new SimpleDateFormat("yyyyMMddHHmmss");
        }
    };
    /**
     * 时间格式化方法
     *
     * @param date
     * @param fromat
     * @return
     */
    public static String getStringByFormat(Date date,
ThreadLocal<SimpleDateFormat> fromat) {
        return fromat.get().format(date);
```

```
    }
}
```

（5）编写支付渠道参数处理工厂类。

在完成支付订单入库操作后，需要进行具体的支付渠道请求的处理。考虑到“支付微服务”接入多个支付渠道的场景，所以这里通过定义一个工厂类来实现具体渠道处理代码的寻找，即业务实现类代码所依赖的 PayChannelServiceFactory 类，代码如下：

```
package com.wudimanong.payment.service;
import lombok.extern.slf4j.Slf4j;
import org.springframework.beans.factory.annotation.Autowired;
import org.springframework.stereotype.Service;
@Slf4j
@Service
public class PayChannelServiceFactory {
    @Autowired
    private PayChannelService aliPayServiceImpl;
    /**
     * 根据渠道代码获取具体的渠道业务层（Service 层）处理类
     *
     * @param channelName
     * @return
     */
    public PayChannelService createPayChannelService(int channelName) {
        switch (channelName) {
            case 1:
                return aliPayServiceImpl;
            default:
                return null;
        }
    }
}
```

本实例只接入支付宝渠道，所以上面的工厂类代码只实现了支付宝渠道处理类的实例化，其他渠道可依次扩展。

> 支付宝渠道接入将在 6.6 节演示。本实现类中涉及数据库持久层（Dao 层）操作的代码依赖将在 6.5.3 节中演示。

4. 开发“渠道支付结果通知”接口的业务层（Service 层）代码

业务层（Service 层）主要用来处理渠道支付结果——包括记录支付通知报文、处理支付订单状态，以及向接入方系统（如电子钱包系统）同步支付结果。

接下来，以支付宝支付结果通知为例，开发“渠道支付结果通知”接口的业务层（Service 层）代码。

（1）定义业务层（Service 层）接口类 PayNotifyService。代码如下：

```
package com.wudimanong.payment.service;
import com.wudimanong.payment.entity.dto.AliPayReceiveDTO;
public interface PayNotifyService {
    /**
     * 定义“支付宝支付结果通知回调”接口
     *
     * @param aliPayReceiveDTO
     * @return
     */
    String aliPayReceive(AliPayReceiveDTO aliPayReceiveDTO);
}
```

（2）实现业务层（Service 层）接口类的方法。代码如下：

```
package com.wudimanong.payment.service.impl;
import com.alibaba.fastjson.JSON;
...
import org.springframework.stereotype.Service;
@Slf4j
@Service
public class PayNotifyServiceImpl implements PayNotifyService {
    /**
     * 渠道参数配置信息持久层（Dao 层）的依赖
     */
    @Autowired
    PayChannelParamDao payChannelParamDao;
    /**
     * 支付订单流水持久层（Dao 层）的依赖
     */
    @Autowired
    PayOrderDao payOrderDao;
    /**
     * 渠道支付通知日志持久层（Dao 层）的依赖
     */
    @Autowired
    PayNotifyDao payNotifyDao;
    @Override
    public String aliPayReceive(AliPayReceiveDTO aliPayReceiveDTO) {
        //对报文进行签名验证
        boolean verifyResult = aliPayReceiveMsgVerify(aliPayReceiveDTO);
```

```
        //如果签名验证失败，则直接返回错误信息
        if (!verifyResult) {
            return "sign verify fail";
        }
        //查询支付订单流水信息
        Map<String, Object> paramMap = new HashMap<>();
        paramMap.put("pay_id", aliPayReceiveDTO.getOut_trade_no());
        List<PayOrderPO> payOrderPOList =
payOrderDao.selectByMap(paramMap);
        if (payOrderPOList == null || payOrderPOList.size() <= 0) {
            return "order not exist";
        }
        //如果签名验证成功，则保存支付结果通知报文信息
        PayOrderPO payOrderPO = payOrderPOList.get(0);
        //通过 MapStruct 工具进行数据对象转换
        PayNotifyPO payNotifyPO =
PayNotifyConvert.INSTANCE.convertPayNotifyPO(payOrderPO);
        payNotifyPO.setMerchantId(aliPayReceiveDTO.getApp_id());
        //设置状态为已处理
        payNotifyPO.setReceiveStatus("2");
        //将支付通知报文转为 JSON 格式进行存储
        payNotifyPO.setFullinfo(JSON.toJSONString(aliPayReceiveDTO));
        payNotifyDao.insert(payNotifyPO);
        //更新支付订单状态（这里放到一个事务中，也可以异步解耦处理）
        payOrderPO.setUpdateTime(new
Timestamp(System.currentTimeMillis()));
        payOrderPO.setStatus("2");
        payOrderPO.setTradeNo(aliPayReceiveDTO.getTrade_no());
        payOrderDao.updateById(payOrderPO);
        //向接入方同步支付结果（逻辑暂不实现）
        return "success";
    }
    /**
     * 支付宝支付通知报文签名验证方法
     *
     * @param aliPayReceiveDTO
     * @return
     */
    private boolean aliPayReceiveMsgVerify(AliPayReceiveDTO
aliPayReceiveDTO) {
        //查询支付宝支付 RSA 公钥信息
        QueryWrapper<PayChannelParamPO> queryWrapper = new QueryWrapper<>();
        queryWrapper.and(wq -> wq.eq("partner",
aliPayReceiveDTO.getApp_id()))
```

```
                .and(wq -> wq.eq("status", "0")).and(wq -> wq.eq("key_type",
"publickey"));
            PayChannelParamPO payChannelParamPO =
payChannelParamDao.selectOne(queryWrapper);
            //如果支付参数信息不存在，则直接返回失败
            if (payChannelParamPO == null) {
                return false;
            }
            //将支付参数对象转换为 Map
            Map<String, String> paramMap =
JSON.parseObject(JSON.toJSONString(aliPayReceiveDTO), Map.class);
            //调用支付宝支付 SDK 验证签名
            boolean signVerified = false;
            try {
                signVerified = AlipaySignature
                        .rsaCheckV1(paramMap, payChannelParamPO.getKeyContext(),
"UTF-8", payChannelParamPO.getSignType());
            } catch (AlipayApiException e) {
                e.printStackTrace();
            }
            //由于模拟时需要支付宝私钥签名，所以这里为了便于测试默认返回签名验证成功
            return true;
        }
    }
```

上述实现类代码的主要逻辑为：

①对支付通知报文进行签名验证，在签名验证方法 aliPayReceiveMsgVerify()中使用了支付宝支付 SDK 所提供的相关方法（参考第 6.6 节内容）。

②在签名验证的过程中，需要使用支付宝支付渠道的公钥信息（公钥信息的生成将在 6.6 节介绍），该公钥信息通过“渠道参数信息表”进行存取。

③在签名验证通过后，会将支付通知报文记录至支付通知信息表，并完成支付订单状态处理等业务逻辑。

④通过接入方在支付时传递的“notify_url”通知地址，向接入方系统同步支付结果。

（3）编写 MapStruct 数据转化类。

在步骤（2）的代码中使用 MapStruct 工具来转换生成 PayNotifyPO 数据对象。转换类的代码如下：

```
package com.wudimanong.payment.convert;
import com.wudimanong.payment.dao.model.PayNotifyPO;
...
```

```
import org.mapstruct.factory.Mappers;
@org.mapstruct.Mapper
public interface PayNotifyConvert {
    PayNotifyConvert INSTANCE = Mappers.getMapper(PayNotifyConvert.class);
    /**
     * 支付结果通知报文日志信息的转换方法
     *
     * @param payOrderPO
     * @return
     */
    @Mappings({})
    PayNotifyPO convertPayNotifyPO(PayOrderPO payOrderPO);
}
```

如果支付系统向接入方系统(如电子钱包系统)同步支付结果，则可以直接采用 HTTP 发送，但是需要考虑接入方接收失败的重复发送问题。如果是内部系统交互，则可以考虑通过可靠消息来实现。

6.5.3 开发 MyBatis 持久层（Dao 层）组件

持久层（Dao 层）的实现，以 MyBatis 及 MyBatis-Plus 组合提供的功能为基础。需要先在程序主类中添加 MyBatis 接口代码的包扫描路径，代码如下：

```
package com.wudimanong.payment;
import org.mybatis.spring.annotation.MapperScan;
...
import org.springframework.cloud.client.discovery.EnableDiscoveryClient;
@EnableDiscoveryClient
@SpringBootApplication
@MapperScan("com.wudimanong.payment.dao.mapper")
public class PaymentApplication {
    public static void main(String[] args) {
        SpringApplication.run(PaymentApplication.class, args);
    }
}
```

在上述代码中，通过@MapperScan 注解定义 MyBatis 持久层（Dao 层）组件代码的包路径。

1. 实现“统一支付”接口的持久层（Dao 层）

“统一支付”接口的持久层（Dao 层）主要以“支付订单信息表”（pay_order）的数据库操作为主。

（1）定义“支付订单信息表”的数据库实体类。代码如下：

```
package com.wudimanong.payment.dao.model;
import com.baomidou.mybatisplus.annotation.TableName;
import java.sql.Timestamp;
import lombok.Data;
@Data
@TableName("pay_order")
public class PayOrderPO {
    /**
     * 自增 ID
     */
    private Integer id;
    /**
     * 业务方订单号（需保证在接入方系统内唯一）
     */
    private String orderId;
    /**
     * 业务方交易类型
     */
    private String tradeType;
    /**
     * 支付订单金额
     */
    private Integer amount;
    /**
     * 支付币种
     */
    private String currency;
    /**
     * 支付订单状态。0-待支付；1-支付中；2-支付成功；3-支付失败
     */
    private String status;
    /**
     * 支付渠道编码
     */
    private String channel;
    /**
     * 渠道支付方式
     */
    private String payType;
    /**
     * “支付微服务”的订单流水号
     */
```

```
    private String payId;
    /**
     * 第三方渠道流水号
     */
    private String tradeNo;
    /**
     * 业务方用户 ID
     */
    private String userId;
    /**
     * 支付订单创建时间
     */
    private Timestamp createTime;
    /**
     * 支付订单更新时间
     */
    private Timestamp updateTime;
}
```

（2）定义“支付订单信息表”的持久层（Dao 层）接口。代码如下：

```
package com.wudimanong.payment.dao.mapper;
import com.baomidou.mybatisplus.core.mapper.BaseMapper;
import com.wudimanong.payment.dao.model.PayOrderPO;
import org.springframework.stereotype.Repository;
@Repository
public interface PayOrderDao extends BaseMapper<PayOrderPO> {
}
```

2. 实现“渠道支付结果通知”接口的持久层（Dao 层）

“渠道支付结果通知”接口的持久层（Dao 层）实现，主要以支付通知信息表（pay_notify）和渠道参数信息表（pay_channel_param）的操作为主。

（1）定义“支付通知信息表”的数据库实体类。代码如下：

```
package com.wudimanong.payment.dao.model;
import com.baomidou.mybatisplus.annotation.TableName;
...
import lombok.NoArgsConstructor;
@Data
@Builder
@AllArgsConstructor
@NoArgsConstructor
@TableName("pay_notify")
public class PayNotifyPO {
```

```
/**
 * 主键 ID
 */
private Integer id;
/**
 * 支付订单号
 */
private String payId;
/**
 * 支付渠道
 */
private Integer channel;
/**
 * 支付状态
 */
private String status;
/**
 * 支付通知原始报文信息
 */
private String fullinfo;
/**
 * 业务方订单号
 */
private String orderId;
/**
 * 报文签名验证结果。0-验证成功；1-签名验证失败
 */
private Integer verify;
/**
 * 渠道支付商户号
 */
private String merchantId;
/**
 * 接收处理状态。1-已接收；2-已处理；3-已同步至业务方
 */
private String receiveStatus;
/**
 * 接入方通知次数
 */
private Integer notifyCount;
/**
 * 接入方最近通知时间
 */
private Timestamp notifyTime;
```

```
    /**
     * 更新时间
     */
    private Timestamp updateTime;
    /**
     * 创建时间
     */
    private Timestamp createTime;
}
```

（2）定义“支付通知信息表”的持久层（Dao 层）接口。代码如下：

```
package com.wudimanong.payment.dao.mapper;
import com.baomidou.mybatisplus.core.mapper.BaseMapper;
import com.wudimanong.payment.dao.model.PayNotifyPO;
import org.springframework.stereotype.Repository;
@Repository
public interface PayNotifyDao extends BaseMapper<PayNotifyPO> {
}
```

（3）定义“渠道参数信息表”的数据库实体类。代码如下：

```
package com.wudimanong.payment.dao.model;
import com.baomidou.mybatisplus.annotation.TableName;
import java.sql.Time;
import java.sql.Timestamp;
import lombok.Data;
@Data
@TableName("pay_channel_param")
public class PayChannelParamPO {
    /**
     * 主键
     */
    private Integer id;
    /**
     * 具体的支付渠道账号
     */
    private String partner;
    /**
     * 报文签名类型
     */
    private String signType;
    /**
     * 密钥类型
     */
    private String keyType;
```

```
    /**
     * 证书文本内容
     */
    private String keyContext;
    /**
     * 证书到期时间
     */
    private Timestamp expireTime;
    /**
     * 状态。0-可用，1-不可用
     */
    private String status;
    /**
     * 更新时间
     */
    private Timestamp updateTime;
    /**
     * 创建时间
     */
    private Timestamp createTime;
    /**
     * 备注信息
     */
    private String remark;
}
```

（4）定义“渠道参数信息表”的持久层（Dao 层）接口。代码如下：

```
package com.wudimanong.payment.dao.mapper;
import com.baomidou.mybatisplus.core.mapper.BaseMapper;
import com.wudimanong.payment.dao.model.PayChannelParamPO;
import org.springframework.stereotype.Repository;
@Repository
public interface PayChannelParamDao extends BaseMapper<PayChannelParamPO> {
}
```

（5）初始化支付宝渠道参数信息配置。

“渠道支付结果通知”逻辑所依赖的渠道支付参数信息，可以通过“渠道参数信息表”实现配置化，以使得“支付微服务”具备一定的平台能力。为此，可以将第 6.6 节中获取的支付宝渠道参数信息先存入“渠道参数信息表”，SQL 代码如下：

```
#初始化支付渠道参数配置
INSERT INTO `pay_channel_param`(`id`, `partner`, `sign_type`, `key_type`,
`key_context`, `expire_time`, `status`, `update_time`, `create_time`, `remark`)
VALUES (1, '2016101800715197', 'RSA2', 'publickey',
'MIIBIjANBgkqhkiG9w0BAQEFAAOCAQ8AMIIBCgKCAQEA4Z0RuCT/DAxYzK4A1qU7yPmhEcO5vFo
```

```
os/r9AI2J94BuvE16gR4rH0Xv6j1i7h/KcSnehdIwh2YNBzKbP+I+KCqyaK4fbbJKND5FOj+nWgv
ug8MII+mjHoTtCbt2h95odeTp+e9nU3zRFZw42018d1hwoGJpZwu8a8C8Dsn9tHMSTGhg1UrjJn3
sP69q8eVTRcIQP+EPCsKohYaolXXmqoeevudSrVg5GIcXyXuuJPFGcKkOQo+Fujxj2JZxQcPYXRx
cqPGVT2Q+bvTRA3BKKtALChWU5JbQTM3zMBdGQSDVfd1ipVnLAubzXB/Np6I23fAWywNKWRCWvLQ
Fq146wwIDAQAB', '2021-03-02 10:07:26', '0', '2020-03-02 10:08:51', '2020-03-02
10:07:26', '');
INSERT INTO `pay_channel_param`(`id`, `partner`, `sign_type`, `key_type`,
`key_context`, `expire_time`, `status`, `update_time`, `create_time`, `remark`)
VALUES (2, '2016101800715197', 'RSA2', 'privatekey',
'MIIEvQIBADANBgkqhkiG9w0BAQEFAASCBKcwggSjAgEAAoIBAQDLBQXRgdVI0QfuTRqa5A2P9/L
dSJ7F6KkOfYsz2VOzJ+QyA6C+Lf
    ...
    jS/bqlllIKid3yM/AoNRS9zxCx3EGktfkZrsRwiw8D04hR14CL0tW2nerd6q0JSR0wtHcXXv
4LXz/Xg=', '2021-03-02 10:09:10', '0', '2020-03-02 10:09:49', '2020-03-02
10:09:10', NULL);
```

以上 SQL 代码中的支付渠道参数信息（如加密方式、密钥类型、公钥、私钥等）需根据具体接入的支付渠道来定。上述配置为 6.6 节中演示接入“支付宝”渠道时所获得的支付接入参数。

如果在统一支付逻辑中实现了配置化路由功能，则可以将其与“渠道参数信息表”结合起来。由于篇幅的关系，本实例就不实现了，读者可以参考“支付渠道路由配置表”的设计自行扩展。

6.6 步骤 4：接入“支付宝”渠道

在接入第三方支付渠道时，接入方需要具备一定的资质。为了方便实验，可以通过“支付宝开放平台”提供的沙箱环境进行测试开发。

6.6.1 申请支付宝沙箱环境

具体步骤如下：

（1）进入支付宝开放平台，单击“开发平台-开发者中心-沙箱环境”，系统会自动创建一个沙箱应用，如图 6-9 所示。

图 6-9

（2）通过官方提供的 RSA 工具生成支付接入所需的 RSA 密钥，如图 6-10 所示。

图 6-10

（3）将 RSA 工具生成的应用公钥配置到步骤（1）创建的沙箱应用中，如图 6-11 所示。

图 6-11

单击“保存设置”按钮。此时，系统除保存应用公钥外，还会自动生成并复制支付宝公钥信息，用于后续支付宝渠道返回报文的签名验证。

（4）将支付宝支付参数信息配置到微服务工程的 application.yml 文件中。代码如下：

```
#支付宝渠道参数配置
channel:
  alipay:
    #沙箱应用 AppId
    appId: 201610180071XXX
    #沙箱应用私钥（公钥需要上传至支付宝后台）
    privateKey:
MIIEvQIBADANBgkqhkiG9w0BAQEFAASCBKcwggSjAgEAAoIBAQDLBQXRgdVI0QfuTR...
    #在上传应用私钥时，支付宝自动配对生成的支付宝公钥
    publicKey: MIIBIjANBgkqhkiG9w0BAQEFAAOCAQ8AMIIBCgKCAQEA4Z0RuC...
    #支付宝沙箱环境支付网关地址
    payUrl: https://XXX.alipaydev.com/gateway.do
```

以上支付参数信息是作者用自己的支付宝账号申请的沙箱应用。读者可以参考支付宝官方网站依据实际情况来配置。

以上配置在 6.5.3 节的“2.”小标题中已经配置到“渠道参数信息表”中了。如果“统一支付”接口也像“渠道支付结果通知”接口那样使用“渠道参数信息表”的逻辑来获取支付参数配置，则步骤（3）可以省略。

6.6.2 开发接入支付宝支付的代码

接下来以接入“支付宝电脑网站支付”为例，开发接入支付宝渠道的具体代码。

（1）在“支付微服务”工程的 pom.xml 文件中，引入支付宝支付服务端 SDK。代码如下：

```
<!--引入支付宝支付服务端 SDK -->
<dependency>
    <groupId>com.alipay.sdk</groupId>
    <artifactId>alipay-sdk-java</artifactId>
    <version>4.9.13.ALL</version>
</dependency>
```

支付宝为了方便开发者接入，提供了支持各类主流语言的服务端 SDK。对于 Java 版本的 SDK，可以直接从 Maven 仓库引入。

（2）编写渠道支付业务层（Service 层）接口类。代码如下：

```
package com.wudimanong.payment.service;
import com.wudimanong.payment.entity.bo.UnifiedPayBO;
import com.wudimanong.payment.entity.dto.UnifiedPayDTO;
public interface PayChannelService {
    /**
     * 定义渠道支付业务层（Service 层）方法
     *
     * @param unifiedPayDTO
     * @return
     */
    UnifiedPayBO pay(UnifiedPayDTO unifiedPayDTO);
}
```

（3）编写支付宝电脑网页支付实现类。

“支付微服务”一般会对接多个支付渠道。在步骤（2）中定义的支付接口是一个通用的定义，不同的支付渠道可以通过不同的实现类来实现。“支付宝电脑网页支付”的实现类代码如下：

```
package com.wudimanong.payment.service.impl;
import com.alibaba.fastjson.JSON;
...
import org.springframework.stereotype.Service;
@Service
public class AliPayServiceImpl implements PayChannelService {
    /**
     * “支付网关”接口的地址
     */
    @Value("${channel.alipay.payUrl}")
    private String payUrl;
    /**
     * 支付宝应用 ID
```

```
     */
    @Value("${channel.alipay.appId}")
    private String appId;
    /**
     * 支付宝应用私钥
     */
    @Value("${channel.alipay.privateKey}")
    private String privateKey;
    /**
     * 支付宝应用公钥
     */
    @Value("${channel.alipay.publicKey}")
    private String publicKey;
    private String format = "json";
    private String charset = "UTF-8";
    private String signType = "RSA2";
    @Override
    public UnifiedPayBO pay(UnifiedPayDTO unifiedPayDTO) {
        //获得初始化的 AlipayClient
        AlipayClient alipayClient = new DefaultAlipayClient(payUrl, appId,
privateKey, format, charset, publicKey,
                signType);
        //创建 API 对应的 request
        AlipayTradePagePayRequest alipayRequest = new
AlipayTradePagePayRequest();
        //在公共参数中设置同步跳转地址和异步支付结果通知地址
        alipayRequest.setReturnUrl(unifiedPayDTO.getReturnUrl());
        alipayRequest.setNotifyUrl(alipayRequest.getNotifyUrl());
        //填充业务参数（参考具体支付产品的请求参数要求）
        BizContent bizContent = BizContent.builder().out_trade_no
(String.valueOf(unifiedPayDTO.getOrderId()))
                .product_code("FAST_INSTANT_TRADE_PAY").total_amount(Double
.valueOf(unifiedPayDTO.getAmount()) / 100)
                .subject(unifiedPayDTO.getSubject()).body(unifiedPayDTO.get
Body())
                .passback_params("merchantBizType%" +
unifiedPayDTO.getTradeType())
                .build();
        alipayRequest.setBizContent(JSON.toJSONString(bizContent));
        //用户支付宝网页跳转携带的“form”表单信息
        String form = "";
        try {
            //调用 SDK 生成支付请求参数
            form = alipayClient.pageExecute(alipayRequest).getBody();
        } catch (AlipayApiException e) {
```

```
                //将支付渠道错误封装为系统可识别的异常码
                throw new
ServiceException(BusinessCodeEnum.BUSI_CHANNEL_FAIL_2000.getCode(),
                        BusinessCodeEnum.BUSI_CHANNEL_FAIL_2000.getDesc(), e);
            }
            return UnifiedPayBO.builder().orderId(unifiedPayDTO.getOrderId()).
extraInfo(form).build();
        }
        /**
         * 此内部类用于封装支付宝请求参数中的业务参数
         */
        @Data
        @Builder
        @NoArgsConstructor
        @AllArgsConstructor
        static class BizContent {
            private String out_trade_no;
            private String product_code;
            private Double total_amount;
            private String subject;
            private String body;
            private String passback_params;
        }
    }
```

上述代码完成了支付宝电脑网页支付方式的接入。“支付微服务”在处理这种支付方式时，会给前端返回一个携带“form”表单数据的提交链接（上述代码的实现为 UnifiedPayBO 对象的 extraInfo 字段）。

前端在接收到“统一支付”接口返回的“form”表单数据后，将用户浏览器重定向至支付宝支付界面（参考第 5.6.5 节中“2.”小标题中的效果）。

上述接入支付渠道的逻辑，是 6.5.2 节“3.”小标题中实现“统一支付”接口的业务层（Service 层）时所依赖的逻辑。

6.6.3 测试“支付宝电脑网页支付”接口

如果存在支付收银台，则可以直接看到支付界面的跳转效果。这里为了方便测试，直接通过 Postman 来调用“统一支付”接口。获取的“form”表单提交数据（返回信息中的 extraInfo 字段）如图 6-12 所示。

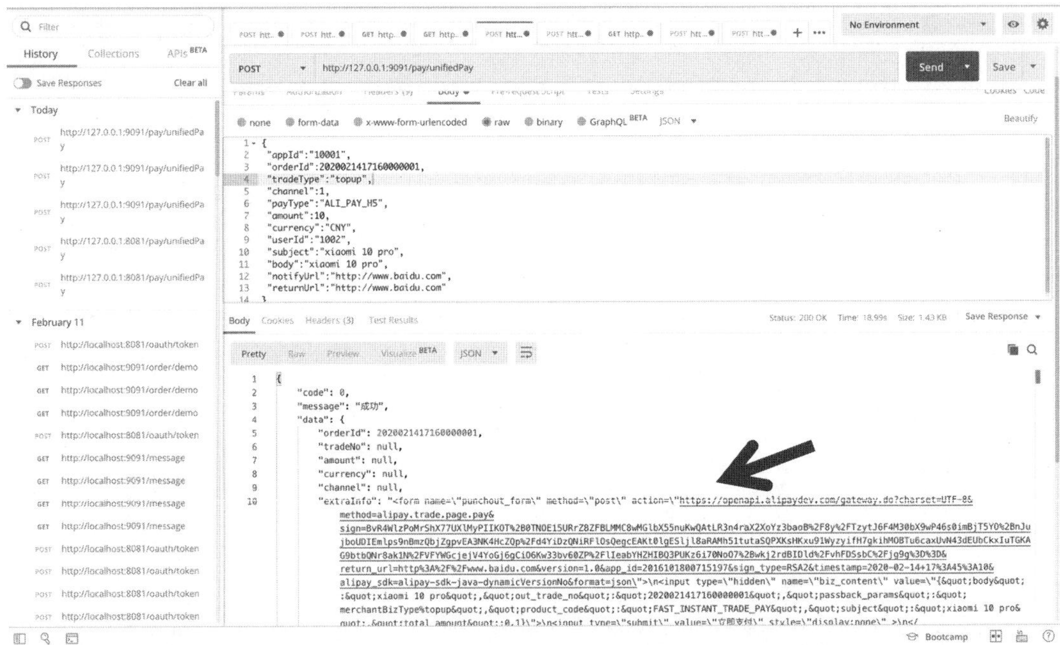

图 6-12

具体的支付请求参数如下：

```
{
 "appId":"10001",
 "orderId":20200214171600000001,
 "tradeType":"topup",
 "channel":1,
 "payType":"ALI_PAY_H5",
 "amount":10,
 "currency":"CNY",
 "userId":"1002",
 "subject":"xiaomi 10 pro",
 "body":"xiaomi 10 pro",
 "notifyUrl":"http://www.baidu.com",
 "returnUrl":"http://www.baidu.com"
}
```

此时，可以复制“form”表单数据到一个 HTML 页面中来演示跳转到支付宝支付界面的效果，具体是指，新建一个叫作 alipay.html 的空页面，然后将“统一支付”接口返回报文中的“extraInfo”字段数据整理、复制到 alipay.html 页面中，代码如下：

```
<form name="punchout_form" method="post"
action="https://openapi.alipaydev.com/gateway.do?charset=UTF-8&method=alipay
```

```
.trade.page.pay&sign=BvR4WlzPoMrShX77UXlMyPIIKOT%2B0TNOE15URrZ8ZFBLMMC8wMGlb
X55nuKwQAtLR3n4raX2XoYz3baoB%2F8y%2FTzytJ6F4M30bX9wP46s0imBjT5YO%2BnJujboUDI
Emlps9nBmzQbjZgpvEA3NK4HcZQp%2Fd4YiDzQNiRFlOsQegcEAKt0lgESljl8aRAMh51tutaSQP
XKsHKxu91WyzyifH7gkihMOBTu6caxUvN43dEUbCkxIuTGKAG9btbQNr8ak1N%2FVFYWGcjejV4Y
oGj6gCiO6Kw33bv60ZP%2FlIeabYHZHIBQ3PUKz6i70NoO7%2Bwkj2rdBIDld%2FvhFDSsbC%2Fj
g9g%3D%3D&return_url=http%3A%2F%2Fwww.baidu.com&version=1.0&app_id=201610180
0715197&sign_type=RSA2&timestamp=2020-02-14+17%3A45%3A10&alipay_sdk=alipay-s
dk-java-dynamicVersionNo&format=json">
    <input type="hidden" name="biz_content"
value="{&quot;body&quot;:&quot;xiaomi 10
pro&quot;,&quot;out_trade_no&quot;:&quot;2020021417160000001&quot;,&quot;pas
sback_params&quot;:&quot;merchantBizType%topup&quot;,&quot;product_code&quot
;:&quot;FAST_INSTANT_TRADE_PAY&quot;,&quot;subject&quot;:&quot;xiaomi 10
pro&quot;,&quot;total_amount&quot;:0.1}">
    <input type="submit" value="立即支付" style="display:none">
   </form>
   <script>
    document.forms[0].submit();
   </script>
```

用浏览器打开该页面，跳转支付宝支付界面的效果如图 6-13 所示。

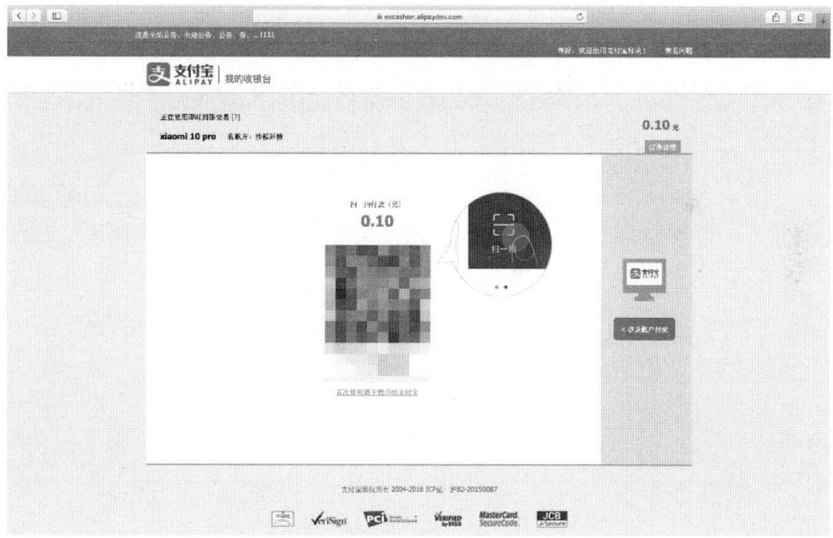

图 6-13

由于沙箱账号的原因，这里不能真正完成用户付款操作，但是实际的支付效果及流程就是这样的。

6.6.4 测试支付宝“渠道支付结果通知”的逻辑

如果用户完成支付，并且“支付微服务”传递给支付宝的通知地址访问正常，则可以实现一个完整的支付流程。

但由于网络、主体资质等客观条件的限制，为了方便测试，这里通过 Postman 来模拟支付宝支付结果通知的回调，请求参数如下：

```
    http://127.0.0.1:9092/notify/aliPayReceive?subject=PC 网站支付交易
&trade_no=2020101221001004580200203978&gmt_create=2020-03-01
21:36:12&notify_type=trade_status_sync&total_amount=0.01&out_trade_no=202002
2916461796172&seller_id=2088201909970555&notify_time=2020-03-01
21:36:12&trade_status=TRADE_SUCCESS&gmt_payment=2020-03-01
21:36:12&passback_params=passback_params123&buyer_id=2088102114562585&app_id
=2016101800715197&notify_id=7676a2e1e4e737cff30015c4b7b55e3kh6&
sign_type=RSA2&sign=oE4ywj/NOF3JPhQg93Zdam/36VadLj9RTqhPXe0OnkpeNeVaTCUL5qhU
2HCdcJvvAzX5dEA8mU3w9pAErbbJ9tUDb8pvXNRtdPfIQxOOFBd5nRuCZ13eQ3Y4IbD+scoDUdO1
9JYoRZdOTaJpmIcc+hiHeb+eaflF4XncbP2dkBXN3AkPURrHbZb6+sRGmYatDziFjpXypkWKB1HN
5FI/BtTlKaCf8U9Ut6XcG0AiOaStMCJLwpO3BvH9ReIo5MFdQb68vNfKnwDKBz6N+rvJFiFdrjSY
4fdp5hJFJzz8IQKVYfTQvqscVCHt9xgnHzL7esoMKXRIos6JGGNkcqbBNQASDF==
```

以上为支付宝“渠道支付结果通知”的模拟报文，其中的订单信息可根据实际测试的信息填充。通过 Postman 工具访问的效果如图 6-14 所示。

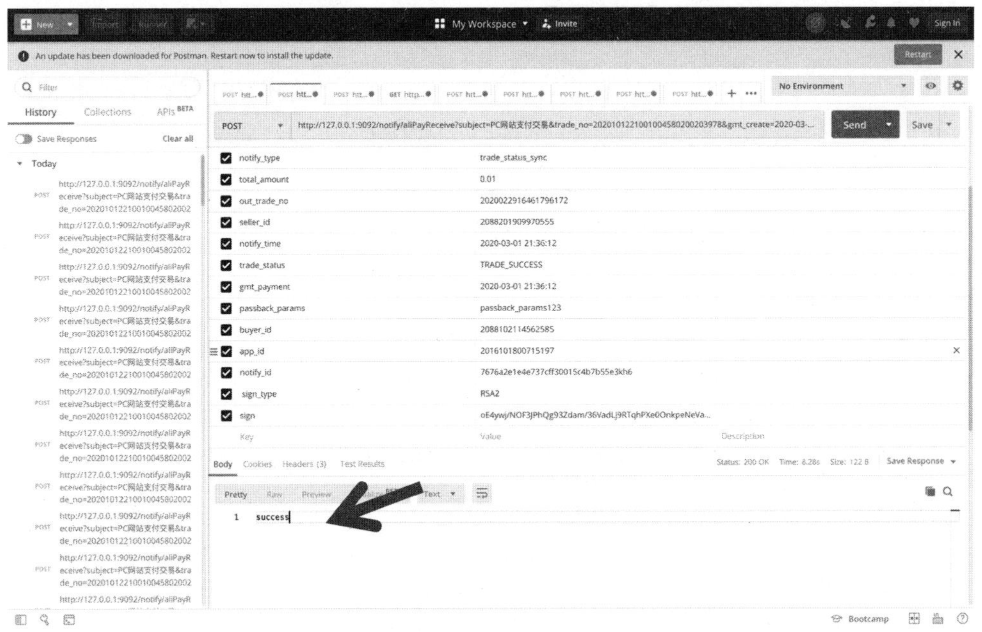

图 6-14

如图 6-14 所示，如果处理成功，则向支付宝响应“success”信息，这样支付宝就会认为通知成功，否则会以一定频率重复通知。

6.7　步骤 5：进行 Spring Cloud 微服务代码单元测试

在日常的编程中，为了保证软件的质量需要做很多测试工作，比较常见的是使用 Postman 这样的 HTTP 测试工具直接进行接口网络测试。但是，这样的测试方式需要启动应用，并且还需要依赖数据库、注册中心及其他中间件服务，测试起来并不方便，并且覆盖率也难以保证。

因此，在一些编程规范的公司中，会要求程序员编写用例覆盖率尽可能高的单元测试代码。本节将演示 Spring Cloud 微服务单元测试代码的编写方法。

6.7.1　认识单元测试

系统开发测试的大致流程是：“编码完成 → UnitTest（单元测试）→ IntegrationTest（集成测试）→ QA 测试 → 发布上线”。开发人员应该进行充分的单元测试，以免将过多的问题留到 QA 测试阶段，从而影响软件的迭代周期。

在 6.5 节实现的业务逻辑中，在服务接口层（Controller 层）、业务层（Service 层）及持久层（Dao 层）之间都是通过依赖注入接口的方式实现彼此调用的——例如，Controller 层依赖 Service 层、Service 层依赖 Dao 层及其他第三方组件。

单元测试分别对局部代码的逻辑进行验证，需要分别对服务接口层（Controller 层）、业务层（Service 层）及持久层（Dao 层）的代码进行单元测试，而对于彼此之间的依赖注入在编写单元测试时需要进行 Mock（模拟）。

在 Spring Cloud 微服务中，可以引入 Mockito 框架来实现对依赖对象的模拟。在微服务工程的 pom.xml 文件中引入如下依赖：

```
<!--单元测试的依赖-->
<dependency>
    <groupId>org.springframework.boot</groupId>
    <artifactId>spring-boot-starter-test</artifactId>
    <scope>test</scope>
</dependency>
```

引入该依赖后会默认引入测试框架 Mockito。使用 Mockito 进行单元测试的一般步骤如图 6-15 所示。

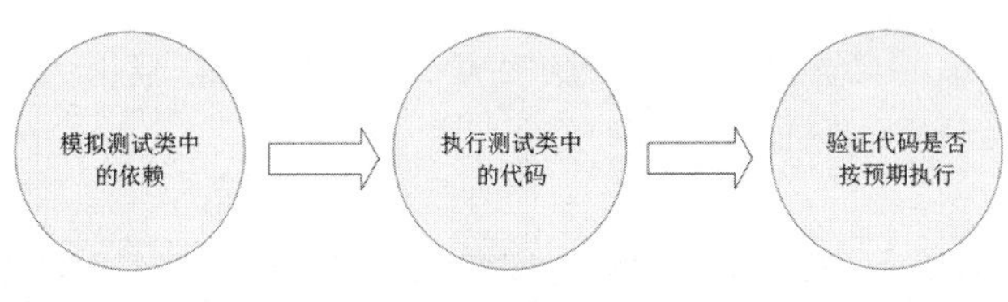

图 6-15

6.7.2 开发 Mockito 单元测试代码

本节以“统一支付”接口的业务层（Service 层）实现类为例，来演示使用 Mockito 开发单元测试代码的方法。

按照工程规范约定，单元测试代码需要放在工程的“src/test”目录下，并根据被测试代码所在目录建立同级子包，如图 6-16 所示。

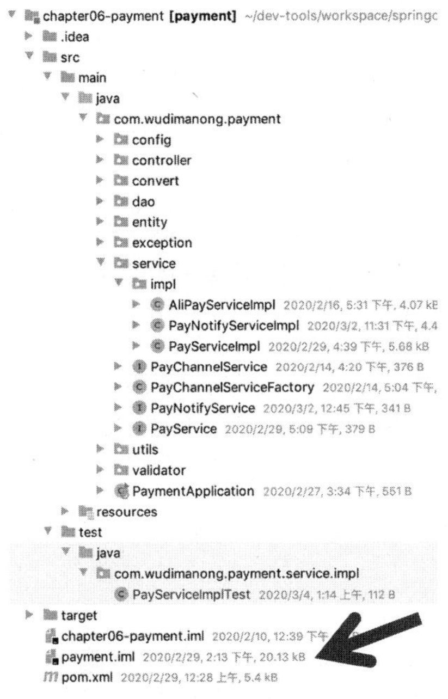

图 6-16

如图 6-16 所示，针对 PayServiceImpl 实现类的单元测试，需要在工程“src/test”包路径中创建对应的包结构，并编写具体的测试类（测试类名为“类名 + Test”）。之后，就可以在测试类中编写具体的单元测试方法了，代码如下：

```
package com.wudimanong.payment.service.impl;
import static org.mockito.ArgumentMatchers.any;
...
```

```
import org.springframework.test.context.junit4.SpringRunner;
@RunWith(SpringRunner.class)
@SpringBootTest(classes = {PayServiceImpl.class})
@ActiveProfiles("test")
public class PayServiceImplTest {
    /**
     * 目标测试类的实例依赖
     */
    @Autowired
    PayServiceImpl payServiceImpl;
    /**
     * 通过 Mockito 框架的@MockBean 注解来模拟 Redis 分布式锁的依赖对象
     */
    @MockBean
    private RedisLockRegistry redisLockRegistry;
    /**
     * 模拟支付订单持久层（Dao 层）的依赖接口
     */
    @MockBean
    PayOrderDao payOrderDao;
    /**
     * 模拟渠道处理工厂类的对象
     */
    @MockBean
    PayChannelServiceFactory payChannelServiceFactory;
    /**
     * 模拟支付宝渠道处理类的对象
     */
    @MockBean
    PayChannelService aliPayServiceImpl;
    /**
     * "统一支付"接口的业务层（Service 层）方法的单元测试方法
     */
    @Test
    public void unifiedPay() {
        //模拟生成请求参数对象
        UnifiedPayDTO unifiedPayDTO = new UnifiedPayDTO();
        unifiedPayDTO.setOrderId("2020021417160000001");
        unifiedPayDTO.setAppId("10001");
        unifiedPayDTO.setTradeType("topup");
        unifiedPayDTO.setChannel(1);
        unifiedPayDTO.setPayType("ALI_PAY_H5");
        unifiedPayDTO.setAmount(10);
        unifiedPayDTO.setCurrency("CNY");
        unifiedPayDTO.setUserId("1002");
```

```
        unifiedPayDTO.setSubject("xiaomi 10 pro");
        unifiedPayDTO.setBody("xiaomi 10 pro");
        unifiedPayDTO.setNotifyUrl("http://www.baidu.com");
        unifiedPayDTO.setReturnUrl("http://www.baidu.com");
        //模拟 Redis 分布式锁的对象行为
        given(redisLockRegistry.obtain(any(String.class))).willReturn(new
Lock() {
            @Override
            public void lock() {
            }
            @Override
            public void lockInterruptibly() throws InterruptedException {
            }
            @Override
            public boolean tryLock() {
                return true;
            }
            @Override
            public boolean tryLock(long time, TimeUnit unit) throws
InterruptedException {
                return true;
            }
            @Override
            public void unlock() {
            }
            @Override
            public Condition newCondition() {
                return null;
            }
        });
        //1  模拟在持久层（Dao 层）依赖方法执行时返回的支付订单数据对象
        given(payOrderDao.selectByMap(any(Map.class))).willReturn(null);
        //2  模拟渠道 Service 工厂类返回的渠道处理实例对象
        given(payChannelServiceFactory.createPayChannelService
(any(Integer.class))).willReturn(aliPayServiceImpl);
        //3  执行单元测试代码
        payServiceImpl.unifiedPay(unifiedPayDTO);
        //4  验证分布式锁获取方法的执行过程
        verify(redisLockRegistry).obtain(any(String.class));
        //5  验证数据库查询方法的执行过程
        verify(payOrderDao).selectByMap(any(Map.class));
        //6  验证支付订单入库逻辑的执行过程
        verify(payOrderDao).insert(any(PayOrderPO.class));
        //7  验证工厂方法的执行过程
```

```
            verify(payChannelServiceFactory).createPayChannelService
(any(Integer.class));
            //8  验证支付方法的执行过程
            verify(aliPayServiceImpl).pay(any(UnifiedPayDTO.class));
        }
    }
```

上述单元测试代码，是针对 PayServiceImpl 类中的 unifiedPay()方法所编写的单元测试方法。说明如下：

（1）在测试类上，通过@RunWith(SpringRunner.class)、@SpringBootTest(classes = {PayServiceImpl.class})注解，指定测试类在 Spring Boot 环境中运行。

（2）在 PayServiceImpl 类中会依赖多个组件——例如 Redis 分布式锁、数据库持久层（Dao 层）等。通过 Mockito 框架提供@MockBean 注解模拟依赖对象。

（3）由于被测试代码中有一些依赖对象的执行方法，所以，在编写单元测试时需要根据逻辑模拟依赖对象的行为。

上述测试代码中的“given(...)willReturn(...)”等子句就是用来模拟依赖对象行为的。

（4）通过模拟依赖对象的行为，被测试代码在执行测试时，可以得到正常的运行的条件，而不需要依赖真的第三方组件或网络，从而尽可能让内部的代码逻辑得到测试。

（5）对于测试的运行结果，通过断言、verify(...)等方式进行验证，以确保测试结果符合预期。

由于测试代码中所依赖的 RedisLockRegistry 类为一个 final Class，而 Mockito 目前的版本虽然支持针对 final Class 类的模拟，但默认是停用的，因此需要在“src/test/resources”目录下手动创建一个名为“org.mockito.plugins.MockMaker”的文件，并在文件中加上如下代码：

```
mock-maker-inline
```

（6）运行单元测试代码，效果如图 6-17 所示。

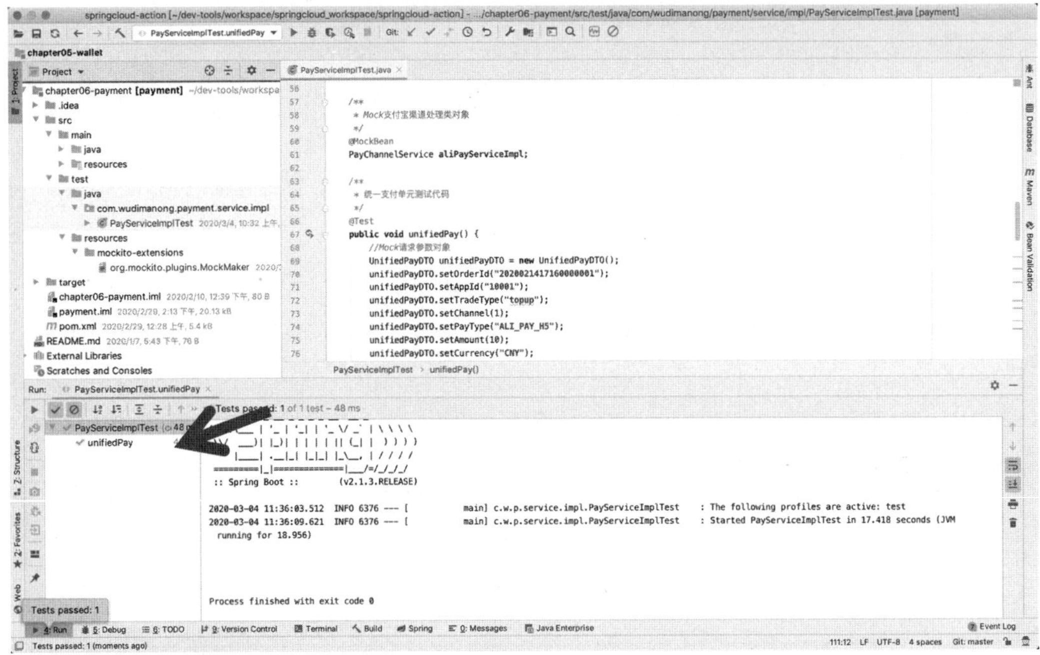

图 6-17

6.8 本章小结

本实例主要演示了 Spring Cloud 微服务架构下“支付微服务”服务端逻辑的构建步骤。对于“电子钱包系统”与“支付系统”交互的演示可同步参考第 5 章的内容。

第 7 章

【实例】A/B 测试系统
——用“Spring Boot Starter 机制 + Caffeine 缓存”实现 A/B 流量切分

在一些以数据驱动为主的互联网公司中，A/B 测试是验证产品设计是否符合预期的重要手段。比如，在互联网产品设计中遇到了不同产品方案或策略的争议——页面上某个按钮的颜色是采用蓝色还是红色，位置是放在左边还是右边等。此时，除依据常规的产品直觉来判断外，还可以采取 A/B 测试这样的科学方法来判断。

从概念上，可以这么理解 A/B 测试：为了验证某个新的产品设计方案、功能点或策略的实际效果，在同一个时段，给两组用户分别展示优化前和优化后的方案，并通过运行过程中上报的系统埋点数据，来分析优化前/后方案在一个或多个评估指标上的差异，从而判断哪个方案更符合预期。

这两组用户分别被称为对照组和实验组。对于用户的分组应足够随机，尽量在统计学上无差别。

为了更好地开展 A/B 测试，并对实验效果及评估指标进行统一的管理，可以建设一个独立的 A/B 测试平台。

本实例将基于 Spring Cloud 微服务构建一个简版的 A/B 测试系统，该系统由 A/B 测试微服务“experiment”和 A/B 测试接入方微服务“experiment-access-demo”组成。

通过本章，读者将学习到以下内容：

- A/B 测试微服务的基本概念。
- Spring Cloud 微服务的构建及组件应用。
- A/B 测试微服务的设计方法及流程。
- 集成高性能本地缓存组件 Caffeine 的方法。
- 编写 Spring Boot Starter“开箱即用”依赖的方法。
- 利用 A/B 测试平台，实现服务端系统的灰度发布。

7.1 功能概述

A/B 测试微服务需要具备以下核心能力。

- 实验管理：提供友好的可视化操作界面，能够让业务系统便捷地创建、查看和修改 A/B 测试信息，实现 A/B 测试的生命周期管理。
- 流量控制：通过流量的自由分配，实现快速地切换新/旧功能的流量接入比例，从而实现灰度发布。
- 指标管理：通过良好的指标设计与管理，提高数据统计分析的效率，从而更好地评估 A/B 测试效果。
- 接入 SDK：A/B 测试微服务能够落地并发挥作用不可缺少的一部分。例如，针对移动 App，需要提供基于 Andriod、iOS 的 SDK 版本；针对前端，需要提供 JS 的 SDK 版本；针对服务端，需要根据实际情况提供多语言版本的服务端“接入 SDK”（如 Java、Go、Python 等）。

7.2 系统设计

A/B 测试的实现理论，主要是基于 Google 发布的论文 *Overlapping Experiment Infrastructure:More, Better, Faster Experimentation*（重叠实验：更多，更好，更快）。所以，在本节的设计中会有部分基于该论文术语的内容，但主要还是从系统实现的角度进行阐述——如系统流程设计、系统结构设计及数据库设计。

7.2.1 系统流程设计

A/B 测试接入方微服务主要通过“接入 SDK”来与 A/B 测试微服务进行交互，来实时获取 A/B 测试及流量配置信息，并通过“接入 SDK”内置的分桶算法来实现流量的分配，如图 7-1 所示。

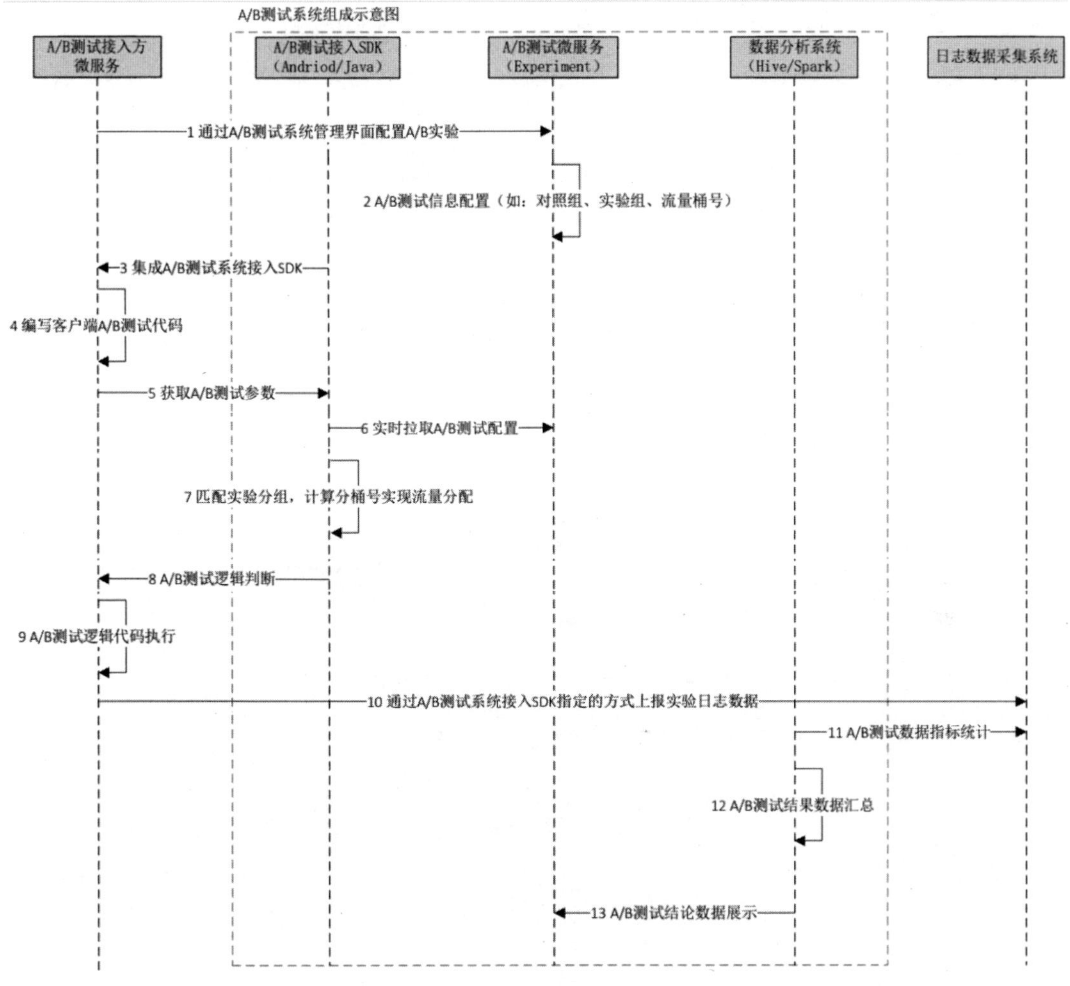

图 7-1

流程说明如下：

（1）A/B 测试接入方微服务通过 A/B 测试微服务提供的操作界面配置 A/B 测试信息——设定好实验组（新逻辑）、对照组（旧逻辑）的流量比例；然后 A/B 测试微服务服务端会根据设置的流量比例，分别计算实验组、对照组对应的流量分桶编号。

（2）A/B 测试接入方微服务通过“接入 SDK”获取 A/B 测试配置信息，并将其缓存至本地进程。

（3）A/B 测试接入方微服务在运行业务逻辑（被 A/B 测试的部分）的过程中，向“接入 SDK”内嵌的流量分桶计算方法传入分流字段（如用户 ID），得到计算出的流量分桶编号。之后，“接入 SDK”会将计算的流量分桶编号与 A/B 测试配置信息中的流量分桶编号列表进行匹配——匹配到哪一组（实验组/对照组），则该流量进入那一组逻辑，以此实现流量控制。

（4）在 A/B 测试逻辑运行的同时，“接入 SDK”会将实验日志同步至数据分析系统。数据分析系统根据设定的指标计算出指标的统计数据。

（5）产品或设计人员通过分析实验数据判断产品设计的真实效果——如果实验组效果更好，则逐步将流量分配至新逻辑；反之则说明新逻辑的产品效果并不符合预期。

7.2.2 系统结构设计

在系统实现上，采用经典的 MVC 分层模式。A/B 测试微服务的结构如图 7-2 所示。

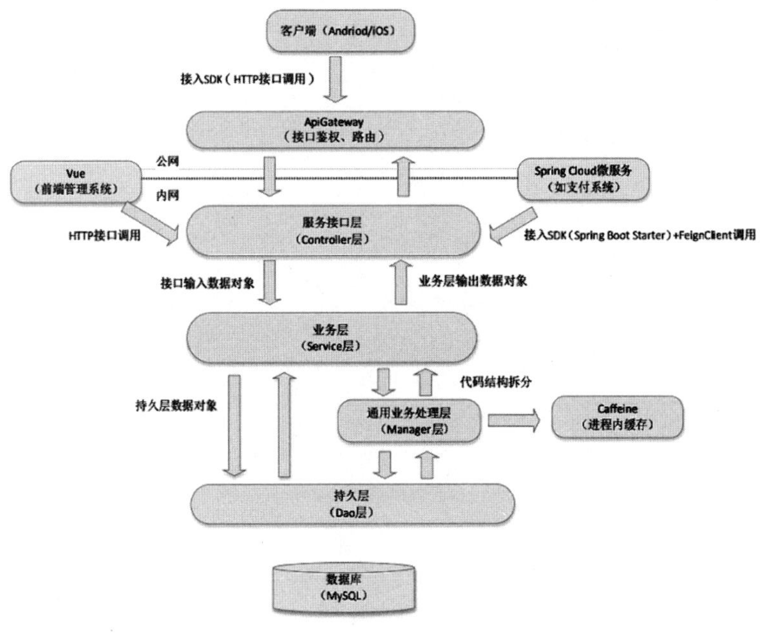

图 7-2

A/B 测试接入方微服务通过“接入 SDK”来接入 A/B 测试微服务，并通过 FeignClient 实现“接入 SDK”与 A/B 测试微服务之间的通信调用。此外，还会通过集成 Caffeine 进程内缓存，来提高 A/B 测试微服务的服务性能。

关于 MVC 分层模式的说明，可参考 6.2.2 节的内容。

7.2.3 数据库设计

下面来定义实现 A/B 测试微服务所需要的表结构。

- 代码 以下表在本书配置资源的“chapter07-experiment/src/main/resources/db.migration”目录下。

1. A/B 测试信息表

“A/B 测试信息表”主要实现 A/B 测试信息的管理，以及 A/B 测试生命周期的控制。具体的 SQL 代码如下：

```
create table abtest_exp_info (
  id int(11) not null auto_increment comment '实验 ID',
  name varchar(255) collate utf8_bin not null comment '实验名称',
  factor_tag varchar(50) collate utf8_bin not null comment '系统对应的业务标签',
  layer_id int(11) not null comment '分层 ID',
  group_field_id int(11) not null comment '分组字段(参数)类型。1-用户; 2-地理位置',
  exp_hyp varchar(255) collate utf8_bin default null comment '实验假设',
  result_expect varchar(255) collate utf8_bin default null comment '实验预期',
  metric_ids varchar(255) collate utf8_bin not null comment '指标 ID 列表',
  start_time datetime default null comment '开始时间',
  end_time datetime default null comment '结束时间',
  status int(4) not null default '0' comment '实验状态。0-新建; 1-已发布; 2-生效中; 3-已暂停; 4-已终止; 5-已结束',
  online tinyint(2) not null default '0' comment '是否已上线。0-未上线; 1-已上线',
  partition_type tinyint(4) not null default '0' comment '分区类型',
  is_sampling tinyint(2) default '0' comment '是否抽样打点。0-否; 1-是',
  sampling_ratio int(4) default '5' comment '抽样率 n%',
  config varchar(1024) collate utf8_bin default null comment '实验配置',
  service_name varchar(50) collate utf8_bin default 'all' comment '系统名称,以逗号分隔, 区分使用端',
  owner varchar(50) collate utf8_bin not null comment '负责人',
  is_delete tinyint(2) not null comment '是否已删除。0-未删除; 1-已删除',
  ext varchar(1000) collate utf8_bin not null default '{}' comment '扩展字段 JSON 格式',
  create_time timestamp not null default current_timestamp comment '记录创建时间, 默认当前时间',
  update_time timestamp not null default current_timestamp on update current_timestamp comment '记录更新时间, 默认当前时间',
  primary key(id)
);
alter table abtest_exp_info comment 'A/B 测试信息表';
```

```
#索引信息
alter table abtest_exp_info add index idx_factor_tag(factor_tag);
alter table abtest_exp_info add index idx_layer(layer_id);
alter table abtest_exp_info add index
idx_partition_type_deleted_status(partition_type,is_delete,status);
```

2. A/B 测试分层表

“A/B 测试分层表”主要用于实现 Google 论文中阐述的分层实验的概念——在多层实验中，通过分层可以实现流量的复用，以验证更多的 A/B 测试场景。具体的 SQL 代码如下：

```
create table abtest_layer (
  id int(11) unsigned not null auto_increment comment '分层ID',
  `name` varchar(255) collate utf8_bin not null comment '名称',
  `desc` varchar(255) collate utf8_bin not null default '' comment '分层描述',
  group_field_id int(11) not null comment '分组字段（参数）类型。1-用户；2-地理
位置',
  bucket_total_num int(11) not null comment '当前层的分桶总数',
  unused_bucket_nos text collate utf8_bin not null comment '未使用的分桶编号
列表',
  partition_type tinyint(4) not null default '0' comment '分区类型',
  update_time datetime not null default current_timestamp on update
current_timestamp comment '更新时间',
  create_time datetime not null default current_timestamp comment '创建时间',
  is_delete tinyint(2) not null comment '是否已删除。0-未删除；1-已删除',
  primary key (id)
);
alter table abtest_layer comment 'A/B测试分层表';
#索引信息
alter table abtest_layer add index
idx_partition_type_group_field(`partition_type`,`group_field_id`);
```

3. A/B 测试分组表

“A/B 测试分组表”是实现流量分配逻辑的关键数据表，主要用于存储 A/B 测试中实验组和对照组的定义，以及流量分桶编号列表。具体的 SQL 代码如下：

```
create table abtest_group (
  id int(11) not null auto_increment comment '分组ID',
  group_type tinyint(4) default null comment '分组类别。0-实验组；1-对照组',
  flow_ratio int(11) not null default '0' comment '分流后，分组流量占比',
  exp_id int(11) default null comment '实验ID',
  name varchar(50) collate utf8_bin default '' comment '分组名称',
  group_partition_type int(11) default null comment '分组类型。0-区间分组；1-
单双号分组',
  group_partition_details text collate utf8_bin comment '分流内包含的编号列表，
用逗号分隔，如：00,09',
```

```
      strategy_detail varchar(1000) collate utf8_bin default null comment '策
略对应的 JSON 格式信息',
      online tinyint(2) not null default '0' comment '是否已上线。0-未上线；1-已
上线',
      create_time timestamp not null default current_timestamp comment '记录创
建时间，默认当前时间',
      update_time timestamp not null default current_timestamp on update
current_timestamp comment '记录更新时间，默认当前时间',
      dilution_ratio int(11) not null default '1' comment '稀释倍率',
      white_list text collate utf8_bin comment '白名单',
      primary key (id)
    );
    alter table abtest_group comment 'A/B 测试分组表';
    #索引信息
    alter table abtest_group add index idx_expid_online(exp_id,online);
```

4. A/B 测试指标表

“A/B 测试指标表”用于指定 A/B 测试需要计算的数据指标，实现 A/B 测试指标的抽象定义及统一管理。具体的 SQL 代码如下：

```
    create table abtest_metric (
      id int(11) not null auto_increment comment '指标 ID',
      `name` varchar(100) collate utf8_bin default null comment '中文名',
      name_en varchar(100) collate utf8_bin default null comment '英文名',
      formula varchar(255) collate utf8_bin default null comment '指标计算公式',
      group_field_id int(11) not null comment '分组因子 ID。0 表示所有分组因子都具
有的指标',
      `desc` varchar(255) collate utf8_bin not null comment '指标描述',
      `status` tinyint(2) not null default '1' comment '是否可用。0-否；1-是',
      create_time datetime not null default current_timestamp comment '创建时间',
      update_time datetime not null default current_timestamp on update
current_timestamp comment '更新时间',
      primary key (id)
    );
    alter table abtest_metric comment 'A/B 测试指标表';
    #索引信息
    alter table abtest_metric add index idx_groupfieldid(group_field_id);
```

7.3 步骤 1：构建 Spring Cloud 微服务工程代码

接下来搭建 A/B 测试微服务所需要的 Spring Cloud 微服务工程，并集成 A/B 测试微服务开发所依赖的组件。

7.3.1 创建 Spring Cloud 微服务工程

1. 创建一个基本的 Maven 工程

利用 2.3.1 节介绍的方法创建一个 Maven 工程，完成后的工程代码结构如图 7-3 所示。

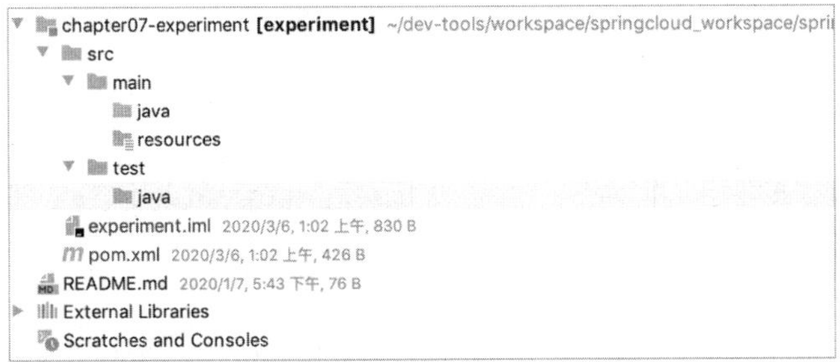

图 7-3

2. 引入 Spring Cloud 依赖，将其改造为微服务项目

（1）引入 Spring Cloud 微服务的核心依赖。

这里可以参考 2.5.2 节中的具体步骤。

（2）在工程代码的 resources 目录新建一个基础性配置文件——bootstrap.yml。配置文件中的代码如下：

```
spring:
  application:
    name: experiment
  profiles:
    active: debug
  cloud:
    consul:
      discovery:
        preferIpAddress: true
        instance-id: ${spring.application.name}:${spring.cloud.client.
ipAddress}:${spring.application.instance_id:${server.port}}:@project.version@
        healthCheckPath: /actuator/health
server:
  port: 9091
```

（3）Spring Boot 并不会默认加载 bootstrap.yml 这个文件，所以需要在 pom.xml 中添加 Maven 资源相关的配置，具体参考 2.5.2 节内容。

（4）创建 A/B 测试微服务“experiment”的入口程序类。代码如下：

```
package com.wudimanong.experiment;
```

```
import org.springframework.boot.SpringApplication;
import org.springframework.boot.autoconfigure.SpringBootApplication;
import org.springframework.cloud.client.discovery.EnableDiscoveryClient;
@EnableDiscoveryClient
@SpringBootApplication
public class ExperimentApplication {
    public static void main(String[] args) {
        SpringApplication.run(ExperimentApplication.class, args);
    }
}
```

至此，A/B 测试微服务所需的 Spring Cloud 微服务工程就构建出来了。

7.3.2 将 Spring Cloud 微服务注入 Consul

参考 2.5.1 节、2.5.3 节的内容，将 A/B 测试微服务“experiment”注入服务注册中心 Consul 中。然后运行所构建的 A/B 测试微服务“experiment”，可以看到该服务已经注册到 Consul 中了，如图 7-4 所示。

打开 Consul 控制台，A/B 测试微服务“experiment”被注册到 Consul 中的效果如图 7-5 所示。

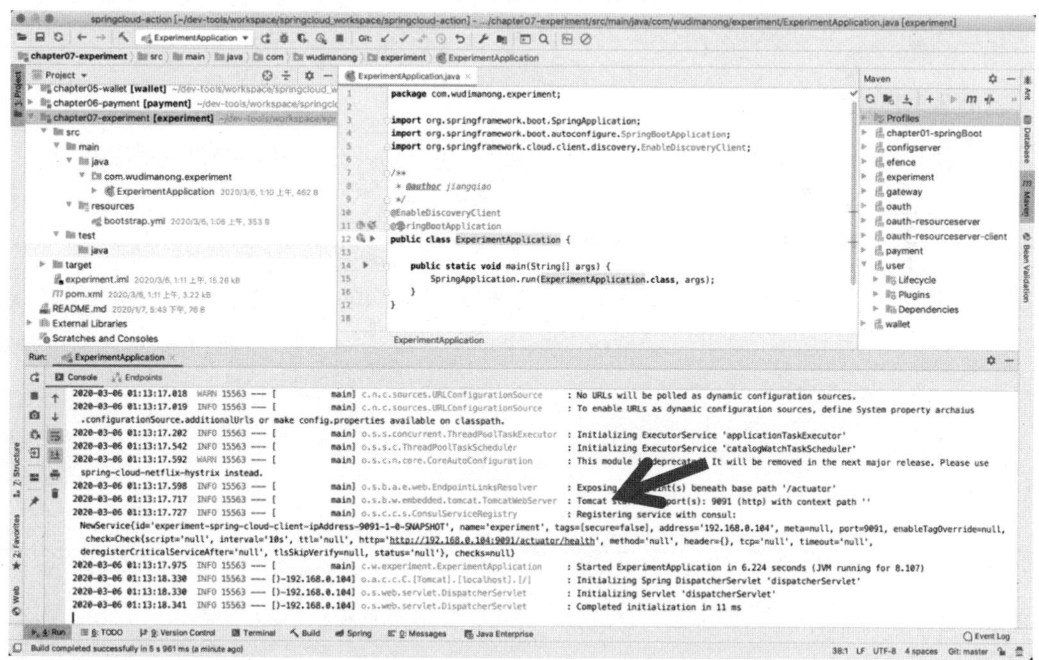

图 7-4

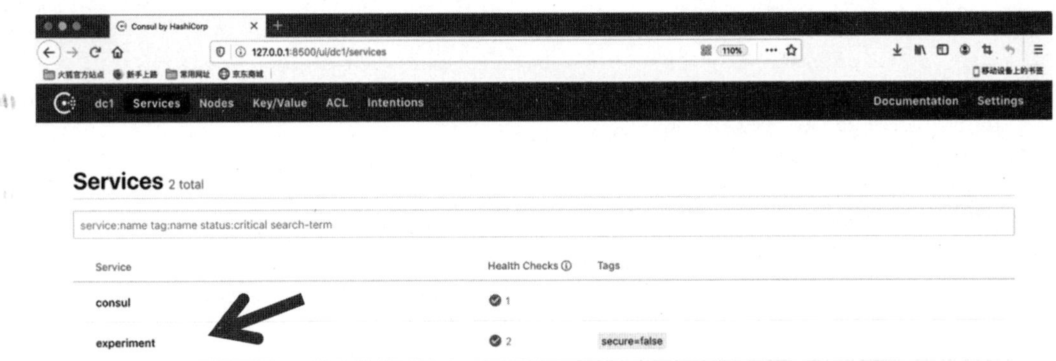

图 7-5

至此，就从技术层面完成了 Spring Cloud 微服务的搭建过程。接下来继续集成开发 A/B 测试微服务“experiment”所依赖的其他组件。

7.3.3 集成 MyBatis，以访问 MySQL 数据库

在本实例中，使用 MyBatis 这个持久层（Dao 层）框架来操作数据库。

1. 引入 MyBatis 框架依赖及 MySQL 数据库驱动程序

具体步骤参考 2.3.3 节内容。

2. 配置项目数据库连接信息

在项目中创建一个新的配置文件 application.yml，添加 MySQL 数据库的连接信息如下：

```
  spring:
    datasource:
      url:
jdbc:mysql://127.0.0.1:3306/abtest?zeroDateTimeBehavior=convertToNull&useUni
code=true&useUnicode=true&characterEncoding=utf-8
      username: root
      password: 123456
      type: com.alibaba.druid.pool.DruidDataSource
      driver-class-name: com.mysql.jdbc.Driver
      separator: //
```

上述配置中涉及的数据库信息，可以参考 1.3.3 节通过 Docker 部署本地 MySQL 的步骤——创建一个名为 abtest 的数据库，并执行 7.2.3 节中所定义的 SQL 脚本。

7.3.4 通过 MyBatis-Plus 简化 MyBatis 的操作

接下来，使用 MyBatis-Plus 框架来简化 MyBatis 的数据库操作。

1. Spring Boot 集成 MyBatis-Plus 框架

具体的集成和配置方法可参考 4.3.5 节内容。

2. 启动项目，验证 MyBatis-Plus 集成效果

启动 A/B 测试微服务“experiment”，其成功集成 MyBatis-Plus 框架的运行效果如图 7-6 所示。

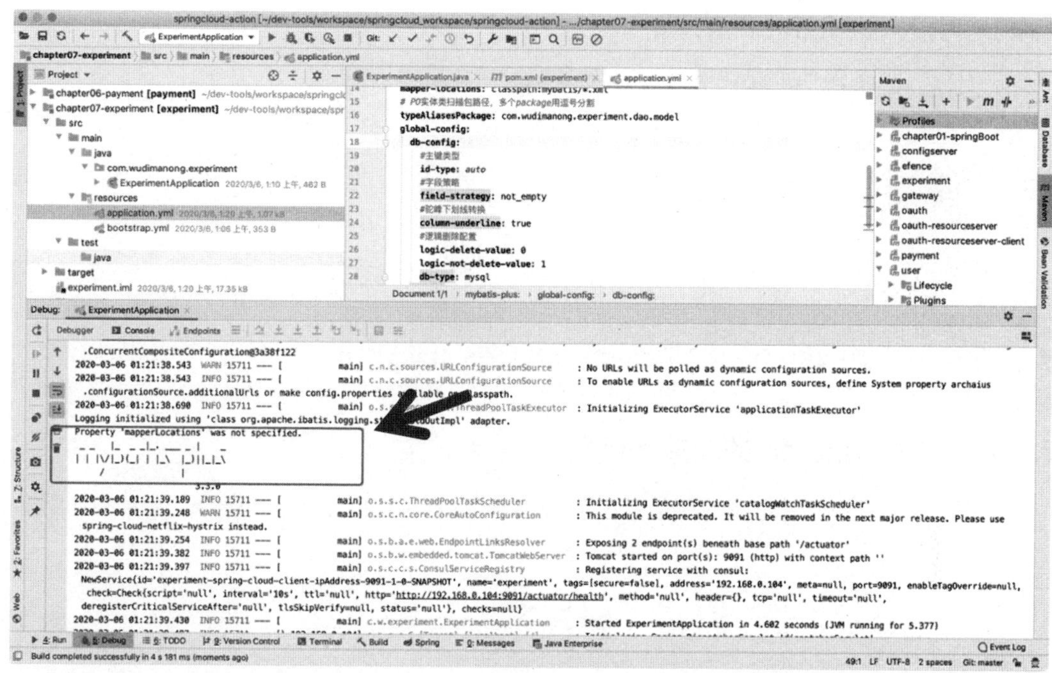

图 7-6

7.4 步骤 2：集成高性能本地缓存 Caffeine

在日常软件开发中，经常会使用缓存技术来降低数据库的访问压力，从而提高服务的性能。主要的缓存方案有：分布式缓存、本地缓存。

（1）分布式缓存，是在分布式服务环境下实现内存全局共享的一种缓存方案。最常见的分布式缓存技术是 Redis。

（2）本地缓存，则是在应用进程内实现内存共享的缓存方案。Java 中常见的本地缓存组件有：EhCache、Guava Cache 和 Caffeine。

- EhCache：ORM 框架 Hibernate 的默认本地缓存组件。
- Guava Cache：一种全内存的本地缓存实现，属于 Google Guava 的一个模块。
- Caffeine：使用 Java 8 对 Guava Cache 的重写版本，基于 LRU 算法实现，支持多种缓存过期策略，用于在 Spring Boot 2.0 及以上版本中取代 Guava Cache。

本实例使用的是 Spring Boot 2.1.3 版本，所以，将通过集成 Caffeine 来提升 A/B 测试配置数据获取的性能。

7.4.1 引入 Caffeine 的依赖

在工程的 pom.xml 文件中引入 Caffeine 的依赖，代码如下：

```
<!--基于 Spring Boot 引入 Caffeine 的依赖-->
<dependency>
    <groupId>org.springframework.boot</groupId>
    <artifactId>spring-boot-starter-cache</artifactId>
</dependency>
<dependency>
    <groupId>com.github.ben-manes.caffeine</groupId>
    <artifactId>caffeine</artifactId>
</dependency>
```

7.4.2 开发 Caffeine 的配置类代码

为了更灵活地控制缓存策略，可以通过系统配置类的方式，实现对 Caffeine 组件的自动配置。代码如下：

```
package com.wudimanong.experiment.config;
import com.github.benmanes.caffeine.cache.Caffeine;
import java.util.ArrayList;
...
import org.springframework.context.annotation.Primary;
//启用缓存
@EnableCaching
@Configuration
public class CacheConfig {
    /**
     * 设置缓存的默认大小
     */
    public static final int DEFAULT_MAXSIZE = 50000;
    /**
     * 设置缓存的默认过期时间（单位：s）
```

```
 */
public static final int DEFAULT_EXPIRE_TIME = 10;
/**
 * 定义缓存的名称、超时时长（s）、最大容量。如果需要修改，则可以在构造方法的参数中修改
 */
public enum Caches {
    //设置一个测试 Caffeine 缓存，缓存有效期 5s
    CAFFEINE_TEST(5, DEFAULT_MAXSIZE),
    //A/B 测试配置信息缓存，缓存有效期 60s
    EXP_CONFIG_INFO(60, DEFAULT_MAXSIZE);
    /**
     * 最大数量
     */
    private int maxSize = DEFAULT_MAXSIZE;
    /**
     * 过期时间（s）
     */
    private int expireTime = DEFAULT_EXPIRE_TIME;
    /**
     * 缓存的构造方法
     */
    Caches(int expireTime, int maxSize) {
        this.expireTime = expireTime;
        this.maxSize = maxSize;
    }
    /**
     * 获取过期时间
     */
    int getExpireTime() {
        return this.expireTime;
    }
    /**
     * 获取缓存大小
     */
    int getMaxSize() {
        return this.maxSize;
    }
}
/**
 * 创建基于 Caffeine 的 Cache Manager（缓存管理器）
 */
@Bean
@Primary
public CacheManager caffeineCacheManager() {
    SimpleCacheManager cacheManager = new SimpleCacheManager();
```

```
        //设置多种不同的缓存策略
        ArrayList<CaffeineCache> caches = new ArrayList<CaffeineCache>();
        for (Caches c : Caches.values()) {
            caches.add(new CaffeineCache(c.name(),
                    Caffeine.newBuilder().recordStats()
                            //从最后一次写入缓存后开始计时，在指定的时间后过期
                            .expireAfterWrite(c.getExpireTime(),
TimeUnit.SECONDS)
                            //缓存的最大容量
                            .maximumSize(c.getMaxSize())
                            .build())
            );
        }
        cacheManager.setCaches(caches);
        return cacheManager;
    }
}
```

上述配置代码的大致逻辑说明如下：

（1）用@EnableCaching 注解开启缓存机制。

（2）定义 Caches 枚举类是为了实现针对具体缓存的策略配置——例如，为了测试 Caffeine 缓存效果，在枚举中添加了一个有效期为 5s 的 CAFFEINE_TEST 缓存；而对于 A/B 测试配置信息，则设置缓存失效时间为 60s 的 EXP_CONFIG_INFO 缓存。

（3）如果有其他的缓存策略设置，则在 Caches 枚举类中进行扩展。

（4）在 Caches 枚举类中定义的缓存策略配置，是通过配置类中的 caffeineCacheManager() 方法创建的 Caffeine 缓存管理器来实现的。

7.4.3 演示 Caffeine 的使用效果

接下来演示本地缓存 Caffeine 的使用效果。

（1）编写一个查询“A/B 测试信息表”的业务层（Service 层）测试类。代码如下：

```
package com.wudimanong.experiment.service;
import org.springframework.cache.annotation.Cacheable;
...
import org.springframework.stereotype.Service;
@Service
public class CaffeineTestService {
    /**
     * “A/B 测试信息表”的持久层（Dao 层）
     */
```

```
        @Autowired
        AbtestExpInfoDao abtestExpInfoDao;
        //以参数 factorTag 为 Key 进行缓存
        @Cacheable(value = "CAFFEINE_TEST", key = "#factorTag", sync = true)
        public AbtestExpInfoPO getExpInfoByFactorTag(String factorTag) {
            //封装查询参数
            AbtestExpInfoPO abtestExpInfoPO = new AbtestExpInfoPO();
            abtestExpInfoPO.setFactorTag(factorTag);
            QueryWrapper<AbtestExpInfoPO> queryWrapper = new
QueryWrapper<>(abtestExpInfoPO);
            //查询 A/B 测试配置信息
            abtestExpInfoPO = abtestExpInfoDao.selectOne(queryWrapper);
            return abtestExpInfoPO;
        }
    }
```

在上方的代码中，通过@Cacheable 注解来使用缓存，其中“value”属性为在 7.4.2 节配置类的枚举中定义的缓存标识；“key”属性的值表示使用 getExpInfoByFactorTag()方法的参数值来作为缓存的键值。

这样，当 getExpInfoByFactorTag()方法被调用时，就会先从缓存中查找数据——如果缓存中存在数据，则直接返回；如果不存在，则在该方法执行完成后将返回结果存入缓存。

以上就是在实际代码中使用 Caffeine 的基本方法。其中，依赖的持久层（Dao 层）代码参考后面 7.5.3 节中的代码。

（2）编写测试接口的服务接口层（Controller 层）。代码如下：

```
    package com.wudimanong.experiment.controller;
    import com.wudimanong.experiment.dao.model.AbtestExpInfoPO;
    ...
    import org.springframework.web.bind.annotation.RestController;
    @Slf4j
    @RestController
    @RequestMapping("/test")
    public class CaffeineTestController {
        /**
         * 注入在步骤（1）中定义的 Service 测试类的实例
         */
        @Autowired
        CaffeineTestService caffeineTestService;
        /**
         * 根据业务标签查询 A/B 测试信息
```

```
     *
     * @param factorTag
     */
    @GetMapping("/findByFactorTag")
    public AbtestExpInfoPO findByFactorTag(@RequestParam("factorTag")
String factorTag) throws InterruptedException {
        Long startTime1 = System.currentTimeMillis();
        AbtestExpInfoPO abtestExpInfoPO =
caffeineTestService.getExpInfoByFactorTag(factorTag);
        long endTime1 = System.currentTimeMillis();
        System.out.println("第 1 次耗时(数据库获取)->" + (endTime1 - startTime1)
+ " 毫秒");
        Long startTime2 = System.currentTimeMillis();
        AbtestExpInfoPO abtestExpInfoPO2 =
caffeineTestService.getExpInfoByFactorTag(factorTag);
        long endTime2 = System.currentTimeMillis();
        System.out.println("第 2 次耗时(从缓存获取)->" + (endTime2 - startTime2)
+ " 毫秒");
        //让线程休眠 5s，以便缓存失效后查看效果
        Thread.sleep(5000);
        Long startTime3 = System.currentTimeMillis();
        AbtestExpInfoPO abtestExpInfoPO3 =
caffeineTestService.getExpInfoByFactorTag(factorTag);
        long endTime3 = System.currentTimeMillis();
        System.out.println("第 3 次耗时(从数据库获取)->" + (endTime3 - startTime3)
+ " 毫秒");
        Long startTime4 = System.currentTimeMillis();
        AbtestExpInfoPO abtestExpInfoPO4 =
caffeineTestService.getExpInfoByFactorTag(factorTag);
        long endTime4 = System.currentTimeMillis();
        System.out.println("第 4 次耗时(从缓存获取)->" + (endTime4 - startTime4)
+ " 毫秒");
        return abtestExpInfoPO4;
    }
}
```

上述测试代码，通过 4 次调用来演示 Caffeine 缓存的运行情况：

- 第 1 次：由于缓存中无数据，所以直接从数据库中查询。
- 第 2 次：由于此时缓存中已有数据，所以直接从缓存中查询，之后线程会休眠 5s。
- 第 3 次：由于缓存的失效时间被设置为 5s，所以再次从数据库中查询。
- 第 4 次：由于缓存中已有数据，所以直接从缓存中查询。

（3）启动微服务程序，分 4 次调用测试接口，查看到的缓存使用情况如图 7-7 所示。

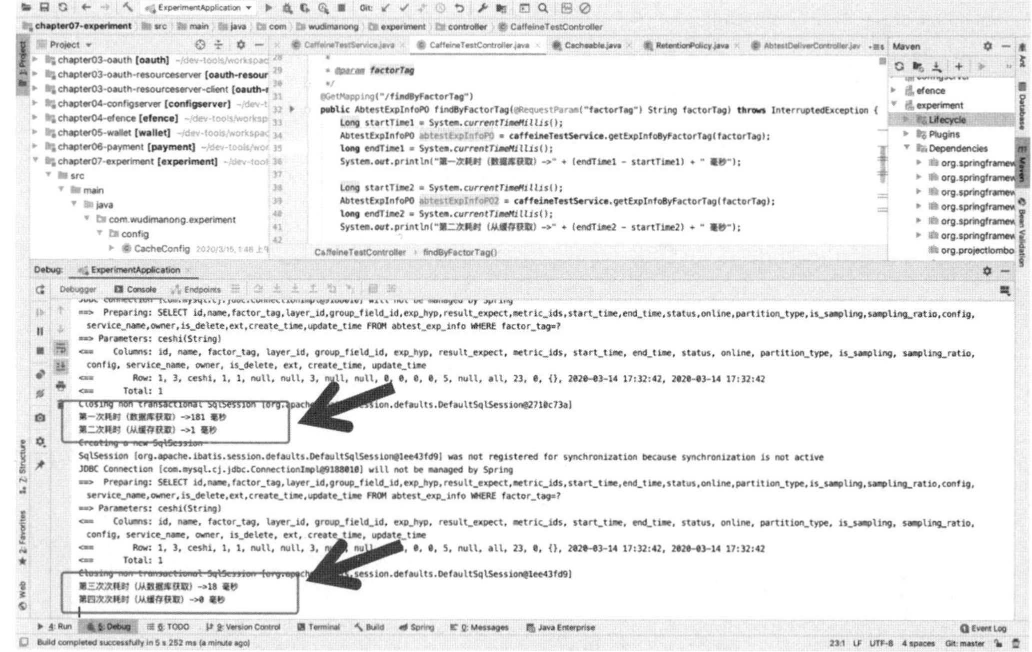

图 7-7

从上述测试程序运行效果来看，缓存逻辑已经生效。

7.5 步骤 3：实现微服务的业务逻辑

接下来将实现 A/B 测试微服务“experiment”的业务逻辑。

7.5.1 定义服务接口层（Controller 层）

在本实例中，A/B 测试微服务“experiment”的服务接口层主要提供两种类型的接口：

- 提供给 A/B 测试接入方微服务 SDK 的数据访问接口。
- 提供给 A/B 测试管理系统的管理性接口。

1. 约定接口数据格式

在定义 Controller 层接口之前，需要先约定接口的请求方式，以及统一的报文格式。具体的规范可以参考 4.1.1 节中“1.”小标题中的内容。

其中涉及的统一报文格式类（ResponseResult）及全局响应码枚举类（GlobalCodeEnum），可以复制 4.1.1 节中“1.”小标题中的代码。

为了定义的接口代码能够被“接入 SDK”复用，这里将 ResponseResult 及 GlobalCodeEnum 等接口格式代码抽象到单独的客户端工程。

（1）创建一个项目名为“experiment-java-client”的客户端工程，如图 7-8 所示。

```
▼ chapter07-experiment-java-client [experiment-java-client] ~/de
  ▶ .idea
  ▼ src
    ▼ main
      ▼ java
          com.wudimanong.experiment.client.entity
        resources
    ▶ test
    experiment-java-client.iml 2020/3/17, 12:07 上午, 830 B
    pom.xml 2020/3/17, 12:02 上午, 438 B
  README.md 2020/1/7, 5:43 下午, 76 B
▶ External Libraries
  Scratches and Consoles
```

图 7-8

（2）在客户端工程的 pom.xml 文件中添加一些基本依赖。代码如下：

```
<?xml version="1.0" encoding="UTF-8"?>
<project xmlns="http://maven.apache.org/POM/4.0.0"
        xmlns:xsi="http://www.w3.org/2001/XMLSchema-instance"
        xsi:schemaLocation="http://maven.apache.org/POM/4.0.0
http://maven.apache.org/xsd/maven-4.0.0.xsd">
    <modelVersion>4.0.0</modelVersion>
    <groupId>com.wudimanong</groupId>
    <artifactId>experiment-java-client</artifactId>
    <version>1.0-SNAPSHOT</version>
    <!--引入 Spring Cloud 父依赖-->
    <parent>
        <groupId>org.springframework.cloud</groupId>
        <artifactId>spring-cloud-starter-parent</artifactId>
        <version>Greenwich.SR1</version>
        <relativePath/>
    </parent>
    <dependencies>
        <!--引入 lombok 开发工具包-->
        <dependency>
            <groupId>org.projectlombok</groupId>
            <artifactId>lombok</artifactId>
        </dependency>
        <!--Jackson 依赖-->
        <dependency>
            <groupId>com.fasterxml.jackson.core</groupId>
            <artifactId>jackson-annotations</artifactId>
        </dependency>
```

```
        <!--参数校验工具-->
        <dependency>
            <groupId>org.springframework.boot</groupId>
            <artifactId>spring-boot-starter-validation</artifactId>
            <version>2.2.1.RELEASE</version>
        </dependency>
    </dependencies>
</project>
```

（3）在 A/B 测试微服务“experiment”主工程的 pom.xml 文件中引入客户端依赖，实现接口定义的复用。代码如下：

```
<!—引入客户端依赖-->
<dependency>
    <groupId>com.wudimanong</groupId>
    <artifactId>experiment-java-client</artifactId>
    <version>1.0-SNAPSHOT</version>
</dependency>
```

2. 定义管理端的“系统登录”接口

在本实例中，A/B 测试管理相关的接口需要与用 Vue.js 开发的前端界面系统进行交互。为了实现系统的登录效果，这里开发一个简单的登录接口供前端调用。

在实际场景中，为了实现管理系统的用户登录及权限，需要设计相对完整的账号功能。由于该内容并不是本章的重点，所以这里只是简单地模拟登录逻辑。

（1）定义管理端“系统登录”接口的 Controller 层。代码如下：

```
package com.wudimanong.experiment.controller;
import com.wudimanong.experiment.client.entity.ResponseResult;
...
import org.springframework.web.bind.annotation.RestController;
@Slf4j
@RestController
public class LoginController {
    @Autowired
    LoginService LoginServiceImpl;
    /**
     * “系统登录”接口
     */
    @PostMapping("/login")
    public ResponseResult<User> login(@RequestBody User user) {
        return ResponseResult.OK(LoginServiceImpl.login(user));
```

```
    }
}
```

该接口的基本逻辑是：利用前端用户输入的账号及密码进行登录验证，如果输入正确则返回用户基本信息，否则提示登录失败。

（2）定义管理端“系统登录”接口的请求/返回参数对象。该接口的输入/输出对象都是 User 对象，代码如下：

```
package com.wudimanong.experiment.entity;
import lombok.Data;
@Data
public class User {
    /**
     * 用户 ID
     */
    private Integer id;
    /**
     * 用户账号
     */
    private String username;
    /**
     * 登录密码
     */
    private String password;
    /**
     * 头像链接
     */
    private String avatar;
    /**
     * 真实姓名
     */
    private String name;
}
```

3. 定义管理端的“实验信息分页查询”接口

管理端的“实验信息分页查询”接口，主要用于实现实验信息分页列表信息的展示，以及条件筛选。

（1）定义管理端“实验信息分页查询”接口的 Controller 层。代码如下：

```
package com.wudimanong.experiment.controller;
import com.wudimanong.experiment.client.entity.ResponseResult;
...
import org.springframework.web.bind.annotation.RestController;
@Slf4j
```

```
@RestController
@RequestMapping("/expInfo")
public class AbtestExpController {
    /**
     * 将依赖注入“实验信息分页查询”接口的业务层（Service 层）实例
     */
    @Autowired
    AbtestExpService abtestExpServiceImpl;
    /**
     * 分页查询实验信息列表
     *
     * @return
     */
    @GetMapping("/getExpInfos")
    public ResponseResult<GetExpInfosBO> getExpInfos(@Validated
GetExpInfosDTO getExpInfosDTO) {
        return
ResponseResult.OK(abtestExpServiceImpl.getExpInfos(getExpInfosDTO));
    }
}
```

（2）定义管理端“实验信息分页查询”接口的请求参数对象。代码如下：

```
package com.wudimanong.experiment.client.entity.dto;
import com.wudimanong.experiment.client.validator.EnumValue;
import javax.validation.constraints.NotNull;
import lombok.Data;
@Data
public class GetExpInfosDTO {
    /**
     * 实验名称模糊匹配
     */
    private String nameLike;
    /**
     * 根据业务标签精准匹配
     */
    private String factorTag;
    /**
     * 状态。0-新建；1-已发布；2-生效中；3-已暂停；4-已终止；5-已结束
     */
    @EnumValue(intValues = {0, 1, 2, 3, 4, 5}, message = "请输入正确的实验状
态值")
    private Integer status;
    /**
     * 页码
     */
```

```
    @NotNull(message = "页码不能为空")
    private Integer pageNo;
    /**
     * 每页条数
     */
    @NotNull(message = "页码大小不能为空")
    private Integer pageSize;
}
```

（3）定义管理端“实验信息分页查询”接口的返回参数对象。代码如下：

```
package com.wudimanong.experiment.client.entity.bo;
import com.wudimanong.experiment.client.entity.AbtestExpInfo;
...
import lombok.NoArgsConstructor;
@Data
@Builder
@AllArgsConstructor
@NoArgsConstructor
public class GetExpInfosBO implements Serializable {
    /**
     * 总记录数
     */
    private Integer total;
    /**
     * 页码
     */
    private Integer pageNo;
    /**
     * 具体数据列表
     */
    List<AbtestExpInfo> list;
}
```

上述返回参数会返回总记录数、当前页面，以及具体的数据列表。

（4）定义管理端“实验信息分页查询”接口的返回参数对象中的数据对象。代码如下：

```
package com.wudimanong.experiment.client.entity;
import com.fasterxml.jackson.annotation.JsonFormat;
...
import lombok.NoArgsConstructor;
@Data
@Builder
@AllArgsConstructor
@NoArgsConstructor
public class AbtestExpInfo implements Serializable {
```

```
/**
 * 实验 ID，使用数据库自增机制
 */
private Integer id;
/**
 * 实验名称
 */
private String name;
/**
 * 系统对应的业务标签
 */
private String factorTag;
/**
 * 分层 ID
 */
private Integer layerId;
/**
 * 分组字段（参数类型）。1-用户；2-地理位置
 */
private Integer groupFieldId;
/**
 * 负责人信息
 */
private String owner;
/**
 * 实验开始时间
 */
@JsonFormat(pattern = "yyyy-MM-dd HH:mm:ss", timezone = "GMT+8")
private Timestamp startTime;
/**
 * 实验结束时间
 */
@JsonFormat(pattern = "yyyy-MM-dd HH:mm:ss", timezone = "GMT+8")
private Timestamp endTime;
/**
 * 实验状态。0-新建；1-已发布；2-生效中；3-已暂停；4-已终止；5-已结束
 */
private Integer status;
/**
 * 实验自定义配置信息（JSON 格式）
 */
private String config;
/**
 * 是否抽样打点。0-否；1-是
 */
```

```
    private Integer isSampling;
    /**
     * 抽样率，例如 5 表示 5%
     */
    private Integer samplingRatio;
    /**
     * 使用端系统名称，例如 pay 表示支付系统
     */
    private String serviceName;
    /**
     * 扩展信息（以 JSON 格式存储）
     */
    private String ext;
    /**
     * 实验假设
     */
    private String expHyp;
    /**
     * 实验预期
     */
    private String resultExpect;
    /**
     * 指标 IDS
     */
    private String metricIds;
    /**
     * 是否已上线
     */
    private Integer online;
    /**
     * 分区类型
     */
    private Integer partitionType;
    /**
     * 创建时间
     */
    @JsonFormat(pattern = "yyyy-MM-dd HH:mm:ss", timezone = "GMT+8")
    private Timestamp createTime;
    /**
     * 更新时间
     */
    @JsonFormat(pattern = "yyyy-MM-dd HH:mm:ss", timezone = "GMT+8")
    private Timestamp updateTime;
}
```

上述数据对象中的字段与“A/B 测试信息表”中的字段基本一致。但为了分层隔离，这里并没有将持久层（Dao 层）数据对象直接输出，而是定义了新的转换数据对象。

4. 定义管理端的“实验创建”接口

“实验创建”接口是 A/B 测试微服务提供的关键接口。通过该接口，接入方可以使用管理系统提供的操作界面来创建 A/B 测试，并进行初始流量的分配。

（1）定义管理端“实验创建”接口的 Controller 层。

在“3.”小标题中创建的 AbtestExpController 类中，增加“实验创建”接口。代码如下：

```
    /**
     * “实验创建”接口
     */
    @PostMapping("/createExp")
    public ResponseResult<CreateExpBO> createExp(@RequestBody @Validated
CreateExpDTO createExpDTO) {
        return
ResponseResult.OK(abtestExpServiceImpl.createExp(createExpDTO));
    }
```

（2）定义管理端“实验创建”接口的请求参数对象。代码如下：

```
    package com.wudimanong.experiment.client.entity.dto;
    import com.wudimanong.experiment.client.validator.EnumValue;
    import javax.validation.constraints.NotNull;
    import lombok.Data;
    @Data
    public class CreateExpDTO {
        /**
         * 接入方服务名
         */
        @NotNull(message = "接入服务名不能为空")
        private String appName;
        /**
         * 实验业务标签
         */
        @NotNull(message = "实验业务标签不能为空")
        @Pattern(regexp = "[A-Za-z]\\w{4,34}_[0-9]{4}$", message = "实验标签不符
合规范")
    private String factorTag;
        /**
         * 实验描述
         */
        @NotNull(message = "实验描述信息不能为空")
        private String desc;
```

```
    /**
     * 分组类型。1-用户；2-地理位置
     */
    @EnumValue(intValues = {1, 2}, message = "分组类型不支持")
    private Integer groupField;
    /**
     * 流量层分层 ID
     */
    private Integer layerId;
    /**
     * 实验负责人
     */
    private String owner;
}
```

（3）定义管理端“实验创建”接口的返回参数对象。代码如下：

```
package com.wudimanong.experiment.client.entity.bo;
import java.io.Serializable;
...
import lombok.NoArgsConstructor;
@Data
@Builder
@AllArgsConstructor
@NoArgsConstructor
public class CreateExpBO implements Serializable {
    /**
     * 是否创建成功
     */
    private Boolean isSuccess;
    /**
     * 返回实验 ID
     */
    private Integer expId;
}
```

5. 定义管理端的“实验流量编辑”接口

通过管理系统的实验流量编辑功能，可以实现对 A/B 测试流量分配的动态调节。这也是 A/B 测试微服务需要具备的核心功能——通过对 A/B 测试流量的动态调整，可以更灵活地验证实验效果。

通过该功能，也可以很好地实现新旧功能在流量上的灰度发布。这也是 A/B 测试系统在实践中应用得比较广泛的场景。

（1）定义管理端“实验流量编辑”接口的 Controller 层。

在“3.”小标题中创建的 AbtestExpController 类中，增加“实验流量编辑”接口。代码如下：

```
/**
 * “实验流量编辑”接口
 */
@PostMapping("/updateFlowRatio")
public ResponseResult<UpdateFlowRatioBO> updateFlowRatio(
        @RequestBody @Validated UpdateFlowRatioDTO updateFlowRatioDTO) {
    return
ResponseResult.OK(abtestExpServiceImpl.updateFlowRatio(updateFlowRatioDTO));
}
```

（2）定义管理端“实验流量编辑”接口的请求参数对象。代码如下：

```
package com.wudimanong.experiment.client.entity.dto;
import java.util.Map;
import javax.validation.constraints.NotNull;
import lombok.Data;
@Data
public class UpdateFlowRatioDTO {
    /**
     * 应用名（主要用于标识接入方身份）
     */
    @NotNull(message = "应用名不能为空")
    private String appName;
    /**
     * 实验标签
     */
    @NotNull(message = "实验标签不能为空")
    private String factorTag;
    /**
     * 实验流量配比（example->{"76":20; "77":40），76 和 77 都表示分组 ID
     */
    private Map<Integer, Integer> flowRatioParam;
}
```

在请求参数对象中，流量占比分配的数据格式为：“分组 ID + 流量占比”的 JSON 格式，由 Map 对象进行接收处理。

（3）定义管理端“实验流量编辑”接口的返回参数对象。代码如下：

```
package com.wudimanong.experiment.client.entity.bo;
import java.util.List;
...
import lombok.NoArgsConstructor;
@Data
```

```
@Builder
@NoArgsConstructor
@AllArgsConstructor
public class UpdateFlowRatioBO {
    /**
     * 接入方应用名称
     */
    private String appName;
    /**
     * 实验标签
     */
    private String factorTag;
    /**
     * 调整后的流量分配情况
     */
    private List<GroupFlowRatioBO> flowRatio;
}
```

（4）定义管理端“实验流量编辑”接口的返回参数对象中流量分配情况的数据对象。代码如下：

```
package com.wudimanong.experiment.client.entity.bo;
import java.io.Serializable;
...
import lombok.NoArgsConstructor;
@Data
@Builder
@AllArgsConstructor
@NoArgsConstructor
public class GroupFlowRatioBO implements Serializable {
    /**
     * 分组 ID
     */
    private Integer groupId;
    /**
     * 分组类型
     */
    private Integer groupType;
    /**
     * 调整后的流量占比
     */
    private Integer flowRatio;
    /**
     * 调整前的流量占比
     */
    private Integer preFlowRatio;
}
```

6. 定义 SDK 端的“实验配置信息获取”接口

“实验配置信息获取”接口，主要是为各版本的“接入 SDK”提供实验配置信息获取服务。从使用场景上看，该接口具备高频访问的特点，所以需要在性能上有更多的考虑。

（1）定义 SDK 端“实验配置信息获取”接口的 Controller 层。代码如下：

```
package com.wudimanong.experiment.controller;
import com.wudimanong.experiment.client.entity.ResponseResult;
...
import org.springframework.web.bind.annotation.RestController;
/**
 * @desc 获取实验配置信息
 */
@Slf4j
@RestController
@RequestMapping("/config")
public class AbtestDeliverController {
    /**
     * 业务层（Service 层）的依赖接口
     */
    @Autowired
    private AbtestDeliverService abtestDeliverServiceImpl;
    /**
     * 根据实验业务标签获取实验配置信息
     */
    @GetMapping("/findByFactorTag")
    public ResponseResult<ConfigBO>
findByFactorTag(@RequestParam("factorTag") String factorTag) {
        return ResponseResult.OK(abtestDeliverServiceImpl.
getExpInfoByFactorTag(factorTag));
    }
}
```

该接口主要通过实验业务标签，以 Get 请求的方式获取 A/B 测试的配置信息——包括实验的基本信息及流量组分桶信息。据此可以实现“接入 SDK”流量路由的相关匹配逻辑。

此外，考虑到实验数据的完整性及传输效率，在接口返回数据对象设计上，需要进行一定特殊的考虑。

（2）定义 SDK 端“实验配置信息获取”接口的返回参数对象。代码如下：

```
package com.wudimanong.experiment.client.entity.bo;
import com.wudimanong.experiment.client.entity.AbtestExp;
...
import lombok.NoArgsConstructor;
```

```
@Data
@Builder
@NoArgsConstructor
@AllArgsConstructor
public class ConfigBO {
    /**
     * 业务系统实验标签
     */
    private String factorTag;
    /**
     * 实验配置信息
     */
    private AbtestExp abtestExp;
}
```

（3）定义 SDK 端“实验配置信息获取”接口的返回参数对象中的实验配置信息数据对象。代码如下：

```
package com.wudimanong.experiment.client.entity;
import java.util.List;
...
import lombok.NoArgsConstructor;
@Data
@Builder
@NoArgsConstructor
@AllArgsConstructor
public class AbtestExp {
    /**
     * 实验 ID
     */
    private Integer expId;
    /**
     * 业务系统实验标签
     */
    private String factorTag;
    /**
     * 分层 ID
     */
    private Integer layerId;
    /**
     * 分组类型 ID，1 表示按用户分组
     */
    private Integer groupFieldId;
    /**
     * 分区类型
     */
```

```
    private String partitionType;
    /**
     * 是否抽样打点
     */
    private Boolean isSampling;
    /**
     * 抽样率
     */
    private Integer samplingRatio;
    /**
     * 实验分组信息列表
     */
    private List<AbtestGroup> abtestGroups;
    /**
     * 实验配置
     */
    private String config;
}
```

（4）编写在步骤（3）中定义的“实验配置信息获取”接口的返回参数对象中“实验分组信息”属性的数据对象。代码如下：

```
package com.wudimanong.experiment.client.entity;
import java.util.List;
...
import lombok.NoArgsConstructor;
@Data
@Builder
@NoArgsConstructor
@AllArgsConstructor
public class AbtestGroup {
    /**
     * 实验 ID
     */
    private Integer expId;
    /**
     * 分组 ID
     */
    private Integer groupId;
    /**
     * 分组类型。0-实验组；1-对照组
     */
    private Integer groupType;
    /**
     * 分组的分区类型
     */
```

```
    private String groupPartitionType;
    /**
     * 原始桶
     */
    private Set<Integer> partitionSerialNums;
    /**
     * 桶是否经过压缩
     */
    private Boolean isUseBase64Nums;
    /**
     * 压缩后的桶
     */
    private String partitionSerialNums64;
    /**
     * 稀释倍率
     */
    private Integer dilutionRatio;
    /**
     * 策略信息
     */
    private List<Strategy> abtestStrategies;
    /**
     * 白名单
     */
    private List<String> whiteList;
}
```

上述代码是“实验分组信息”及“流量配置”的数据结构——描述了具体的分组类型，以及分组内流量分桶编号的列表。

考虑到流量“分桶编号列表”在接口传输上存在数据量大的问题，所以，在实现上需要设置阈值，来判断是否采用 Base 64 方式对分桶编号列表进行压缩后传输。

（5）编写在步骤（4）中定义的数据对象中“策略信息”属性的数据对象。代码如下：

```
package com.wudimanong.experiment.client.entity;
import lombok.AllArgsConstructor;
...
import lombok.NoArgsConstructor;
@Data
@Builder
@AllArgsConstructor
@NoArgsConstructor
public class Strategy {
    /**
     * 策略 Key
```

```
     */
    private String key;
    /**
     * 权重值
     */
    private String weight;
}
```

7.5.2 开发业务层（Service 层）的代码

接下来开发 A/B 测试微服务业务层（Service 层）的代码。

1. 定义业务异常处理机制

关于业务层（Service 层）的业务异常处理机制，可以参考 4.4.2 节中“1.”小标题中的内容。

定义的业务层（Service 层）异常码的枚举类，具体代码如下：

```
package com.wudimanong.experiment.client.entity;
public enum BusinessCodeEnum {
    /**
     * 登录用户权限相关的错误码（以 1000 开头，根据业务扩展）
     */
    BUSI_LOGIN_FAIL_1000(1000, "用户密码错误"),
    /**
     * A/B 测试业务逻辑相关的错误码（以 2000 开头，根据业务扩展）
     */
    BUSI_LOGICAL_FAIL_2000(2000, "factor 已存在实验"),
    BUSI_LOGICAL_LAYER_IS_NOT_EXIST(2001, "分层信息不存在"),
    BUSI_LOGICAL_OVER_AVAILABLE_FLOW(2002, "超出可用流量"),
    BUSI_LOGICAL_EXP_IS_NOT_EXIST(2003, "实验信息不存在");
    /**
     * 编码
     */
    private Integer code;
    /**
     * 描述
     */
    private String desc;
    BusinessCodeEnum(Integer code, String desc) {
        this.code = code;
        this.desc = desc;
    }
    /**
     * 根据编码获取枚举类型
     */
```

```
    public static BusinessCodeEnum getByCode(String code) {
        //判断编码是否为空
        if (code == null) {
            return null;
        }
        //循环处理
        BusinessCodeEnum[] values = BusinessCodeEnum.values();
        for (BusinessCodeEnum value : values) {
            if (value.getCode().equals(code)) {
                return value;
            }
        }
        return null;
    }
    public Integer getCode() {
        return code;
    }
    public String getDesc() {
        return desc;
    }
}
```

此枚举类可在具体的业务层（Service 层）逻辑处理中根据业务进行扩展。例如，将实验系统内部相关的业务异常码的值约定为 2000～2999。

2. 引入 MapStruct 实体映射工具

具体参考 4.4.2 节中“2.”小标题中的内容。

3. 开发管理端“系统登录”接口的业务层（Service 层）代码

接下来开发管理端“系统登录”接口的业务层（Service）代码。

（1）定义业务层（Service 层）接口类 LoginService。代码如下：

```
package com.wudimanong.experiment.service;
import com.wudimanong.experiment.entity.User;
public interface LoginService {
    /**
     * “用户登录”接口
     */
    User login(User user);
}
```

（2）实现业务层（Service 层）接口类的方法。代码如下：

```
package com.wudimanong.experiment.service.impl;
import com.wudimanong.experiment.client.entity.BusinessCodeEnum;
```

```
...
import org.springframework.stereotype.Service;
@Slf4j
@Service
public class LoginServiceImpl implements LoginService {
    @Override
    public User login(User user) {
        //为方便测试，这里以“硬编码”的方式设置固定的账号和密码（正式系统可扩展用户登录的逻辑）
        if (user.getUsername().equals("admin") &&
user.getPassword().equals("123456")) {
            user.setId(123);
            user.setName("无敌码农");
            user.setPassword("");
            //通过 GitHub 存储小头像

            user.setAvatar("https://github.com/manongwudi/repos/
blob/master/static/images/avator-wudimanong.jpg");
        } else {
            throw new
ServiceException(BusinessCodeEnum.BUSI_LOGIN_FAIL_1000.getCode(),
                    BusinessCodeEnum.BUSI_LOGIN_FAIL_1000.getDesc());
        }
        return user;
    }
}
```

上述代码，只是以“硬编码”的方式实现了简单的登录逻辑。如有需要，读者可以自行完善用户登录逻辑。

4. 开发管理端“实验信息分页查询”接口的业务层（Service 层）代码

接下来开发管理端“实验信息分页查询”接口的业务层（Service 层）代码。

（1）定义业务层（Service 层）接口类 AbtestExpService。代码如下：

```
package com.wudimanong.experiment.service;
import com.wudimanong.experiment.client.entity.bo.GetExpInfosBO;
import com.wudimanong.experiment.client.entity.dto.GetExpInfosDTO;
import org.springframework.validation.annotation.Validated;
public interface AbtestExpService {
    /**
     * “实验信息分页查询”接口
     */
    GetExpInfosBO getExpInfos(GetExpInfosDTO getExpInfosDTO);
}
```

（2）实现业务层（Service 层）接口类的方法。代码如下：

```
package com.wudimanong.experiment.service.impl;
import com.baomidou.mybatisplus.core.conditions.query.QueryWrapper;
...
import org.springframework.stereotype.Service;
@Service
@Slf4j
public class AbtestExpServiceImpl implements AbtestExpService {
    /**
     * 注入“A/B 测试信息表”的持久层（Dao 层）依赖
     */
    @Autowired
    AbtestExpInfoDao abtestExpInfoDao;
    /**
     * 实验信息分页查询逻辑的实现方法
     */
    @Override
    public GetExpInfosBO getExpInfos(GetExpInfosDTO getExpInfosDTO) {
        //基于 MyBatis-Plus 的语法拼装查询条件
        QueryWrapper<AbtestExpInfoPO> queryWrapper = new QueryWrapper<>();
        if (getExpInfosDTO.getNameLike() != null
&& !"".equals(getExpInfosDTO.getNameLike())) {
            queryWrapper.like("name", getExpInfosDTO.getNameLike());
        }
        if (getExpInfosDTO.getFactorTag() != null
&& !"".equals(getExpInfosDTO.getFactorTag())) {
            queryWrapper.eq("factor_tag", getExpInfosDTO.getFactorTag());
        }
        if (getExpInfosDTO.getStatus() != null) {
            queryWrapper.eq("status", getExpInfosDTO.getStatus());
        }
        //过滤已删除的数据
        queryWrapper.eq("is_delete", 0);
        //进行 ID 降序排序
        queryWrapper.orderByDesc("id");
        //MyBatisPlus 分页支持（设置页码及每页记录数据）
        Page<AbtestExpInfoPO> page = new Page<>(getExpInfosDTO.getPageNo(),
getExpInfosDTO.getPageSize());
        //执行分页查询，返回分页查询结果
        IPage<AbtestExpInfoPO> resultPage =
abtestExpInfoDao.selectPage(page, queryWrapper);
        //获取具体的分页数据
        List<AbtestExpInfoPO> abtestExpInfoPOList =
resultPage.getRecords();
```

```
        //通过 MapStruct 数据复制工具，将“持久层的（Dao 层）列表对象”转换为“业务层（Service 层）输出对象”
        List<AbtestExpInfo> abtestExpInfoList = AbtestExpConvert.INSTANCE.convertAbtestInfo(abtestExpInfoPOList);
        //构造返回输出数据的对象
        GetExpInfosBO getExpInfosBO = GetExpInfosBO.builder().total(Long.valueOf(resultPage.getTotal()).intValue())
                .pageNo(Long.valueOf(resultPage.getCurrent()).intValue()).list(abtestExpInfoList).build();
        return getExpInfosBO;
    }
}
```

上方代码的主要逻辑是：利用 MyBatis-Plus 提供的数据库操作语法，实现“A/B 测试信息表”的分页查询功能。

（3）编写针对 MyBatis-Plus 分页插件的配置类。

在步骤（2）中使用了 MyBatis-Plus 默认集成的分页插件。为了正常使用该分页插件，还需要编写一个 Spring 配置类，具体代码如下：

```
package com.wudimanong.experiment.config;
import com.baomidou.mybatisplus.extension.plugins.PaginationInterceptor;
import org.springframework.context.annotation.Bean;
import org.springframework.context.annotation.Configuration;
@Configuration
public class MybatisPlusConfig {
    /**
     * 创建分页插件
     */
    @Bean
    public PaginationInterceptor paginationInterceptor() {
        return new PaginationInterceptor();
    }
}
```

（4）编写 MapStruct 组件的数据转化接口。

在步骤（2）中，使用了 MapStruct 组件来实现“持久层（Dao 层）数据对象”到“业务层（Service 层）输出的数据对象”的转换，所以还需要编写一个数据转换接口，代码如下：

```
package com.wudimanong.experiment.convert;
import com.wudimanong.experiment.client.entity.AbtestExpInfo;
...
import org.mapstruct.factory.Mappers;
/**
```

```
 * @author jiangqiao
 */
@Mapper
public interface AbtestExpConvert {
    AbtestExpConvert INSTANCE = Mappers.getMapper(AbtestExpConvert.class);
    /**
     *将实验信息列表“持久层（Dao 层）数据对象”转换为“业务层（Service 层）输出的数据对象”
     */
    @Mappings({})
    List<AbtestExpInfo> convertAbtestInfo(List<AbtestExpInfoPO> abtestExpInfoPOList);
}
```

5. 开发管理端“实验创建”接口的业务层（Service 层）代码

管理端“实验创建”接口的业务层（Service 层）逻辑涉及 A/B 测试微服务的流量分桶、流量分配等关键逻辑，在具体实现上有一定的复杂度。为方便读者理解，下面尽可能多地通过注释及文字来进行说明。

（1）定义业务层（Service 层）方法。

在“4.”小标题下步骤（1）中定义的业务层（Service 层）AbtestExpService 接口类中，增加“实验创建”业务层接口方法。代码如下：

```
/**
 * 创建实验
*/
CreateExpBO createExp(CreateExpDTO createExpDTO);
```

（2）实现业务层（Service 层）方法。

在“4.”小标题下步骤（2）中定义的业务层（Service 层）实现类 AbtestExpServiceImpl 中，增加“实验创建”接口的业务层（Service 层）方法的实现。代码如下：

```
/**
 * “实验创建”接口业务层（Service 层）方法的实现（核心方法）
*/
@Transactional(rollbackFor = Exception.class)
@Override
public CreateExpBO createExp(CreateExpDTO createExpDTO) {
    //1 验证实验是否已存在
    QueryWrapper<AbtestExpInfoPO> queryWrapper = new QueryWrapper<>();
    queryWrapper.eq("factor_tag", createExpDTO.getFactorTag());
    AbtestExpInfoPO abtestExpInfoPO = abtestExpInfoDao.selectOne(queryWrapper);
```

```
        if (abtestExpInfoPO != null) {
            throw new
ServiceException(BusinessCodeEnum.BUSI_LOGICAL_FAIL_2000.getCode(),
                    BusinessCodeEnum.BUSI_LOGICAL_FAIL_2000.getDesc());
        }
        //2 判断分层信息是否存在，如果不存在，则创建默认流量分层
        AbtestLayerPO abtestLayerPO = null;
        if (createExpDTO.getLayerId() != null) {
            //如果流量层不存在，则抛出异常返回失败
            if (!isExistLayer(createExpDTO.getLayerId())) {
                throw new ServiceException
(BusinessCodeEnum.BUSI_LOGICAL_LAYER_IS_NOT_EXIST.getCode(),
                        BusinessCodeEnum.
BUSI_LOGICAL_LAYER_IS_NOT_EXIST.getDesc());
            }
        } else {
            //生成实验分层
            abtestLayerPO = createAbtestLayer(createExpDTO);
            //持久化分层信息
            abtestLayerDao.insert(abtestLayerPO);
        }
        //3 生成实验基本信息
        abtestExpInfoPO = createAbtestInfo(createExpDTO, abtestLayerPO);
        //持久化实验基本信息
        abtestExpInfoDao.insert(abtestExpInfoPO);
        //4 创建实验分组信息并持久化
        List<AbtestGroupPO> groupInfos =
createAbtestGroupList(abtestExpInfoPO);
        //批量持久化分组信息
        abtestGroupDao.batchInsert(groupInfos);
        //5 流量分配初始化
        //筛选流量分配占比超过 0 的分组
        Map<Integer, Integer> flowRatioMap = groupInfos.stream().filter(o ->
o.getFlowRatio() > 0)
                .collect(Collectors.toMap(AbtestGroupPO::getId,
AbtestGroupPO::getFlowRatio));
        if (flowRatioMap.size() > 0) {
            //调用“流量桶分配”方法（在调用之前需要将 GroupList 中的流量占比设置为 0，以确
保正常的初始分配）
            updateFlowRatio(abtestExpInfoPO, abtestLayerPO,
                    groupInfos.stream().peek(group ->
group.setFlowRatio(0)).collect(
                            Collectors.toList()), flowRatioMap);
        }
```

```
        return CreateExpBO.builder().isSuccess(true).expId
(abtestExpInfoPO.getId()).build();
    }
```

上述方法实现了 A/B 测试所涉及的核心逻辑，大致流程为：

①根据实验名称判断是否已存在相同实验，如果存在，则抛出重复异常。

②判断接口所传递的实验分层信息是否存在。如果存在，则沿用已指定的分层流量数据；如果不存在，则创建新的分层流量信息。

③生成实验基本信息，并通过持久层（Dao 层）组件将其插入数据库。

④创建默认的实验分组数据，并将其插入数据库。

⑤根据实验分组初始流量分配占比，进行流量分配计算，并更新流量分配信息。

上述步骤均在一个数据库事务中进行管理，使用@Transactional 注解实现。由于逻辑相对复杂，涉及的代码量也比较大，因此在该方法的实现中会对部分逻辑进行单独的方法抽象。

（3）定义步骤（2）中涉及的“创建实验分层”的私有方法。

在“4.”小标题下步骤（2）中定义的业务层（Service 层）实现 AbtestExpServiceImpl 类中，增加“创建实验分层”的私有方法。代码如下：

```
    /**
     * 创建实验分层信息的方法
    */
    private AbtestLayerPO createAbtestLayer(CreateExpDTO createExpDTO) {
        AbtestLayerPO abtestLayerPO = new AbtestLayerPO();
        //名称及描述信息设置
        abtestLayerPO.setName(createExpDTO.getFactorTag());
        abtestLayerPO.setDesc(createExpDTO.getDesc());
        //设置流量分组类型 ID
        abtestLayerPO.setGroupFieldId(createExpDTO.getGroupField());
        //初始化流量分桶（这里是最核心的逻辑，通过编写 BucketUtils 工具类实现）
        //设置每个分层的默认分桶总数
        abtestLayerPO.setBucketTotalNum(BucketUtils.BUCKET_TOTAL_NUM);
        //通过 BucketUtils 工具类的方法，实现流量分桶的初始化
        abtestLayerPO
                .setUnusedBucketNos(StringUtils.join(BucketUtils.getShuffledBu
cketNoList().stream().toArray(), ","));
        abtestLayerPO.setIsDelete(0);
        return abtestLayerPO;
    }
```

在上述创建实验分层信息的方法中涉及的流量分桶逻辑，是通过 BucketUtils 工具类中的“流

量分桶计算”方法来实现的，具体参考下面步骤（7）的内容。

（4）定义步骤（2）中涉及的“创建实验基本信息”的私有方法。

在“4.”小标题下步骤（2）中定义的业务层（Service 层）实现 AbtestExpServiceImpl 类中，增加“创建实验基本信息”的私有方法。代码如下：

```
    /**
     * 创建实验基本信息的方法
     */
    private AbtestExpInfoPO createAbtestInfo(CreateExpDTO createExpDTO,
AbtestLayerPO abtestLayerPO) {
        AbtestExpInfoPO abtestExpInfoPO = new AbtestExpInfoPO();
        abtestExpInfoPO.setName(createExpDTO.getDesc());
        abtestExpInfoPO.setFactorTag(createExpDTO.getFactorTag());
        //设置分层信息
        abtestExpInfoPO
                .setLayerId(createExpDTO.getLayerId() == null ?
abtestLayerPO.getId() : createExpDTO.getLayerId());
        abtestExpInfoPO.setGroupFieldId(createExpDTO.getGroupField());
        //设置默认的抽样率
        abtestExpInfoPO.setIsSampling(1);
        //默认设置抽样率为 5
        abtestExpInfoPO.setSamplingRatio(5);
        //设置指标等信息
        abtestExpInfoPO.setMetricIds("");
        abtestExpInfoPO.setOwner(createExpDTO.getOwner());
        abtestExpInfoPO.setServiceName(createExpDTO.getAppName());
        //默认设置为已发布状态
        abtestExpInfoPO.setStatus(1);
        abtestExpInfoPO.setIsDelete(0);
        //设置为未上线
        abtestExpInfoPO.setOnline(0);
        return abtestExpInfoPO;
    }
```

（5）定义步骤（2）中涉及的“创建实验分组信息”的私有方法。

在“4.”小标题下步骤（2）中定义的业务层（Service 层）实现 AbtestExpServiceImpl 类中，增加“创建实验分组信息”的私有方法。代码如下：

```
    /**
     * 设置初始流量的默认占比
     */
    public static final Integer defaultGroupInitFlowRatio = 50;
```

```
    /**
     * 创建实验分组信息的方法
    */
    private List<AbtestGroupPO> createAbtestGroupList(AbtestExpInfoPO
abtestExpInfoPO) {
        //生成流量分组信息
        List<AbtestGroupPO> groupInfos = new ArrayList<>();
        //1 生成实验组
        AbtestGroupPO testGroup = new AbtestGroupPO();
        testGroup.setExpId(abtestExpInfoPO.getId());
        //设置流量占比
        testGroup.setFlowRatio(defaultGroupInitFlowRatio);
        //设置分组名称
        testGroup.setName("实验组");
        //分组类型为 0 表示实验组
        testGroup.setGroupType(0);
        //设置默认分流内包含的分桶编号
        testGroup.setGroupPartitionDetails("");
        //设置策略明细
        testGroup.setStrategyDetail("");
        testGroup.setOnline(abtestExpInfoPO.getOnline());
        testGroup.setDilutionRatio(0);
        groupInfos.add(testGroup);
        //2 生成对照组
        AbtestGroupPO controlGroup = new AbtestGroupPO();
        controlGroup.setExpId(abtestExpInfoPO.getId());
        //设置流量占比
        controlGroup.setFlowRatio(defaultGroupInitFlowRatio);
        //设置分组名称
        controlGroup.setName("对照组");
        //分组类型为 1 表示对照组
        controlGroup.setGroupType(1);
        //设置默认分流内包含的分桶编号
        controlGroup.setGroupPartitionDetails("");
        //设置策略明细
        controlGroup.setStrategyDetail("");
        controlGroup.setOnline(abtestExpInfoPO.getOnline());
        controlGroup.setDilutionRatio(0);
        groupInfos.add(0, controlGroup);
        return groupInfos;
    }
```

上述方法的逻辑是：创建实验分组（实验组、对照组）的基本信息，并设置它们的流量占比。

（6）定义步骤（2）中涉及的“流量桶分配”的方法。

在“4.”小标题下步骤（2）中定义的业务层（Service 层）实现 AbtestExpServiceImpl 类中，增加“流量桶分配”的方法。代码如下：

```
    /**
     * 流量桶分配（流量桶分配是最核心的公共方法，其算法也是整个 A/B 测试微服务的核心）
     *
     * @param expInfo
     * @param layer
     * @param groupList
     * @param flowRatioMap
     */
    @Transactional(rollbackFor = Exception.class)
    public void updateFlowRatio(AbtestExpInfoPO expInfo, AbtestLayerPO layer, 
List<AbtestGroupPO> groupList,
            Map<Integer, Integer> flowRatioMap) {
        //获取需要进行流量分配的分组信息
        try {
            groupList.stream().filter(group -> 
flowRatioMap.containsKey(group.getId())).sorted(
                    Comparator.comparing(group -> flowRatioMap.get(group.getId()) 
- group.getFlowRatio()))
                    .map(group -> (Function<List<Integer>, List<Integer>>) unused 
-> {
                        Result bucketResult;
                        try {
                            Request bucketRequest = new Request();
                            bucketRequest.setCurrBucketRatio
(group.getFlowRatio());
                            bucketRequest.setDestBucketRatio
(flowRatioMap.get(group.getId()));
                            
bucketRequest.setCurrUnusedBucketNoListOfLayer(unused);
                            //将分桶数据转换为 List<Integer>
                            List<Integer> currUsedBucketNoListOfGroup = 
(group.getGroupPartitionDetails() != null && !""
                                    .equals(group.getGroupPartitionDetails())) ? 
Arrays
                                    .asList(group.getGroupPartitionDetails().split
(",")).stream()
                                    .map(o -> 
Integer.valueOf(o)).collect(Collectors.toList()) : new ArrayList<>();
                            bucketRequest.setCurrUsedBucketNoListOfGroup
(currUsedBucketNoListOfGroup);
                            //执行重新分桶洗牌逻辑
```

```
                        bucketResult =
BucketUtils.bucketReallocate(bucketRequest);
                        //设置已分配好的分桶编号
                        group.setGroupPartitionDetails(
                                StringUtils.join(bucketResult.
getBucketNoListOfGroup(), ","));
                        //更新当前时间
                        group.setUpdateTime(new Timestamp
(System.currentTimeMillis()));
                        //更新流量占比
                        group.setFlowRatio(flowRatioMap.get(group.getId()));
                        abtestGroupDao.updateById(group);
                    } catch (Exception e) {
                        throw new ServiceException(BusinessCodeEnum.
BUSI_LOGICAL_OVER_AVAILABLE_FLOW.getCode(),

BusinessCodeEnum.BUSI_LOGICAL_OVER_AVAILABLE_FLOW.getDesc(), e);
                    }
                    //返回当前分层中未被分配的流量分桶编号
                    return bucketResult.getUnusedBucketNoListOfLayer();
                }).reduce(Function::andThen)
                .ifPresent(func -> {
                    //更新分层信息中未使用的分桶编号信息
                    List<Integer> layerUnusedBucketNos =
                            (layer.getUnusedBucketNos() != null
&& !"".equals(layer.getUnusedBucketNos())) ? Arrays
                                    .asList(layer.getUnusedBucketNos().split("
,"))
                                    .stream().map(o ->
Integer.valueOf(o)).collect(Collectors.toList())
                                    : new ArrayList<>();
                    String unused =
StringUtils.join(func.apply(layerUnusedBucketNos).toArray(), ",");
                    layer.setUnusedBucketNos(unused);
                    layer.setUpdateTime(new
Timestamp(System.currentTimeMillis()));
                    abtestLayerDao.updateById(layer);
                });
        } catch (Exception e) {
            throw new ServiceException
(BusinessCodeEnum.BUSI_LOGICAL_OVER_AVAILABLE_FLOW.getCode(),
                BusinessCodeEnum.BUSI_LOGICAL_OVER_AVAILABLE_FLOW.getDesc(),
e);
        }
    }
```

上述方法是本层实现逻辑中最为核心的方法，其核心逻辑是：

- 通过输入的实验基本信息、流量分层信息、现有流量分组信息，以及上一层方法设定的流量目标占比，进行流量分桶的分配计算。
- 根据计算结果，更新流量分层中的可用流量分桶编号，以及具体实验分组中所持有的流量分桶。

（7）定义在步骤（3）、步骤（6）中“流量分桶计算”逻辑所依赖的 BucketUtils 工具类。代码如下：

```
    package com.wudimanong.experiment.utils;
    import java.util.Base64;
    import java.util.BitSet;
    import java.util.Collections;
    import java.util.LinkedList;
    import java.util.List;
    public class BucketUtils {
        /**
         * 每个分层的 bucket 总数
         */
        public static final Integer BUCKET_TOTAL_NUM = 1000;
        /**
         * 原始分桶编号
         */
        private static final List<Integer> ORIGINAL_BUCKET_NOS = new
LinkedList<>();
        // 初始化原始分桶编号
        static {
            for (int index = 0; index < BUCKET_TOTAL_NUM; index++) {
                ORIGINAL_BUCKET_NOS.add(index);
            }
        }
        /**
         * 洗牌，获取bucket 的分桶编号
         */
        public static List<Integer> getShuffledBucketNoList() {
            List<Integer> currentBucketNos = new LinkedList<>(ORIGINAL_BUCKET_
NOS);
            //这里通过集合对象提供的 shuffle()方法（使用指定的随机源对指定列表进行置换，所
有置换发生的可能性都是大致相等的）来进行分桶编号洗牌
            Collections.shuffle(currentBucketNos);
            return currentBucketNos;
        }
        /**
```

```
     * 核心流量分桶调整逻辑（输入：当前百分比、目标百分比、未分配分桶编号、已分配分桶编号；输出：当前百分比、未分配分桶编号、调整后已分配的分桶编号）
     */
    public static BucketAllocate.Result
bucketReallocate(BucketAllocate.Request request) throws Exception {
        //流量占比值为 0~100
        if (request.getCurrBucketRatio() < 0 || request.getDestBucketRatio()
< 0 || request.getCurrBucketRatio() > 100
                || request.getDestBucketRatio() > 100) {
            throw new Exception("flowRatio value is invalid", null);
        }
        //定义流量计算结果对象
        BucketAllocate.Result result = new BucketAllocate.Result();
        //计算目标占比与当前占比的差（单位：%）
        int gapPercent = request.getDestBucketRatio() -
request.getCurrBucketRatio();
        if (gapPercent == 0) {
            //1 如果目标占比与当前占比一致，则分组内分桶数量不变
            result.setBucketRatio(request.getDestBucketRatio());
            result.setUnusedBucketNoListOfLayer
(request.getCurrUnusedBucketNoListOfLayer());
            result.setBucketNoListOfGroup
(request.getCurrUsedBucketNoListOfGroup());
        } else if (gapPercent > 0) {
            //2 如果目标占比大于当前占比，则需要扩充分组内的分桶数量
            //2-1 这是一个核心公式，计算当前分组需要扩充的分桶数量
            int needAddBucketNumOfGroup = gapPercent * BUCKET_TOTAL_NUM / 100;
            //2-2 检查当前流量层未使用分桶数量是否满足扩充需要
            List<Integer> currUnusedBucketNoListOfLayer =
request.getCurrUnusedBucketNoListOfLayer();
            int unusedBucketNumOfLayer = currUnusedBucketNoListOfLayer ==
null ? 0: currUnusedBucketNoListOfLayer.size();
            if (needAddBucketNumOfGroup > unusedBucketNumOfLayer) {
                throw new Exception("needAddBucketNumOfGroup >
unusedBucketNumOfLayer", null);
            }
            //2-3 调整流量分层中未使用的分桶编号，并将其分配至相应分组
            //继续计算流量分层中应该持有的未分配桶数量
            int unusedBucketRemainNum = unusedBucketNumOfLayer -
needAddBucketNumOfGroup;
            //根据比例将之前流量分层中未被使用的分桶，按照新的占比进行重新分配
            List<Integer> currUsedBucketNoListOfGroup =
request.getCurrUsedBucketNoListOfGroup();
            List<Integer> destUnusedBucketNoListOfLayer = new LinkedList<>();
```

```
            List<Integer> destUsedBucketNoListOfGroup = new LinkedList<>
(currUsedBucketNoListOfGroup);
            for (int index = 0; index < unusedBucketNumOfLayer; index++) {
                Integer currBucketNo =
currUnusedBucketNoListOfLayer.get(index);
                if (index < unusedBucketRemainNum) {
                    destUnusedBucketNoListOfLayer.add(currBucketNo);
                } else {
                    destUsedBucketNoListOfGroup.add(currBucketNo);
                }
            }
            //填充计算结果
            result.setBucketRatio(request.getDestBucketRatio());
            result.setUnusedBucketNoListOfLayer
(destUnusedBucketNoListOfLayer);
            result.setBucketNoListOfGroup(destUsedBucketNoListOfGroup);
        } else {
            //3 如果目标占比小于当前占比，则需要收缩分组内分桶数量
            //3-1 计算当前分组内需要收缩的分桶数量
            int needMinusBucketNumOfGroup = -1 * gapPercent * BUCKET_TOTAL_NUM
/ 100;
            //3-2 检查“当前分组内已使用的分桶数量”是否大于“需要收缩的分桶数量”
            List<Integer> currUsedBucketNoListOfGroup =
request.getCurrUsedBucketNoListOfGroup();
            int usedBucketNumOfGroup = currUsedBucketNoListOfGroup == null ?
0 : currUsedBucketNoListOfGroup.size();
            if (needMinusBucketNumOfGroup > usedBucketNumOfGroup) {
                throw new Exception("needMinusBucketNumOfGroup >
usedBucketNumOfGroup", null);
            }
            //3-3 调整当前分组内已使用的分桶编号，并将其回收至流量分层中未使用分桶编号池中
            int usedBucketRemainNum = usedBucketNumOfGroup -
needMinusBucketNumOfGroup;
            List<Integer> currUnusedBucketNoListOfLayer =
request.getCurrUnusedBucketNoListOfLayer();
            List<Integer> destUnusedBucketNoListOfLayer = new
LinkedList<>(currUnusedBucketNoListOfLayer);
            List<Integer> destUsedBucketNoListOfGroup = new LinkedList<>();
            for (int index = 0; index < usedBucketNumOfGroup; index++) {
                Integer currBucketNo = currUsedBucketNoListOfGroup.get
(index);
                if (index < usedBucketRemainNum) {
                    destUsedBucketNoListOfGroup.add(currBucketNo);
                } else {
                    destUnusedBucketNoListOfLayer.add(currBucketNo);
```

```
                }
            }
            //填充计算结果
    result.setBucketRatio(request.getDestBucketRatio());

            result.setUnusedBucketNoListOfLayer(destUnusedBucketNoListOfLa
yer);
            result.setBucketNoListOfGroup(destUsedBucketNoListOfGroup);
        }
        return result;
    }
}
```

上述代码是流量分桶最核心的逻辑。大致结构为：设置流量桶的默认总数为 1000；在创建实验分层时，通过 getShuffledBucketNoList()方法实现流量分桶的洗牌逻辑（由 Collections.shuffle()方法实现）。

其中，bucketReallocate()方法是流量占比重新分配的关键逻辑，其基本算法逻辑如下：

- 如果“目标流量占比”与“当前分组流量占比”的差为 0，则无须进行流量重新分配，直接返回当前流量分配情况即可。
- 如果“目标流量占比”与“当前分组流量占比”的差大于 0，则需要对当前分组流量进行扩充，具体需要扩充的流量分桶数的计算公式为：$\frac{流量占比差\times总的流量桶}{100}$。之后根据扩充分桶数计算结果，从流量分层的可用分桶数中抽取一定的流量分桶编号。
- 如果“目标流量占比”与“当前分组流量占比”的差小于 0，则需要对当前分组流量进行收缩，具体需要收缩的流量分桶计算公式为：$\frac{-1\times流量占比差\times总的流量桶}{100}$。之后根据需要收缩的分桶数计算结果，从当前分组流量的可用分桶数抽取一定的分桶编号，并将其放回实验分层未使用的分桶编号池中。

定义 bucketReallocate()方法的输入/输出参数对象，代码如下：

```
package com.wudimanong.experiment.utils;
import java.util.List;
import lombok.Data;
public class BucketAllocate {
    /**
     * 定义分桶分配方法的参数对象
     */
    @Data
    public static class Request {
        // 当前的分桶占比（举例：25）
        private int currBucketRatio;
        // 目标的分桶占比
```

```
        private int destBucketRatio;
        // 在分层中，当前的未分配分桶编号
        private List<Integer> currUnusedBucketNoListOfLayer;
        // 在分组内，已经分配的分桶编号
        private List<Integer> currUsedBucketNoListOfGroup;
    }
    /**
     * 定义分桶操作的辅助输出对象
     */
    @Data
    public static class Result {
        // 分桶占比
        private int bucketRatio;
        // 在分层中，当前的未分配分桶编号
        private List<Integer> unusedBucketNoListOfLayer;
        // 在分组内，已经分配的分桶编号
        private List<Integer> bucketNoListOfGroup;
    }
}
```

6. 开发管理端“实验流量编辑”接口的业务层（Service 层）代码

“实验流量编辑”接口的主要功能是：重新分配流量分桶编号，从而实现流量的调节功能。

（1）定义业务层（Service 层）方法。

在“4.”小标题下步骤（1）中定义的业务层（Service 层）AbtestExpService 接口类中，增加“实验流量编辑”接口的业务层（Service 层）方法。代码如下：

```
/**
 * 修改实验流量占比
*/
UpdateFlowRatioBO updateFlowRatio(UpdateFlowRatioDTO updateFlowRatioDTO);
```

（2）实现业务层（Service 层）方法。

在“4.”小标题下步骤（2）中定义的业务层（Service 层）实现 AbtestExpServiceImpl 类中，增加“实验流量编辑”接口的业务层（Service 层）方法的实现。代码如下：

```
/**
 * 实验流量编辑，修改实验流量的占比
*/
@Override
public UpdateFlowRatioBO updateFlowRatio(UpdateFlowRatioDTO
updateFlowRatioDTO) {
    //获取实验基本信息
    QueryWrapper<AbtestExpInfoPO> queryWrapper = new QueryWrapper<>();
```

```
        queryWrapper.eq("factor_tag", updateFlowRatioDTO.getFactorTag());
        queryWrapper.eq("service_name", updateFlowRatioDTO.getAppName());
        AbtestExpInfoPO abtestExpInfoPO = abtestExpInfoDao.selectOne(queryWrapper);
        if (abtestExpInfoPO == null) {
            throw new ServiceException
(BusinessCodeEnum.BUSI_LOGICAL_EXP_IS_NOT_EXIST.getCode(),
                    BusinessCodeEnum.BUSI_LOGICAL_EXP_IS_NOT_EXIST.getDesc());
        }
        //获取实验分层信息
        AbtestLayerPO abtestLayerPO = abtestLayerDao.selectById
(abtestExpInfoPO.getLayerId());
        if (abtestLayerPO == null) {
            throw new ServiceException
(BusinessCodeEnum.BUSI_LOGICAL_LAYER_IS_NOT_EXIST.getCode(),

BusinessCodeEnum.BUSI_LOGICAL_LAYER_IS_NOT_EXIST.getDesc());
        }
        //获取实验分组信息
        Map<String, Object> param = new HashMap<>();
        param.put("exp_id", abtestExpInfoPO.getId());
        List<AbtestGroupPO> groupList = abtestGroupDao.selectByMap(param);
        //筛选需要调整的分组（当目标分组流量与当前分组流量占比不一致时，才需要调整）
        List<AbtestGroupPO> oldGroupList = groupList.stream().filter(group
-> !Objects.equals(group.getFlowRatio(),

updateFlowRatioDTO.getFlowRatioParam().getOrDefault(group.getId(), 0)))
                .collect(Collectors.toList());
        //调用公共流量调整方法进行流量调节（与实验创建共用）
        updateFlowRatio(abtestExpInfoPO, abtestLayerPO, oldGroupList,
updateFlowRatioDTO.getFlowRatioParam());
        //封装流量调节后的结果数据
        List<GroupFlowRatioBO> groupFlowRatioResult = groupList.stream()
                .map(group ->
GroupFlowRatioBO.builder().groupId(group.getId()).groupType(group.getGroupTy
pe())
                        .preFlowRatio(group.getFlowRatio())
                        .flowRatio(updateFlowRatioDTO.getFlowRatioParam().get(gr
oup.getId())).build())
                .collect(Collectors.toList());
        //封装返回参数对象
        UpdateFlowRatioBO updateFlowRatioBO =
UpdateFlowRatioBO.builder().appName(updateFlowRatioDTO.getAppName())
                .factorTag(updateFlowRatioDTO.getFactorTag()).flowRatio(groupF
lowRatioResult).build();
        return updateFlowRatioBO;
    }
```

7. 开发 SDK 端"实验配置信息获取"接口的业务层（Service 层）代码

接下来开发 SDK 端"实验配置信息获取"接口的业务层（Service 层）代码。

（1）定义业务层（Service 层）接口类 AbtestDeliverService。代码如下：

```
package com.wudimanong.experiment.service;
import com.wudimanong.experiment.client.entity.bo.ConfigBO;
public interface AbtestDeliverService {
    /**
     * 根据业务标签获取实验配置信息
     */
    ConfigBO getExpInfoByFactorTag(String factorTag);
}
```

（2）实现业务层（Service 层）接口类的方法。代码如下：

```
package com.wudimanong.experiment.service.impl;
import com.baomidou.mybatisplus.core.conditions.query.QueryWrapper;
...
import org.springframework.stereotype.Service;
@Slf4j
@Service
public class AbtestDeliverServiceImpl implements AbtestDeliverService {
    /**
     * "A/B 测试信息表"的持久层（Dao 层）接口
     */
    @Autowired
    AbtestExpInfoDao abtestExpInfoDao;
    /**
     * "A/B 测试分组表"的持久层（Dao 层）接口
     */
    @Autowired
    AbtestGroupDao abtestGroupDao;
    /**
     * 根据业务系统标识，获取实验配置信息信息（以参数 factorTag 为 Key，使用 Caffeine
进行缓存）
     */
    @Override
    @Cacheable(value = "EXP_CONFIG_INFO", key = "#factorTag")
    public ConfigBO getExpInfoByFactorTag(String factorTag) {
        //根据业务系统参数，查询实验的基本信息
        AbtestExpInfoPO abtestExpInfoPO = new AbtestExpInfoPO();
        abtestExpInfoPO.setFactorTag(factorTag);
        QueryWrapper<AbtestExpInfoPO> queryWrapper = new
QueryWrapper<>(abtestExpInfoPO);
        abtestExpInfoPO = abtestExpInfoDao.selectOne(queryWrapper);
```

```
            //如果实验信息不存在，则返回空配置
            if (abtestExpInfoPO == null) {
                return null;
            }
            //根据实验 ID，查询分组列表信息
            QueryWrapper<AbtestGroupPO> groupQueryWrapper = new
QueryWrapper<>();
            groupQueryWrapper.eq("exp_id", abtestExpInfoPO.getId());
            List<AbtestGroupPO> groupPOList =
abtestGroupDao.selectList(groupQueryWrapper);
            //将实验配置信息的“持久层（Dao 层）数据对象”转换为“接口的返回参数对象”
            ConfigBO configBO =
AbtestExpConvert.INSTANCE.convertConfig(abtestExpInfoPO, groupPOList);
            return configBO;
        }
    }
```

上述代码，主要是通过实验业务标签，查询实验信息及实验分组信息——按照返回数据对象的约定，对实验基本信息、分组流量等信息进行封装。

（3）编写数据对象转换接口类 AbtestExpConvert。

在“4.”小标题下步骤（4）中定义的数据对象转换接口类 AbtestExpConvert 中增加转换方法。代码如下：

```
    /**
     * 将根据实验及分组信息，转换“实验配置信息获取”接口的输出数据对象
    */
    default ConfigBO convertConfig(AbtestExpInfoPO abtestExpInfoPO,
List<AbtestGroupPO> groupPOList) {
        AbtestExp abtestExp = new AbtestExp();
        abtestExp.setExpId(abtestExpInfoPO.getId());
        abtestExp.setFactorTag(abtestExpInfoPO.getFactorTag());
        abtestExp.setLayerId(abtestExpInfoPO.getLayerId());
        abtestExp.setGroupFieldId(abtestExpInfoPO.getGroupFieldId());
        abtestExp.setIsSampling(abtestExpInfoPO.getIsSampling() == 1 ? true :
false);
        abtestExp.setSamplingRatio(abtestExpInfoPO.getSamplingRatio());
        //通过函数式编程的方法转换并设置分组信息
        abtestExp.setAbtestGroups(map(groupPOList, AbtestGroupPO::mapGroup));
        return
ConfigBO.builder().factorTag(abtestExp.getFactorTag()).abtestExp(abtestExp).
build();
    }
    /**
     * 实验分组信息转换的函数式方法
```

```
    */
    default <T, R> List<R> map(List<T> list, Function<T, R> func) {
        if (CollectionUtils.isEmpty(list)) {
            return Collections.emptyList();
        }
        return list.stream().map(func).collect(Collectors.toList());
    }
```

在某些场景下，如果 MapStruct 转换接口不能完全满足要求，则可以通过 Java 8 所支持的 default 类型的方法编写数据转换代码。这样不仅可以确保软件分层清晰，也可以实现相对灵活的编程方式。

（4）定义步骤（3）中实验分组信息转换逻辑依赖的函数式（Java 8 特性）方法——map()。

在步骤（3）的转换逻辑中，在将“流量分组信息的数据库持久层（Dao 层）数据对象”转换为“实验配置信息获取接口的输出对象”时，使用了 AbtestGroupPO 持久层（Dao 层）数据对象（参考 7.5.3 节中“2.”小标题的内容）中定义的函数方法，具体代码如下：

```
    /**
     * 将“持久层（Dao 层）数据对象”转换为“接口层数据对象”
     */
    public static AbtestGroup mapGroup(AbtestGroupPO abtestGroupPO) {
        //分流内包含的桶编号列表
        List<Integer> partitionSerialNums = null;
        if (abtestGroupPO.getGroupPartitionDetails() != null
&& !"".equals(abtestGroupPO.getGroupPartitionDetails())) {
            partitionSerialNums =
Arrays.asList(abtestGroupPO.getGroupPartitionDetails().split(",")).stream()
                    .map(o -> Integer.valueOf(o)).collect(Collectors.toList());
        } else {
            partitionSerialNums = new ArrayList<>();
        }
        //判断桶数量大小是否大于 100，以此决定传输时是否启用压缩
        boolean useBase64Nums = partitionSerialNums.size() > 100;
        AbtestGroup group = new AbtestGroup();
        group.setExpId(abtestGroupPO.getExpId());
        group.setGroupId(abtestGroupPO.getId());
        group.setGroupType(abtestGroupPO.getGroupType());
        //是否使用 Base64 方式进行压缩
        group.setIsUseBase64Nums(useBase64Nums);
        //如果不需要进行 Base64 方式压缩，则直接设置分桶编号列表
        group.setPartitionSerialNums(useBase64Nums ? Collections.emptySet() :
new HashSet<>(partitionSerialNums));
        //如果需要进行Base64 方式压缩，则将分桶编号列表信息压缩后设置到“partitionSerialNums64”
字段中
```

```
        group.setPartitionSerialNums64(useBase64Nums ?
BucketUtils.bucketsToBitStr(partitionSerialNums) : "");
        group.setDilutionRatio(abtestGroupPO.getDilutionRatio());
        //设置分组实验策略信息
    group.setAbtestStrategies(JSON.parseObject(abtestGroupPO.getStrategyDeta
il(), List.class));
        group.setWhiteList(
                Objects.nonNull(abtestGroupPO.getWhiteList()) ?
Arrays.asList(abtestGroupPO.getWhiteList().split(","))
                        : null);
        return group;
    }
```

上述逻辑会根据流量分桶编号是否大于 100，来判断是否需要对分桶数据进行压缩，而具体的压缩算法是在“5.”小标题下步骤（7）中创建的 BucketUtils 工具类中增加“压缩”方法来实现的。代码如下：

```
/**
 * 将桶编号列表进行Base64 方式压缩
 */
public static String bucketsToBitStr(Iterable<Integer> bucketNos) {
    BitSet bitSet = new BitSet(BUCKET_TOTAL_NUM);
    bucketNos.forEach(bitSet::set);
    return Base64.getUrlEncoder().encodeToString(bitSet.toByteArray());
}
```

以上就是 SDK 端“实验配置信息获取”接口所涉及的业务层（Service 层）实现代码。

为了提升该接口功能的服务性能，在业务层（Service 层）实现方法定义上，通过 @Cacheable 注解使用 Caffeine 进程内缓存（缓存时效为 60s）来提升性能。有关 Caffeine 缓存的配置可参考 7.4 节的内容。

7.5.3 开发 MyBatis 持久层（Dao 层）组件

持久层（Dao 层）的实现，以“MyBatis + MyBatis-Plus”组合提供的功能为基础。

需要先在程序主类中添加 MyBatis 接口代码的包扫描路径，代码如下：

```
package com.wudimanong.experiment;
import org.mybatis.spring.annotation.MapperScan;
...
import org.springframework.cloud.client.discovery.EnableDiscoveryClient;
@EnableDiscoveryClient
@SpringBootApplication
```

```
@MapperScan("com.wudimanong.experiment.dao.mapper")
public class ExperimentApplication {
    public static void main(String[] args) {
        SpringApplication.run(ExperimentApplication.class, args);
    }
}
```

在上述代码中，通过@MapperScan 注解定义了 MyBatis 持久层（Dao 层）组件的代码包路径。

1. 开发“实验信息分页查询”接口的持久层（Dao 层）代码

“实验信息分页查询”的持久层（Dao 层）实现，主要以“A/B 测试信息表”（abtest_exp_info）的数据库操作为主。

（1）定义“A/B 测试信息表”的数据库实体类。代码如下：

```
package com.wudimanong.experiment.dao.model;
import com.baomidou.mybatisplus.annotation.IdType;
...
import lombok.Data;
@Data
@TableName("abtest_exp_info")
public class AbtestExpInfoPO {
    /**
     * 实验 ID，使用数据库自增机制
     */
    @TableId(value = "id", type = IdType.AUTO)
    private Integer id;
    /**
     * 实验名称
     */
    private String name;
    /**
     * 接入系统对应的业务标签
     */
    private String factorTag;
    /**
     * 分层 ID
     */
    private Integer layerId;
    /**
     * 分组字段（参数类型）。 1-用户; 2-地理位置
     */
    private Integer groupFieldId;
    /**
```

```
     * 实验假设
     */
    private String expHyp;
    /**
     * 实验预期
     */
    private String resultExpect;
    /**
     * 指标 IDS
     */
    private String metricIds;
    /**
     * 实验开始时间
     */
    private Timestamp startTime;
    /**
     * 实验结束时间
     */
    private Timestamp endTime;
    /**
     * 实验状态。0-新建；1-已发布；2-生效中；3-已暂停；4-已终止；5-已结束
     */
    private Integer status;
    /**
     * 是否已上线
     */
    private Integer online;
    /**
     * 分区类型
     */
    private Integer partitionType;
    /**
     * 是否抽样打点。 0-否，1- 是
     */
    private Integer isSampling;
    /**
     * 抽样率，例如 5 表示 5%
     */
    private Integer samplingRatio;
    /**
     * 实验自定义配置信息
     */
    private String config;
    /**
     * 接入端系统名称，例如 pay 表示支付系统
```

```
     */
    private String serviceName;
    /**
     * 负责人信息
     */
    private String owner;
    /**
     * 是否已删除。0-未删除；1-已删除
     */
    private Integer isDelete;
    /**
     * 扩展信息
     */
    private String ext;
    /**
     * 创建时间
     */
    private Timestamp createTime;
    /**
     * 更新时间
     */
    private Timestamp updateTime;
}
```

（2）定义“A/B 测试信息表”的持久层（Dao 层）接口。代码如下：

```
package com.wudimanong.experiment.dao.mapper;
import com.baomidou.mybatisplus.core.mapper.BaseMapper;
import com.wudimanong.experiment.dao.model.AbtestExpInfoPO;
import org.springframework.stereotype.Repository;
@Repository
public interface AbtestExpInfoDao extends BaseMapper<AbtestExpInfoPO> {
}
```

2. 开发“实验创建”接口的持久层（Dao 层）代码

“实验创建”接口的持久层（Dao 层）实现，涉及“A/B 测试信息表”（abtest_exp_info）、“A/B 测试分层表”（abtest_layer），以及“A/B 测试分组表”（abtest_group）的数据库操作。其中，“A/B 测试信息表”的持久层（Dao 层）在“1.”小标题中已经定义。接下来定义“A/B 测试分层表”，以及“A/B 测试分组表”的持久层（Dao 层）。

（1）定义“A/B 测试分层表”的数据库实体类。代码如下：

```
package com.wudimanong.experiment.dao.model;
import com.baomidou.mybatisplus.annotation.IdType;
...
```

```
import lombok.Data;
@Data
@TableName("abtest_layer")
public class AbtestLayerPO {
    /**
     * 分层 ID，使用数据库自增机制
     */
    @TableId(value = "id", type = IdType.AUTO)
    private Integer id;
    /**
     * 分层名称
     */
    private String name;
    /**
     * 分层描述
     */
    @TableField(value = "`desc`")
    private String desc;
    /**
     * 流量分组字段类型。1-用户；2-地理位置
     */
    private Integer groupFieldId;
    /**
     * 当前流量层的分桶总数
     */
    private Integer bucketTotalNum;
    /**
     * 未被使用的分桶编号列表
     */
    private String unusedBucketNos;
    /**
     * 分区类型（功能待扩展字段）
     */
    private Integer partitionType;
    /**
     * 更新时间
     */
    private Timestamp updateTime;
    /**
     * 创建时间
     */
    private Timestamp createTime;
    /**
     * 是否已删除。0-未删除；1-已删除
     */
```

```
    private Integer isDelete;
}
```

（2）定义“A/B 测试分层表”的持久层（Dao 层）接口。代码如下：

```
package com.wudimanong.experiment.dao.mapper;
import com.baomidou.mybatisplus.core.mapper.BaseMapper;
import com.wudimanong.experiment.dao.model.AbtestLayerPO;
import org.springframework.stereotype.Repository;
@Repository
public interface AbtestLayerDao extends BaseMapper<AbtestLayerPO> {
}
```

（3）定义“A/B 测试分组表”的数据库实体类。代码如下：

```
package com.wudimanong.experiment.dao.model;
import com.alibaba.fastjson.JSON;
...
import lombok.Data;
@Data
@TableName("abtest_group")
public class AbtestGroupPO {
    /**
     * 分组 ID
     */
    @TableId(value = "id", type = IdType.AUTO)
    private Integer id;
    /**
     * 分组类别。0-实验组；1-对照组
     */
    private Integer groupType;
    /**
     * 分组后流量占比
     */
    private Integer flowRatio;
    /**
     * 实验 ID
     */
    private Integer expId;
    /**
     * 分组名称
     */
    private String name;
    /**
     * 分组类型。0-区间分组；1-单双号分组
     */
    private Integer groupPartitionType;
```

```
    /**
     * 分流内包含的桶编号列表
     */
    private String groupPartitionDetails;
    /**
     * 策略对应的 JSON 格式信息
     */
    private String strategyDetail;
    /**
     * 是否已上线
     */
    private Integer online;
    /**
     * 创建时间
     */
    private Timestamp createTime;
    /**
     * 更新时间
     */
    private Timestamp updateTime;
    /**
     * 分流的稀释系数，实际分流为 flowRatio / dilutionRatio
     */
    private Integer dilutionRatio;
    /**
     * 白名单
     */
    private String whiteList;
}
```

（4）定义“A/B 测试分组表”的持久层（Dao 层）接口。代码如下：

```
package com.wudimanong.experiment.dao.mapper;
import com.baomidou.mybatisplus.core.mapper.BaseMapper;
...
import org.springframework.stereotype.Repository;
@Repository
public interface AbtestGroupDao extends BaseMapper<AbtestGroupPO> {
    /**
     * 批量插入方法
     */
    int batchInsert(List<AbtestGroupPO> list);
}
```

上述持久层（Dao 层）接口的实现，并非完全基于 MyBatis-Plus 的默认操作方法。在业务层（Service 层）的 A/B 测试分组信息的持久化逻辑中需要用到“批量插入”功能，而 MyBatis-Plus 针对“批量插入”功能并没有提供相应地支持。因此，下面继续通过 MyBatis 的 XML 映射方式定义数据库“批量插入”的方法。

在 A/B 测试微服务工程的“src/resources/mybatis/”目录下创建 SQL 映射文件——AbtestGroupDao.xml。代码如下：

```
    <?xml version="1.0" encoding="UTF-8"?>
    <!DOCTYPE mapper PUBLIC "-//mybatis.org//DTD Mapper 3.0//EN"
"http://mybatis.org/dtd/mybatis-3-mapper.dtd">
    <mapper namespace="com.wudimanong.experiment.dao.mapper.AbtestGroupDao">
        <!--批量插入（插入后返回主键 ID）-->
        <insert id="batchInsert" useGeneratedKeys="true" keyProperty="id"

parameterType="com.wudimanong.experiment.dao.model.AbtestGroupPO">
            INSERT INTO

abtest_group(group_type,flow_ratio,exp_id,`name`,group_partition_details,str
ategy_detail,online,create_time,update_time,dilution_ratio,white_list)
            VALUES
            <foreach collection="list" item="cost" index="index" separator=",">

(#{cost.groupType,jdbcType=INTEGER},#{cost.flowRatio,jdbcType=INTEGER},#{cos
t.expId,jdbcType=INTEGER},#{cost.name,jdbcType=VARCHAR},#{cost.groupPartitio
nDetails,jdbcType=VARCHAR},#{cost.strategyDetail,jdbcType=VARCHAR},#{cost.on
line,jdbcType=INTEGER},#{cost.createTime,jdbcType=TIMESTAMP},#{cost.updateTi
me,jdbcType=TIMESTAMP},#{cost.dilutionRatio,jdbcType=INTEGER},#{cost.whiteLi
st,jdbcType=VARCHAR})
            </foreach>
        </insert>
    </mapper>
```

3. 开发“实验流量编辑”接口及“实验配置信息获取”接口的持久层（Dao 层）代码

“实验流量编辑”接口涉及的持久层（Dao 层）操作组件及数据对象，沿用“2.”小标题中定义的数据库表的持久层（Dao 层）组件即可。

“实验配置信息获取”接口涉及的持久层（Dao 层）操作组件及数据对象，沿用“1.”小标题及“2.小标题”中定义的数据库表的持久层（Dao 层）组件即可。

7.6 步骤 4：基于 Spring Boot Starter 方式编写“接入 SDK”

在 A/B 测试微服务中，“接入 SDK”的便捷性是很重要的一个方面。一般来说，针对不同的 A/B 测试场景应该提供不同端类型及编程语言类型的“接入 SDK”。

本节以将 Spring Cloud 微服务接入 A/B 测试微服务为例，基于 Spring Boot Starter 方式来编写 A/B 测试微服务的“接入 SDK”。

7.6.1 创建 Spring Boot Starter 工程代码

提供 Spring Boot Starter 方式的“接入 SDK”，可以较大限度地降低将 Spring Cloud 微服务接入 A/B 测试微服务进行 A/B 测试的难度。接下来，构建 A/B 测试微服务“接入 SDK”的工程代码。

1. 创建一个基本的 Maven 工程

利用 2.3.1 节介绍的方法创建一个 Maven 工程，完成后的工程代码结构如图 7-9 所示。

```
▼ chapter07-experiment-java-starter [experiment-java-starter] ~/dev-tools/worl
  ▼ src
    ▼ main
        java
        resources
    ▼ test
        java
    chapter07-experiment-java-starter.iml 2020/4/29, 10:38 上午, 80 B
    experiment-java-starter.iml 2020/4/29, 10:39 上午, 830 B
    pom.xml 2020/4/29, 10:37 上午, 439 B
  README.md 2020/1/7, 5:43 下午, 76 B
▶ External Libraries
  Scratches and Consoles
```

图 7-9

2. 引入 Spring Boot Starter 依赖，并配置自动加载机制

（1）引入 Spring Cloud 的父依赖。

在“1.”小标题中创建的 Maven 工程的 pom.xml 文件中引入 Spring Cloud 的父依赖。代码如下：

```
<?xml version="1.0" encoding="UTF-8"?>
<project xmlns="http://maven.apache.org/POM/4.0.0"
         xmlns:xsi="http://www.w3.org/2001/XMLSchema-instance"
```

```
        xsi:schemaLocation="http://maven.apache.org/POM/4.0.0
http://maven.apache.org/xsd/maven-4.0.0.xsd">
      <modelVersion>4.0.0</modelVersion>
      <groupId>com.wudimanong</groupId>
      <artifactId>experiment-java-starter</artifactId>
      <version>1.0-SNAPSHOT</version>
      <!--引入 Spring Cloud 的父依赖-->
      <parent>
         <groupId>org.springframework.cloud</groupId>
         <artifactId>spring-cloud-starter-parent</artifactId>
         <version>Greenwich.SR1</version>
         <relativePath/>
      </parent>
   </project>
```

为了使得“接入 SDK”与其他微服务所依赖的 Spring Boot 版本一致，需要引入相同的 Spring Cloud 父依赖。

（2）在 pom.xml 文件中引入其他的依赖。代码如下：

```
   <!--引入 A/B 测试微服务“experiment”本身的 Client 依赖-->
   <dependency>
      <groupId>com.wudimanong</groupId>
      <artifactId>experiment-java-client</artifactId>
      <version>1.0-SNAPSHOT</version>
   </dependency>
   <!--Spring Boot 的基本依赖-->
   <dependency>
      <groupId>org.springframework.cloud</groupId>
      <artifactId>spring-cloud-starter-openfeign</artifactId>
   </dependency>
   <!--引入“openfeign”的依赖，后续接入的 SDK 将通过微服务的服务发现机制连接 A/B 测试微
服务来获取配置-->
   <dependency>
      <groupId>org.springframework.cloud</groupId>
      <artifactId>spring-cloud-starter-openfeign</artifactId>
   </dependency>
   <!--引入配置注解依赖-->
   <dependency>
      <groupId>org.springframework.boot</groupId>
      <artifactId>spring-boot-configuration-processor</artifactId>
      <optional>true</optional>
   </dependency>
```

（3）实现 Spring Boot Starter 组件的自动化配置机制。

在“接入 SDK”工程的“/src/main/resources/META-INF/”目录下创建 spring.factories 文件——它是 Spring Boot SPI 机制的入口。

在 spring.factories 文件中配置 Spring Boot 自动配置类，以实现 Spring Boot Starter 组件的自动化配置。代码如下：

```
org.springframework.boot.autoconfigure.EnableAutoConfiguration=com.wudim
anong.experiment.starter.autoconfigure.ExperimentAutoConfiguration
```

在上方配置中，所依赖的自动化配置类的实现可以参考 7.6.2 节的内容。

7.6.2 开发“接入 SDK”的代码

1. 编写“接入 SDK”的自动配置类

（1）编写在“接入 SDK”工程的“/src/resources/META-INF/spring.factories”文件中设置的自动注入的配置类 ExperimentAutoConfiguration。代码如下：

```
package com.wudimanong.experiment.starter.autoconfigure;
import com.wudimanong.experiment.starter.ExperimentTemplate;
...
import org.springframework.context.annotation.Configuration;
@Configuration
@ConditionalOnProperty(value = "experiment.enable", havingValue = "true")
@EnableConfigurationProperties({ExperimentProperties.class})
public class ExperimentAutoConfiguration {
    @EnableFeignClients(clients = ExperimentFeignClient.class)
    @EnableDiscoveryClient
    @ConditionalOnMissingBean({ExperimentFeignSource.class,
ExperimentFeignClient.class})
    public class ExperimentSourceConfiguration {
        @Bean
        @ConditionalOnMissingBean(ExperimentFeignSource.class)
        public ExperimentFeignSource
experimentFeignSource(ExperimentFeignClient experimentFeignClient) {
            return new ExperimentFeignSource(experimentFeignClient);
        }
    }
    @Bean
    @ConditionalOnBean(ExperimentFeignSource.class)
```

```
    public ExperimentTemplate experimentTemplate(ExperimentFeignSource
experimentFeignSource) {
        return new ExperimentTemplate(experimentFeignSource);
    }
}
```

在上述代码中，使用 Spring Boot 注解来实现对依赖注入行为的控制——例如，通过@ConditionalOnProperty 注解控制自动配置类是否加载。如果在系统配置参数中设置了 experiment.enable 属性，且设置为 true，则 Spring 容器就会自动加载该配置类。

（2）编写步骤（1）中涉及的配置属性类 ExperimentProperties。

通过@EnableConfigurationProperties 注解，可以将系统配置参数直接映射成 Java 对象。例如，通过定义 ExperimentProperties 类，将“接入 SDK”所依赖的配置参数直接映射成 Java 对象，代码如下：

```
package com.wudimanong.experiment.starter.properties;
import lombok.Data;
import
org.springframework.boot.context.properties.ConfigurationProperties;
@Data
@ConfigurationProperties("experiment")
public class ExperimentProperties {
    private String enable;
}
```

上述代码将以“experiment”为前缀的属性直接映射到 ExperimentProperties 类的对象中。对于需要依赖外部配置的 Starter 组件来说，这样的方式可以更高效地实现配置参数的获取及控制。

2. 开发访问 A/B 测试微服务“experiment”的 FeignClient 代码

（1）开发调用 A/B 测试微服务“实验配置信息”接口的 FeignClient 代码。

在编写“接入 SDK”的过程中获取“实验配置信息”，需要访问 A/B 测试微服务的接口。所以，在“1.”小标题中的自动配置类中，通过@EnableFeignClients 及@EnableDiscoveryClient 注解引入 A/B 测试微服务的 FeignClient 接口 ExperimentFeignClient。代码如下：

```
package com.wudimanong.experiment.starter.feign;
import com.wudimanong.experiment.client.entity.ResponseResult;
import com.wudimanong.experiment.client.entity.bo.ConfigBO;
...
import org.springframework.web.bind.annotation.RequestParam;
```

```
    @FeignClient(value = "experiment", configuration =
ExperimentFeignConfiguration.class, fallbackFactory =
ExperimentFeignFallbackFactory.class)
    public interface ExperimentFeignClient {
        /**
         * 获取“实验配置信息”的 FeignClient 接口的定义
         */
        @GetMapping("/config/findByFactorTag")
        ResponseResult<ConfigBO> findByFactorTag(@RequestParam("factorTag")
String factorTag);
    }
```

上述代码演示了 Spring Cloud 微服务 FeignClient 接口的编写方式，并通过定义 findByFactorTag()接口方法，实现对 A/B 测试系统中“实验配置信息获取”微服务接口的远程访问。

（2）编写步骤（1）中@FeignClient 注解的“configuration”和“fallbackFactory”属性所指定的配置类及降级处理类。

涉及的配置类 ExperimentFeignConfiguration 的代码如下：

```
    package com.wudimanong.experiment.starter.feign;
    import org.springframework.context.annotation.Bean;
    import org.springframework.context.annotation.Configuration;
    @Configuration
    public class ExperimentFeignConfiguration {
        /**
         * 构建 Fallback 工厂类
         */
        @Bean
        ExperimentFeignFallbackFactory experimentFeignFallbackFactory() {
            return new ExperimentFeignFallbackFactory();
        }
    }
```

涉及的降级处理类 ExperimentFeignFallbackFactory 的代码如下：

```
    package com.wudimanong.experiment.starter.feign;
    import com.wudimanong.experiment.client.entity.ResponseResult;
    import com.wudimanong.experiment.client.entity.bo.ConfigBO;
    import feign.hystrix.FallbackFactory;
    import lombok.extern.slf4j.Slf4j;
    @Slf4j
    public class ExperimentFeignFallbackFactory implements
FallbackFactory<ExperimentFeignClient> {
```

```
        @Override
        public ExperimentFeignClient create(Throwable cause) {
            return new ExperimentFeignClient() {
                @Override
                public ResponseResult<ConfigBO> findByFactorTag(String factorTag) {
                    log.info("A/B 测试微服务调用的降级逻辑处理...");
                    log.error(cause.getMessage());
                    return null;
                }
            };
        }
    }
```

3. 实现“接入 SDK”配置类的核心逻辑

在“1.”小标题中初步定义了支持 Spring Boot Starter 方式的自动配置类，接下来实现该配置类的核心逻辑。

（1）定义 ExperimentFeignSource 类。

在配置类 ExperimentAutoConfiguration 中定义的内部类 ExperimentSourceConfiguration，会通过@ConditionalOnMissingBean 注解来约定：如果缺失 ExperimentFeignSource 对象，则实例化该对象。ExperimentFeignSource 类的代码如下：

```
    package com.wudimanong.experiment.starter.feign;
    import com.wudimanong.experiment.client.entity.bo.ConfigBO;
    import java.util.Optional;
    public class ExperimentFeignSource {
        /**
         * A/B 测试微服务“experiment”的 FeignClient 接口
         */
        private ExperimentFeignClient experimentFeignClient;
        public ExperimentFeignSource(ExperimentFeignClient
experimentFeignClient) {
            this.experimentFeignClient = experimentFeignClient;
        }
        /**
         * 获取“实验配置信息”
         */
        public ConfigBO getDeliverConfig(String factorTag) {
            return Optional.of(experimentFeignClient.findByFactorTag
(factorTag).getData()).get();
        }
    }
```

上述代码主要实现了，通过 FeignClient 接口访问远程 A/B 测试微服务“experiment”的逻辑。

（2）定义 ExperimentTemplate 类。

在配置类 ExperimentAutoConfiguration 中，会通过@ConditionalOnBean 注解在 ExperimentFeignSource 类的实例存在的情况下去实例化"接入 SDK"的模板类 ExperimentTemplate。模板类 ExperimentTemplate 的代码如下：

```
package com.wudimanong.experiment.starter;
import com.wudimanong.experiment.client.entity.AbtestExp;
...
import com.wudimanong.experiment.utils.BucketUtils;
import java.util.Optional;
public class ExperimentTemplate {
    /**
     * 设置 A/B 测试微服务"experiment"的 FeignClient 接口的依赖
     */
    private ExperimentFeignSource experimentFeignSource;
    public ExperimentTemplate(ExperimentFeignSource experimentFeignSource) {
        this.experimentFeignSource = experimentFeignSource;
    }
    /**
     * 根据指定 ID 获取分流结果
     */
    public AbtestInfo get(String factorTag, String currIdStr) {
        //获取实验结果匹配信息
        MatchResult matchResult = math(factorTag, currIdStr);
        //生成返回结果对象数据
        AbtestInfo abtestInfo = AbtestInfo.builder().factorTag
(matchResult.getAbtestExp().getFactorTag())
                .paramId(currIdStr).result(matchResult).build();
        return abtestInfo;
    }
    /**
     * "实验业务标签 + 分流 ID"的匹配方法
     */
    private MatchResult math(String factorTag, String currIdStr) {
        //以 FeignClient 接口调用的方式，获取"实验的配置信息"
        ConfigBO result = experimentFeignSource.getDeliverConfig
(factorTag);
        if (result == null) {
            //如果获取"实验配置信息"失败，则返回空配置
            return MatchResult.builder().build();
        }
        //获取"实验配置信息"中的 A/B 测试关键配置信息
        AbtestExp abtestExp = result.getAbtestExp();
        //计算当前层流量分桶编号（核心逻辑）
```

```
        Long currBucketNo = AbtestUtils.getBucketNo(currIdStr,
abtestExp.getLayerId());
        //匹配“实验配置信息”中的流量分组信息（以 lambda 语法的方式，通过流过滤匹配来
计算当前流量桶号应该匹配到哪个分组）
        AbtestGroup destGroup = Optional.ofNullable
(abtestExp.getAbtestGroups().stream()
                .filter(lt -> lt.getIsUseBase64Nums() ?
BucketUtils.bitStr2buckets(lt.getPartitionSerialNums64())
                        .contains(currBucketNo.intValue())
                        : lt.getPartitionSerialNums().contains
(currBucketNo.intValue())).findFirst().get()).get();
        return
MatchResult.builder().destGroup(destGroup).abtestExp(abtestExp).retrieveType
(RetrieveType.BUCKET)
                .build();
    }
}
```

上述代码是“接入 SDK”的核心逻辑：

- 通过接入方传递的 A/B 测试业务标签及流量计算参数（例如 UID），来计算本次请求流量的分桶编号。
- 匹配“实验配置信息”中的流量分组——如果匹配实验组，则表示流量应该走新逻辑；如果匹配对照组，则说明逻辑应该走旧逻辑。从而实现流量的路由及分配。

（3）定义步骤（2）中 match()方法所依赖的 AbtestUtils 工具类。代码实现如下：

```
package com.wudimanong.experiment.starter.utils;
public class AbtestUtils {
    /**
     * 计算当前层流量分桶编号的方法
     */
    public static Long getBucketNo(String currIdStr, Integer layerId) {
        //将分流标识 ID 与流量分层 ID 拼装
        String destKey = currIdStr + layerId;
        //取 MD5 哈希值
        String md5Hex = Md5Utils.md5Hex(destKey, "UTF-8");
        //获取 Hash 数值类型
        Long hash = Long.parseLong(md5Hex.substring(md5Hex.length() - 16,
md5Hex.length() - 1), 16);
        if (hash < 0) {
            hash *= -1;
        }
        // 取模
        return (hash % 1000L);
    }
}
```

在上述代码中，通过计算“流量参数 ID + 实验分层 ID”的 MD5 值，并对 1000 取模，来均匀地得到 1000 以内的分桶编号数值。MD5 工具类的代码如下：

```
package com.wudimanong.experiment.starter.utils;
import java.security.MessageDigest;
public class Md5Utils {
    private static final String hexDigits[] = {"0", "1", "2", "3", "4", "5",
"6", "7", "8", "9","a", "b", "c", "d", "e", "f"};
    /**
     * 获取MD5哈希值的方法
     */
    public static String md5Hex(String origin, String charsetname) {
        String resultString = null;
        try {
            resultString = new String(origin);
            MessageDigest md = MessageDigest.getInstance("MD5");
            if (charsetname == null || "".equals(charsetname)) {
                resultString =
byteArrayToHexString(md.digest(resultString.getBytes()));
            } else {
                resultString =
byteArrayToHexString(md.digest(resultString.getBytes(charsetname)));
            }
        } catch (Exception exception) {
        }
        return resultString;
    }
    private static String byteArrayToHexString(byte b[]) {
        StringBuffer resultSb = new StringBuffer();
        for (int i = 0; i < b.length; i++) {
            resultSb.append(byteToHexString(b[i]));
        }
        return resultSb.toString();
    }
    private static String byteToHexString(byte b) {
        int n = b;
        if (n < 0) {
            n += 256;
        }
        int d1 = n / 16;
        int d2 = n % 16;
        return hexDigits[d1] + hexDigits[d2];
    }
}
```

（4）在“experiment-java-client”工程的 BucketUtils 工具类中，增加“分桶列表压缩”方法。

回到 match()方法，在进行具体的流量分桶编号匹配时，A/B 测试微服务为了接口方便传输，会对分桶编号列表大于 100 的数据进行 Base64 方式的压缩。

所以，在获取分桶编号列表配置进行计算时，需要考虑数据是否被 Base64 方式压缩——如果压缩过，则需要进行解压缩操作。具体在“experiment-java-client”工程的工具类 BucketUtils 中增加如下方法：

```
/**
 * 将以 Base64 方式压缩的分桶数据解压缩成 Set<Integer>类型
*/
public static Set<Integer> bitStr2buckets(String str) {
    BitSet bitSet = BitSet.valueOf(Base64.getUrlDecoder().decode(str));
    return bitSet.stream().boxed().collect(Collectors.toSet());
}
```

（5）回到 match()方法，定义该方法的返回数据对象 MatchResult。代码如下：

```
package com.wudimanong.experiment.starter.entity;
import com.wudimanong.experiment.client.entity.AbtestExp;
...
import lombok.Data;
import lombok.NoArgsConstructor;
@Data
@Builder
@AllArgsConstructor
@NoArgsConstructor
public class MatchResult {
    /**
     * 目标分桶结果
     */
    private AbtestGroup destGroup;
    /**
     * 原始实验配置信息
     */
    private AbtestExp abtestExp;
    /**
     * 匹配类型（有分桶编号、白名单两种类型）
     */
    private RetrieveType retrieveType;
    public enum RetrieveType {
        BUCKET, WHITE_LIST
    }
}
```

（6）定义 AbtestInfo 输出数据类。

回到模板类 ExperimentTemplate，为方便接入方使用该类，并没有直接将 match()方法暴露出去，而是通过 get()方法对其进行了包装，并重新定义了输出对象 AbtestInfo，代码如下：

```
package com.wudimanong.experiment.starter.entity;
import com.wudimanong.experiment.client.entity.Strategy;
...
import lombok.NoArgsConstructor;
@Data
@Builder
@AllArgsConstructor
@NoArgsConstructor
public class AbtestInfo {
    /**
     * 实验 Tag
     */
    private String factorTag;
    /**
     * 分流参数
     */
    private String paramId;
    /**
     * 匹配结果
     */
    private MatchResult result;
    /**
     * 策略信息
     */
    private List<Strategy> abtestStrategies;
    /**
     * 判断匹配分组是否为对照组
     */
    public Boolean isControl() {
        return this.result.getDestGroup().getGroupType().equals(1);
    }
    /**
     * 判断匹配分组是否为实验组
     */
    public Boolean isAbtest() {
        return this.result.getDestGroup().getGroupType().equals(1);
    }
}
```

至此，基于 Spring Boot Starter“开箱即用”方式，完成了 A/B 测试微服务“接入 SDK”代码的编写。

7.7 步骤 5：接入 A/B 测试微服务，实现灰度发布

接下来新构建一个 Spring Cloud 测试微服务，并将其接入 A/B 测试微服务，实现微服务新/旧功能的灰度发布。

7.7.1 创建 A/B 测试接入方微服务示例工程代码

1. 创建一个基本的 Maven 工程

利用 2.3.1 节介绍的方法创建一个 Maven 工程，完成后的工程代码结构如图 7-10 所示。

```
▼ chapter07-experiment-access-demo [experiment-access-demo] ~/dev-tools/wc
  ▼ src
    ▼ main
        java
        resources
    ▼ test
        java
    chapter07-experiment-access-demo.iml 2020/5/9, 5:26 下午, 80 B
    experiment-access-demo.iml 2020/5/9, 5:28 下午, 830 B
    pom.xml 2020/5/9, 5:26 下午, 438 B
```

图 7-10

2. 引入 Spring Cloud 依赖，将其改造为微服务项目

（1）引入 Spring Cloud 微服务的核心依赖。

这里可以参考 2.5.2 节中的具体步骤。

（2）在工程代码的 resources 目录新建一个基础性配置文件——bootstrap.yml。配置文件中的代码如下：

```
spring:
  application:
    name: experiment-access-demo
  profiles:
    active: debug
  cloud:
    consul:
      discovery:
```

```
        preferIpAddress: true
        instance-id: ${spring.application.name}:${spring.cloud.client.
ipAddress}:${spring.application.instance_id:${server.port}}:@project.version@
        healthCheckPath: /actuator/health
server:
  port: 8081
```

（3）Spring Boot 并不会默认加载 bootstrap.yml 这个文件，所以需要在 pom.xml 中添加 Maven 资源相关的配置，具体参考 2.5.2 节内容。

（4）创建接入测试微服务的入口程序类。代码如下：

```
package com.wudimanong.access.demo;
import org.springframework.boot.SpringApplication;
import org.springframework.boot.autoconfigure.SpringBootApplication;
import org.springframework.cloud.client.discovery.EnableDiscoveryClient;
@EnableDiscoveryClient
@SpringBootApplication
public class AccessDemo {
    public static void main(String[] args) {
        SpringApplication.run(AccessDemo.class, args);
    }
}
```

（5）在微服务工程的 pom.xml 文件中，引入 A/B 测试微服务“接入 SDK”的依赖。代码如下：

```
<!--引入 A/B 测试微服务“接入 SDK”的依赖-->
 <dependency>
     <groupId>com.wudimanong</groupId>
     <artifactId>experiment-java-starter</artifactId>
     <version>1.0-SNAPSHOT</version>
 </dependency>
```

7.7.2 通过接口调用的方式创建 A/B 测试

在正常情况下，A/B 测试的创建及分组流量分配的操作，是通过前端管理系统来完成的。这里为了方便测试，通过接口调用的方式来完成。

（1）运行 A/B 测试微服务“experiment”工程，如图 7-11 所示。

（2）通过 Postman 调用 A/B 测试微服务的管理端“实验创建”接口，来模拟 A/B 测试的创建过程，如图 7-12 所示。

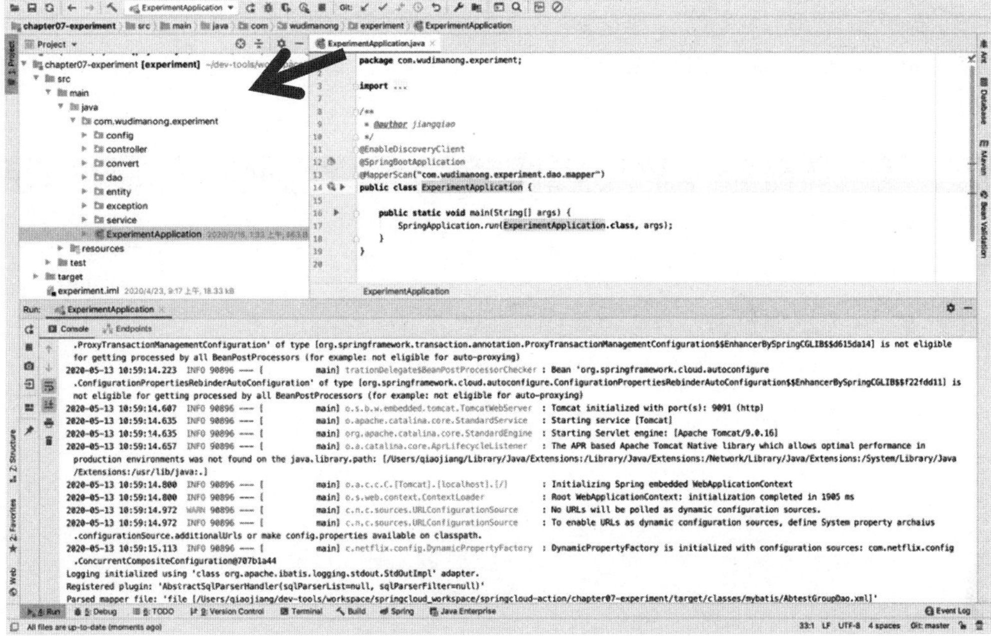

图 7-11

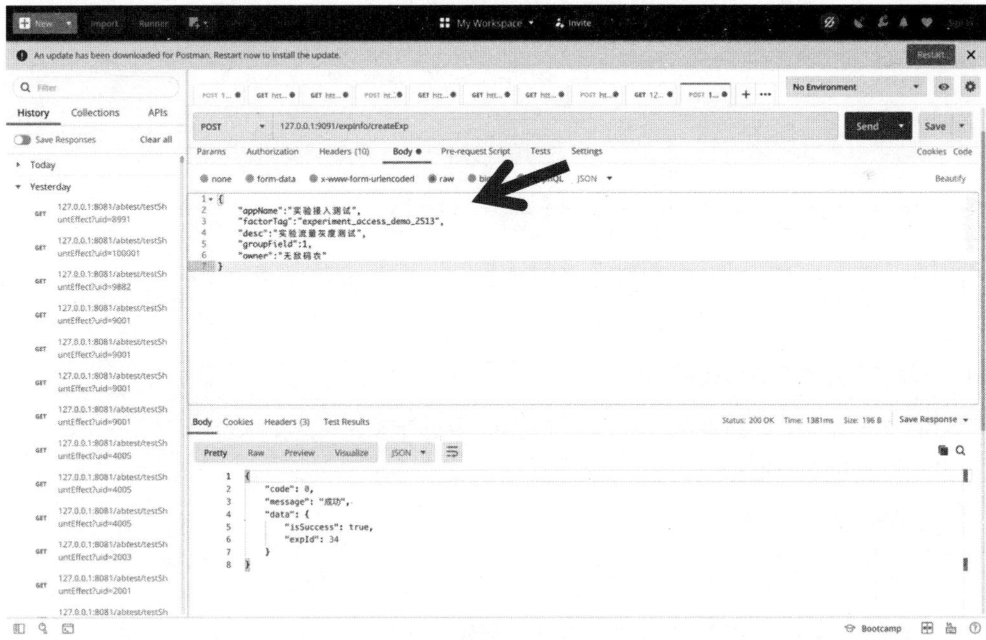

图 7-12

（3）在 A/B 测试创建成功后，默认会将实验组和对照组的流量占比各设置为 50%。调用 A/B 测试微服务的 SDK 端“实验配置信息获取”接口，可以查看步骤（1）中所创建 A/B 测试的配置信息，如图 7-13 所示。

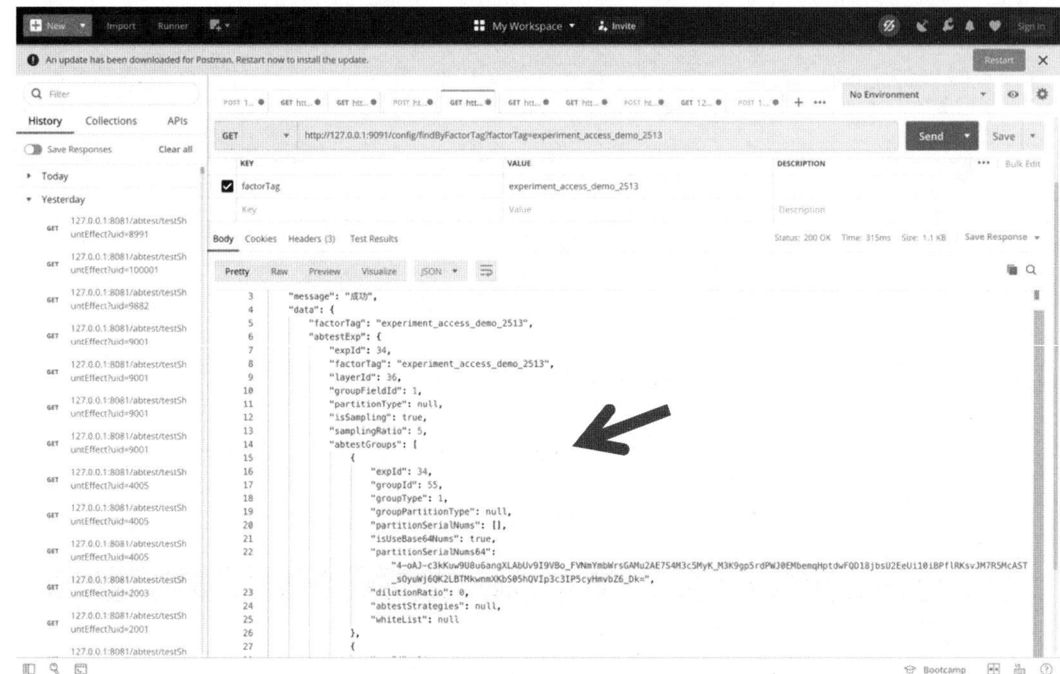

图 7-13

7.7.3 开发 A/B 测试代码，实现灰度流量切分

接下来，在 A/B 测试接入方微服务中编写 A/B 测试的分流逻辑，根据用户 ID 实现灰度流量切分的效果。

1. 开发接入方 A/B 测试代码

（1）在接入方微服务工程的“/src/resources”目录下创建配置文件 application.yml，并开启“接入 SDK”的开关。代码如下：

```
#开启“接入 SDK”的开关
experiment:
  enable: true
```

（2）编写一个测试接口的 Controller 层。代码如下：

```
package com.wudimanong.access.demo.controller;
```

```
import com.wudimanong.access.demo.entity.TestShuntEffectBO;
...
import org.springframework.web.bind.annotation.RestController;
@Slf4j
@RestController
@RequestMapping("/abtest")
public class AbtestController {
    @Autowired
    AbtestService abtestServiceImpl;
    @GetMapping("/testShuntEffect")
    public TestShuntEffectBO testShuntEffect(TestShuntEffectDTO
testShuntEffectDTO) {
        return abtestServiceImpl.testShuntEffect(testShuntEffectDTO);
    }
}
```

定义测试接口的请求参数对象。代码如下：

```
package com.wudimanong.access.demo.entity;
import lombok.Data;
@Data
public class TestShuntEffectDTO {
    private Long uid;
}
```

定义测试接口的返回参数对象。代码如下：

```
package com.wudimanong.access.demo.entity;
import lombok.Builder;
import lombok.Data;
@Data
@Builder
public class TestShuntEffectBO {
    private Long uid;
    private Boolean isNewLogic;
    private Integer testGroupCounter;
    private Integer controlGroupCounter;
}
```

（3）开发测试接口的业务层（Service 层）代码。

定义业务层（Service 层）接口类，代码如下：

```
package com.wudimanong.access.demo.service;
import com.wudimanong.access.demo.entity.TestShuntEffectBO;
import com.wudimanong.access.demo.entity.TestShuntEffectDTO;
public interface AbtestService {
    /**
```

```
         * A/B 测试接入示例方法
         */
        TestShuntEffectBO testShuntEffect(TestShuntEffectDTO
testShuntEffectDTO);
    }
```

实现业务层（Service 层）接口类的方法，代码如下：

```
    package com.wudimanong.access.demo.service.impl;
    import com.wudimanong.access.demo.entity.TestShuntEffectBO;
    ...
    import org.springframework.stereotype.Service;
    @Slf4j
    @Service
    public class AbtestServiceImpl implements AbtestService {
        /**
         * 定义记录“对照组”被调用次数的全局计数器
         */
        public static AtomicInteger testGroupCounter = new AtomicInteger(0);
        /**
         * 定义记录“实验组”被调用次数的全局计数器
         */
        public static AtomicInteger controlGroupCounter = new AtomicInteger(0);
        /**
         * 注入“接入 SDK”模板类的依赖
         */
        @Autowired
        ExperimentTemplate experimentTemplate;
        @Override
        public TestShuntEffectBO testShuntEffect(TestShuntEffectDTO
testShuntEffectDTO) {
            boolean isNewLogic = isNewLogic(testShuntEffectDTO.getUid());
            if (isNewLogic) {
                //给“对照组”增加计数次数
                controlGroupCounter.getAndIncrement();
                log.info("执行新逻辑->实验组执行");
            } else {
                log.info("执行老逻辑->对照组执行");
                //给“实验组”增加计数次数
                testGroupCounter.getAndIncrement();
            }
            return
TestShuntEffectBO.builder().uid(testShuntEffectDTO.getUid()).isNewLogic(isNe
wLogic)
                    .testGroupCounter(testGroupCounter.get()).controlGroupCount
er(controlGroupCounter.get()).build();
```

```
    }
    /**
     * 判断是否执行新逻辑（实验组）
     *
     * @param uid
     * @return
     */
    private Boolean isNewLogic(Long uid) {
        //获取“实验配置信息”
        AbtestInfo abtestInfo =
experimentTemplate.get("experiment_access_demo_2513", String.valueOf(uid));
        //判断流量应该匹配的分组，如果为实验组，则流量走新逻辑，否则走老逻辑
        if (abtestInfo.isAbtest()) {
            return true;
        }
        return false;
    }
}
```

通过上述步骤，完成了接入方 A/B 测试代码的编写。

2. 演示 A/B 测试效果，实现灰度流量切分

（1）启动接入方微服务（需要同时启动 A/B 测试微服务“experiment”），如图 7-14 所示。

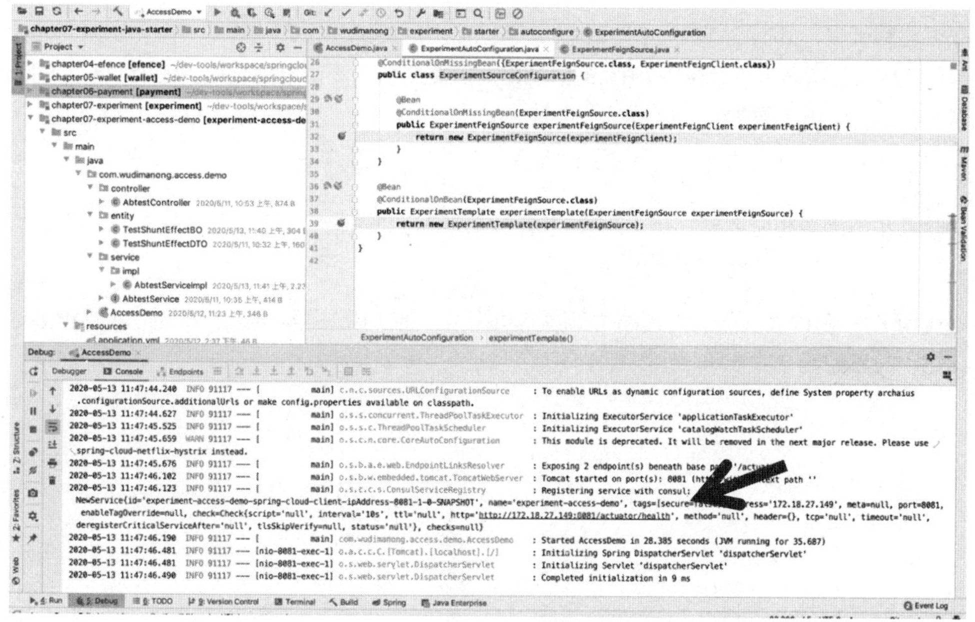

图 7-14

（2）测试“实验组”“对照组”流量各占 50%情况下的 A/B 测试效果。

按照默认的流量分配设置，接下来将分别以 UID 为 1001、2002、3001、4001、5001、6001、7001、8001、9001、1010 的用例，来模拟访问接入方服务，并验证其分流效果是否会以 50%的比例执行。最终测试结果如图 7-15 所示。

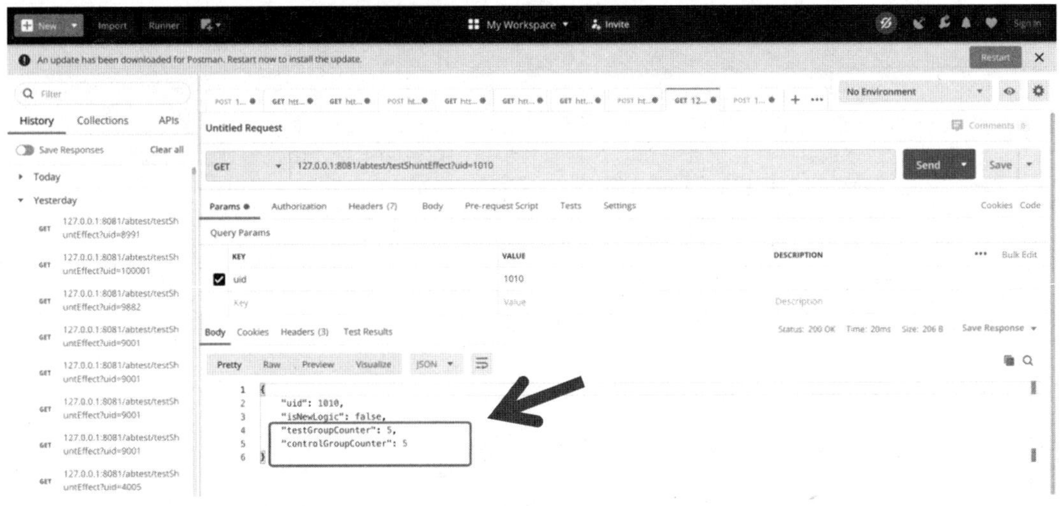

图 7-15

在上述测试中，分别以 10 个不同的 UID 来模拟访问测试接口。从最后一次调用返回的计数器结果来看，这 10 次请求的流量以 50%的分配比例被路由到了不同的逻辑层，这说明分流的实际效果满足流量比例设置的要求。

（3）测试将“实验组”流量占比调到 100%后的 A/B 测试效果。

为验证动态流量切分效果，接下来通过接口调用的方式调整流量分配比例——将实验组的流量占比调到 100%，将对照组流量调到 0%。

在调用 A/B 测试微服务的管理端“实验流量编辑”接口后，再调用 SDK 端“实验配置信息获取”接口，效果如图 7-16 所示。

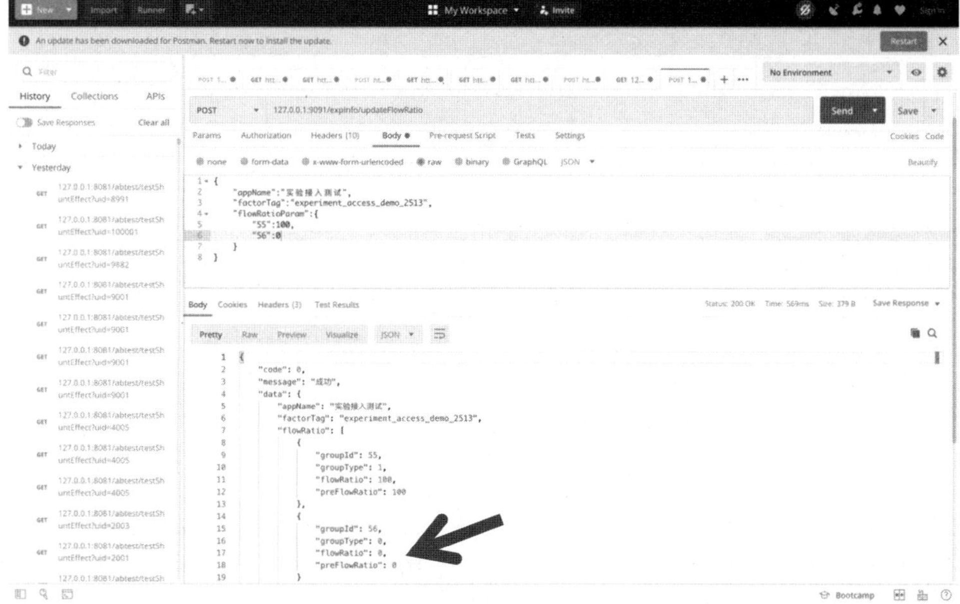

图 7-16

可以看到，调整后的实验组流量占比达到了 100%，而对照组流量则调整为 0%。

接下来，重新执行步骤（2），调用 10 次测试接口的分流效果如图 7-17 所示。

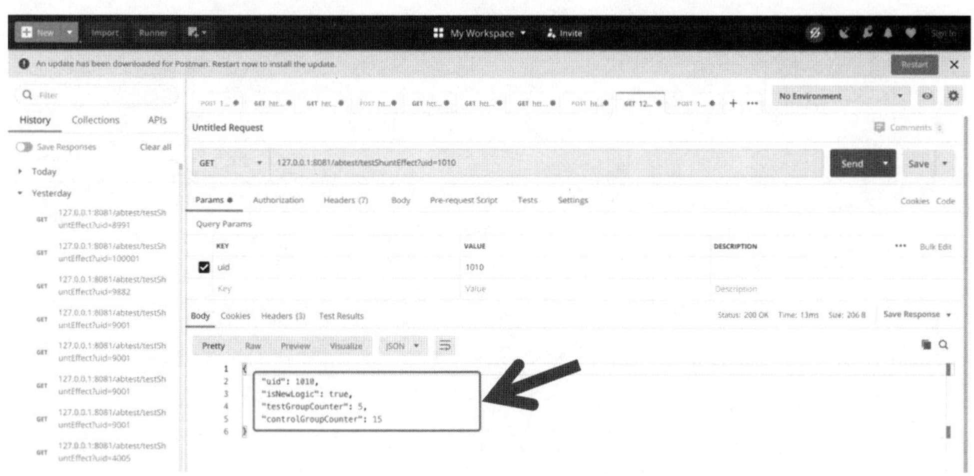

图 7-17

可以看到，这 10 次请求的流量已经全部分到实验组逻辑，这可以证明流量动态调整效果达到预期。

7.8 本章小结

本章主要演示了在 Spring Cloud 微服务架构下 A/B 测试微服务的构建。通过流程和系统的设计，展示了以数据驱动为核心的 A/B 测试的关键逻辑。在具体的实现过程中，重点讲解了如何基于 Spring Boot 框架特性编写具备“开箱即用”能力的 SDK 组件，这是 Spring Cloud 微服务日常开发工作中会遇到的场景。

本章最后演示了在实际场景下将一个微服务接入 A/B 测试微服务，实现功能逻辑的灰度流量切分方法，而这也是在互联网系统研发中确保服务平稳迭代的常用手段。

第 8 章

【实例】分布式任务调度系统

——用“ZooKeeper + ElasticJob”处理分布式任务

在实际业务中，任务处理是微服务体系中常见的一类功能。例如，支付系统会通过定时任务来处理支付对账；员工系统会通过任务来扫描即将过生日的员工，并批量发送关怀邮件。

由于微服务具备分布式部署的特点，所以，处理同一个任务的应用往往会被部署在多个服务器节点上。如果一个任务同时被多个节点调度执行，则可能引起业务上的错误和混乱。

因此，任务处理功能需要具备分布式调度能力，以确保某一个任务同一个时刻只会被调度到一个节点上执行。要实现此类功能，需要将任务处理功能接入独立的分布式调度系统。

本实例将通过主流的分布式任务调度方案，来介绍在 Spring Cloud 中构建分布式任务调度系统的方法。

通过本章，读者将学习到以下内容：

- 分布式任务调度系统的基本概念。
- 构建基于 ElasticJob 框架的分布式任务调度系统的方法。
- 基于 Spring Boot Starter 方式编写 ElasticJob 的“接入 SDK”。
- 将 Spring Cloud 微服务接入分布式任务调度系统的方法。

8.1 功能概述

任务一般可分为以下两种。

- 周期任务：根据设置的周期不间断运行的任务。例如，将支付账单下载任务设置为每天上午 10 点执行；将支付状态检查任务设置为每 1 分钟执行一次等。
- 定时任务：在固定时间点执行的任务。例如，在某个确定的时间点发送推送消息之类。

一般来说，微服务场景下的任务处理，除需要支持不同的任务类型外，还需要具备分布式任务的调度能力——既能支持任务处理节点的高可用、多副本部署，又能实现分布式任务调度（不允许同一个任务在同一时刻被多个节点重复执行）。

分布式任务调度系统还应该支持基于 Cron 表达式的配置，以便根据需求的变化随时调整任务的执行时间或周期。

实现任务处理的方式有很多：如 Linux 系统中的 Crontab、Java 语言中的 Timer 类，以及可在 Spring 中集成的 Spring Task 和 Quartz 等框架。但这些方式都不支持分布式任务处理。在微服务体系中，要实现分布式任务处理，需要引入一套独立的分布式任务调度系统，以简化任务的开发，并实现海量分布式任务的集中调度和管理。

8.2 步骤 1：构建分布式任务调度系统

本实例将基于分布式任务调度框架 ElasticJob，来实现独立的分布式任务调度系统。

分布式任务调度系统的实现，需要分布式任务调度系统的支持。在 Spring Cloud 微服务中，可以通过接入由“Zookeeper + ElasticJob”组成的分布式任务调度系统，来实现微服务任务处理的分布式调度。

8.2.1 认识分布式任务调度框架 ElasticJob

ElasticJob 是一款应用广泛的分布式任务调度框架，能够满足较大规模分布式任务调度需求。ElasticJob 已经成为 Apache 顶级开源项目“Apache ShardingSphere”的一个子项目，这也使得它具备了更强的可持续维护能力。

在引入 ElasticJob 后，开发人员可以更专注于任务本身的业务逻辑，而像分布式协调、管理调度及性能等非业务功能层面的逻辑，则由 ElasticJob 框架及 ZooKeeper 服务进行处理。

ElasticJob 的开源版本主要由以下两个相互独立的子项目组成。

- ElasticJob-Lite：一种无中心化的、依赖分布式协调服务 ZooKeeper 来实现分布式任务调度的解决方案。
- ElasticJob-Cloud：一种中心化的、依托容器调度平台 Mesos 的分布式任务调度系统。从架构方式看，基于容器平台的调度，能够实现任务进程级的瞬时调度——不用像 ElasticJob-Lite 方案那样常驻内存，从而更节约系统资源。

ElasticJob-Cloud 依赖 Mesos 容器平台环境，所以，实施起来存在一定门槛，且当前主流的容器平台是 Kubernetes。此外，这种方案在实际场景中的落地案例也不多，大家了解即可。

本实例将采用 ElasticJob-Lite 作为分布式任务调度的实现方案，其架构如图 8-1 所示。

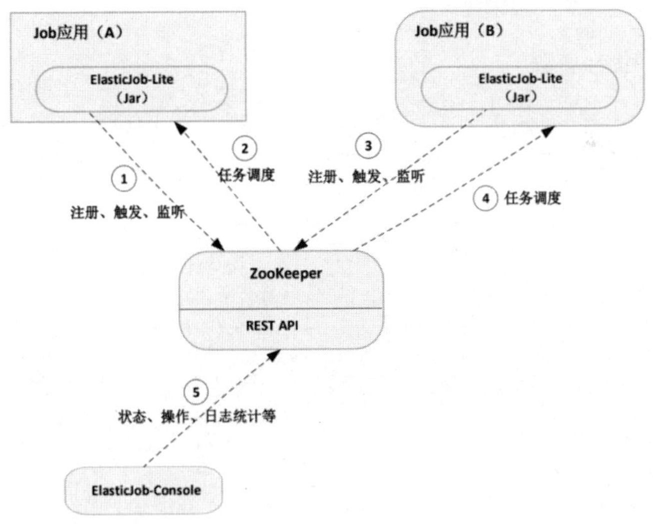

图 8-1

ElasticJob-Lite 的架构说明如下：

（1）微服务任务处理系统，通过集成 ElasticJob-Lite 来定义分布式任务。

（2）集成 ElasticJob-Lite 的任务处理系统，会将任务定义信息注册到分布式协调服务 ZooKeeper 中，并通过监听 ZooKeeper 的事件来完成任务的触发、调度执行等逻辑。

（3）ElasticJob-Console 控制台通过 REST API 与 ZooKeeper 连接，来实现管理 ElasticJob 任务信息、执行分布式调度节点的操作，以及执行日志查询等功能。

8.2.2 搭建 ZooKeeper 分布式协调服务

ZooKeeper 是一个开源的分布式协调服务，是 ElasticJob 实现分布式任务调度所依赖的核心服务。接下来搭建用于实验的 ZooKeeper 集群环境。

1. 下载 ZooKeeper 安装包

（1）通过“wget”命令将 ZooKeeper 的安装包下载至服务器的指定目录（如/opt）：

```
$ wget https://XXX/zookeeper/zookeeper-3.7.0/apache-zookeeper-3.7.0-bin.
tar.gz
```

（2）将安装包解压缩至指定目录：

```
$ tar -zxvf apache-zookeeper-3.7.0-bin.tar.gz -C /opt/zookeeper/node1
```

出于对安装路径合理规划的考虑，在“/opt”目录下创建一级子目录“/zookeeper”，并在该目录中再分别创建 3 个二级子目录：“node1”“node2”“node3”。之后，将 ZooKeeper 的安装包分别解压缩至这 3 个二级子目录中。

2. 安装 ZooKeeper 集群

为了保证 ZooKeeper 集群节点选举机制的正常运转，1 个 ZooKeeper 集群至少需要 3 个节点。如果条件有限，则可以通过在 1 台服务器中同时运行 3 个节点来组建一个 ZooKeeper 集群。

在“1.”小标题中所创建的“node1”“node2”“node3”目录，就是用于在一台机器中同时运行 3 个 ZooKeeper 节点时的安装程序运行配置目录。

（1）分别进入 ZooKeeper 安装包的解压缩目录，复制“/conf”目录中的样本配置：

```
$ cp zoo_sample.cfg zoo.cfg
```

（2）编辑“zoo.cfg”配置文件。代码如下：

```
...
#此处为 ZooKeeper 数据存储路径的设置
dataDir=/opt/zookeeper/node1/data
...
```

```
    #设置客户端连接端口。由于部署在 1 台机器上，所以分别将 node1、node2、node3 中的该参数设置为 2181、2182、2183
    clientPort=2181
    ...
    #分设置 ZooKeeper 集群节点组成，其格式为：server.{服务器编号}={IP 地址}:{Leader 选举端口}:{ZooKeeper 服务器的通信端口}
    server.1=127.0.0.1:2888:3888
    server.2=127.0.0.1:2889:3889
    server.3=127.0.0.1:2890:3890
```

参照上述配置，依次修改“node2”“node3”解压缩目录中的 zoo.cfg 配置文件。

（3）配置 ZooKeeper 服务节点的编号。

在各个节点的“data”数据目录中，分别创建名为“myid”的文件。根据节点设置，分别在该文件中写入 1、2、3 标号，以对应步骤（2）中“zoo.cfg”配置文件中用来指定节点 IP 及端口的“*server.{服务器编号}*”配置参数。

3. 安装 JDK

完成前面的步骤，实际上就完成了 ZooKeeper 集群的基本设置。但由于 ZooKeeper 是由 Java 编写的，所以，要正常运行它还需要在服务器中安装 JDK（1.8 版本）。

以 Ubantu 为例，安装 JDK 的具体命令如下：

```
    # sudo apt-get install openjdk-8-jdk
```

其他系统环境 JDK 的安装方法，可参考对应系统环境的安装方式。

4. 运行 ZooKeeper 集群

分别进入 ZooKeeper 安装程序解压缩目录的“/bin”目录下，执行启动 ZooKeeper 的脚本。命令如下：

```
$ ./node1/apache-zookeeper-3.7.0-bin/bin/zkServer.sh start
...
Starting zookeeper ... STARTED
```

执行成功后，可查看 ZooKeeper 集群的具体运行状态。命令如下。

```
$ ./node1/apache-zookeeper-3.7.0-bin/bin/zkServer.sh status
...
Mode: follower
$ ./node2/apache-zookeeper-3.7.0-bin/bin/zkServer.sh status
...
```

```
Mode: leader
$ ./node3/apache-zookeeper-3.7.0-bin/bin/zkServer.sh status
...
Mode: follower
```

如果集群运行成功，则在每个节点执行上述状态查询命令时，都会返回该节点所处的角色。此时 node2 为 leader 节点，剩下两个为 follower 节点。

8.2.3 部署 ElasticJob 的 Console 管理控制台

接下来，通过部署 ElasticJob 的 Console 控制台来连接 ZooKeeper 集群，实现对 ElasticJob 分布式任务的可视化管理。

1.下载 ElasticJob-UI 的安装包

ElasticJob-UI 既可以通过编译源码的方式进行安装，也可以通过编译好的安装包进行安装。提示，源码及编译好的安装包的下载地址以官网为准。

之后，将下载或编译好的 ElasticJob-UI 安装包解压缩至服务器中的指定目录中备用。

2. 添加 MySQL 驱动程序

由于软件许可的原因，一些数据库的 JDBC 驱动程序不能直接被 ElasticJob 引入，需要手动添加相关的驱动程序。

（1）下载 MySQL 驱动程序。提示，MySQL 驱动程序的下载地址以官网为准。命令如下：

```
#MySQL 驱动程序下载
$ wget https://{下载地址}/get/Downloads/Connector-J/mysql-connector-java-8.0.23.tar.gz
```

（2）解压缩下载的 MySQL 驱动程序，找到“mysql-connector-java-8.0.23.jar”文件，将其复制到 ElasticJob-UI 安装路径的“../ext-lib”目录下。

3. 启动 ElasticJob-UI 服务

进入 ElasticJob-UI 安装文件的“./bin”目录下，执行如下启动命令：

```
# ./bin/start.sh
Starting the ShardingSphere-ElasticJob-UI ...
Please check the STDOUT file:
/opt/elasticjob-console/apache-shardingsphere-elasticjob-3.0.0-RC1-lite-ui-bin/logs/stdout.log
```

通过查看控制台输出日志判断服务是否启动成功，命令如下：

```
# tail -f logs/stdout.log
...
 [INFO ] 10:52:16.414 [main] o.s.j.e.a.AnnotationMBeanExporter - Registering beans for JMX exposure on startup
[INFO ] 10:52:16.428 [main] o.a.coyote.http11.Http11NioProtocol - Starting ProtocolHandler ["http-nio-8088"]
[INFO ] 10:52:16.433 [main] o.a.tomcat.util.net.NioSelectorPool - Using a shared selector for servlet write/read
[INFO ] 10:52:16.491 [main] o.s.b.c.e.t.TomcatEmbeddedServletContainer - Tomcat started on port(s): 8088 (http)
[INFO ] 10:52:16.495 [main] o.a.s.elasticjob.lite.ui.Bootstrap - Started Bootstrap in 7.471 seconds (JVM running for 8.548)
```

可以看到，ElasticJob-UI 的服务已经在 8088 端口运行成功。

4. 配置 ZooKeeper 连接，实现对分布式任务的管理

在 ElasticJob-UI 运行成功后，通过浏览器打开控制台。地址为："***部署服务器 IP 地址*** + 8088"。例如：

```
http://10.211.55.12:8088/#/
```

此时系统会进入登录页面，如图 8-2 所示。

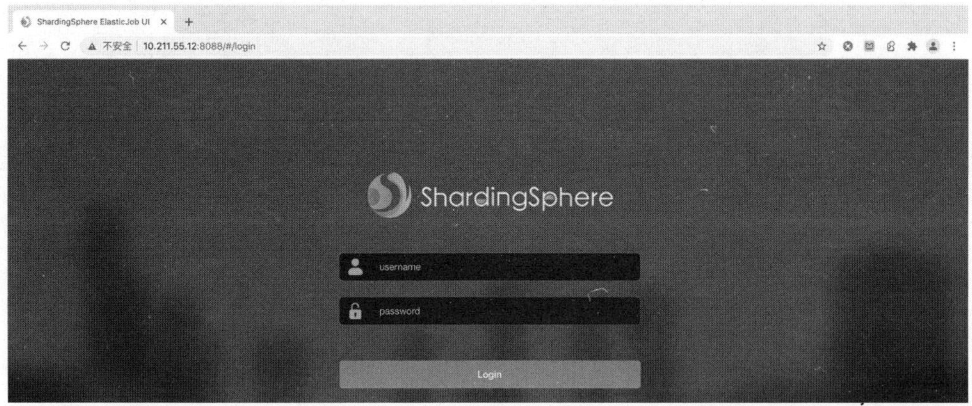

图 8-2

这里默用户和密码都为“root”，可以通过 ElasticJob-UI 安装目录中的“./conf/application.properties”配置文件进行修改。

登录后的效果如图 8-3 所示。

图 8-3

ElasticJob-UI 本身并不与任何环境绑定，但它可以通过连接 ZooKeeper 注册中心，实现对 ElasticJob 分布式任务的管理。

5. **添加注册中心**

下面用 ElasticJob-UI 可以同时管理多个 ZooKeeper 集群，并通过给同一个 ZooKeeper 集群配置不同的命名空间，来实现对不同任务服务的隔离。

以支付对账类的任务系统为例，在注册中心界面中配置了名称为“支付对账任务系统”、命名空间为“check-schedule”的注册中心，并指定了 ZooKeeper 集群的地址，如图 8-4 所示。

图 8-4

此时，如果单击“注册中心配置”菜单，进入注册中心配置列表后，单击连接已配置的注册中心，则控制台操作切换到对应的 ZooKeeper 集群上，如图 8-5 所示。

图 8-5

之后，在“作业操作”中可以看到该注册中心中的作业信息，如图 8-6 所示。

图 8-6

至此，完成了“ZooKeeper + ElasticJob”分布式任务调度系统的构建。后续将据此实现 Spring Cloud 微服务的分布式任务处理。

8.3 步骤 2：实现 Spring Cloud 微服务分布式任务处理

接下来，搭建 Spring Cloud 微服务工程代码结构，并集成 ElasticJob 分布式任务调度框架，以实现微服务的分布式任务处理。这里要实现的分布式任务处理是支付对账。

8.3.1 创建 Spring Cloud 微服务工程

1. 创建一个基本的 Maven 工程结构

利用 2.3.1 节介绍的方法创建一个 Maven 工程，完成后的工程代码结构如图 8-7 所示。

```
▼ chapter08-check-schedule [check-schedule] ~/dev-tools/workspace/springc
  ▼ src
    ▼ main
        java
        resources
    ▶ test
    chapter08-check-schedule.iml 2021/4/6, 6:24 下午, 80 B
    check-schedule.iml 2021/4/6, 6:25 下午, 830 B
    pom.xml 2021/4/6, 6:24 下午, 430 B
```

图 8-7

2. 引入 Spring Cloud 依赖，将其改造为微服务项目

（1）引入 Spring Cloud 微服务的核心依赖。

这里可以参考 2.5.2 节中的具体步骤。

（2）在工程代码的 resources 目录新建一个基础性配置文件——bootstrap.yml。配置文件中的代码如下：

```
spring:
  application:
    name: check-schedule
  profiles:
    active: debug
  cloud:
    consul:
      discovery:
        preferIpAddress: true
        instance-id:
${spring.application.name}:${spring.cloud.client.ipAddress}:${spring.applica
tion.instance_id:${server.port}}:@project.version@
        healthCheckPath: /actuator/health
server:
  port: 9091
```

（3）Spring Boot 并不会默认加载 bootstrap.yml 这个文件，所以需要在 pom.xml 中添加 Maven 资源相关的配置，具体参考 2.5.2 节内容。

（4）创建微服务的入口程序类。代码如下：

```
package com.wudimanong.schedule;
import org.springframework.boot.SpringApplication;
import org.springframework.boot.autoconfigure.SpringBootApplication;
import org.springframework.cloud.client.discovery.EnableDiscoveryClient
@EnableDiscoveryClient
@SpringBootApplication
public class CheckScheduleApplication {
```

```
    public static void main(String[] args) {
        SpringApplication.run(CheckScheduleApplication.class, args);
    }
}
```

至此，Spring Cloud 任务调度示例微服务工程就构建出来了。

8.3.2 编写 ElasticJob 的“接入 SDK”

ElasticJob 官方提供的 Spring Boot 集成方式，对分布式任务开发来说并不友好。在实际应用中，可以基于 Spring Boot Starter 方式编写自定义的“接入 SDK”，来更方便、优雅地定义分布式任务。

编写一个好用的 Spring Boot Starger 方式的“接入 SDK”，可以极大地方便 Spring Boot 项目快速接入一个基础组件。同样，编写 ElasticJob 的“接入 SDK”，目的也是为了方便微服务能够更快速地实现分布式任务处理。

1. 创建一个基本的 Maven 工程结构

利用 2.3.1 节介绍的方法创建一个 Maven 工程，完成后的工程代码结构如图 8-8 所示。

```
▼ chapter08-elasticjob-springboot-starter [elasticjob-springboot-starter] ~/dev-tools/wor
  ▶ .idea
  ▼ src
    ▼ main
        java
        resources
    ▼ test
        java
    chapter08-elasticjob-springboot-starter.iml 2021/4/25, 5:02 下午, 80 B
    pom.xml 2021/4/25, 5:01 下午, 445 B
```

图 8-8

2. 引入 Spring Boot 父依赖

（1）在 Maven 工程的 pom.xml 文件中，引入 Spring Boot 父依赖。代码如下：

```
<?xml version="1.0" encoding="UTF-8"?>
<project xmlns="http://maven.apache.org/POM/4.0.0"
         xmlns:xsi="http://www.w3.org/2001/XMLSchema-instance"
         xsi:schemaLocation="http://maven.apache.org/POM/4.0.0
http://maven.apache.org/xsd/maven-4.0.0.xsd">
    <modelVersion>4.0.0</modelVersion>
    <groupId>com.wudimanong</groupId>
    <artifactId>elasticjob-springboot-starter</artifactId>
```

```
    <version>1.0-SNAPSHOT</version>
    <!--引入 Spring Boot 父依赖-->
    <parent>
        <groupId>org.springframework.boot</groupId>
        <artifactId>spring-boot-starter-parent</artifactId>
        <version>2.1.3.RELEASE</version>
        <relativePath/>
    </parent>
</project>
```

"接入 SDK"工程引入的 Spring Boot 父依赖的版本，应与本书其他章节引入的 Spring Cloud 父依赖所对应的 Spring Boot 的版本一致。

（2）引入 Spring Boot 的基础依赖。代码如下：

```
<!--Spring Boot 的基础依赖-->
<dependency>
    <groupId>org.springframework.boot</groupId>
    <artifactId>spring-boot-starter</artifactId>
</dependency>
<dependency>
    <groupId>org.springframework.boot</groupId>
    <artifactId>spring-boot-starter-logging</artifactId>
</dependency>
<dependency>
    <groupId>org.springframework.boot</groupId>
    <artifactId>spring-boot-starter-test</artifactId>
    <scope>test</scope>
</dependency>
<!--配置注解依赖-->
<dependency>
    <groupId>org.springframework.boot</groupId>
    <artifactId>spring-boot-configuration-processor</artifactId>
    <optional>true</optional>
</dependency>
```

（3）引入对接 ElasticJob 分布式任务调度框架所需要的依赖。代码如下：

```
<!--引入对接 ElasticJob 分布式任务调度框架所需要的依赖-->
<dependency>
    <groupId>org.apache.shardingsphere.elasticjob</groupId>
    <artifactId>elasticjob-lite-core</artifactId>
    <version>3.0.0-RC1</version>
    <!--排除 ZooKeeper 依赖（JAR 包）冲突-->
```

```
        <exclusions>
            <exclusion>
                <groupId>org.apache.curator</groupId>
                <artifactId>curator-framework</artifactId>
            </exclusion>
            <exclusion>
                <groupId>org.apache.curator</groupId>
                <artifactId>curator-recipes</artifactId>
            </exclusion>
        </exclusions>
    </dependency>
    <!--单独引入 ZooKeeper 连接相关的依赖-->
    <dependency>
        <groupId>org.apache.curator</groupId>
        <artifactId>curator-framework</artifactId>
        <version>5.1.0</version>
    </dependency>
    <dependency>
        <groupId>org.apache.curator</groupId>
        <artifactId>curator-recipes</artifactId>
        <version>5.1.0</version>
    </dependency>
```

3. 开发“接入 SDK”的代码

（1）在“1.”小标题创建的“接入 SDK”工程的“/src/main/resources/META-INF/”目录中创建 spring.factories 文件。具体的配置内容如下：

```
    org.springframework.boot.autoconfigure.EnableAutoConfiguration=com.wudim
anong.starter.task.ElasticAutoConfiguration
```

该文件是 Spring Boot SPI 机制的入口。在该文件中，通过配置 Spring Boot 自动配置类实现了 Spring Boot Starter 组件的自动加载。

（2）编写步骤（1）中配置的自动配置类的代码。代码如下：

```
    package com.wudimanong.elasticjob.starter;
    import
org.apache.shardingsphere.elasticjob.reg.zookeeper.ZookeeperConfiguration;
    ...
    import org.springframework.context.annotation.Configuration;
    @Configuration
    @ConditionalOnProperty({"elasticjob.zk.serverLists", "elasticjob.zk.namespace"})
    public class ElasticAutoConfiguration {
```

```
        /**
         * ZooKeeper 注册中心的配置
         */
        @Bean
        @ConfigurationProperties("elasticjob.zk")
        public ZookeeperConfiguration
getConfiguration(@Value("${elasticjob.zk.serverLists}") String serverLists,
                @Value("${elasticjob.zk.namespace}") String namespace) {
            return new ZookeeperConfiguration(serverLists, namespace);
        }

        /**
         * 初始化注册信息
         */
        @Bean(initMethod = "init")
        public ZookeeperRegistryCenter
zookeeperRegistryCenter(ZookeeperConfiguration configuration) {
            return new ZookeeperRegistryCenter(configuration);
        }
        /**
         * 配置处理自定义 ElasticJob 任务的逻辑类
         */
        @Bean
        public ElasticJobBeanPostProcessor
elasticJobBeanPostProcessor(ZookeeperRegistryCenter center) {
            return new ElasticJobBeanPostProcessor(center);
        }
    }
```

上述自动配置类的主要逻辑是：

①通过@ConditionalOnProperty 注解，判断应用是否配置了 ZooKeeper 的链接地址、命名空间等信息。如果存在 ZooKeeper 的链接地址及命名空间等配置信息，则自动初始化该配置类。

②自动配置类会根据配置的 ZooKeeper 连接信息，来初始化 ElasticJob 框架所定义的 ZooKeeper 配置类。

③通过@Bean(initMethod="init")注解，在 Bean 初始化时创建 ZooKeeper 的注册中心连接。

④通过@Bean 注解，实例化自定义 ElasticJob 任务的配置类 ElasticJobBeanPostProcessor。

（3）编写处理自定义 ElasticJob 任务的配置类 ElasticJobBeanPostProcessor。代码如下：

```
package com.wudimanong.elasticjob.starter;
import java.util.ArrayList;
import java.util.List;
...
```

```
    import org.springframework.beans.factory.config.BeanPostProcessor;
    public class ElasticJobBeanPostProcessor implements BeanPostProcessor,
DisposableBean {
        /**
         * ZooKeeper 注册器
         */
        private ZookeeperRegistryCenter zookeeperRegistryCenter;
        /**
         * 已注册任务列表
         */
        private List<ScheduleJobBootstrap> schedulers = new ArrayList<>();
        /**
         * 构造注册中心
         */
        public ElasticJobBeanPostProcessor(ZookeeperRegistryCenter
zookeeperRegistryCenter) {
            this.zookeeperRegistryCenter = zookeeperRegistryCenter;
        }
        @Override
        public Object postProcessBeforeInitialization(Object o, String s) throws
BeansException {
            return o;
        }
        /**
         * Spring IOC 容器扩展接口方法，在 Bean 初始化后执行
         */
        @Override
        public Object postProcessAfterInitialization(Object o, String s) throws
BeansException {
            Class<?> clazz = o.getClass();
            //只处理自定义的 ElasticTask 注解
            if (!clazz.isAnnotationPresent(ElasticTask.class)) {
                return o;
            }
            if (!(o instanceof ElasticJob)) {
                return o;
            }
            ElasticJob job = (ElasticJob) o;
            //获取注解定义
            ElasticTask annotation = clazz.getAnnotation(ElasticTask.class);
            //定义任务名称
            String jobName = annotation.jobName();
            //定义 Cron 表达式
            String cron = annotation.cron();
            //设置任务参数
```

```
        String jobParameter = annotation.jobParameter();
        //设置任务描述信息
        String description = annotation.description();
        //设置数据分片数
        int shardingTotalCount = annotation.shardingTotalCount();
        //设置数据分片参数
        String shardingItemParameters = annotation.shardingItemParameters();
        //设置是否禁用分片项
        boolean disabled = annotation.disabled();
        //设置重启任务定义信息是否覆盖
        boolean overwrite = annotation.overwrite();
        //设置是否开启故障转移
        boolean failover = annotation.failover();
        //设置是否开启错过任务重新执行
        boolean misfire = annotation.misfire();
        //配置分片策略类
        String jobShardingStrategyClass = annotation.jobShardingStrategyClass();

        //根据自定义注解的配置设置 ElasticJob 任务的配置
        JobConfiguration coreConfiguration = JobConfiguration
                .newBuilder(jobName, shardingTotalCount).cron(cron).
jobParameter(jobParameter).overwrite(overwrite)
                .failover(failover).misfire(misfire).description(description)
                .shardingItemParameters(shardingItemParameters).disabled(di
sabled)
                .jobShardingStrategyType(jobShardingStrategyClass)
                .build();
        //创建任务调度对象
        ScheduleJobBootstrap scheduleJobBootstrap = new
ScheduleJobBootstrap(zookeeperRegistryCenter, job,
                coreConfiguration);
        //触发任务调度
        scheduleJobBootstrap.schedule();
        //将创建的任务对象加入集合，便于统一销毁
        schedulers.add(scheduleJobBootstrap);return job;
    }
    /**
     * 任务销毁方法
     */
    @Override
    public void destroy() {
        schedulers.forEach(jobScheduler -> jobScheduler.shutdown());
    }
}
```

上述代码的逻辑是：根据自定义注解@ElasticTask 获取任务的定义信息，并通过 ElasticJob 框架原生的任务配置、创建及启动的方法，实现自定义注解任务配置到 ElasticJob 原生任务配置定义方式的映射。

（4）定义注解@ElasticTask。代码如下：

```
package com.wudimanong.elasticjob.starter;
import java.lang.annotation.ElementType;
...
import org.springframework.stereotype.Component;
@Component
@Target(ElementType.TYPE)
@Retention(RetentionPolicy.RUNTIME)
public @interface ElasticTask {
    /**
     * 定义任务名称
     */
    String jobName();
    /**
     * 定义 Cron 时间表达式
     */
    String cron();
    /**
     * 定义任务参数信息
     */
    String jobParameter() default "";
    /**
     * 定义任务描述信息
     */
    String description() default "";
    /**
     * 定义任务分片数
     */
    int shardingTotalCount() default 1;
    /**
     * 定义任务分片参数
     */
    String shardingItemParameters() default "";
    /**
     * 设置是否禁用数据分片功能
     */
    boolean disabled() default false;
    /**
     * 配置分片策略
     */
```

```
    String jobShardingStrategyClass() default "";
    /**
     * 配置是否故障转移功能
     */
    boolean failover() default false;
    /**
     * 配置是否在运行重启时重置任务定义信息（包括 Cron 时间片设置）
     */
    boolean overwrite() default false;
    /**
     * 配置是否开启错误任务重新执行
     */
    boolean misfire() default true;
}
```

注解@ElasticTask 根据 ElasticJob 支持的基本功能，实现了对任务名称、Cron 表达式、任务描述、任务参数、分片处理设置等基本任务属性的配置。

在定义注解@ElasticTask 的过程中，可以根据实际的需要屏蔽或暴露一些 ElasticJob 所支持的原生功能，从而简化接入的复杂度。

至此，基于 Spring Boot Starter 方式完成了 ElasticJob“接入 SDK”的编写。后面通过该“接入 SDK”可以方便地将微服务接入 ElasticJob 分布式任务调度系统，实现微服务分布式任务处理功能。

8.3.3 定义微服务分布式任务

通过“8.3.2”节定义的“接入 SDK”，可以降低将 Spring Cloud 微服务接入 ElasticJob 分布式任务调度系统的复杂度，以及微服务处理分布式任务的复杂度。

通过“接入 SDK”将 Spring Cloud 微服务接入 ElasticJob 分布式调度系统的步骤如下。

1. 引入 ElasticJob“接入 SDK”的依赖

（1）在第 8.3.1 节创建的微服务工程的 pom.xml 中，引入 8.3.2 节编写的 ElasticJob“接入 SDK”的依赖。代码如下：

```
<!--引入自定义封装的 ElasticJob“接入 SDK”的依赖-->
 <dependency>
     <groupId>com.wudimanong</groupId>
     <artifactId>elasticjob-springboot-starter</artifactId>
     <version>1.0-SNAPSHOT</version>
 </dependency>
```

```
<!--引入 ZooKeeper 客户端的依赖 -->
<dependency>
    <groupId>org.apache.curator</groupId>
    <artifactId>curator-framework</artifactId>
    <version>5.1.0</version>
</dependency>
<dependency>
    <groupId>org.apache.curator</groupId>
    <artifactId>curator-recipes</artifactId>
    <version>5.1.0</version>
</dependency>
```

（2）按照 ElasticJob“接入 SDK”依赖所定义的配置项，在项目的“src/main/resources”目录中创建 application.yml 配置文件，并配置 ElasticJob 分布式任务调度所使用的 ZooKeeper 服务的连接信息。代码如下：

```
#配置 ZooKeeper IP 地址及命名空间配置
elasticjob:
  zk:
    serverLists: 10.211.55.12:2181,10.211.55.12:2182,10.211.55.12:2183
    namespace: check-schedule
```

2. 通过在“接入 SDK”中定义的注解，快速定义一个分布式任务

通过在 ElasticJob“接入 SDK”中定义的注解@ElasticTask，快速定义一个微服务分布式任务。代码如下：

```
package com.wudimanong.schedule.job;
import com.wudimanong.elasticjob.starter.ElasticTask;
import java.util.Date;
import org.apache.shardingsphere.elasticjob.api.ShardingContext;
import org.apache.shardingsphere.elasticjob.simple.job.SimpleJob;

@ElasticTask(jobName = "testJob", cron = "*/5 * * * * ?", description = "
自定义 Task", overwrite = true)
public class TestJob implements SimpleJob {
    @Override
    public void execute(ShardingContext shardingContext) {
        System.out.println("跑任务->" + new Date());
    }
}
```

可以看到，通过注解@ElasticTas 可以直接定义分布式任务的信息：名称、Cron 时间片、参数及数据分片等。

在启动微服务应用时，通过注解@ElasticTask 定义的分布式任务将被自动注册到 ZooKeeper

中，并通过 ElasticJob 实现任务的分布式调度，效果如图 8-9 所示。

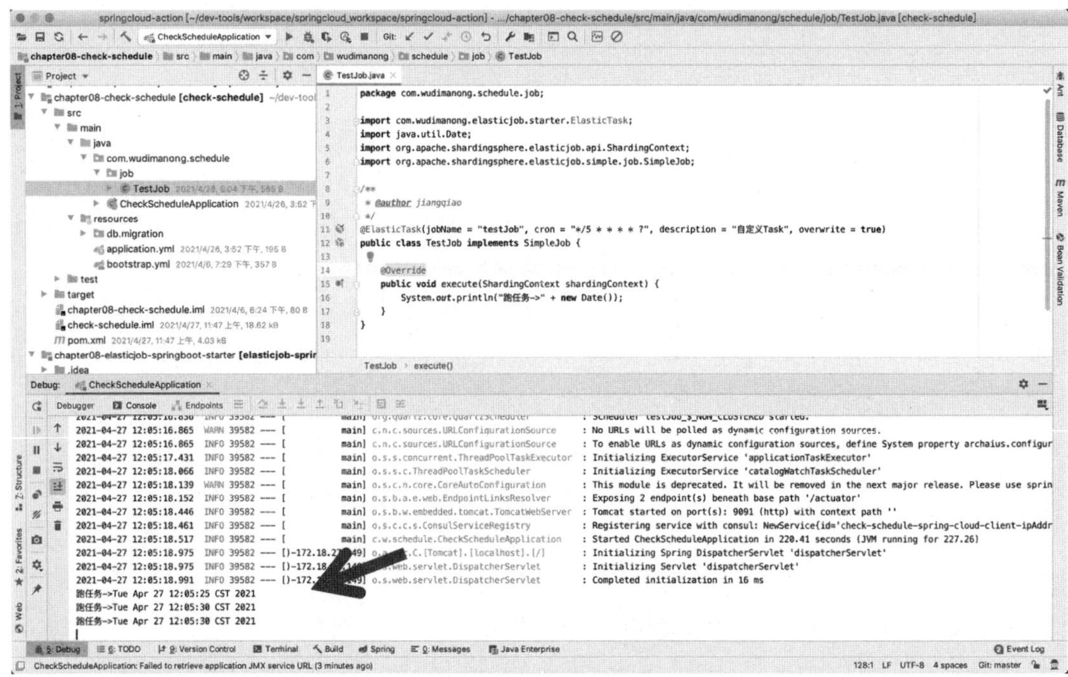

图 8-9

从图 8-9 可以看到，通过注解@ElasticTask 定义的分布式任务，已经按照所设定的 cron 时间片，被调度运行起来了。

此时，通过 ElasticJob 的 Console 控制台也能看到定义的分布式任务信息，如图 8-10 所示。

图 8-10

至此，完成了基于“ZooKeeper + ElasticJob”分布式任务调度系统实现微服务分布式任务处理的基本过程。

8.4 本章总结

本章演示了构建 Spring Cloud 微服务分布式任务处理的主流方法——通过搭建基于“ZooKeeper + ElasticJob”的分布式任务调度系统，实现 Spring Cloud 微服务的分布式任务处理；并通过 Spring Boot Starter 方式的 ElasticJob“接入 SDK”快速开发微服务分布式任务。

本章所构建的分布式任务调度系统，可以作为整个 Spring Cloud 微服务体系实现分布式任务集中管理的统一调度平台。

第 9 章

搭建微服务 DevOps 发布系统

——用“GitLab + Harbor + Kubernetes”构建 Spring Cloud 微服务 CI/CD 自动化发布体系

在实施微服务架构后，单体系统变成了数量众多的微服务应用。在这种变化的冲击下，原先的开发、测试和运维部署流程都会面临一定的挑战。

提高工程研发效率，确保开发、测试和运维部署流程的顺畅，是微服务架构能够真正落地，并产生实际收益的关键。

实现上述目标，需要基于 DevOps 方法构建基于 CI/CD（持续集成/持续交付）流程的自动化发布系统。在该系统中，开发人员、测试人员及运维人员可以随时随地构建代码，并将其发布至指定的运行环境。

关于 DevOps 的具体实践，一般会根据自身的发展阶段和实际需要来选择落地方案：具备条件的公司，可以研发功能丰富的可视化发布系统；条件有限的创业公司，可以通过开源或现有的技术组件（如 GitLab、Jenkins 等）来实现操作相对简陋的自动化发布系统。

本章将以 Spring Cloud 微服务体系为背景，通过 GitLab 自带的 CI/CD 机制，并基于 Kubernetes 容器化技术，来实现具备相对完整 CI/CD 流程的自动化发布系统。

通过本章，读者将学习到以下内容：

- 持续集成（CI）/持续交付（CD）的概念及基本流程。

- 自动化发布系统的设计及构建方法。
- 基于 GitLab 的代码管理及 CI/CD 配置。
- 常见的 Docker 镜像仓库及选型。
- 基于 Kubernetes 的容器编排技术实战。
- 基于“GitLab + Habor + Kubernetes”CI/CD 流程的微服务容器化部署。

9.1 CI/CD 概述

DevOps 并不是在微服务架构流行后才产生的概念，而是业界在多年软件开发实践中积累的理论和工具的集合。

本章所要讨论的 DevOps 发布系统，实际上是通过搭建 CI/CD 流水线，来建立一套应用程序构建、测试、打包及发布的高效自动化方法。

1. 持续集成（CI）

> CI（持续集成）/CD（持续交付）并不是某一种具体的技术，而是一种软件工程文化加一系列操作原则和具体实践的集合。

持续集成（CI）的主要目标是，通过建立一致的自动化构建方法来打包程序代码，使得开发团队成员可以更频繁地构建代码、更早地进行代码集成，以及时地发现和解决代码中的问题、提高协作开发效率及软件交付质量。

可持续集成（CI）的基本流程如图 9-1 所示。

图 9-1

从实现流程上来说，CI 的主要过程是：将开发人员提交的代码，以高度自动化的方式打包成可以在具体基础架构环境（例如 Docker 镜像）中运行的程序包。这个过程可以由一组工具，如 GitLab Runner（CI Pipeline）、Sonar（代码检测工具）等，去完成。在具体构建 CI 流程时，根据实际需要集成即可。

2. 持续交付（CD）

持续交付（CD）的主要逻辑是：将“在 CI 流程中构建的程序镜像”从镜像仓库自动发布到具体的基础架构环境（如测试/生产 Kubernetes 集群）。

实现 CD 的工具主要有 GitLab Runner（CD Pipeline）、Helm（Kubernetes 软件包管理工具）等。

CD 的核心是：通过输入的各种用户参数（如 yaml 文件、环境配置参数等）自动生成具体的发布指令（如 Helm 指令），并根据参数中设置的信息来配置程序的具体运行环境。

可持续交付（CD）的基本运行流程如图 9-2 所示。

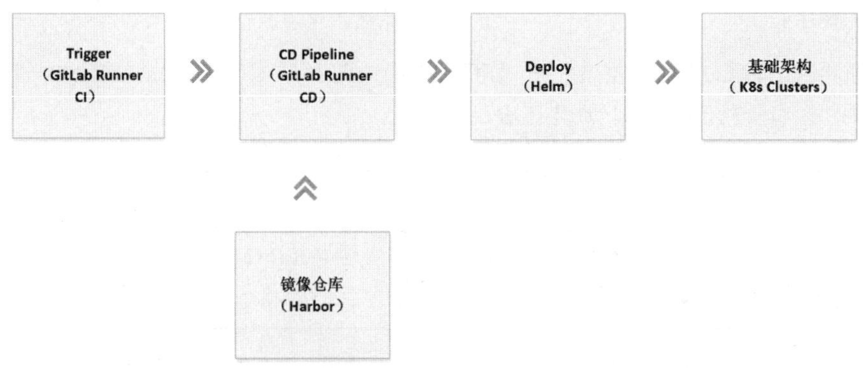

图 9-2

本节描述了 CI/CD 的基本概念及流程，后面节将依据这些内容来构建具体的 DevOps 发布系统。

9.2 了解 DevOps 发布系统的设计流程

本章所设计的 DevOps 发布系统，主要是利用 GitLab 提供的 CI 机制，来实现在代码发生提交或合并等事件时自动触发预设的 CI/CD 流程。具体的系统结构如图 9-3 所示。

- CI 阶段主要包括基本的代码编译、构建和打包，并将打包好的应用 Docker 镜像发布至镜像仓库中。
- CD 阶段则是从镜像仓库拉取应用 Docker 镜像，并根据设置的 CD 流程将应用发布至指定的 Kubernetes 集群中。
- DevOps 发布系统主要由 GitLab、Harbor 镜像仓库及 Kubernetes 集群组成。其中，GitLab 主要承担代码版本管理，以及 CI/CD 流程的定义和触发；Harbor 负责应用 Docker

镜像的存储和分发；Kubernetes 集群则是应用容器运行的基础架构环境。

在后面的实战中将逐步构建这些组件，并使用它们来搭建起 DevOps 发布系统。

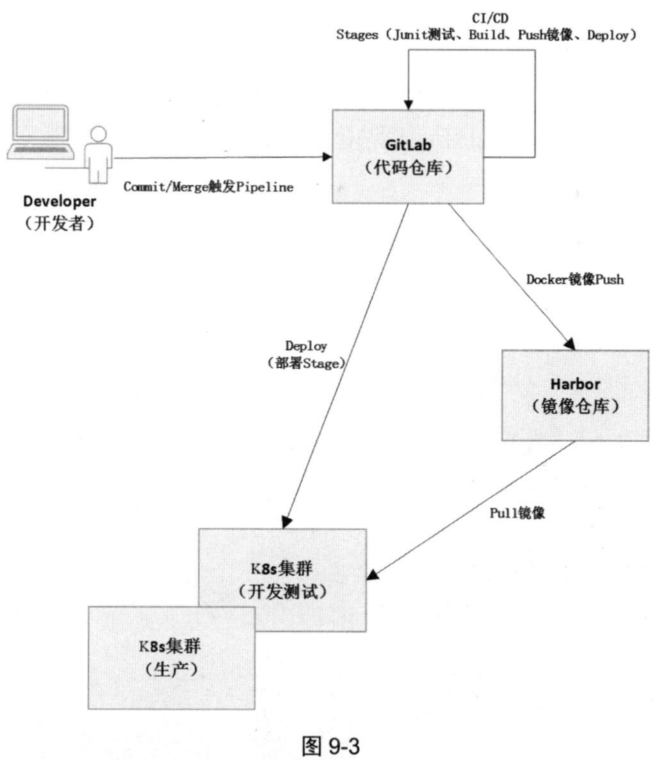

图 9-3

9.3　基础知识 1：GitLab 代码仓库

GitLab 是目前主流的开源代码管理仓库。除基本的版本管理外，GitLab 还提供了可持续集成 GitLab CI 功能：只要在代码仓库根目录创建一个“.gitlab-ci.yml”文件，并指派一个 Runner，即可实现在有代码合并或提交请求时自动触发构建、打包等操作。而这些的具体操作阶段（STAGES），还可以在“.gitlab-ci.yml”文件中进行定制，这也是本实例实现 DevOps 发布系统的基础。

9.3.1　部署 GitLab 代码仓库

在本章所演示的 DevOps 发布系统中，GitLab 扮演了关键角色。接下来将演示如何在 Linux 服务器中部署 GitLab（版本 13.2.2）。

（1）准备一台 Linux 服务器（版本 Ubantu 20.04 LTS），并安装必要的系统环境依赖。命令如下：

```
#刷新本地包索引
sudo apt update
#安装 Postfix 以发送邮件通知
sudo apt install ca-certificates curl openssh-server postfix
```

对于 Postfix 的安装，如果出现提示，则选择“Internet 站点”，并在下一个界面中输入服务器的域名，以配置系统发送邮件的方式。

（2）安装 GitLab。命令如下：

```
#切换到临时目录
cd /tmp/
#下载 GitLab 安装脚本
curl -LO https://packages.gitlab.com/install/repositories/gitlab/gitlab-ce/script.deb.sh
#运行脚本执行安装操作，该脚本会设置本服务器为 GitLab 维护的存储库，并使用与其他系统软件包相同的包管理工具来管理 GitLab
sudo bash /tmp/script.deb.sh
#使用 apt 命令安装实际的 GitLab 应用
sudo apt install gitlab-ce
```

如果出现如图 9-4 所示效果，则说明安装成功。

```
(Reading database ... 71038 files and directories currently installed.)
Preparing to unpack .../gitlab-ce_13.2.2-ce.0_amd64.deb ...
Unpacking gitlab-ce (13.2.2-ce.0) ...
Setting up gitlab-ce (13.2.2-ce.0) ...
It looks like GitLab has not been configured yet; skipping the upgrade script.
```

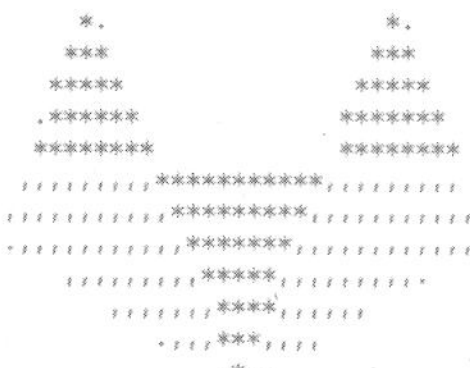

```
Thank you for installing GitLab!
GitLab was unable to detect a valid hostname for your instance.
Please configure a URL for your GitLab instance by setting `external_url`
configuration in /etc/gitlab/gitlab.rb file.
Then, you can start your GitLab instance by running the following command:
  sudo gitlab-ctl reconfigure

For a comprehensive list of configuration options please see the Omnibus GitLab readme
https://gitlab.com/gitlab-org/omnibus-gitlab/blob/master/README.md
```

图 9-4

（3）修改 GitLab 配置文件，指定服务器的 IP 地址及自定义端口号。命令如下：

```
#编辑配置文件
vim /etc/gitlab/gitlab.rb
```

需要将顶部的“external_url”配置项设置为实际的域名。假设以本机的 80 端口作为默认访问地址，则更改配置如下：

```
## GitLab URL
##! URL on which GitLab will be reachable.
##! For more details on configuring external_url see:
##!
https://docs.gitlab.com/omnibus/settings/configuration.html#configuring-the-external-url-for-gitlab
external_url 'http://10.211.55.11:80'
```

保存并关闭配置文件，然后运行重新配置 GitLab 的命令：

```
sudo gitlab-ctl reconfigure
```

（4）重启 GitLab 服务。命令如下：

```
#执行重启命令
sudo gitlab-ctl restart
```

（5）访问 GitLab 页面。

输入服务访问地址“http://10.211.55.11”，如图 9-5 所示。

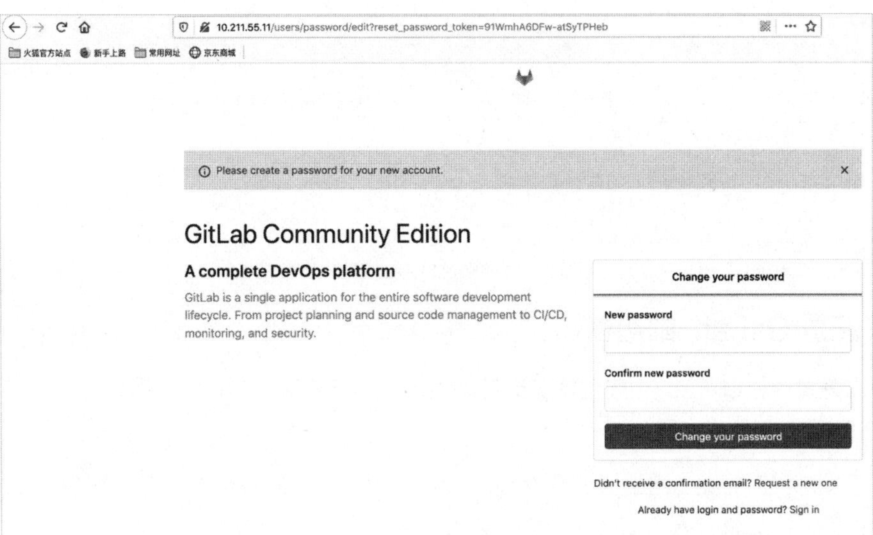

图 9-5

首次登录会要求修改“root”管理员的密码，这里将其设置为“12345678”。

在密码设置成功后，以“root”管理员身份登录 GitLab，登录后的界面如图 9-6 所示。

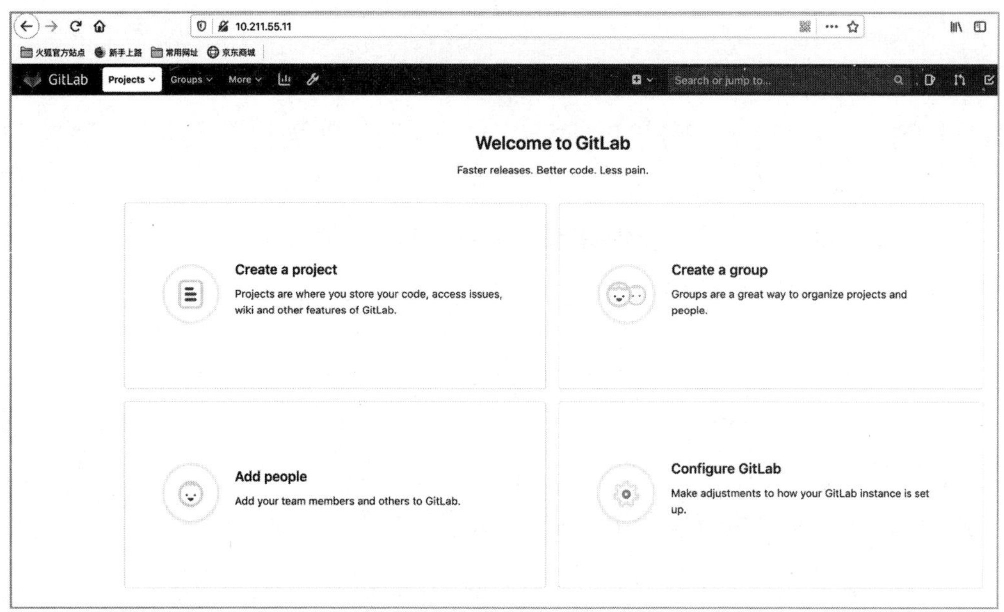

图 9-6

至此完成了 GitLab 服务的安装。

（6）为了防止服务器重启，可以设置 GitLab 开机自启动。命令如下：

```
#设置 GitLab 开机自启动
sudo systemctl enable gitlab-runsvdir.service
```

9.3.2 配置 GitLab 邮箱通知

在日常开发中，GitLab 中代码的变更可以通过邮件的方式通知给开发人员。接下来演示配置 GitLab 邮箱通知的具体方式。

（1）设置邮箱服务器。这里通过设置 QQ 邮箱的 POP3/SMTP 服务作为邮件服务器，完成后保存好授权码，如图 9-7 所示。

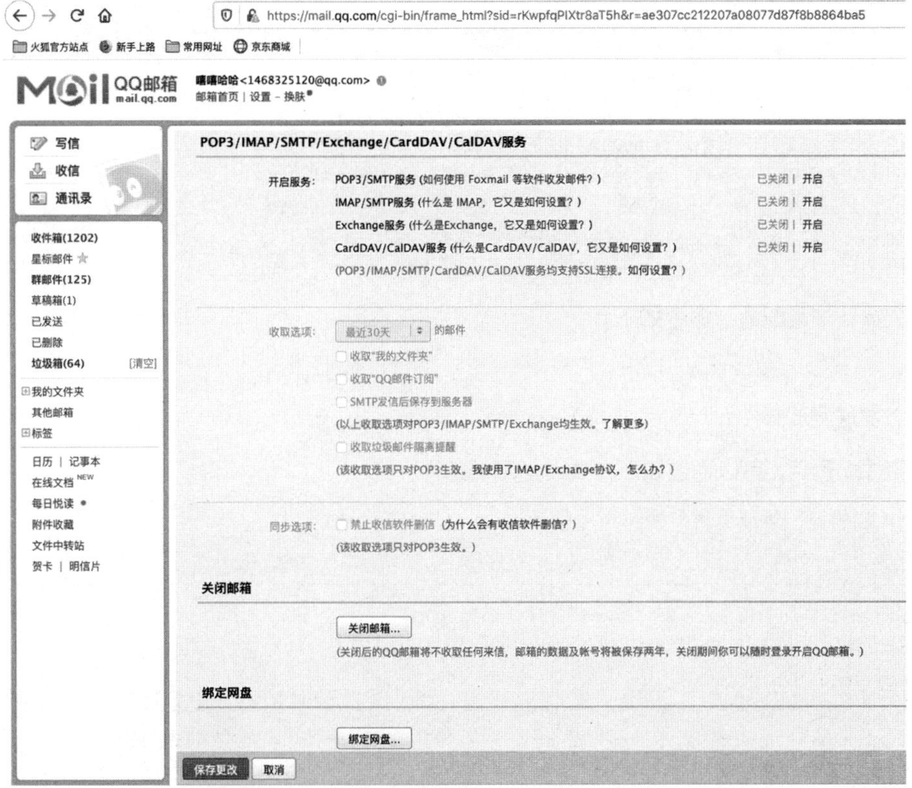

图 9-7

在正常情况下，GitLab 的邮件通知功能会使用到公司邮箱服务。这里为了便于实验，暂时使用 QQ 邮箱提供的邮箱服务功能：单击“设置”→“账户”，找到如图 9-7 所示的开启 POP3/SMTP 邮箱服务界面，单击“开启”按钮，验证成功后记住授权码信息。

（2）修改 GitLab 的配置文件，设置邮箱信息。

打开配置文件：

```
sudo vim /etc/gitlab/gitlab.rb
```

修改相关配置如下：

```
#配置邮箱来源与邮箱显示名称
gitlab_rails['gitlab_email_enabled'] = true
gitlab_rails['gitlab_email_from'] = '146832512*@qq.com'
gitlab_rails['gitlab_email_display_name'] = 'Wudimanong-GitLab'
...
#SMTP 配置
gitlab_rails['smtp_enable'] = true
```

```
gitlab_rails['smtp_address'] = "smtp.qq.com"
gitlab_rails['smtp_port'] = 465
gitlab_rails['smtp_user_name'] = "146832512*@qq.com"
gitlab_rails['smtp_password'] = "eeiowlzvwsjehdix"
gitlab_rails['smtp_domain'] = "smtp.qq.com"
gitlab_rails['smtp_authentication'] = "login"
gitlab_rails['smtp_enable_starttls_auto'] = true
gitlab_rails['smtp_tls'] = true
gitlab_rails['gitlab_email_from'] = '146832512*@qq.com'
```

（3）重新加载配置。命令如下：

```
sudo gitlab-ctl reconfigure
```

（4）发送测试邮件。

在完成配置后，可以通过如下方式发送测试邮件：

```
#进入控制台，然后发送邮件
sudo gitlab-rails console
--------------------------------------------------------------------------------
 GitLab:       13.2.2 (64fc0138d55) FOSS
 GitLab Shell: 13.3.0
 PostgreSQL:   11.7
--------------------------------------------------------------------------------
Notify.test_emLoading production environment (Rails 6.0.3.1)
irb(main):001:0>
#发送一封测试邮件
Notify.test_email('wudimanong@qq.com', 'wudimanong', 'Hello
World').deliver_now
```

发送成功后，可以进入邮箱查看是否收到邮件。

9.3.3 设置 GitLab 的 CI/CD 功能

从 GitLab 8.0 开始，GitLab CI/CD 就已经被集成在 GitLab 中了。

在 9.3.1 节中已经成功安装了 GitLab。如果此时在 GitLab 上创建 Project，则能看到 CI/CD 相关的菜单页，如图 9-8 所示。

从图 9-8 中可以看到，项目在进行 commit 后并没有触发自动构建流程。这说明，虽然 GitLab 集成了 CI/CD 的功能，但还需要配置 CI/CD 才能够自动生效。所以下面进行一定的设置。

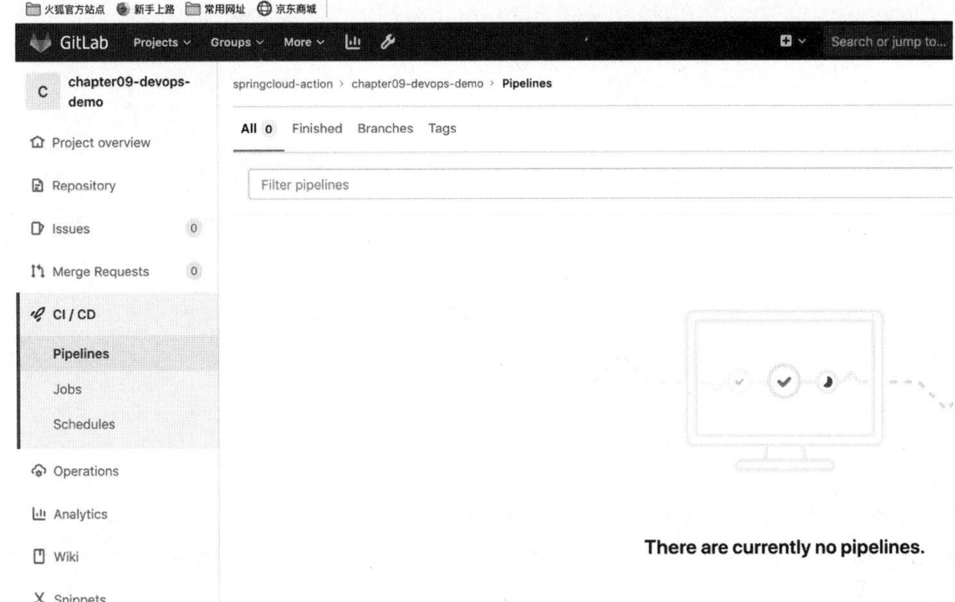

图 9-8

1. GitLab CI/CD 的相关概念

在介绍 GitLab CI/CD 配置前，先来介绍 GitLab CI/CD 的相关概念。

- Pipeline。

一次 Pipeline 相当于一次构建任务，其中可以包含多个执行流程，如安装依赖、运行测试、编译、测试服务器部署、生产服务器部署等流程。

任何代码提交或者“Merge Request”分支合并的动作都可以触发 Pipeline。

- Stage。

Stage 表示构建阶段，实质上就是上面提到的 Pipeline 流程。我们可以在一次 Pipeline 中定义多个 Stage。

所有的 Stage 都会按照顺序执行：在一个 Stage 完成后，下一个 Stage 才会开始。只有当所有的 Stage 完成后，Pipeline 构建任务才会成功；任何一个 Stage 失败，后面的 Stage 都不会执行，从而导致本次 Pipeline 任务失败。

- Job。

Job 表示 Stage 中具体的执行工作。例如定义了一个叫作“Deploy”的 Stage，在该 Stage 中部署了两套不同的测试环境，那么就可以针对这个“Deploy”Stage 定义两个 Job。在同一个

Stage 中的 Job 会并行执行，只有 Stage 中的所有 Job 都执行成功，该 Stage 才算成功；否则该 Stage 算执行失败，从而使本次 Pipeline 失败。

2. 谁来执行构建任务

在理解了上面的概念后，会产生一个问题——具体由谁来执行构建任务呢?

实际上，虽然 GitLab 已经自带了 CI/CD 功能，但是 GitLab CI 本身却不会执行 Pipeline，而是由一个被叫作“GitLab Runner”的独立组件来单独运行构建任务。这主要是因为：GitLab CI 属于 GitLab 的一部分，如果由 GitLab CI 来运行构建任务，则会导致 GitLab 性能大幅下降。

因此，GitLab CI 主要用来管理各个项目的构建状态，而具体的构建工作这种耗费资源的事情则由 GitLab Runner 来执行。

3. 部署 GitLab Runner

由于 GitLab Runner 是独立的组件，可以被安装在不同的机器上，所以在构建任务期间它并不会影响 GitLab 的性能。因此，要让 GitLab CI/CD 功能正常运行起来，还需要独立部署 GitLab Runner 组件。

（1）在 Linux 服务器上部署 GitLab Runner。命令如下：

```
#通过远程命令添加镜像仓库
curl -L
https://packages.gitlab.com/install/repositories/runner/gitlab-runner/script
.deb.sh | sudo bash
#安装最新版本（13.2.2）的 GitLab Runner
sudo apt-get install gitlab-runner
```

（2）启动 GitLab Runner。命令如下：

```
#将 GitLab Runner 服务设置为开机自启动
sudo systemctl enable gitlab-runner
#启动 GitLab Runner
systemctl start gitlab-runner
```

（3）在完成启动后，查看 GitLab Runner 的运行状态。命令如下：

```
systemctl status gitlab-runner
```

如果显示如下信息，则说明 GitLab Runner 服务启动成功。

```
gitlab-runner.service - GitLab Runner
     Loaded: loaded (/etc/systemd/system/gitlab-runner.service; enabled;
vendor preset: enabled)
     Active: active (running) since Sun 2020-08-02 10:46:10 UTC; 6min ago
   Main PID: 34871 (gitlab-runner)
      Tasks: 8 (limit: 2275)
```

```
      Memory: 5.2M
      CGroup: /system.slice/gitlab-runner.service
            └─34871 /usr/lib/gitlab-runner/gitlab-runner run
--working-directory /home/gitlab-runner --config
/etc/gitlab-runner/config.toml --service gitlab-runner --syslog --user
gitlab-runner
   ...
```

4. 将 GitLab Runner 注册到 GitLab

前面提到过，GitLab Runner 作为独立运行的组件，可以被单独部署在别的机器上。所以，如果要让 GitLab 通过 GitLab Runner 来执行构建任务，则需要 GitLab Runner 将自身信息注册到 GitLab 中。

需要说明的是，如果 GitLab Runner 被注册到某个具体的项目中，则该 Runner 只会对该项目生效，这样每创建一个新的项目都需要单独注册 Runner，比较麻烦。因此，可以配置对整个项目组都生效的 GitLab Runner。

（1）通过 Group 主页→Settings→CI/CD→Runners Expand，获取针对整个项目组的 gitlab-ci 的 Token，如图 9-9 所示。

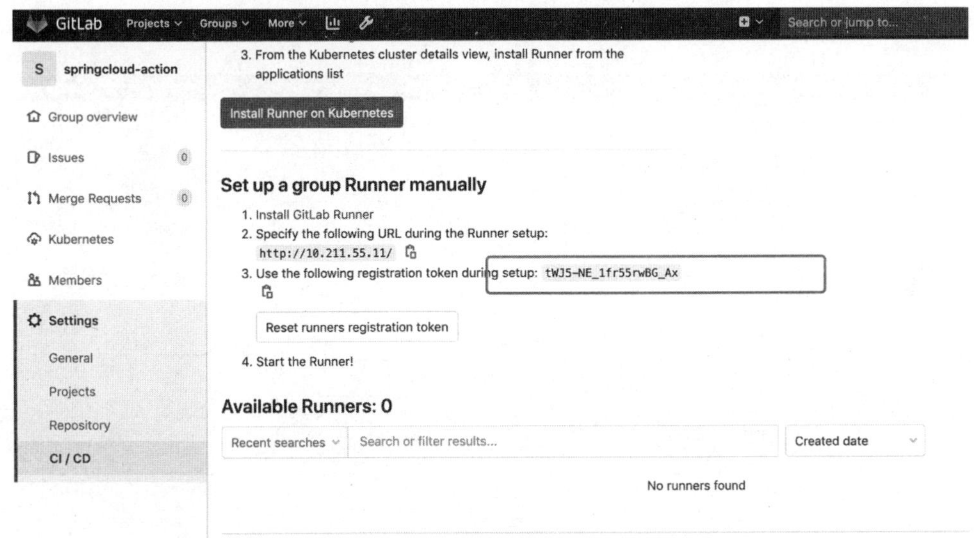

图 9-9

（2）通过交互式命令方式实现 GitLab Runner 与 GitLab CI/CD 的绑定，命令如下：

```
   [root@localhost src]# sudo gitlab-ci-multi-runner register
   Runtime platform                                    arch=amd64 os=linux pid=7335
revision=de7731dd version=12.1.0
```

```
    Running in system-mode.
    #输入 GitLab 的服务 URL
    Please enter the gitlab-ci coordinator URL (e.g. https://gitlab.com/):
    http://10.211.55.11
    #输入 gitlab-ci 的 Token（如图 9-9 所示）
    Please enter the gitlab-ci token for this runner:
    tWJ5-NE_1fr55rwBG_Ax
    #集成服务中对于这个 Runner 的描述
    Please enter the gitlab-ci description for this runner:
    [wudimanong-gitlab]: Runner01
    #给 Runner 输入一个 tag。该 tag 非常重要，在后续的使用过程中需要使用该 tag 来指定 GitLab
Runner
    Please enter the gitlab-ci tags for this runner (comma separated):
    Runner01
    Registering runner... succeeded                    runner=NzwN_gw6
    #选择执行器。GitLab Runner 实现了很多执行器，这里选择 shell
    Please enter the executor: docker, parallels, shell, ssh, docker+machine,
docker-ssh+machine, custom, docker-ssh, virtualbox, kubernetes:
    kubernetes
    Runner registered successfully. Feel free to start it, but if it's running
already the config should be automatically reloaded!
```

（3）刷新如图 9-9 所示界面，就能看到刚才注册的 Group Runner 了，如图 9-10 所示。

图 9-10

在 GitLab Runner 注册的过程中设置了 gitlab-ci 的 tag，而要想项目的构建任务被该 Runner 执行，则需要在其项目根目录的 CI 配置中指定该 tag，否则项目的构建任务是无法被 GitLab Runner 执行的。

5. 将 GitLab Runner 指定到 Group 级别

在实际应用场景中，一般会将 GitLab Runner 直接指定到 Group 级别。这样，该 Group 下的所有项目的构建任务都能够被 GitLab Runner 正常执行（无论其是否指定了 Runner 的 tag）。

要实现这样的效果，需要对 GitLab Runner 进行一定的设置。

（1）单击进入 GitLab 代码空间中的 Settings→CI/CD，找到 Group Runners，单击可用 Runners 列表后面的 Edit 按钮，如图 9-11 所示。

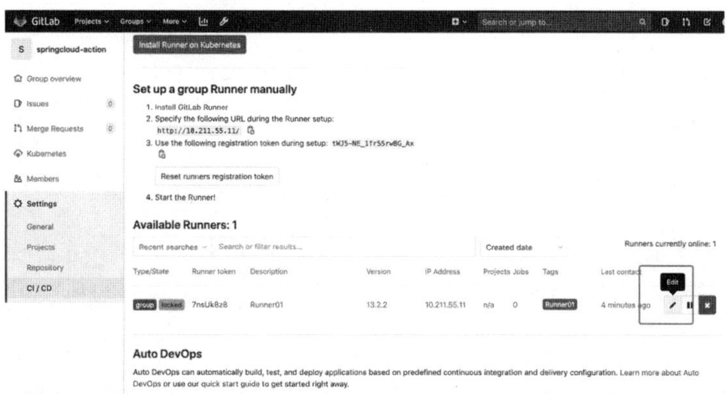

图 9-11

（2）如图 9-12 所示勾选“Indicates whether this runner can pick jobs without tags”选项，然后保存变更。

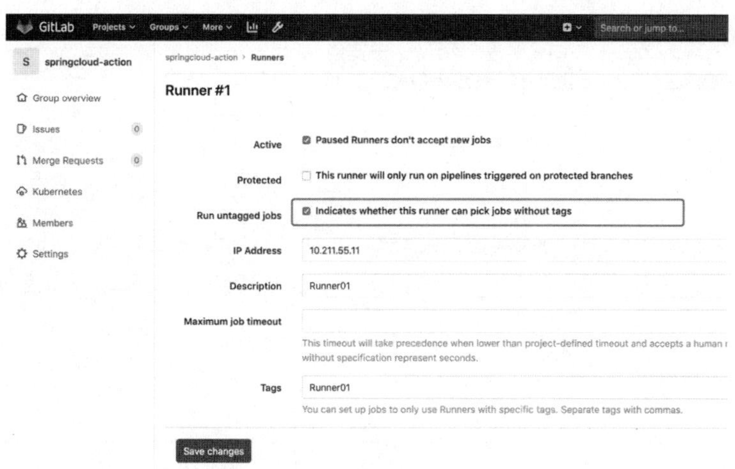

图 9-12

这样，在该 Group 下创建的所有项目不需要指定特定的 Runner tag 就能自动触发 Pipeline 构建任务了。

9.3.4 安装 Maven 及 Docker 环境

在基于 GitLab 的 CI/CD 流程中，是通过相应的编译、构建工具来完成软件代码的编译打包的。

而在 Spring Cloud 微服务体系中，目前主流的构建方式是：通过“Maven + Docker”将 Spring Cloud 应用打包成 Docker 镜像，并将镜像部署至指定环境。这就要求在 GitLab 的服务器上也要安装 Maven 及 Docker 环境。

在 GitLab 的服务器上安装 Maven 及 Docker 环境的命令如下：

```
#安装 Maven 环境
apt install maven
#安装 Docker 环境
apt install docker.io
#启动 Docker 服务
service docker start
#设置 Docker 开机自启动
systemctl enable docker
```

在 Maven 项目的构建过程中涉及依赖 Jar 包的下载。在默认情况下，Maven 会直接从中央仓库拉取，但这样会导致构建速度的降低。所以，一般会在公司内部架设一个 Maven 私有仓库（如 Nexus）。由于篇幅的原因，这里就不具体演示了，但在实际应用场景中需要注意。

9.4 基础知识 2：Docker 镜像仓库

本章所演示的 DevOps 发布系统对应的基础架构环境是 Kubernetes 集群，其中应用发布的主要载体是 Docker 容器镜像。

在 CI 流程中，开发人员在将代码提交到 GitLab 仓库后会自动触发 Pipeline。而该动作的主要逻辑是：将代码编译构建后打包成 Docker 镜像，并将其上传至对应的 Docker 镜像仓库中。

而之后的 CD 流程，实际上就是从镜像仓库拉取 Docker 容器镜像，并将其部署到指定的 Kubernetes 集群环境中。

从整个 CI/CD 流程上来看，镜像仓库是非常关键的衔接点，是整个 DevOps 发布系统必不可少的基础组件。

下面就来演示如何安装部署 Docker 镜像仓库。

9.4.1 Docker 镜像简介

在部署 Docker 镜像仓库前，先简单介绍下什么是 Docker 镜像。

从本质上说，Docker 镜像可以被看作是一种特殊的文件系统，它封装了“容器运行时”所需的程序、依赖库、资源、配置文件及环境参数等依赖。这样，无论你是在本地还是在服务器端的任何一台机器，只要解压缩打包好的 Docker 镜像，就能将这个应用所需要的执行环境完整地重现出来。

正是因为 Docker 镜像提供了这种深入到操作系统级别的运行环境一致性的能力，使得它革命性地解决了应用打包过程中的环境依赖问题。这也是 Docker 项目能够快速流行起来的关键原因。

借助 Docker 镜像，可以高效地打通“开发–测试–部署”流程中的每一个环节，从而大大提高了软件系统的迭代效率。目前，基于 Docker 镜像的发布方式，已经逐渐成为软件发布的主流方式。

9.4.2 选择 Docker 镜像仓库

镜像仓库作为 Docker 技术的核心组件之一，其主要作用就是负责镜像内容的存储和分发。

从使用范围来说，Docker 镜像仓库分为“公有镜像仓库”和“私有镜像仓库”。

- 公有镜像仓库可以被任何人使用。Docker 公司维护的在线存储库 Docker Hub，以及部分云服务厂商（如阿里云）提供的在线 Docker 镜像库等，都属于公有镜像仓库。
- 私有镜像仓库是指，部署在公司或组织内部，用于自身应用 Docker 镜像存储、分发的镜像仓库。在构建公司内部使用的自动化发布系统的过程中，从安全的角度出发，应用的打包镜像一般情况下只会被存储在私有镜像仓库中。CI/CD 流程的主要环节也是通过向私有镜像仓库上传镜像和拉取镜像。

在目前企业级私有镜像仓库构建方案中，比较流行的是：开源的企业级 Docker 镜像仓库 Harbor、商业镜像仓库 JFrog Artifactory。

这两种 Docker 镜像仓库各自都有一定的市场。在作者所工作过的公司中，使用 Harbor 和 JFrog Artifactory 作为私有镜像仓库的都有。就成熟度和功能完整性来说，JFrog Artifactory 作为商业级解决方案更具优势；但从社区活跃度及开源特性来说，Harbor 的使用范围则要更广泛一些。在本章的实例中，将采用 Harbor 作为 DevOps 发布系统的私有镜像仓库。

9.4.3 部署 Harbor 私有镜像仓库

Harbor 是 VMware 公司开源的企业级 Docker Registry 项目，其目标是帮助用户快速搭建一个企业级的 Docker 镜像仓库。

Harbor 提供了包括管理用户界面、角色访问控制、日志审计等在内的企业级特性，加上其开源，使得其获得了不少互联网公司的青睐。

1. 部署 Harbor 镜像仓库

接下来演示部署 Harbor 镜像仓库的具体步骤。

（1）准备一台 Linux 服务器（版本：Ubantu 20.04 LTS），并安装 Docker。命令如下：

```
    #更新 apt 源，并添加 HTTPS 支持
    sudo apt-get update && sudo apt-get install apt-transport-https
ca-certificates curl software-properties-common -y

    #使用 utc 源添加 GPG Key
    curl -fsSL https://mirrors.ustc.edu.cn/docker-ce/linux/ubuntu/gpg | sudo
apt-key add

    #添加 Docker-CE 稳定版源地址
    sudo add-apt-repository "deb [arch=amd64]
https://mirrors.ustc.edu.cn/docker-ce/linux/ubuntu $(lsb_release -cs) stable"

    #更新源
    sudo apt-get update
    #安装最新版 Docker
    sudo apt install -y docker-ce
```

（2）安装 docker-compose 环境。

Harbor 在设计上是以 docker-compose 的规范来组织其各个组件的，并且通过 docker-compose 进行相关服务组件的启动和暂停。所以，安装 Harbor 需要安装 docker-compose 环境。命令如下：

```
    #运行此命令下载 docker-compose 的稳定版本
    sudo curl -L
"https://github.com/docker/compose/releases/download/1.26.2/docker-compose-$
(uname -s)-$(uname -m)" -o /usr/local/bin/docker-compose
```

如果下载速度过慢，也可以通过国内的源进行下载，例如：

```
    #通过 DaoCloud 下载 docker-compose 的稳定版本
```

```
curl -L https://get.daocloud.io/docker/compose/releases/download/1.26.2/docker-compose-`uname -s`-`uname -m` > /usr/local/bin/docker-compose
```

在下载成功后，给二进制文件应用赋予可执行权限，如下：

```
sudo chmod +x /usr/local/bin/docker-compose
```

（3）下载 Harbor 安装包。

Harbor 有离线和在线两种安装方式，这里选择下载在线安装包，命令如下：

```
#下载 Harbor 的在线安装包
wget
https://github.com/goharbor/harbor/releases/download/v2.0.2/harbor-online-installer-v2.0.2.tgz
```

可以根据 GitHub 官方发布的信息来选择合适的版本。

（4）解压缩并安装 Harbor。命令如下：

```
#将在线安装包解压缩到指定目录
tar zxvf harbor-online-installer-v2.0.2.tgz -C /usr/local/
```

在解压缩后，编辑 harbor.yml 配置文件（可以复制 harbor/harbor.yml.tmpl 模板文件），主要修改点如下：

```
#1  修改 hostname 为本机 IP 地址
hostname: 10.211.55.11
...
#2  修改 HTTP 端口号
http:
  # port for http, default is 80. If https enabled, this port will redirect to https port
  port: 8088

#3  为便于测试这里注释掉了 HTTPS 配置
#HTTPS:
  # https port for harbor, default is 443
  #  port: 443
  # The path of cert and key files for nginx
  # certificate: /your/certificate/path
  # private_key: /your/private/key/path
```

在完成配置文件修改后执行安装脚本：

```
./install.sh
```

如果出现以下提示，则表示安装成功。

```
...
Creating harbor-log ... done
Creating harbor-portal ... done
Creating redis        ... done
Creating registry      ... done
Creating harbor-db     ... done
Creating registryctl   ... done
Creating harbor-core   ... done
Creating nginx            ... done
Creating harbor-jobservice ... done
✔ ----Harbor has been installed and started successfully.----
```

如果上述步骤都执行正常，则访问 Harbor 用户地址“http://10.211.55.11:8088”，登录界面如图 9-13 所示。

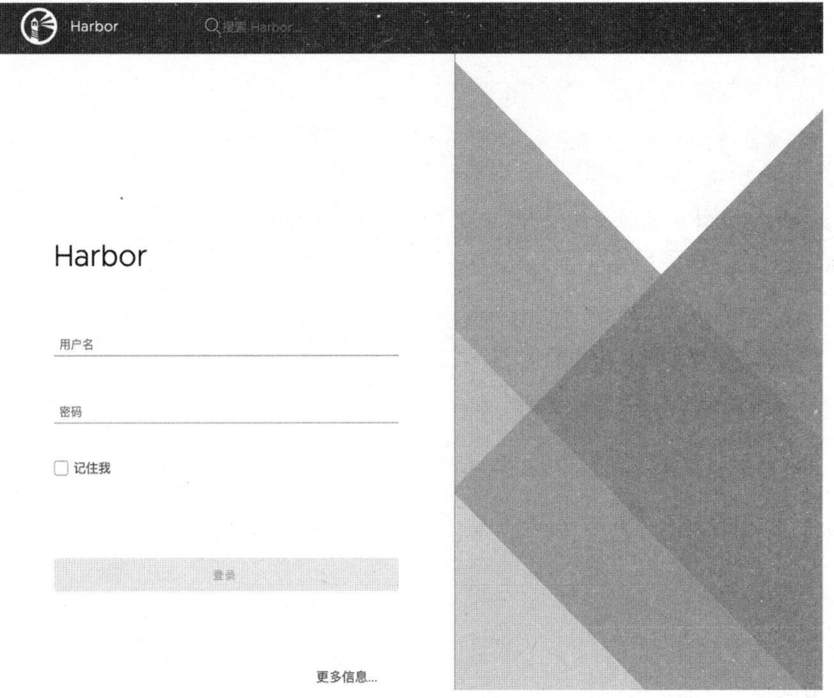

图 9-13

使用默认的“账号：admin”“密码：Harbor12345”登录验证，登录成功后的主界面如图 9-14 所示。

图 9-14

在安装成功后，编辑系统配置文件，设置开机自启动，具体如下：

```
#编辑系统文件
vi /etc/rc.local

#添加如下内容
#harbor start
cd /usr/local/harbor && docker-compose up -d
```

2. 测试镜像推送

在完成上述步骤后，通过 Docker 命令来测试镜像仓库是否能被正常使用。步骤如下。

（1）登录另外一台服务器，确保 Docker 已经安装好，然后本地配置仓库地址，具体如下：

```
#编辑 Docker 本地仓库配置文件
vim /etc/docker/daemon.json
```

添加如下内容：

```
{"insecure-registries": ["10.211.55.11:8088"]}
```

（2）重启 Docker 服务，命令如下：

```
systemctl restart docker
```

（3）使用 Docker 命令登录 Harbor 仓库，具体如下：

```
docker login -u admin -p Harbor12345 10.211.55.11:8088
```

如果显示如下信息，则说明登录成功：

```
...
Login Succeeded
```

（4）测试镜像推送是否正常，具体如下：

```
#从公有镜像仓库拉取一个镜像
$ docker pull alpine
Using default tag: latest
latest: Pulling from library/alpine
df20fa9351a1: Pull complete
Digest:
sha256:185518070891758909c9f839cf4ca393ee977ac378609f700f60a771a2dfe321
Status: Downloaded newer image for alpine:latest
docker.io/library/alpine:latest
```

（5）给拉取的镜像打上 tag，并执行 push 命令：

```
#Docker tag
$ docker tag alpine:latest 10.211.55.11:8088/library/alpine

#执行push命令
$ docker push 10.211.55.11:8088/library/alpine
The push refers to repository [10.211.55.11:8088/library/alpine]
50644c29ef5a: Pushed
latest: digest:
sha256:a15790640a6690aa1730c38cf0a440e2aa44aaca9b0e8931a9f2b0d7cc90fd65 size:
528
```

（6）在镜像推送成功后，进入 Harbor 后台验证其是否上传成功，如图 9-15 所示。

图 9-15

从图 9-15 中可以看到，镜像已经成功被上传至 Harbor 镜像仓库。

9.5 基础知识 3：Kubernetes 容器编排技术

在第 5.7 节中介绍了 Docker 容器技术，并演示了如何将 Spring Cloud 微服务应用以 Docker

容器的方式进行部署。但在实际的生产实践中，Docker 作为单一的容器技术工具并不能很好地定义容器的“组织方式”和“管理规范”，难以独立地支撑起生产级的大规模容器化部署。

正因为如此，容器技术的发展就迅速走向了以 Kubernetes 为代表的“容器编排”的技术路线。这也是为什么现在很少看到在生产环境中直接使用 Docker 部署应用的原因。

这并不是说 Kubernetes 与 Docker 一点关系也没有，要知道 Docker 最大的技术成功在于它定义了 Docker 镜像，从而解决了困扰开发者多年的应用打包问题。所以，虽然 Kubernetes 并不完全依赖 Docker 的“容器运行时”技术，但它目前所运行的绝大部分应用都还是以 Docker 镜像为载体的。

本节将介绍 Kubernetes 容器编排技术的基本原理，并搭建一套功能完整的 Kubernetes 集群，以此作为 DevOps 发布系统中 Spring Cloud 微服务应用运行的基础架构环境。

9.5.1 Kubernetes 简介

前面简要介绍了 Kubernetes 与 Docker 技术之间的关系，并提到了“容器编排”的概念。相对于 Docker 单一容器而言，Kubernetes 容器编排技术可以很好地实现大规模容器的组织和管理，从而使容器技术实现了从“容器”到“容器云”的飞跃。那么 Kubernetes 技术是从何而来的？又真正解决了什么问题呢？

Kubernetes 是由 Google 与 RedHat 公司共同主导的开源容器编排项目，它起源于 Google 公司的 Borg 系统，所以它在超大规模集群管理方面的经验明显优于其他容器编排技术。加上 Kubernetes 在社区管理方面的开放性，使得它很快打败了 Docker 公司推出的容器编排解决方案（Compose + Swarm），成了容器编排领域事实上的标准。

在功能上，Kubernetes 是一种综合的、基于容器构建分布式系统的基础架构环境。它不仅能够实现基本的拉取镜像和运行容器，还可以提供路由网关、水平扩展、监控、备份、灾难恢复等一系列运维能力。更重要的是，Kubernetes 可以按照用户的意愿和整个系统的规则，高度自动化地处理容器之间的各种关系实现“编排”能力。

此外，Kubernetes 的出现也重新定义了微服务架构的技术方向。目前通常所说的“云原生”及 Service Mesh（服务网格）等概念，很大程度上依赖 Kubernetes 所提供的能力。

由于篇幅有限，这里就不展开介绍了，感兴趣的读者可以参考其他专业书籍或技术资料。

9.5.2 搭建 Kubernetes 集群

要实现基于 Kubernetes 集群环境的 DevOps 发布系统，就要有一个集群。接下来，演示搭建 Kubernetes 集群的具体步骤。

1. 系统环境准备

要搭建 Kubernetes 集群，首先需要准备机器。最直接的办法是到公有云（如阿里云等）申请几台虚拟机。如果条件允许，用几台本地物理服务器来组建集群最好不过了。但是这些机器需要满足以下几个条件：

- 64 位 Linux 操作系统，且内核版本要求 3.10 及以上，能满足安装 Docker 所需的要求。
- 机器之间要保持网络互通。这是未来容器之间网络互通的前提条件。
- 要有外网访问权限，因为在部署过程中需要拉取相应的镜像，要求能够访问 gcr.io、quay.io 这两个 docker registry（有小部分镜像需要从这里拉取）。
- 单机可用资源建议两核 CPU、8G 内存或以上。如果小一点也可以，但是能调度的 Pod 数量就比较有限了。
- 磁盘空间要求在 30GB 以上，主要用于存储 Docker 镜像及相关日志文件。

本次实验准备了两台虚拟机，其具体配置如下：

- 两核 CPU、2GB 内存、30GB 的磁盘空间。
- Unbantu 20.04 LTS 的 Sever 版本，其 Linux 内核为 5.4.0。
- 内网互通，外网访问权限不受控制。

2. 了解 Kubernetes 集群部署工具 Kubeadm

Kubernetes 的部署一直是困扰初学者进入 Kubernetes 世界的一个大障碍。

在早期，Kubernetes 的部署主要依赖社区维护的各种脚本，但这其中涉及二进制编译、配置文件，以及 kube-apiserver 授权配置文件等诸多运维工作。

目前，各大云服务厂商常用的 Kubernetes 部署方式是“使用 SaltStack、Ansible 等运维工具自动化地执行这些烦琐的步骤”。但即使这样，部署过程对于初学者来说依然非常烦琐。

正是基于这样的痛点，在志愿者的推动下，Kubernetes 社区发起了 Kubeadm 这个独立的一键部署工具。使用 Kubeadm，通过几条简单的指令即可快速部署一个 Kubernetes 集群。

接下来使用 Kubeadm 来部署一个 Kubernetes 集群。

3. 安装 Kubeadm 及 Docker 环境

（1）编辑操作系统的安装源配置文件，添加 Kubernetes 镜像源。命令如下：

```
#添加 Kubernetes 官方镜像源 apt-key
root@kubenetesnode01:~#curl -s
https://packages.cloud.google.com/apt/doc/apt-key.gpg | apt-key add -
#添加 Kubernetes 的官方镜像源地址
root@kubernetesnode01:~# vim /etc/apt/sources.list
#add kubernetes source
deb http://apt.kubernetes.io/ kubernetes-xenial main
```

上方操作添加的是 Kubernetes 的官方镜像源。如果因为网络访问不到 apt.kubernetes.io，也可以换成国内的 Ubantu 镜像源，如阿里云的镜像源地址：

```
#添加阿里云的 Kubernetes 镜像源 apt-key
root@kubenetesnode01:~# curl -s
https://mirrors.aliyun.com/kubernetes/apt/doc/apt-key.gpg | apt-key add -
#添加阿里云的 Kubernetes 镜像源地址
root@kubernetesnode01:~# vim /etc/apt/sources.list
deb https://mirrors.aliyun.com/kubernetes/apt/ kubernetes-xenial main
```

（2）更新 apt 资源列表。命令如下：

```
root@kubernetesnode01:~# apt-get update
Hit:1 http://cn.archive.ubuntu.com/ubuntu focal InRelease
Hit:2 http://cn.archive.ubuntu.com/ubuntu focal-updates InRelease
Hit:3 http://cn.archive.ubuntu.com/ubuntu focal-backports InRelease
Hit:4 http://cn.archive.ubuntu.com/ubuntu focal-security InRelease
Get:5 https://packages.cloud.google.com/apt kubernetes-xenial InRelease
[8,993 B]
Get:6 https://packages.cloud.google.com/apt kubernetes-xenial/main amd64
Packages [37.7 kB]
Fetched 46.7 kB in 7s (6,586 B/s)
Reading package lists... Done
```

（3）通过“apt-get”命令安装 Kubeadm。命令如下：

```
root@kubernetesnode01:~# apt-get install -y docker.io kubeadm
Reading package lists... Done
Building dependency tree
Reading state information... Done
The following additional packages will be installed:
bridge-utils cgroupfs-mount conntrack containerd cri-tools dns-root-data
dnsmasq-base ebtables kubectl kubelet kubernetes-cni libidn11 pigz runc socat
ubuntu-fan
...
```

这里直接使用 Ubantu 的 docker.io 安装源。在上述安装过程中，kubeadm、kubelet、kubectl、kubernetes-cni 这几个 Kubernetes 核心组件的二进制文件也都被自动安装好。

（4）修改 Docker 操作系统限制。

在完成前面的操作步骤后，系统中会自动安装好 Docker 引擎。但在具体部署 Kubernetes 前，还需要对 Docker 的配置信息进行一些调整。

① 编辑系统“/etc/default/grub”文件，在配置项 GRUB_CMDLINE_LINUX 中添加如下参数：

```
GRUB_CMDLINE_LINUX=" cgroup_enable=memory swapaccount=1"
```

在完成编辑后保存文件。执行更新命令，并重启服务器，命令如下：

```
root@kubernetesnode01:/opt/kubernetes-config# update-grub
root@kubernetesnode01:/opt/kubernetes-config# reboot
```

上方修改主要解决的是可能出现的“Docker 警告 WARNING: No swap limit support”问题。

② 编辑“/etc/docker/daemon.json”文件，添加如下内容：

```
{
  "exec-opts": ["native.cgroupdriver=systemd"]
}
```

③ 在完成保存后，执行重启 Docker 的命令，如下：

```
root@kubernetesnode01:/opt/kubernetes-config# systemctl restart docker
```

④ 查看 Docker 的 Cgroup 信息，如下：

```
root@kubernetesnode01:/opt/kubernetes-config# docker info | grep Cgroup
 Cgroup Driver: systemd
```

上方修改主要解决的是“Docker cgroup driver. The recommended driver is "systemd"”问题。

以上只是作者在安装过程中遇到的具体问题的解决方法。如果读者在安装过程中遇到其他问题，请自行查阅相关资料。

⑤ 禁用虚拟内存。

由于 Kubernetes 是禁用虚拟内存的，所以要先禁用 swap，否则会在 Kubeadm 初始化 Kubernetes 时报错，命令如下：

```
root@kubernetesnode01:/opt/kubernetes-config# swapoff -a
```

该命令只是临时禁用 swap，如果要保证系统重启后仍然生效，则需要编辑“edit /etc/fstab”文件，注释掉 swap 那一行。

（5）启动系统 Docker 服务。

命令如下：

```
root@kubenetesnode02:~# systemctl enable docker.service
```

4. 部署 Kubernetes 的 Master 节点

在 Kubernetes 中，Master 节点是集群的控制节点，它由 3 个紧密协作的独立组件组合而成，分别是：负责 API 服务的 kube-apiserver、负责调度的 kube-scheduler，以及负责容器编排的 kube-controller-manager。整个集群的持久化数据由 kube-apiserver 处理后保存在 Etcd 中。

要部署 Master 节点，则可以直接通过 Kubeadm 进行一键部署。但如果希望部署一个相对完整的 Kubernetes 集群，则可以通过配置文件来开启一些实验性功能。具体方法如下：

（1）在安装服务器中新建“/opt/kubernetes-config”目录，并创建一个给 Kubeadm 使用的 YAML 文件（kubeadm.yaml），具体内容如下：

```
apiVersion: kubeadm.k8s.io/v1beta2
kind: ClusterConfiguration
controllerManager:
 extraArgs:
     horizontal-pod-autoscaler-use-rest-clients: "true"
     horizontal-pod-autoscaler-sync-period: "10s"
     node-monitor-grace-period: "10s"
apiServer:
  extraArgs:
     runtime-config: "api/all=true"
kubernetesVersion: "v1.18.1"
```

在上方 YAML 文件中：

- horizontal-pod-autoscaler-use-rest-clients:"true" 表示将来部署的 kuber-controller-manager 能够使用自定义资源（Custom Metrics）进行自动水平扩展。感兴趣的读者可以自行查阅相关资料。
- v1.18.1 是 Kubeadm 要部署的 Kubernetes 版本号。

需要注意的是，在执行过程中如果由于国内网络限制问题导致无法下载相应的 Docker 镜像，则可以根据报错信息在国内网站（如阿里云）上找到相关镜像，然后再将这些镜像重新打上 tag 之后再进行安装。具体如下：

```
    #从阿里云 Docker 仓库拉取 Kubernetes 组件镜像
    docker pull
registry.cn-hangzhou.aliyuncs.com/google_containers/kube-apiserver-amd64:v1.
18.1
    docker pull
registry.cn-hangzhou.aliyuncs.com/google_containers/kube-controller-manager-
amd64:v1.18.1
    docker pull
registry.cn-hangzhou.aliyuncs.com/google_containers/kube-scheduler-amd64:v1.
18.1
    docker pull
registry.cn-hangzhou.aliyuncs.com/google_containers/kube-proxy-amd64:v1.18.1
    docker pull
registry.cn-hangzhou.aliyuncs.com/google_containers/etcd-amd64:3.4.3-0
    docker pull registry.cn-hangzhou.aliyuncs.com/google_containers/pause:3.2
    docker pull
registry.cn-hangzhou.aliyuncs.com/google_containers/coredns:1.6.7
```

在下载完成后，再将这些 Docker 镜像重新打上 tag，具体命令如下：

```
    #给镜像打上 tag
    docker tag registry.cn-hangzhou.aliyuncs.com/google_containers/pause:3.2
k8s.gcr.io/pause:3.2
    docker tag
registry.cn-hangzhou.aliyuncs.com/google_containers/coredns:1.6.7
k8s.gcr.io/coredns:1.6.7
    docker tag
registry.cn-hangzhou.aliyuncs.com/google_containers/etcd-amd64:3.4.3-0
k8s.gcr.io/etcd:3.4.3-0
    docker tag
registry.cn-hangzhou.aliyuncs.com/google_containers/kube-scheduler-amd64:v1.
18.1 k8s.gcr.io/kube-scheduler:v1.18.1
    docker tag
registry.cn-hangzhou.aliyuncs.com/google_containers/kube-controller-manager-
amd64:v1.18.1 k8s.gcr.io/kube-controller-manager:v1.18.1
    docker tag
registry.cn-hangzhou.aliyuncs.com/google_containers/kube-apiserver-amd64:v1.
18.1 k8s.gcr.io/kube-apiserver:v1.18.1
    docker tag
registry.cn-hangzhou.aliyuncs.com/google_containers/kube-proxy-amd64:v1.18.1
k8s.gcr.io/kube-proxy:v1.18.1
```

此时通过 Docker 命令就可以查看这些 Docker 镜像信息了，命令如下：

```
    root@kubernetesnode01:/opt/kubernetes-config# docker images
```

```
    REPOSITORY   TAG     IMAGE ID      CREATED          SIZE
    k8s.gcr.io/kube-proxy
v1.18.1           4e68534e24f6       2 months ago       117MB
    registry.cn-hangzhou.aliyuncs.com/google_containers/kube-proxy-amd64
v1.18.1           4e68534e24f6       2 months ago       117MB
    k8s.gcr.io/kube-controller-manager
v1.18.1           d1ccdd18e6ed       2 months ago       162MB
    registry.cn-hangzhou.aliyuncs.com/google_containers/kube-controller-mana
ger-amd64  v1.18.1           d1ccdd18e6ed       2 months ago       162MB
    k8s.gcr.io/kube-apiserver
v1.18.1           a595af0107f9       2 months ago       173MB
    registry.cn-hangzhou.aliyuncs.com/google_containers/kube-apiserver-amd64
v1.18.1           a595af0107f9       2 months ago       173MB
    k8s.gcr.io/kube-scheduler
v1.18.1           6c9320041a7b       2 months ago       95.3MB
    registry.cn-hangzhou.aliyuncs.com/google_containers/kube-scheduler-amd64
v1.18.1           6c9320041a7b       2 months ago       95.3MB
    k8s.gcr.io/pause
3.2               80d28bedfe5d       4 months ago       683kB
    registry.cn-hangzhou.aliyuncs.com/google_containers/pause
3.2               80d28bedfe5d       4 months ago       683kB
    k8s.gcr.io/coredns
1.6.7             67da37a9a360       4 months ago       43.8MB
    registry.cn-hangzhou.aliyuncs.com/google_containers/coredns
1.6.7             67da37a9a360       4 months ago       43.8MB
    k8s.gcr.io/etcd
3.4.3-0           303ce5db0e90       8 months ago       288MB
    registry.cn-hangzhou.aliyuncs.com/google_containers/etcd-amd64
3.4.3-0           303ce5db0e90       8 months ago       288MB
```

（2）执行 Kubeadm 的部署命令完成 Kubernetes Master 节点的部署。具体命令及执行结果如下：

```
    root@kubernetesnode01:/opt/kubernetes-config# kubeadm init --config
kubeadm.yaml --v=5
    ...
    Your Kubernetes control-plane has initialized successfully!

    To start using your cluster, you need to run the following as a regular user:

      mkdir -p $HOME/.kube
      sudo cp -i /etc/kubernetes/admin.conf $HOME/.kube/config
      sudo chown $(id -u):$(id -g) $HOME/.kube/config

    You should now deploy a pod network to the cluster.
    Run "kubectl apply -f [podnetwork].yaml" with one of the options listed at:
```

```
    https://kubernetes.io/docs/concepts/cluster-administration/addons/

  Then you can join any number of worker nodes by running the following on each
as root:

  kubeadm join 10.211.55.6:6443 --token jfulwi.so2rj5lukgsej2o6 \
      --discovery-token-ca-cert-hash
sha256:d895d512f0df6cb7f010204193a9b240e8a394606090608daee11b988fc7fea6
```

从上面部署执行结果中可以看到，在部署成功后 Kubeadm 会生成如下信息：

```
  kubeadm join 10.211.55.6:6443 --token d35pz0.f50zacvbdarqn2vi \
      --discovery-token-ca-cert-hash
sha256:58958a3bf4ccf4a4c19b0d1e934e77bf5b5561988c2274364aaadc9b1747141d
```

其中“kubeadm join”是用来给该 Master 节点添加更多 Worker 节点（工作节点）的命令，后面具体部署 Worker 节点时将会使用到它。

此外，Kubeadm 还会提示第一次使用 Kubernetes 集群所需要配置的命令：

```
  mkdir -p $HOME/.kube
  sudo cp -i /etc/kubernetes/admin.conf $HOME/.kube/config
  sudo chown $(id -u):$(id -g) $HOME/.kube/config
```

需要这些配置命令的原因在于，Kubernetes 集群默认是用加密方式访问的，所以这几条命令就是将刚才部署生成的 Kubernetes 集群的安全配置文件保存到当前用户的“.kube”目录下。之后，kubectl 会默认使用该目录下的授权信息访问 Kubernetes 集群。如果不这么做，则在每次访问集群前都需要设置环境变量“export KUBECONFIG”来告诉 kubectl 这个安全配置文件的位置。

（3）使用“kubectl get”命令查看当前 Kubernetes 集群节点的状态，执行效果如下：

```
  root@kubernetesnode01:/opt/kubernetes-config# kubectl get nodes
  NAME               STATUS     ROLES    AGE   VERSION
  kubernetesnode01   NotReady   master   35m   v1.18.4
```

从以上命令输出的结果中可以看到，Master 节点的状态为“NotReady”。为了查找具体原因，可以通过“kuberctl describe”命令来查看该节点（Node）对象的详细信息，命令如下：

```
  root@kubernetesnode01:/opt/kubernetes-config# kubectl describe node
kubernetesnode01
```

该命令可以非常详细地获取节点对象的状态、事件等详情。这种方式也是在调试 Kubernetes 集群时最重要的排查手段。根据显示的如下信息，可以看到节点处于“NodeNotReady”状态的原因在于尚未部署任何网络插件。

```
...
Conditions
...
Ready False... KubeletNotReady runtime network not ready: NetworkReady=false
reason:NetworkPluginNotReady message:docker: network plugin is not ready: cni
config uninitialized
...
```

为了进一步验证这一点，还可以通过 kubectl 检查这个节点上各个 Kubernetes 系统 Pod 的状态。命令及执行效果如下：

```
root@kubernetesnode01:/opt/kubernetes-config# kubectl get pods -n
kube-system
NAME                                        READY   STATUS     RESTARTS   AGE
coredns-66bff467f8-l4wt6                    0/1     Pending    0          64m
coredns-66bff467f8-rcqx6                    0/1     Pending    0          64m
etcd-kubernetesnode01                       1/1     Running    0          64m
kube-apiserver-kubernetesnode01             1/1     Running    0          64m
kube-controller-manager-kubernetesnode01 1/1     Running    0          64m
kube-proxy-wjct7                            1/1     Running    0          64m
kube-scheduler-kubernetesnode01             1/1     Running    0          64m
```

命令中“kube-system”表示 Kubernetes 项目预留的系统 Pod 空间（Namespace）。需要注意，它并不是 Linux Namespace，而是 Kuebernetes 划分的不同工作空间单位。

从命令输出结果可以看到，coredns 等依赖网络的 Pod 都处于“Pending”（调度失败）状态，这说明该 Master 节点的网络尚未部署就绪。

5. 部署 Kubernetes 网络插件

由于没有部署网络插件，所以前面部署的 Master 节点显示“NodeNotReady”状态。接下来部署网络插件。

在 Kubernetes“一切皆容器”的设计理念指导下，网络插件也会以独立 Pod 的方式运行在系统中。所以，部署网络插件很简单，只需要执行“kubectl apply”命令即可。例如，以 Weave 网络插件为例：

```
root@kubernetesnode01:/opt/kubernetes-config# kubectl apply -f https://
cloud.weave.works/k8s/net?k8s-version=$(kubectl version | base64 | tr -d '\n')
serviceaccount/weave-net created
clusterrole.rbac.authorization.k8s.io/weave-net created
clusterrolebinding.rbac.authorization.k8s.io/weave-net created
role.rbac.authorization.k8s.io/weave-net created
rolebinding.rbac.authorization.k8s.io/weave-net created
daemonset.apps/weave-net created
```

在部署完成后，通过“kubectl get”命令重新检查 Pod 的状态：

```
root@kubernetesnode01:/opt/kubernetes-config# kubectl get pods -n kube-system
NAME                                          READY   STATUS    RESTARTS   AGE
coredns-66bff467f8-l4wt6                      1/1     Running   0          116m
coredns-66bff467f8-rcqx6                      1/1     Running   0          116m
etcd-kubernetesnode01                         1/1     Running   0          116m
kube-apiserver-kubernetesnode01               1/1     Running   0          116m
kube-controller-manager-kubernetesnode01      1/1     Running   0          116m
kube-proxy-wjct7                              1/1     Running   0          116m
kube-scheduler-kubernetesnode01               1/1     Running   0          116m
weave-net-746qj                               2/2     Running   0          14m
```

可以看到，此时所有的系统 Pod 都成功启动了。刚才部署的 Weave 网络插件在 kube-system 下新建了一个名为“weave-net-746qj”的 Pod，这个 Pod 就是容器网络插件在每个节点上的控制组件。

至此，Kubernetes 的 Master 节点就部署完成了。如果只需要一个单节点的 Kubernetes，则现在就可以使用了。但是在默认情况下，Kubernetes 的 Master 节点是不能运行用户 Pod 的，所以需要通过额外的操作进行调整，在本节的最后会介绍。

6. 部署 Kubernetes 的 Worker 节点

为了构建一个完整的 Kubernetes 集群，还需要继续部署 Worker 节点。

实际上，Kubernetes 的 Worker 节点和 Master 节点几乎是相同的，它们都运行着一个 Kubelet 组件，主要的区别在于“kubeadm init”过程中：在 Kubelet 组件启动后，Master 节点还会自动启动 kube-apiserver、kube-scheduler 及 kube-controller-manager 这 3 个系统 Pod。

与部署 Master 节点一样，在具体部署 Worker 节点前，也需要在所有 Worker 节点上执行前面小标题“3. 安装 Kubeadm 及 Decker 环境”中的所有步骤。之后在 Worker 节点执行在“4.”小标题中部署 Master 节点时所生成的“kubeadm join”命令，具体如下：

```
root@kubenetesnode02:~# kubeadm join 10.211.55.6:6443 --token jfulwi.so2rj5lukgsej2o6     --discovery-token-ca-cert-hash sha256:d895d512f0df6cb7f010204193a9b240e8a394606090608daee11b988fc7fea6 --v=5

...
This node has joined the cluster:
* Certificate signing request was sent to apiserver and a response was received.
* The Kubelet was informed of the new secure connection details.
```

```
    Run 'kubectl get nodes' on the control-plane to see this node join the cluster.
```

为了便于在 Worker 节点执行 kubectl 相关命令，还需要进行如下配置：

```
    #创建配置目录
    root@kubenetesnode02:~# mkdir -p $HOME/.kube
    #将Master节点中“$/HOME/.kube/”目录中的config文件复制至Worker节点对应目录下
    root@kubenetesnode02:~# scp root@10.211.55.6:$HOME/.kube/config
$HOME/.kube/
    #权限配置
    root@kubenetesnode02:~# sudo chown $(id -u):$(id -g) $HOME/.kube/config
```

之后就可以在 Worker 或 Master 节点执行节点状态查看命令“kubectl get nodes”，具体如下：

```
    root@kubernetesnode02:~# kubectl get nodes
    NAME                 STATUS      ROLES     AGE     VERSION
    kubenetesnode02      NotReady    <none>    33m     v1.18.4
    kubernetesnode01     Ready       master    29h     v1.18.4
```

具体节点描述信息如下。节点状态显示，此时 Work 节点还处于“NotReady”状态。

```
    root@kubernetesnode02:~# kubectl describe node kubenetesnode02
    ...
    Conditions:
    ...
    Ready False ... KubeletNotReady runtime network not ready: NetworkReady=false
reason:NetworkPluginNotReady message:docker: network plugin is not ready: cni
config uninitialized
    ...
```

根据描述信息发现，Worker 节点 NotReady 的原因也在于网络插件没有部署。继续执行小标题“5. 部署 Kubernetes 网络插件”中的步骤即可。

但要注意，在部署网络插件时会同时部署 kube-proxy。其中涉及从 k8s.gcr.io 仓库获取镜像的动作。如果无法访问外网则可能导致网络部署异常，这里可以参考前面布署 Master 节点时的做法：在通过国内镜像仓库下载后，通过 tag 的方式进行标识。具体如下：

```
    #从阿里云拉取必要的镜像
    docker pull
registry.cn-hangzhou.aliyuncs.com/google_containers/kube-proxy-amd64:v1.18.1
    docker pull registry.cn-hangzhou.aliyuncs.com/google_containers/pause:3.2
    #给镜像重新打上tag
    docker tag
registry.cn-hangzhou.aliyuncs.com/google_containers/kube-proxy-amd64:v1.18.1
k8s.gcr.io/kube-proxy:v1.18.1
    docker tag registry.cn-hangzhou.aliyuncs.com/google_containers/pause:3.2
k8s.gcr.io/pause:3.2
```

如果一切正常，则继续查看节点状态，命令如下：

```
root@kubenetesnode02:~# kubectl get node
NAME                 STATUS   ROLES    AGE     VERSION
kubenetesnode02      Ready    <none>   7h52m   v1.18.4
kubernetesnode01     Ready    master   37h     v1.18.4
```

可以看到，此时 Worker 节点的状态已经变成“Ready”。不过细心的读者可能会发现，Worker 节点的 ROLES 并不像 Master 节点那样显示“master”，而是显示“<none>”。这是因为，新安装的 Kubernetes 环境 Node 节点有时会丢失 ROLES 信息。遇到这种情况，可以手工进行添加，具体命令如下：

```
root@kubenetesnode02:~# kubectl label node kubenetesnode02
node-role.kubernetes.io/worker=worker
```

再次运行节点状态命令，就能看到正常的显示了。命令及效果如下：

```
root@kubenetesnode02:~# kubectl get node
NAME                 STATUS   ROLES    AGE   VERSION
kubenetesnode02      Ready    worker   8h    v1.18.4
kubernetesnode01     Ready    master   37h   v1.18.4
```

至此，部署完了具有一个 Master 节点和一个 Worker 节点的 Kubernetes 集群。作为实验环境，它已经具备了 Kubernetes 集群的基本功能。

7. 部署 Dashboard 可视化插件

在 Kubernetes 社区中有一个很受欢迎的 Dashboard 项目。它给用户提供了一个可视化的 Web 界面，来查看当前集群中的各种信息。

该插件也是以容器化方式进行部署的，操作也非常简单，可在 Master、Worker 节点或其他能够安全访问 Kubernetes 集群的 Node 节点上进行部署。

具体命令如下：

```
root@kubenetesnode02:~# kubectl apply -f
https://raw.githubusercontent.com/kubernetes/dashboard/v2.0.3/aio/deploy/recommended.yaml
```

在部署完成后，就可以查看 Dashboard 项目对应的 Pod 的运行状态了，执行效果如下：

```
root@kubenetesnode02:~# kubectl get pods -n kubernetes-dashboard
NAME                                         READY   STATUS    RESTARTS AGE
dashboard-metrics-scraper-6b4884c9d5-xfb8b   1/1     Running      0     12h
kubernetes-dashboard-7f99b75bf4-9lxk8        1/1     Running      0     12h
```

除此之外，还可以查看 Dashboard 的服务（Service）信息，命令如下：

```
root@kubenetesnode02:~# kubectl get svc -n kubernetes-dashboard
```

```
    NAME                        TYPE        CLUSTER-IP      EXTERNAL-IP   PORT(S)    AGE
    dashboard-metrics-scraper   ClusterIP   10.97.69.158    <none>        8000/TCP   13h
    kubernetes-dashboard        ClusterIP   10.111.30.214   <none>        443/TCP    13h
```

需要注意的是，由于 Dashboard 是一个 Web 服务，所以从安全角度出发，Dashboard 默认只能通过 Proxy 的方式在本地访问。具体方式为：在本地机器安装 Kubectl 管理工具，并将“Master 节点$HOME/.kube/”目录中的 config 文件复制至本地主机相同目录下，之后运行“kubectl proxy”命令，具体如下：

```
    qiaodeMacBook-Pro-2:.kube qiaojiang$ kubectl proxy
    Starting to serve on 127.0.0.1:8001
```

在本地代理启动后，访问 Kubernetes Dashboard 地址，具体如下：

```
    http://localhost:8001/api/v1/namespaces/kubernetes-dashboard/services/ht
tps:kubernetes-dashboard:/proxy/
```

如果访问正常，则会看到如图 9-16 所示界面。

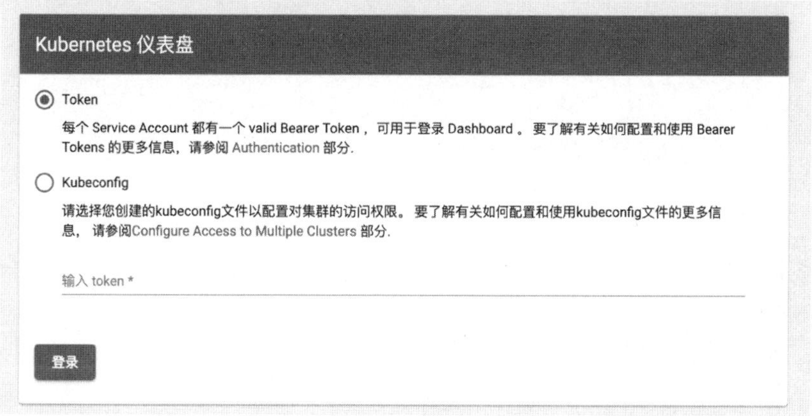

图 9-16

如图 9-16 所示，Dashboard 访问需要进行身份认证，主要有 Token 及 Kubeconfig 两种方式。这里选择 Token 方式。Token 的生成步骤如下：

（1）创建一个服务账号。

在命名空间 kubernetes-dashboard 中创建一个名为“admin-user”的服务账户，具体步骤为：

① 在本地目录中创建文件“dashboard-adminuser.yaml”，具体内容如下：

```
    apiVersion: v1
    kind: ServiceAccount
```

```
metadata:
  name: admin-user
  namespace: kubernetes-dashboard
```

② 执行创建命令：

```
qiaodeMacBook-Pro-2:.kube qiaojiang$ kubectl apply -f
dashboard-adminuser.yaml
Warning: kubectl apply should be used on resource created by either kubectl
create --save-config or kubectl apply
serviceaccount/admin-user configured
```

（2）创建 ClusterRoleBinding。

在使用 Kubeadm 工具配置完 Kubernetes 集群后，集群中已经存在 ClusterRole，可以使用它为上一步创建的 ServiceAccount 创建 ClusterRoleBinding。具体步骤为：

① 在本地目录中创建文件“dashboard-clusterRoleBingding.yaml”，具体内容如下：

```
apiVersion: rbac.authorization.k8s.io/v1
kind: ClusterRoleBinding
metadata:
  name: admin-user
roleRef:
  apiGroup: rbac.authorization.k8s.io
  kind: ClusterRole
  name: cluster-admin
subjects:
- kind: ServiceAccount
  name: admin-user
  namespace: kubernetes-dashboard
```

② 执行创建命令：

```
qiaodeMacBook-Pro-2:.kube qiaojiang$ kubectl apply -f
dashboard-clusterRoleBingding.yaml
clusterrolebinding.rbac.authorization.k8s.io/admin-user created
```

（3）执行获取 Bearer Token 的命令，具体如下：

```
qiaodeMacBook-Pro-2:.kube qiaojiang$ kubectl -n kubernetes-dashboard
describe secret $(kubectl -n kubernetes-dashboard get secret | grep admin-user
| awk '{print $1}')
Name:         admin-user-token-xxq2b
Namespace:    kubernetes-dashboard
Labels:       <none>
Annotations:  kubernetes.io/service-account.name: admin-user
              kubernetes.io/service-account.uid:
213dce75-4063-4555-842a-904cf4e88ed1
```

```
Type:  kubernetes.io/service-account-token

Data
====
ca.crt:     1025 bytes
namespace:  20 bytes
token:
eyJhbGciOiJSUzI1NiIsImtpZCI6IlplSHRwcXhNREs0SUJPcTZIYU1kT0pidlFuOFJaVXYzLWx0
c1BOZzZZY28ifQ.eyJpc3MiOiJrdWJlcm5ldGVzL3NlcnZpY2VhY2NvdW50Iiwia3ViZXJuZXRlc
y5pby9zZXJ2aWNlYWNjb3VudC9uYW1lc3BhY2UiOiJrdWJlcm5ldGVzLWRhc2hib2FyZCIsImt1Y
mVybmV0ZXMuaW8vc2VydmljZWFjY291bnQvc2VjcmV0Lm5hbWUiOiJhZG1pbi11c2VyLXRva2VuL
Xh4cTJiIiwia3ViZXJuZXRlcy5pby9zZXJ2aWNlYWNjb3VudC9zZXJ2aWNlLWFjY291bnQubmFtZ
SI6ImFkbWluLXVzZXIiLCJrdWJlcm5ldGVzLmlvL3NlcnZpY2VhY2NvdW50L3NlcnZpY2UtYWNjb
3VudC51aWQiOiIyMTNkY2U3NS00MDYzLTQ1NTUtODQyYS05MDRjZjRlODhlZDEiLCJzdWIiOiJze
XN0ZW06c2VydmljZWFjY291bnQ6a3ViZXJuZXRlcy1kYXNoYm9hcmQ6YWRtaW4tdXNlciJ9.MIjS
ewAk4aVgVCU6fnBBLtIH7PJzcDUozaUoVGJPUu-TZSbRZHotugvrvd8Ek_f5urfyYhj14y1BSe1E
Xw3nINmo4J7bMI94T_f4HvSFW1RUznfWZ_uq24qKjNgqy4HrSfmickav2PmGv4TtumjhbziMreQ3
jfmaPZvPqOa6Xmv1uhytLw3G6m5tRS97kl0i8A1lqnOWu7COJX0TtPkDrXiPPX9IzaGrp3Hd0pKH
WrI_-orxsI5mmFj0cQZt1ncHarCssVnyHkWQqtle41jV2HAO-bgY1j0E1pOPTlzpmSSbmAmedXZy
m77N10YNaIqtWvFjxMzhFqeTPNo539V1Gg
```

在获取后将回到前面的认证方式选择界面，将获取的 Token 信息填入即可正式进入 Dashboard 的系统界面。看到的 Kubernetes 集群的详细可视化信息如图 9-17 所示。

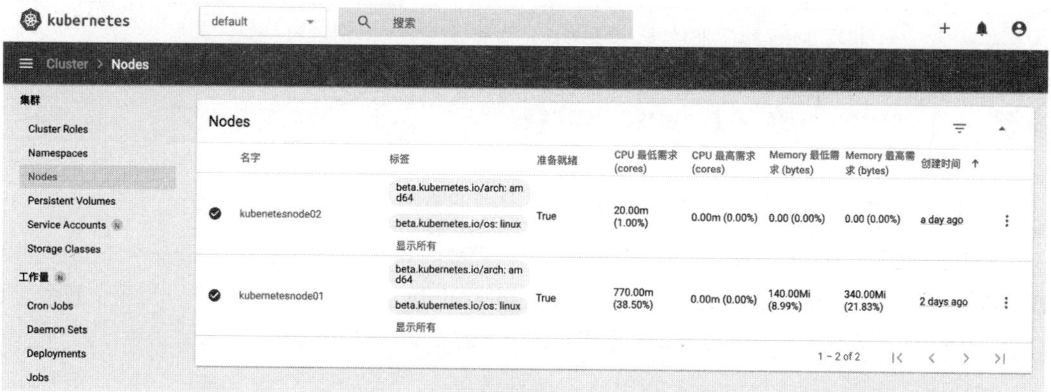

图 9-17

至此，完成了 Kubernetes 可视化插件的部署，并通过本地 Proxy 的方式进行了登录。

在实际的生产环境中，如果觉得每次通过本地 Proxy 的方式进行访问不够方便，则可以使用 Ingress 的方式配置在集群外访问 Dashboard。感兴趣的读者可以自行尝试下。

8. Kubernetes 容器存储插件

使用 Kubernetes 容器技术，很多时候都需要用数据卷（Volume）把外面宿主机上的目录或者文件挂载进容器的 Mount Namespace 中，从而让容器和宿主机共享这些目录或者文件，进而使得容器里的应用可以在这些数据卷中新建或写入文件。

容器最典型的特征之一是“无状态”：在某台机器上启动的容器，无法看到其他机器上的容器在它们数据卷中所写入的文件。

持久化存储，就是用来保存容器存储状态的重要手段。存储插件会在容器中挂载一个基于网络或其他机制的远程数据卷，使得在容器中创建的文件实际上是保存在远程存储服务器上的，与当前宿主机没有直接的绑定关系。这样，无论在哪个宿主机上启动新容器，都可以设置挂载指定的持久化数据卷。

由于 Kubernetes 本身的松耦合设计，所以绝大多数存储项目（例如 Ceph、ClusterFS、NFS 等）都可以为 Kubernetes 提供持久化存储能力。目前 Kubernetes 社区中热度较高的存储方案是 Rook 项目，它是一个基于 Ceph 的 Kubernetes 存储插件。

相比 Ceph 的简单封装，Rook 在自己的实现中加入了横向扩展、灾备、监控等许多企业级功能，所以其更适合作为生产级别可用的存储插件。由于篇幅的关系这里就不再具体演示，感兴趣的读者可以自己尝试部署。

9. 通过 Taint/Toleration 调整 Master 执行 Pod 的策略

在前面提到，Kubernetes 集群的 Master 节点在默认情况下是不能运行用户 Pod 的。能有这样的效果，Kubernetes 依靠的正是 Taint/Toleration 机制。该机制的原理是：一旦某个节点被加上“Taint”，则表示该节点被打上了“污点”，所有的 Pod 就不能再在这个节点上运行。

而 Master 节点之所以不能运行用户 Pod 就在于：其在运行成功后会为自身节点打上“Taint”，从而达到禁止其他用户 Pod 运行在 Master 节点上的效果（不影响已经运行的 Pod）。可以通过命令查看 Master 节点上的相关信息，命令及执行效果如下：

```
root@kubenetesnode02:~# kubectl describe node kubernetesnode01
Name:               kubernetesnode01
Roles:              master
...
```

```
    Taints:              node-role.kubernetes.io/master:NoSchedule
    ...
```

可以看到，Master 节点默认被加上了“node-role.kubernetes.io/master:NoSchedule”这样的“污点”。其中，值“NoSchedule”意味着这个 Taint 只会在调度新的 Pod 时产生作用，而不会影响在该节点上已经运行的 Pod。如果在实验中只想要一个单节点的 Kubernetes，则可以删除 Master 节点上的这个 Taint，具体命令如下：

```
    root@kubernetesnode01:~# kubectl taint nodes --all
node-role.kubernetes.io/master-
```

在上方命令中，通过在“nodes --all node-role.kubernetes.io/master”键后面加一个短横线“-”，表示移除所有以该键为键的 Taint。

至此，一个基本的 Kubernetes 集群就部署完成了。

有了 Kubeadm 这样的原生管理工具，Kubernetes 的部署被大大简化了。其中证书、授权及各个组件配置等最麻烦的操作，Kubeadm 都帮我们完成了。

9.5.3 Kubernetes 的技术原理

前面简单介绍了 Kubernetes 的基本概念，并通过 Kubeadm 具体演示了 Kubernetes 集群的部署方法，其中涉及 apiserver、controller-manager、scheduler、etcd 这样的 Kubernetes 核心组件，也涉及 Pod、Service 这样的容器编排概念。

本节将从“Kubernetes 的系统架构”及“容器编排核心概念”两个方面来阐述 Kubernetes 的基本技术原理。

1. Kubernetes 的系统架构

Kubernetes 的系统架构如图 9-18 所示。

从图 9-18 中可以看到，Kubernetes 架构上主要由 Master 和 Node 两种节点组成，这两种节点分别对应着控制节点和计算节点。其中，Master 节点（即控制节点）是整个 Kubernetes 集群的大脑，主要负责编排、管理和调度用户提交的作业，并能根据集群系统资源的整体使用情况将作业任务自动分发到可用 Node 节点（即计算节点）。

（1）Master 节点。

Master 节点主要由以下 3 个紧密协作的独立组件组合而成。

- kube-apiserver：Kubernetes 集群 API 服务的入口，主要提供资源访问操作、认证、授权、访问控制，以及 API 注册和发现等功能。
- kube-scheduler：负责 Kubernetes 的资源调度，能按照预定的调度策略将 Pod 调度到相应的机器上。

- kube-controller-manager：负责容器编排及 Kubernetes 集群状态的维护，例如故障检测、自动扩展、滚动更新等。

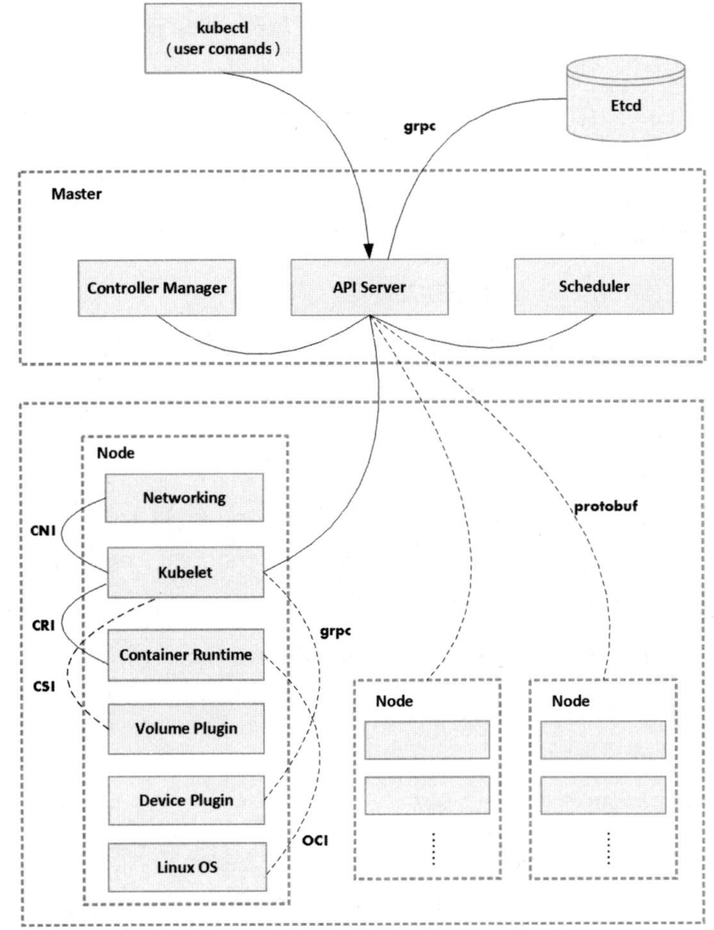

图 9-18

在工作状态下，上述组件还会产生许多需要进行持久化的数据。这些数据会被 kube-apiserver 处理后统一被保存到 Etcd 中。所以，kube-apiserver 不仅是外部访问 Kubernetes 集群的入口，也是维护整个 Kubernetes 集群状态的信息枢纽。

（2）Node 节点。

Node 节点最核心的部分是 kubelet 组件。该组件的核心功能如下：

- 通过 CRI（Container Runtime Interface）远程接口与“容器运行时”（如 Docker）进行交互，对容器的生命周期进行维护。其中，CRI 接口会定义“容器运行时”的各项核心操作，例如启动容器所需的命令及参数等。
- 通过 GRPC 协议同 Device Plugin 插件进行交互，实现 Kubernetes 对宿主机物理设备的管理。
- 通过 CNI（Container Networking Interface）来调用网络插件为容器配置网络，以及通过 CSI（Container Storage Interface）调用存储插件来为容器配置持久化存储。

在 Kubernetes 中，Kubelet 组件会通过 CRI 接口与“容器运行时”进行交互。在容器运行时，kubelet 组件则通过 OCI 容器运行时规范与底层 Linux 操作系统（如 Namespace、Cgroups 等）进行交互。

这里所说的“容器运行时”对应的英文是 Container Runtime，是一种实现容器技术的规范。所有实现了 CRI 接口规范的容器项目，都可以作为 Kubernetes 的“容器运行时”存在。其原因在于，Kubernetes 从设计之初就没有把 Docker 作为整个架构的核心，而只是将其作为最底层的一个“容器运行时”来实现。Kubernetes 要重点解决的是：对于运行在大规模集群中的各种任务，根据其关系进行作业编排及管理。这也是为什么 Kubernetes 被称作容器编排技术，而不仅仅只是容器技术的原因。这种能力也正是 Kubernetes 项目能够流行的关键原因。

2. Kubernetes 容器编排

“处理任务之间的各种关系，实现容器编排”是 Kubernetes 技术的核心，那么“容器编排”到底是一个什么概念？在 Kubernetes 中是如何实现容器编排的呢？

所谓“容器编排”，通俗点举例就是，如果两个应用的调用关系比较紧密，则在运行时将它们部署在同一台机器上，从而提升服务之间的通信效率。而“能够自动将具有此类关系的应用，以容器的方式部署在同一台机器上”的技术就是容器编排。

这里所说的“紧密关系”只是一种形象的说法。在实际的技术场景中，这种“紧密关系”可以被划分为很多类型，例如 Web 应用与数据库之间的访问关系、负载均衡和它后端服务之间的代理关系、门户应用与授权组件之间的调用关系等。

对于 Kubernetes 来说，这样的关系描述显然还是过于具体。因为，Kubernetes 的设计目标是“不仅能够支持前面提到的所有类型的关系，还能够支持未来可能出现的更多种类的关系”。这就要求，Kubernetes 能从更宏观的角度来定义任务之间的各种关系，并且为将来“支持更多种类的关系”留有余地。

具体来说，Kubernetes 对容器间的访问进行了分类：如果这些应用需要进行频繁的交互和访问，或者它们之间存在直接通过本地文件进行信息交换的情况，则在 Kubernetes 中可以将这些容器划分为一个 Pod；Pod 中的容器将共享同一个 Network Namespace、同一组数据卷，从而实现高效率通信。

Pod 是 Kubernetes 中最基础的编排对象，是 Kubernetes 最小的调度单元，也是 Kubernetes 实现容器编排的载体。Pod 本质上是一组共享了某些系统资源的容器集合。在 Kubernetes 中，围绕 Pod 所延伸的核心概念如图 9-19 所示。

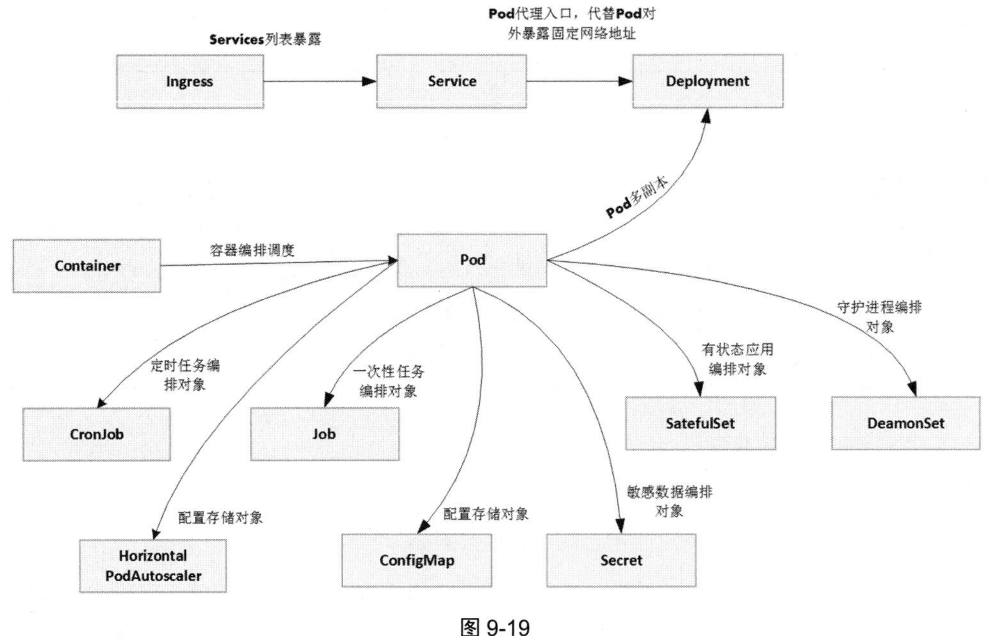

图 9-19

如图 9-19 所示，在 Kubernetes 中，Pod 解决了容器间紧密协作（即编排）的问题。而 Pod 要实现一次启动多个 Pod 副本，则需要 Deployment 这个 Pod 多实例管理器。在有了这样一组 Pod 后，我们需要通过一个固定网络地址以负载均衡的方式访问它们，于是有了 Service。

根据不同的编排场景，Pod 又衍生出了：描述一次性运行任务的 Job 编排对象、描述每个宿主机上必须有且只能运行一个副本的守护进程服务 DaemonSet、描述定义任务的 CronJob 编排对象，以及针对有状态应用的 StatefulSet 对象等。以上这些正是 Kubernetes 定义容器间关系和形态的主要方法。

9.6 自动化发布 Spring Cloud 微服务

通过前面的操作，我们搭建了 GitLab 代码管理仓库、Harbor 容器镜像仓库，以及 Kubernetes 集群等 DevOps 发布系统所依赖的基础环境。

本节将具体演示如何集成 Spring Cloud 微服务与 GitLab CI/CD 机制，实现将 Spring Cloud 微服务自动化发布到 Kubernetes 集群的 DevOps 发布系统。

9.6.1 创建 Spring Cloud 微服务的示例工程

在演示 Spring Cloud 微服务的自动化发布前，需要先创建一个受 GitLab 代码仓库管理的 Spring Cloud 微服务示范工程。

1. 创建一个基本的 Maven 工程结构

利用 2.3.1 节介绍的方法创建一个 Maven 工程，完成后的工程代码结构如图 9-20 所示。

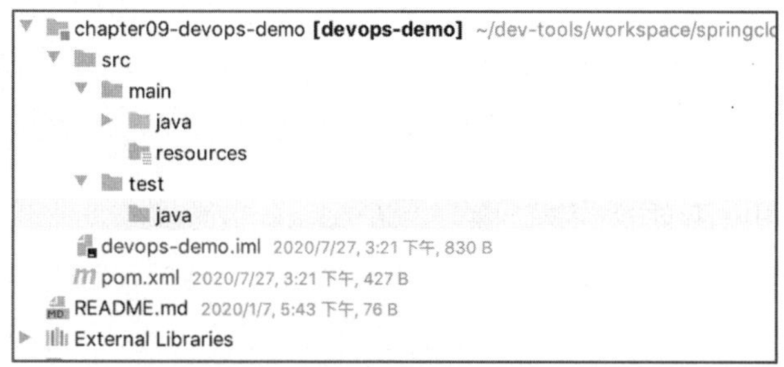

图 9-20

2. 引入 Spring Cloud 依赖，将其改造为微服务项目

（1）引入 Spring Cloud 微服务的核心依赖。

这里可以参考 2.5.2 节中的具体步骤。

（2）在工程代码的 resources 目录新建一个基础性配置文件——bootstrap.yml。配置文件中的代码如下：

```
spring:
  application:
    name: devops-demo
  profiles:
    active: debug
  cloud:
    consul:
```

```
      discovery:
        preferIpAddress: true
        instance-id: ${spring.application.name}:${spring.cloud.client.
ipAddress}:${spring.application.instance_id:${server.port}}:@project.version@
        healthCheckPath: /actuator/health
  server:
    port: 9092
```

（3）Spring Boot 并不会默认加载 bootstrap.yml 这个文件，所以需要在 pom.xml 中添加 Maven 资源相关的配置，具体参考 2.5.2 节内容。

（4）创建 Spring Cloud 示范微服务的入口程序类。代码如下：

```
package com.wudimanong.devops;
import org.springframework.boot.SpringApplication;
import org.springframework.boot.autoconfigure.SpringBootApplication;
import org.springframework.cloud.client.discovery.EnableDiscoveryClient;
@EnableDiscoveryClient
@SpringBootApplication
public class DevopsDemoApplication {
    public static void main(String[] args) {
        SpringApplication.run(DevopsDemoApplication.class, args);
    }
}
```

至此，Spring Cloud 示范微服务就构建完成了。

（5）为了便于后续验证测试，这里编写一个简单的 HTTP 接口。代码如下：

```
package com.wudimanong.devops.controller;
import org.springframework.web.bind.annotation.GetMapping;
import org.springframework.web.bind.annotation.RequestMapping;
import org.springframework.web.bind.annotation.RestController;
@RestController
@RequestMapping("/devops")
public class DevopsTestController {
    @GetMapping("/test")
    public String devopsTest() {
        return "自动化发布示范工程测试接口返回->OK!";
    }
}
```

完成后运行微服务程序。如果该接口能够被正常访问，则说明示例工程基本搭建成功了。后面的内容将演示如何将该服务以自动化发布的方式部署到 Kubernetes 集群中。

9.6.2　配置 Spring Cloud 项目的 Docker 打包插件

1. 配置 Docker 镜像打包插件

在进行具体的 CI/CD 流程定义前，需要在 Spring Cloud 项目工程中配置 Docker 镜像的 Maven 打包插件，使其能够通过 Maven 构建命令将微服务应用打包成 Docker 镜像。

（1）在 Spring Cloud 工程的 pom.xml 文件中配置 Docker 打包插件。具体如下：

```
<!--添加 Docker 镜像的 Maven 打包插件-->
<plugin>
    <groupId>com.spotify</groupId>
    <artifactId>dockerfile-maven-plugin</artifactId>
    <version>1.4.13</version>
    <executions>
        <execution>
            <id>build-image</id>
            <phase>package</phase>
            <goals>
                <goal>build</goal>
            </goals>
        </execution>
    </executions>
    <configuration>
        <!--指定 Dockerfile 文件位置-->
        <dockerfile>docker/Dockerfile</dockerfile>
        <!--指定 Docker 镜像仓库路径-->
        <repository>${docker.repository}/springcloud-action/${app.name}
</repository>
        <buildArgs>
            <!--向 Dockerfile 传递构建参数-->
            <JAR_FILE>target/${project.build.finalName}.jar</JAR_FILE>
        </buildArgs>
    </configuration>
</plugin>
```

上述代码添加了“dockerfile-maven-plugin”插件。该插件是之前“docker-maven-plugin”插件的替代品，支持将 Maven 项目打包为 Docker 镜像。

（2）创建 Docker 镜像打包文件。

在步骤(1)中，Docker 镜像的具体构建方式，是通过在<configuration>标签中指定 Dockerfile 文件来实现的。在微服务工程中新建一个名为“docker”的目录，并创建 Dockerfile 文件。代码如下：

```
FROM openjdk:8u191-jre-alpine3.9
ENTRYPOINT ["/usr/bin/java", "-jar", "/app.jar"]
ARG JAR_FILE
ADD ${JAR_FILE} /app.jar
EXPOSE 8080
```

2. 配置 Docker 镜像的仓库地址

（1）在“1.”小标题的步骤（1）的插件配置中，关于 Docker 镜像仓库路径的配置是根据 Harbor 仓库的地址及存储项目空间来确定的。具体可在 pom.xml 文件中定义属性，代码如下：

```
<properties>
    <!--定义 Docker 镜像仓库地址-->
    <docker.repository>10.211.55.11:8088</docker.repository>
    <!--定义项目名称，作为镜像名称生成的组成部分-->
    <app.name>chapter09-devops-demo</app.name>
</properties>
```

（2）在步骤（1）的配置中，指定了 Harbor 仓库的地址及应用名称。在实际操作时，可以根据自己所搭建的实际环境来确定。而在镜像仓库路径的配置中，除地址外，还指定了镜像仓库的项目空间“springcloud-action”，这是为了便于针对不同类型的项目镜像进行分类而在 Harbor 仓库中提前创建好的，具体如图 9-21 所示。

图 9-21

在创建成功后，单击 Harbor 项目列表就能看到对应的项目空间，如图 9-22 所示。

图 9-22

9.6.3　准备 GitLab CI/CD 服务器的 Kubernetes 环境

在 9.6.2 节中，我们已经在 Spring Cloud 项目中集成了打包 Docker 镜像的 Maven 插件。此时，如果要在 GitLab CI/CD 服务器（需要安装 Docker 及 Maven 环境）中执行 CI 构建流程，则可以将 Spring Cloud 微服务应用打包成 Docker 镜像，并上传至 Harbor 镜像仓库中。

但如果要将 Harbor 仓库中的微服务镜像部署至 Kubernetes 集群，则还需要 GitLab CI/CD 服务器具备与 Kubernetes 集群交互的能力。

接下来演示在 GitLab CI/CD 服务器中安装 Kubernetes 的客户端 kubectl，并配置其与 Kubernetes 集群通信的参数。

（1）在 GitLab CI/CD 服务器（Unbantu 系统）中安装 kubectl。命令如下：

```
#在 Unbantu 系统中使用 snap 安装 kubectl
# snap install kubectl -classic

#安装后检查 kubectl 的版本信息
# kubectl version -client
```

（2）将 9.5.2 节中安装的 Kubernetes 集群 Master 节点中的“./kube/config”文件复制到 GitLab Runner 服务器对应的“./kube/”目录中。可通过类似以下的命令进行复制：

```
#在 GitLab CI/CD 服务器上远程复制 Kubernetes Master 集群的配置文件
#scp root@10.211.55.6:$HOME/.kube/config $HOME/.kube/
```

在“./kube/config”配置文件中，主要定义了 Kubernetes 集群的访问地址及访问证书等信息，内容如下：

```
apiVersion: v1
clusters:
- cluster:
    certificate-authority-data: XXXX
    server: https://10.211.55.12:6443
  name: kubernetes
contexts:
- context:
    cluster: kubernetes
    user: kubernetes-admin
  name: kubernetes-admin@kubernetes
current-context: kubernetes-admin@kubernetes
kind: Config
preferences: {}
users:
- name: kubernetes-admin
  user:
    client-certificate-data: XXXX
    client-key-data: XXXX
```

需要注意，当前环境连接的 Kubernetes 集群 context 为“kubernetes-admin@kubernetes”，后面在具体编写 Kubernetes 部署 CD 指令时将用到它。

（3）配置 Harbor 私有镜像仓库的登录授权信息。

要保证 Kubernetes 集群中各个节点能够正常从 Harbor 私有镜像仓库拉取镜像，则需要提前在各节点配置 Harbor 私有镜像仓库的登录授权信息，例如：

```
$ docker login 10.211.55.11:8088 -u admin -p Harbor12345
WARNING! Using --password via the CLI is insecure. Use --password-stdin.
WARNING! Your password will be stored unencrypted in
/home/ubuntu/.docker/config.json.
Configure a credential helper to remove this warning. See
https://docs.docker.com/engine/reference/commandline/login/#credentials-
store

Login Succeeded
```

如上所示，在某个 Kubernetes 节点以登录 Harbor 仓库的方式获取了授权信息，那么后续 Pod 运行在该节点时就能正常访问该私有镜像仓库，从而拉取镜像、运行容器。

9.6.4 编写 Kubernetes 的发布部署文件

将应用发布至 Kubernetes 集群，主要是通过编写 YAML 配置文件来实现的。

在定义 CI/CD 流程前，先在 Spring Cloud 微服务工程目录中编写一个 Kubernetes 发布文件，具体是：在项目中创建 kubernetes 目录，并在其中创建文件“deploy.yaml”，内容如下：

```
---
#指定 API 的版本
apiVersion: apps/v1
#指定资源 API 对象类型，这里为 Deployment
kind: Deployment
#定义资源的元数据/属性
metadata:
  #资源的名称（在同一个 Namespace 中必须唯一），这里采用动态传值的方式
  name: __APP_NAME__
#指定该资源的内容
spec:
  #Pod 的副本个数
  replicas: __REPLICAS__
  selector:
    matchLabels:
      app: __APP_NAME__
  #定义升级策略为“滚动升级”
  strategy:
    type: RollingUpdate
  #定义 Pod 模板
  template:
    metadata:
      labels:
        app: __APP_NAME__
    spec:
      imagePullSecrets:
        - name: wudimanong-ecr
      #定义容器
      containers:
       #容器的名字
        - name: __APP_NAME__
          #容器使用的镜像仓库地址，这里采用动态传值的方式
          image: __IMAGE__
          #资源管理
          resources:
            requests:
              memory: "1000M"
            limits:
              memory: "1000M"
          volumeMounts:
            - name: time-zone
              mountPath: /etc/localtime
            - name: java-logs
```

```
          mountPath: /opt/logs
        ports:
          - containerPort: __PORT__
        #指定容器中的环境变量
        env:
          - name: SPRING_PROFILES_ACTIVE
            value: __PROFILE__
          - name: JAVA_OPTS
            value: -Xms1G -Xmx1G -Dapp.home=/opt/
    #定义一组挂载设备
    volumes:
      - name: time-zone
        hostPath:
          path: /etc/localtime
      - name: java-logs
        hostPath:
          path: /data/app/deployment/logs
```

这样的 YAML 文件，对应到 Kubernetes 中就是一个 API Object（即 API 对象）。在为这个对象的各个字段填好定义的值，并将其提交到 Kubernetes 后，Kubernetes 负责创建出这些对象所定义的容器和相关类型的 API 资源。

9.6.5 定义 Spring Cloud 微服务的 GitLab CI/CD 流程

在 GitLab 完成关于 CI/CD 机制的配置后，要实现 CI/CD 流程的自动化，则还需要在受 GitLab 管理的微服务工程的根目录中添加“.gitlab-ci.yml”文件，并在其中定义具体的 CI/CD 构建阶段及指令。具体如下：

```
#环境参数信息
variables:
  #Docker 镜像仓库地址和账号密码信息
  DOCKER_REPO_URL: "10.211.55.11:8088"
  DOCKER_REPO_USERNAME: admin
  DOCKER_REPO_PASSWORD: Harbor12345
  #Kubernetes 相关信息配置(空间与服务端口)
  K8S_NAMESPACE: "wudimanong"
  PORT: "8080"

#定义 CI/CD 阶段
stages:
  - test
  - build
  - push
  - deploy
```

```
#执行单元测试阶段
maven-test:
  stage: test
  script:
    - mvn clean test

#代码编译打包镜像阶段
maven-build:
  stage: build
  script:
    - mvn clean package -DskipTests

#将打包的 Docker 镜像上传至私有镜像仓库
docker-push:
  stage: push
  script:
    #对打包的镜像添加 tag
    - docker tag $DOCKER_REPO_URL/$CI_PROJECT_PATH
$DOCKER_REPO_URL/$CI_PROJECT_PATH/$CI_BUILD_REF_NAME:${CI_COMMIT_SHA:0:8}
    #登录私有镜像仓库
    - docker login $DOCKER_REPO_URL -u $DOCKER_REPO_USERNAME -p
$DOCKER_REPO_PASSWORD
    #上传应用镜像至镜像仓库
    - docker push $DOCKER_REPO_URL/$CI_PROJECT_PATH/$CI_BUILD_REF_NAME:
${CI_COMMIT_SHA:0:8}
    - docker rmi
$DOCKER_REPO_URL/$CI_PROJECT_PATH/$CI_BUILD_REF_NAME:${CI_COMMIT_SHA:0:8}
    - docker rmi $DOCKER_REPO_URL/$CI_PROJECT_PATH

#将应用发布至 Kubernetes 测试集群（这里指定为手动确认方式）
deploy-test:
  stage: deploy
  when: manual
  script:
    - kubectl config use-context kubernetes-admin@kubernetes
    - sed -e "s/__REPLICAS__/1/; s/__PORT__/$PORT/;
s/__APP_NAME__/$CI_PROJECT_NAME/; s/__PROFILE__/test/;
s/__IMAGE__/$DOCKER_REPO_URL\/${CI_PROJECT_PATH//\//\\/}\/${CI_BUILD_REF_NA
ME//\//\\/}:${CI_COMMIT_SHA:0:8}/" kubernetes/deploy.yaml | kubectl -n
${K8S_NAMESPACE} apply -f -
```

如上所述，在“.gitlab-ci.yml”文件中定义了“test”“build”“push”“deploy”这 4 个阶段。这几个阶段的具体说明如下。

- test：执行单元测试代码。

- build：执行打包指令，将应用打包为 Docker 镜像。
- push：将 build 阶段打包的本地 Docker 镜像经过 tag 处理后上传至 Harbor 镜像仓库，并在成功后清理本地镜像文件。
- deploy：执行 Kubernetes 指令，根据 Kubernetes 发布部署文件的配置，将容器镜像部署至 Kubernetes 集群。

在以上 CI/CD 阶段的定义中，在具体部署至 Kubernetes 集群时，为了便于应用的统一管理，在发布指令中指定了 Kubernetes 集群的 Namespace，但在默认情况下，在 Kubernetes 集群中是不存在这样的 Namespace 的，所以需要提前手工创建，命令如下：

```
#连接 Kubernetes 集群创建 Kubernetes Namespace
# kubectl create namespace wudimanong
namespace/wudimanong created
```

此外，在 Kubernetes 发布过程中，由于 Kubernetes 集群中的节点需要从 Harbor 仓库拉取镜像，所以，为了确保访问成功，还需要在集群各节点进行如下配置：

```
#编辑 Docker 本地仓库的配置文件
vim /etc/docker/daemon.json
```

添加 Harbor 镜像仓库的地址，具体内容如下：

```
{"insecure-registries": ["10.211.55.11:8088"]}
```

重启 Docker 服务，命令如下：

```
systemctl restart docker
```

至此，完成了通过 GitLab CI/CD 机制构建自动化发布系统的全部准备工作了，后面将具体演示 CI/CD 自动化发布的流程。

9.6.6 将微服务应用自动发布到 Kubernetes 集群中

本节演示将 Spring Cloud 微服务应用自动发布到 Kubernetes 集群。

（1）因为微服务应用是需要注册到 Consul 服务注册中心的，所以需要找一台与 Kubernetes 集群网络互通的服务器部署一个 Consul 服务。具体如下：

```
#部署一个 Consul 服务
#docker run --name consul -p 8500:8500 -v /tmp/consul/conf/:/consul/conf/
-v /tmp/consul/data/:/consul/data/ -d consul
```

如上所示，在与 Kubernetes 集群同网络环境下，基于 Docker 部署了一个 Consul 服务。在部署完成后，可以测试是否能够正常访问它。

（2）创建微服务部署 Kubernetes 集群环境的配置文件。

如果 Consul 服务正常，则在 Spring Cloud 示例项目中添加指定的 Consul 地址。由于本次部署的 Kubernetes 环境为 test 环境，所以可以创建新的配置文件“application-test.yml”，具体如下：

```
spring:
  cloud:
    consul:
      #配置测试环境 Consul 注册中心的 IP 地址
      host: 10.211.55.12
      port: 8500
```

如果不指定 Consul 地址，则 Spring Cloud 应用会自动连接本地地址 127.0.0.1:8500。但在 Kubernetes 环境下，需要根据具体环境指定对应的 Consul 服务地址。

（3）提交代码，触发 GitLab 的 Pipeline 自动构建。

此时如果提交代码，则 GitLab Pipeline 将会被自动触发，并根据在“.gitlab-ci.yml”文件中设置的 CI/CD 阶段开始执行构建命令，如图 9-23 所示。

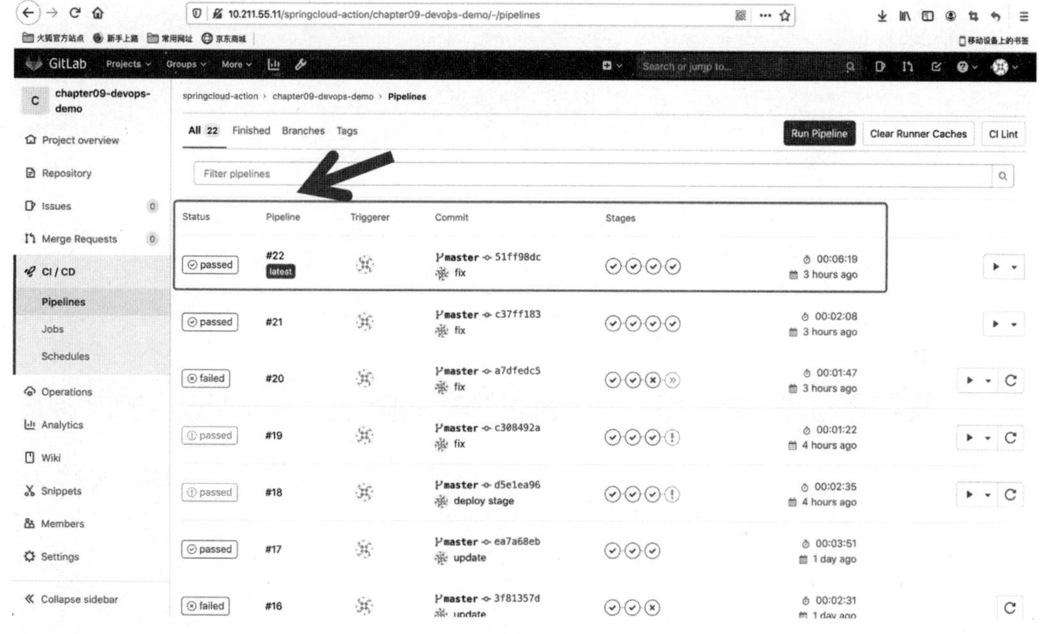

图 9-23

单击构建详情按钮（passed），可以看到具体各个阶段的执行情况，如图 9-24 所示。

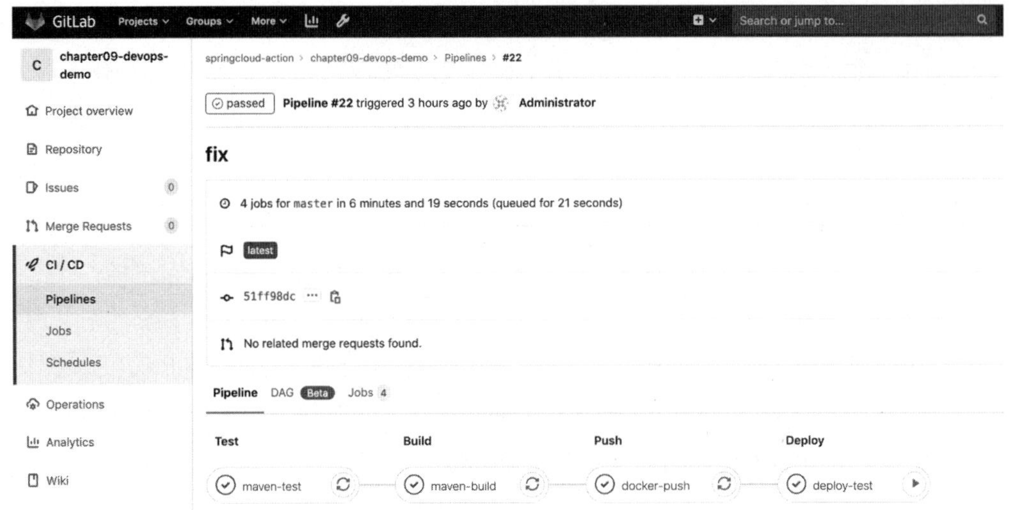

图 9-24

（4）连接 Kubernetes 集群，查看 Pod 是否运行正常。指令执行效果如下所示。

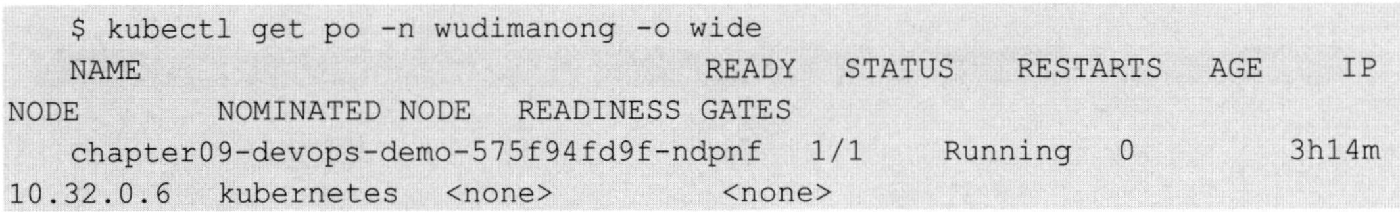

```
$ kubectl get po -n wudimanong -o wide
NAME                                        READY   STATUS    RESTARTS   AGE     IP
NODE         NOMINATED NODE   READINESS GATES
chapter09-devops-demo-575f94fd9f-ndpnf   1/1     Running   0          3h14m
10.32.0.6   kubernetes   <none>           <none>
```

通过 Pod 查看命令可以看到应用 Pod 已经成功运行。

（5）登录 Consul 控制台查看微服务的注册情况，如图 9-25 所示。

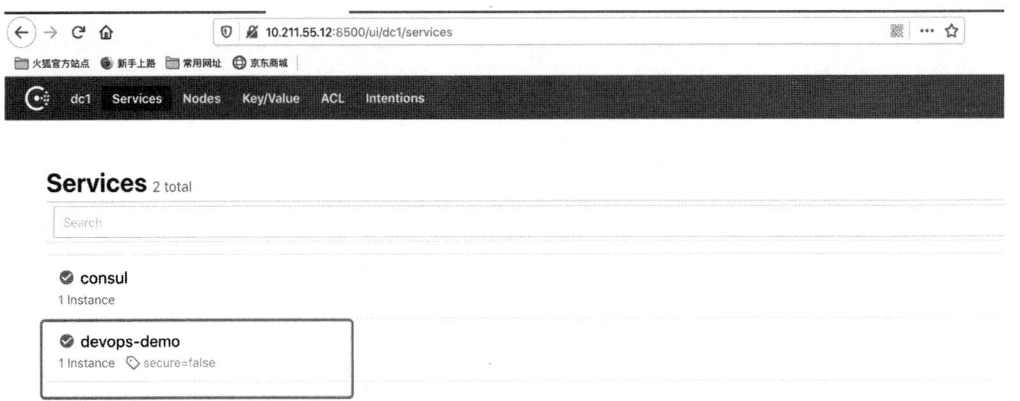

图 9-25

可以看到，在 Kubernetes 集群中运行的微服务应用已经成功注册到 Consul 注册中心了，这说明应用已经部署成功。

9.7 本章小结

本章所演示的 DevOps 发布系统是一个非常综合的实例，其中不仅涉及 CI/CD 这些基本 DevOps 概念，还从实战的角度详细介绍了容器、镜像仓库及 Kubernetes 容器编排等目前 DevOps 领域非常流行的容器化技术。

更重要的是，本章以实践的方式详细演示了如何将这些技术组件组合在一起，构建出一套流程完整的 DevOps 自动化发布体系。这样的发布方式是目前 Spring Cloud 微服务软件发布的主流方式。

通过本章的实践学习，读者可以从整个软件生命周期的角度去理解 Spring Cloud 微服务构建的流程及方法。

第 10 章

搭建微服务监控系统

——用“Prometheus + Grafana + SkyWalking”实现度量指标监控及分布式链路追踪

对于线上系统来说，完善的监控系统是确保服务持续稳定的重要手段。而在微服务场景中，由于服务之间的调用链路较长，且基于 Kubernetes 容器化部署的特点，所以整体服务的稳定性更需要依靠一套强大的监控报警系统来保证。一般来说，可以将监控层次从下向上依次分为基础层监控、中间层监控、应用层监控和业务层监控。

本章将依据这几个监控层次，演示在以 Kubernetes 为基础设施的云原生环境下构建针对 Spring Cloud 微服务的监控系统。

通过本章，读者将学习到以下内容：

- 监控系统的概念与基本原理。
- 搭建 Kubernetes 微服务监控体系的方法。
- 将 Spring Cloud 微服务接入 Prometheus 指标监控系统的方法。
- 分布式链路追踪的概念，以及基于 SkyWalking 实现分布式链路追踪系统。

10.1 认识监控系统

监控系统是运维体系乃至整个软件产品生命周期中最重要的一环。完善的监控可以帮助技术人员在事前及时发现故障，在事后快速追查定位问题。

在以微服务为代表的云原生架构体系中，系统分层多样，服务之间调用链路复杂，所以系统需要监控的目标非常多。如果没有一套完善的监控系统，则难以保证整体服务的持续稳定。

10.1.1 监控对象及分层

在实际场景中，按照监控的对象及系统层次结构，从下向上可依次划分为：基础层监控、中间层监控、应用层监控和业务层监控，如图 10-1 所示。

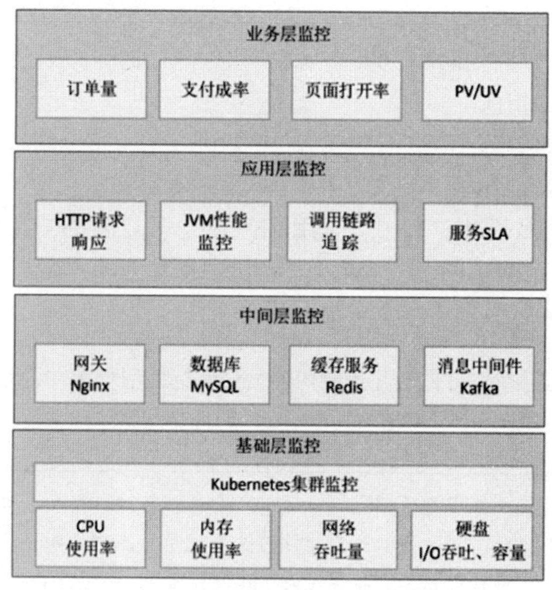

图 10-1

1. 基础层监控

基础层监控，就是对主机服务器（包括宿主机、容器）及其底层资源进行监控，以保证应用运行所依赖的基础环境稳定运行。基础层监控主要有以下两个方向。

- 资源利用：对 I/O 利用率、CPU 利用率、内存使用率、磁盘使用率、网络负载等进行监控，避免因应用本身或其他特殊情况引起硬件资源消耗，进而出现服务故障。

- 网络通信：对服务器之间的网络状态进行监控。网络通信是互联网的重要基石，如果主机之间的网络出现延迟过大、丢包率高等网络问题，则会严重影响业务。

在基于 Kubernetes 的云原生基础设施中，在基础层不仅要对宿主机本身进行监控，还要对 Kubernetes 集群状态，以及其容器资源使用情况进行监控。

2. 中间层监控

中间层监控是指，对诸如 Nginx、Redis、MySQL、RocketMQ、Kafka 等应用服务所依赖的中间件进行监控。它们的稳定也是保证应用持续可用的关键。

3. 应用层监控

应用层监控是指对业务应用进行监控。一般来说，其关注点主要体现在以下几个方面。

- HTTP 接口请求访问：包括接口响应时间、吞吐量等。
- JVM 监控指标：对于 Java 服务，还会重点关注 GC 时间、线程数、FGC/YGC 耗时等与 JVM 性能相关的指标。
- 资源消耗：部署应用会消耗一定的资源，例如对内存、CPU 的消耗。
- 服务的健康状态：当前服务是否存活、运行是否稳定等。
- 调用链路：在微服务架构中，由于调用链路较长，所以需要监控服务之间调用链路的稳定性，避免上下游服务之间的局部链路故障引发系统全局性“雪崩”。

4. 业务层监控

业务层监控也是监控系统关注的一个重要内容。在实际业务场景中，常常会对应用产生的业务数据进行监控，例如，网站系统所关注的 PV、UV，后端交易系统所关注订单量、成功率等。

业务指标也是体现系统稳定性的核心要素。对于任何出现了问题的系统，最先受影响的肯定是业务指标。对于核心业务指标的设定，因具体的业务和场景而异。因此，对于业务层的监控，需要构建具备业务特点的业务监控系统。

10.1.2 常见的监控指标及类型

在指标类监控系统中，通过统计指标可以直观地了解到整个系统的运行情况。在出现问题后，各个指标会首先出现波动，这些波动会反映出系统是哪些方面出了问题，从而可以据此排查问题出现的原因。

下面来看一下统计指标的类型，以及常见的统计指标都有哪些。这是进一步理解指标类监控系统的基础。

1. 指标类型

从整体上看，常见的指标有以下 4 类类型。

（1）计数器（Counter）。

计数器是一种具有累加特性的指标类型，它的值一般为 Double 或者 Long 类型。例如，常见的统计指标“QPS”的值就是通过计数器并配合一些统计函数计算出来的。

（2）测量仪（Gauge）。

测量仪是指在某个时间点对某个数值进行测量的指标类型。测量仪和计数器都可以用来查询特定指标在某个时间点的数值。

和计数器不同，测量仪的值可以随意变化，可以增加，也可以减少。比如，获取 Java 线程池中活跃的线程数，测量仪使用的是 ThreadPoolExecutor 中的 getActiveCount()方法。

常见的统计指标（如 CPU 使用率、内存占用量等）都是通过测量仪来统计的。

（3）直方图（Histogram）。

直方图是一种将多个数值聚合在一起的数据结构，可以表示数据的分布情况。

以常见的响应耗时举例，把响应耗时数据分为多个桶（Bucket），每个桶代表一个耗时区间，例如 0～100ms、100ms～500ms。以此类推，通过这样的形式，可以直观地看到一个时间段内请求耗时的分布情况。

（4）摘要（Summary）。

摘要与直方图类似，表示的也是一段时间内的数据结果。但摘要一般用于标识分位值，分位值其实就是业界常说的“TP90”“TP99”等术语。

例如，有 100 个耗时数值，如果将所有的数值从低到高排列，取第 90%的位置，则这个位置的值就是“TP90”的值；如果这个位置对应的值是 80ms，则代表小于或等于 90%位置的请求耗时都小于或等于 80ms。

2. 常见的监控指标

下面再来看一些工作中常见的监控指标。通过这些指标可以了解系统运行的基本情况。

（1）QPS。

QPS（Query Per Second），指每秒查询的数量。这是一个非常常见的监控指标，它不仅指“查询”这个特殊的条件，也与请求量挂钩。通过这个值，可以查看某个接口的请求量，例如，在

1s 内进行了一次接口调用，则可以认为在这 1s 内 QPS 增加了 1。如果系统经过了压测，则可以通过估算出的 QPS 的峰值来预估系统的容量。

（2）TPS。

TPS（Transaction Per Second），指每秒传输的事务处理的个数。一个事务是指：从客户机向服务器发送请求，到客户机收到服务器响应的全过程。它包括“用户请求服务器”“服务器自己内部处理”及“服务器返回给用户”这 3 个过程。

一般来说，对系统性能的评价，都是以 TPS 来衡量的。系统的整体处理能力，取决于处理能力最弱模块的 TPS。

（3）SLA。

SLA（Service Level Agreement，服务等级协议），一般用于服务商和用户之间的协定，它规定了服务的性能和可用性。依据这种可量化的协定，在服务商与用户之间可以制定出详细的规则，来保证所提供服务的可用性。例如，在使用云服务时，有些服务厂商会与用户签订一个 SLA 协议，约定如果服务达不到相应的 SLA 等级则会进行相应赔偿。

关于 SLA，最常使用的说法是：“4 个 9”“5 个 9”等。其中，“4 个 9”指的是 99.99%，“5 个 9”指的是 99.999%，以此类推。

SLA 并不是一个固定的数值，“几个 9”只是代表系统可以保持稳定的时间，SLA 会因为成功数与请求数的不同而动态变化，例如可能是 95%，也可能是 98%。

如果要计算某个接口的 SLA 情况，则可以指定一段时间区间，然后根据以下公式来计算：

总计成功数 / 总计请求数 = 百分比（100%）

算出 SLA 能有什么作用呢？例如，要保证某个服务 1 年内的 SLA 是“5 个 9”，那么 1 年就是时间单位，由此可以算出服务不可用时间如下：

```
1 年 = 365 天 = 8760 小时

3 个 9 = 8760 × (1 - 99.9%) = 8.76 小时

4 个 9 = 8760 × (1 - 99.99%) = 0.876 小时 = 0.876 × 60 = 52.6 分钟

5 个 9 = 8760 × (1 - 99.999%) = 0.0876 小时 = 0.0876 × 60 = 5.26 分钟
```

由此可见，想要保证越多的“9”，就要提示服务的持续可用时间，减少服务错误的时间。因此，SLA 等级协议的概念对系统稳定有着重要的意义，“几个 9”也成了业界关于服务可用性评价的公认标准。

不同的监控对象，需要关注的指标会不同，需要在实际的业务场景中根据具体的监控对象进行统计。

10.1.3　主流的监控系统及选型

选择开源的监控系统，是快速构建微服务监控体系的最佳方案。

1. 度量指标监控

在度量指标监控领域，目前最知名的开源项目是 Prometheus。它是一款由前 Google 员工发布的开源指标监控系统，支持从基础层到业务层的全域监控。

Prometheus 目前已被 CNCF（云原生基金会）托管，现在已经成为 Kubernetes 集群的标准监控解决方案。

按照监控对象及分层划分，Prometheus 适用的范围如图 10-2 所示。

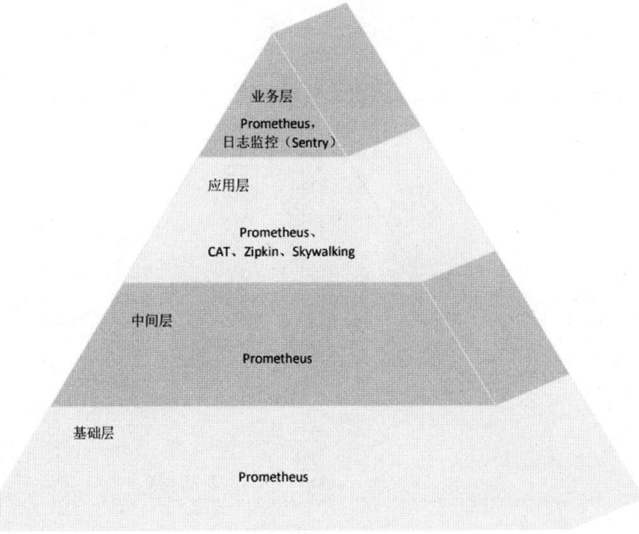

图 10-2

从图 10-2 可以看出，Prometheus 作为一款监控系统，其适用范围十分广，无论是对基础层 CPU、内存、I/O 等主机硬件资源的监控，还是对 Kubernetes 集群及容器资源的监控都可以很好地覆盖。

Prometheus 社区还提供了大量专门针对各类基础软件（如 MySQL、Redis 等）的官方或第三方的 Exporter（采集客户端），能很好地支持对中间层软件系统的指标监控。

对于应用层（微服务系统）的指标监控，Prometheus 也能通过相应的采集客户端实现大部分通用指标（如 QPS、JVM 内存使用情况）的采集监控。

除此之外，Prometheus 也可以实现业务层指标的监控，但是由于业务监控指标复杂多样，所以在具体实践的过程中，还没办法通过某种通用的采集客户端来进行采集，而是依赖对业务代码进行侵入式的数据埋点。

Prometheus 广泛的适用范围（特别是对 Kubernetes 云原生基础设施的天然集成），使得目前采用 Prometheus 作为指标监控系统的公司越来越多，社区生态也越发繁荣。从某种程度上说，Prometheus 已经成为度量指标监控领域的首选开源产品。

2. 分布式链路追踪

除度量指标监控外，在分布式链路追踪（Tracing）及应用性能管理（APM）方面，也有像 CAT、Zipkin、SkyWalking 等开源项目。它们之间的对比分析如图 10-3 所示。

	CAT	Zipkin	Apache Skywalking
社区支持	美团开源，国内流行	Twitter开源，国外主流	Apache支持，国内社区活跃
典型案例	大众点评、携程	京东	华为、小米
APM支持	支持	不支持	支持
源头产品	eBay CAL	Google Dapper	Google Dapper
同类产品	暂无	Spring Cloud Sleuth	Naver Pinpoint
调用链可视	有	有	有
聚合报表	非常丰富	少	较丰富
埋点方式	侵入	侵入	非侵入，运行期字节码增强
VM指标	好	无	有
告警支持	有	无	有
多语言支持	Java/.Net	丰富	Java/.Net/NodeJS/PHP/Go（需手动埋点）
优点	企业级产品，APM报表丰富	社区生态好	社区活跃，非侵入式集成、Apache背书
不足	用户体验一般，社区活跃度不高	APM报表能力弱，代码侵入	开源时间不长，文档支持一般

图 10-3

在上述开源的链路监控系统中，APM（Application Performance Management，应用性能管理）是非常重要的一种能力，但是 Zipkin 并不支持。因此，不建议采用它。

CAT 是国内企业大规模实践出来的产物。从 APM 能力和报表的丰富性来说，它是非常优秀的。但它的缺点是，社区活跃度不高。

SkyWalking 是一款优秀的开源“APM + 分布式链路追踪”系统，支持无侵入式的接入方式。目前 SkyWalking 在国内互联网公司中使用得非常广泛，并且社区活跃度也很高。所以，对于分布式链路系统的选型，会更倾向于 SkyWalking。

综上所述，在本章 Spring Cloud 微服务监控体系的构建中，会采用 Prometheus 和 SkyWalking 来构建微服务的度量指标监控系统，以及分布式链路追踪系统。

10.2 【实战】构建微服务度量指标监控系统

从监控对象及系统分层的角度来看，需要监控的范围非常广。但从微服务监控的角度来看，如果服务都是部署在 Kubernetes 环境中，则需要关注的监控对象主要就是 Kubernetes 集群本身，以及运行在集群中的各类应用容器；而监控内容主要是容器资源的使用情况（例如 CPU 使用率、内存使用率、网络、I/O 等指标）。

这并不是说基础层的物理机、虚拟机或者中间层软件的监控不需要关注，只是这部分工作一般会有专门的人员去做。如果使用的是云服务，则云服务厂商大都已经提供了相关支持。

对于基础物理层、中间件的监控并不是本书的重点，所以就不做过多的介绍，大家对此有一个全局的认识即可。

回到 Kubernetes 微服务监控体系的话题。虽然在 Kubernetes 的早期版本中，监控体系曾经非常复杂，社区中也有各种各样的方案。但是这套体系发展到今天，已经完全演变成了以 Prometheus 为主的统一方案。

本节将基于 Prometheus 来构建针对 Kubernetes 的微服务监控系统。

10.2.1 认识 Prometheus

经过行业多年的实践和沉淀，监控系统按实现方式主要分为以下几类：

- 基于时间序列的 Metrics（度量指标）监控。
- 基于调用链的 Tracing（链路）监控。

- 基于 Logging（日志）的监控。
- 健康性检查（Healthcheck）。

在以上几种监控方式中，Metrics（度量指标）监控是最主要的一种监控方式。

Metrics（度量指标）本质上就是在离散的时间点上产生的数值点[Time,Value]。而一组数值点的序列也被称为“时间序列”，因此，Metrics（度量指标）监控也常被称为“时间序列监控”。

Prometheus 是一款基于时间序列的开源 Metrics（度量指标）监控系统，它可以很方便地进行统计指标的存储、查询和告警。Prometheus 的系统结构如图 10-4 所示。

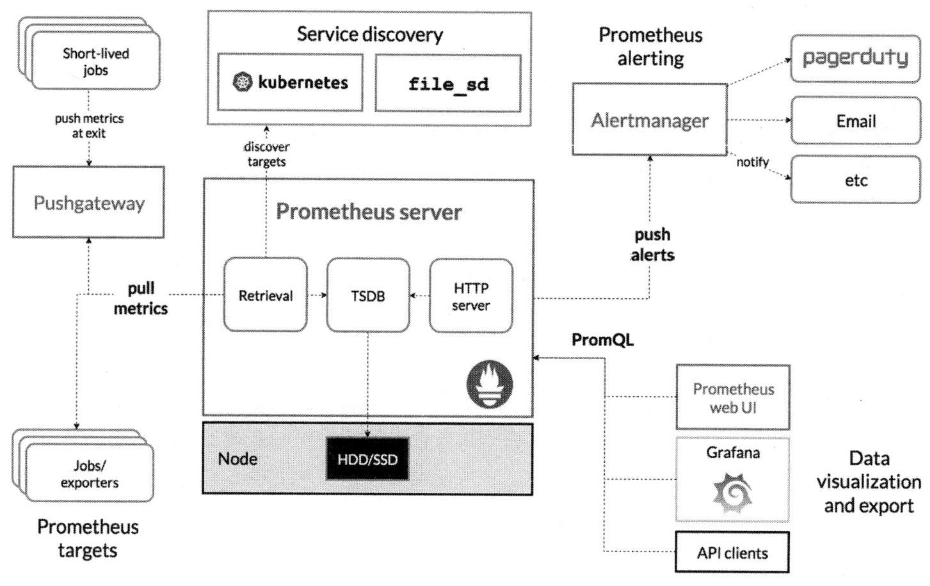

图 10-4

Prometheus 主要是使用 Pull（拉取）的模式去采集被监控对象的 Metrics（度量指标）数据，然后将收到的 Metrics（度量指标）数据进行聚合计算，并将计算结果存储在时间序列数据库（TSDB）中，以便后续实现各种维度的检索。

常见的时间序列数据库有 OpenTSDB、InfluxDB 等。

除 Pull（拉取）的模式外，Prometheus 中的 PushGateway 组件也允许被监控对象以 Push（推送）的模式向 Prometheus 发送 Metrics（度量指标）数据。而 Alertmanager 组件，则可以根据 Metrics（度量指标）信息灵活地设置报警。

Prometheus 还提供了一套完整的查询语言 PromQL。通过 Prometheus 提供的 HTTP 查询接口，使用者可以很方便地使用 PromQL 将 Metrics（度量指标）数据与 Grafana 这样的可视化工具结合起来，从而灵活地定制系统关键 Metrics（度量指标）的监控 Dashboard（看板）。

接下来，将按照监控系统的一般性原理，从数据收集、指标存储、指标查询及规则告警这几个方面进一步阐述 Prometheus 的核心工作方式。

1. 数据采集

Prometheus 是通过在监控对象中运行采集程序，然后暴露 Metrics（度量指标）采集接口，并由 Prometheus 服务以 Pull（拉取）模式来实现数据采集的。

例如，在 Spring Cloud 中引入“spring-boot-starter-actuator”及“micrometer-registry-Prometheus”依赖后，就会自动在微服务应用中开启一些基础指标（如 HTTP 层调用次数、本机的 CPU、内存资源使用等）的采集接口。

对于宿主机及其他对象的监控，Prometheus 维护了一组 Node Exporter 工具。这些工具会以后台进程的形式运行在监控对象所在的服务器中，并代替被监控对象向 Prometheus 暴露 Metrics（度量指标）的采集接口。

Prometheus 采集的 Metrics（度量指标）数据，除了负载、CPU、内存、磁盘、网络等常规信息外，针对不同的监控对象，还可以使用特定的组件去采集更有针对性的指标——例如“MySQL server exporter”这样的 Node Exporter，还可以采集像“collect.binlog_size”这样的 Metrics（度量指标）。

不同的 Node Exporter 组件能提供的具体指标数据，可参考 Prometheus 官方文档中各种 exporters 列表的详细信息。

2. 指标存储

Prometheus 提供了两种方式来存储 Metrics（度量指标）数据。

- 本地存储：Prometheus 在采集到样本后，会以时间序列的方式将样本保存在内存中，并定时同步到磁盘中。这种存储方式的优势是运维简单，但无法支持海量 Metrics（度量指标）数据的持久化，并存在丢失数据的风险。
- 远端存储：为了解决本地单节点存储的限制，Prometheus 提供了远程读写的接口，用户可以自己选择用合适的时序数据库来实现 Prometheus 的存储扩展。

一般来说，Prometheus 可以通过以下两种方式来实现与远端存储系统的对接：

- Prometheus 按照标准的格式将 Metrics（度量指标）数据写入远端存储系统。
- Prometheus 按照标准格式提供访问接口，远端存储系统通过 Prometheus 提供的接口来读取 Prometheus 采集的 Metrics（度量指标）数据。

Prometheus 与远端存储系统的交互如图 10-5 所示。

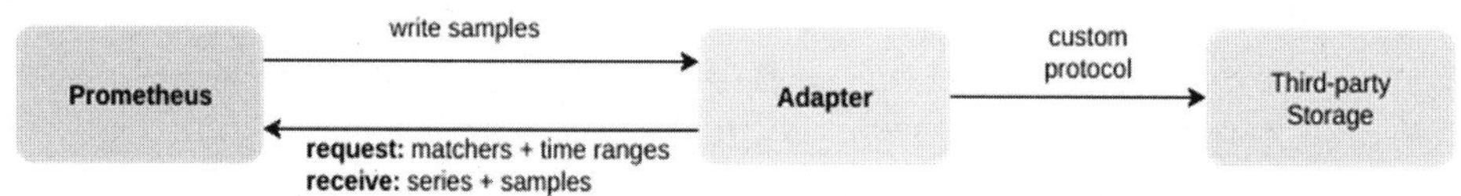

图 10-5

关于 Prometheus 与远端存储系统对接的具体方法，感兴趣的读者可以自行查阅相关资料。

3. 指标查询

在 Prometheus 中，通过 PromeQL 查询语言可以方便地查询各种维度指标数据，进而实现监控数据的可视化图形绘制、监控内容告警的设置等功能。例如：

```
http_error_requests{job="apiserver"}[5m]
```

上面这段 PromeQL 语句可以查询出在“http_error_requests”这个统计指标中，HTTP 请求错误的次数（Job 值等于“apiserver”，每 5 分钟统计一次）。利用这样的数据，就可以绘制出柱状图或折线图来将监控指标数据可视化。

PromQL 还提供了很多计算函数来实现更多维度指标数据的查询，例如：

```
rate(http_error_requests[5m])
```

该 PromeQL 函数统计的是：指标“http_error_requests”的当前值与前 5 分钟相比较的增值。

关于 PromeQL 更多的用法，可以参考官网提供的相关教程。

4. 规则告警

Prometheus 实现规则告警的主要逻辑是，定期执行 PromeQL 告警语句，如果符合告警条件，则进行告警通知。

可以通过进一步学习 Alertmanager 组件，来了解更详细的告警规则及通知方式的配置。

接下来演示部署 Prometheus 构建 Kubernetes 微服务监控体系的具体步骤。

10.2.2　步骤 1：部署 Prometheus Operator

在实际的应用场景中，针对不同的监控对象，Prometheus 的部署方式有所不同。例如，要监控的对象是底层的物理机，或者以物理机方式部署的数据库等中间件系统，则一般将 Prometheus 监控系统部署在物理机环境中。

如果是针对 Kubernetes 集群的监控，则目前主要是通过 Promethues-Operator 的方式将 Promethues 直接部署在 Kubernetes 集群中，从而以更原生的方式实施对 Kubernetes 集群及其运行容器的监控。

这里所说的 Promethues-Operator 是指专门针对 Kubernetes 的 Promethues 封装包，以此来简化 Prometheus 在 Kubernetes 环境中的部署和配置。

通过 Promethues-Operator 在 Kubernetes（Kubernetes 集群的部署可参考 9.5.2 节的内容）中部署 Promethues 的步骤如下。

1. 安装 Helm

由于在安装过程中会使用到 Kubernetes 的包管理工具 Helm，所以需要先在 Kubernetes 集群中安装 Helm。

Helm 是 Kubernetes 的一种包管理工具，与 Java 中的 Maven、NodeJs 中的 Npm、Ubuntu 的 apt，以及 CentOS 的 yum 类似，主要用来简化 Kubernetes 对应用的部署和管理。

（1）从 Github 下载相应版本的 Helm 安装包，并将安装包保存在某个安装了 kubectl 的节点中。

（2）解压缩安装包，并将可执行文件“helm”复制到系统文件夹“/usr/local/bin”中。命令如下：

```
$ tar -zxvf helm-v3.4.0-rc.1-linux-amd64.tar.gz

#将安装包中的可执行文件“helm”复制到文件夹“/usr/local/bin”中
```

```
$ mv linux-amd64/helm /usr/local/bin/
```

（3）执行“helm version”命令。具体如下：

```
$ helm version

version.BuildInfo{Version:"v3.4.0-rc.1",
GitCommit:"7090a89efc8a18f3d8178bf47d2462450349a004", GitTreeState:"clean",
GoVersion:"go1.14.10"}
```

如果能看到 Helm 的版本信息，则说明 Helm 客户端安装成功了。

（4）添加“helm charts”官方仓库的地址。

由于一些公共 Kubernetes 包是在远程仓库中管理的，所以还需要添加“helm charts”官方仓库地址，命令如下：

```
$ helm repo add stable https://×××.helm.sh/stable
```

Helm 中的 Kubernetes 安装包又被称为“charts”。

（5）查看本地 Helm 仓库是否添加成功。命令如下：

```
$ helm repo list

NAME    URL
stable  https://×××.helm.sh/stable
```

（6）查看 Helm 仓库。命令如下：

```
$ helm search repo stable

NAME                          CHART VERSION   APP VERSION    DESCRIPTION
stable/acs-engine-autoscaler    2.1.3           2.1.1          Scales
worker nodes within agent pools
stable/aerospike                0.1.7           v3.14.1.2      A Helm
chart for Aerospike in Kubernetes
stable/anchore-engine           0.1.3           0.1.6          Anchore
container analysis and policy evaluatio...
stable/artifactory              7.0.3           5.8.4          Universal
Repository Manager supporting all maj...
stable/artifactory-ha           0.1.0           5.8.4        Universal
Repository Manager supporting all maj...
stable/aws-cluster-autoscaler   0.3.2                        Scales worker
nodes within autoscaling groups.
```

```
   stable/bitcoind                  0.1.0         0.15.1      Bitcoin is an
innovative payment network and a ...
   stable/buildkite                 0.2.1         3           Agent for
Buildkite
   ...
```

如上所示，通过“Helm search”命令可以查看到各种stable版本的Kubernetes安装包（charts列表）了。

2. 通过 Helm 查找 Prometheus-Operator 安装包

在安装 Prometheus-Operator 之前，先用“helm”命令来搜索 Prometheus 的安装包，命令如下：

```
$ helm search repo prometheus
```

查找结果如图 10-6 所示：

NAME	CHART VERSION	APP VERSION	DESCRIPTION
stable/prometheus	11.12.1	2.20.1	DEPRECATED Prometheus is a monitoring system an...
stable/prometheus-adapter	2.5.1	v0.7.0	DEPRECATED A Helm chart for k8s prometheus adapter
stable/prometheus-blackbox-exporter	4.3.1	0.16.0	DEPRECATED Prometheus Blackbox Exporter
stable/prometheus-cloudwatch-exporter	0.8.4	0.8.0	DEPRECATED A Helm chart for prometheus cloudwat...
stable/prometheus-consul-exporter	0.1.6	0.4.0	DEPRECATED A Helm chart for the Prometheus Cons...
stable/prometheus-couchdb-exporter	0.1.2	1.0	DEPRECATED A Helm chart to export the metrics f...
stable/prometheus-mongodb-exporter	2.8.1	v0.10.0	DEPRECATED A Prometheus exporter for MongoDB me...
stable/prometheus-mysql-exporter	0.7.1	v0.11.0	DEPRECATED A Helm chart for prometheus mysql ex...
stable/prometheus-nats-exporter	2.5.1	0.6.2	DEPRECATED A Helm chart for prometheus-nats-exp...
stable/prometheus-node-exporter	1.11.2	1.0.1	DEPRECATED A Helm chart for prometheus node-exp...
stable/prometheus-operator	9.3.2	0.38.1	DEPRECATED Provides easy monitoring definitions...
stable/prometheus-postgres-exporter	1.3.1	0.8.0	DEPRECATED A Helm chart for prometheus postgres...
stable/prometheus-pushgateway	1.4.3	1.2.0	DEPRECATED A Helm chart for prometheus pushgateway
stable/prometheus-rabbitmq-exporter	0.5.6	v0.29.0	DEPRECATED Rabbitmq metrics exporter for promet...
stable/prometheus-redis-exporter	3.5.1	1.3.4	DEPRECATED Prometheus exporter for Redis metrics
stable/prometheus-snmp-exporter	0.0.6	0.14.0	DEPRECATED Prometheus SNMP Exporter
stable/prometheus-to-sd	0.3.1	0.5.2	DEPRECATED Scrape metrics stored in prometheus ...
stable/elasticsearch-exporter	3.7.0	1.1.0	Elasticsearch stats exporter for Prometheus
stable/helm-exporter	0.3.3	0.4.0	DEPRECATED Exports helm release stats to promet...
stable/karma	1.7.0	v0.72	A Helm chart for Karma - an UI for Prometheus A...
stable/stackdriver-exporter	1.3.0	0.6.0	Stackdriver exporter for Prometheus
stable/weave-cloud	0.3.7	1.4.0	Weave Cloud is a add-on to Kubernetes which pro...
stable/kube-state-metrics	2.9.1	1.9.7	Install kube-state-metrics to generate and expo...
stable/kuberhealthy	1.2.7	v1.0.2	DEPRECATED. Please use https://comcast.github.i...
stable/mariadb	7.3.14	10.3.22	DEPRECATED Fast, reliable, scalable, and easy t...

图 10-6

从图 10-6 可以看到，在 Helm 仓库中可以搜索到版本为 0.38.1 的“stable/prometheus-operator”的安装包。接下来就可以通过 Helm 进行安装了。

3. 通过 Helm 安装 Prometheus-Operator

（1）通过 Helm 安装 Prometheus-Operator。命令如下：

```
#创建 K8s 的命名空间
$ kubectl create ns monitoring

#通过 helm 安装 Prometheus-Operator
$ helm install promethues-operator stable/prometheus-operator -n monitoring
```

（2）查看运行的 Kubernetes Pods 信息。命令如下：

```
$ kubectl get po -n monitoring

NAME                                                      READY   STATUS
RESTARTS   AGE
alertmanager-promethues-operator-promet-alertmanager-0   2/2     Running
0          5m42s
prometheus-promethues-operator-promet-prometheus-0       3/3     Running
1          5m31s
promethues-operator-grafana-5df74d9cb4-5d475             2/2     Running
0          6m53s
promethues-operator-kube-state-metrics-89d8c459f-449k4   1/1     Running
0          6m53s
promethues-operator-promet-operator-79f8b5f7ff-pfpbl     2/2     Running
0          6m53s
promethues-operator-prometheus-node-exporter-6ll4z       1/1     Running
0          6m53s
promethues-operator-prometheus-node-exporter-bvdb4       1/1     Running
0          6m53s
```

如上所示，Prometheus 的相关组件已经以 Pod 的方式运行在 Kubernetes 集群中了。

（3）通过 Helm 命令查看具体的发布记录。命令如下：

```
$ helm list -n monitoring

NAME   NAMESPACE   REVISION   UPDATED   STATUS   CHART   APP VERSION
promethues-operator monitoring  1  2020-10-26 10:15:45.664673683 +0000 UTC
deployed    prometheus-operator-9.3.2  0.38.1
```

10.2.3 步骤 2：演示 Prometheus 的 Metrics（度量指标）监控效果

在 10.2.2 节中，已经将 Prometheus 部署到 Kubernetes 集群中了。此时，Prometheus 实际上就已经发挥作用，可以开始采集 Kubernetes 集群的相关运行指标了。可以通过 Promethues 内置的监控界面进行查看，步骤如下。

（1）在 Kubernetes 中查看内置监控界面所在的 Pod 节点。命令如下：

```
$ kubectl -n monitoring get svc
```

（2）设置 Promethues 内置的监控界面服务的访问端口。

使用“nodeport”的方式设置 Promethues 内置监控界面服务的集群外访问端口（例如 30444）。命令如下：

```
$ kubectl  patch svc promethues-operator-promet-prometheus -n monitoring -p
'{"spec":{"type":"NodePort","ports":[{"port":9090,"targetPort":9090,"nodePor
t":30444}]}}'
service/promethues-operator-promet-prometheus patched
```

（3）查看 Promethues 的监控可视化界面。

在浏览器中输入“Kubernetes 集群的 IP 地址 + 映射端口”，看到的界面效果如图 10-7 所示。

图 10-7

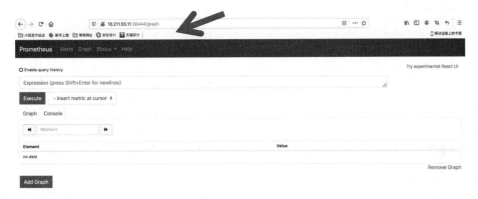

（4）使用 PromeQL 的查看指标“http_requests_total”，效果如图 10-8 所示。

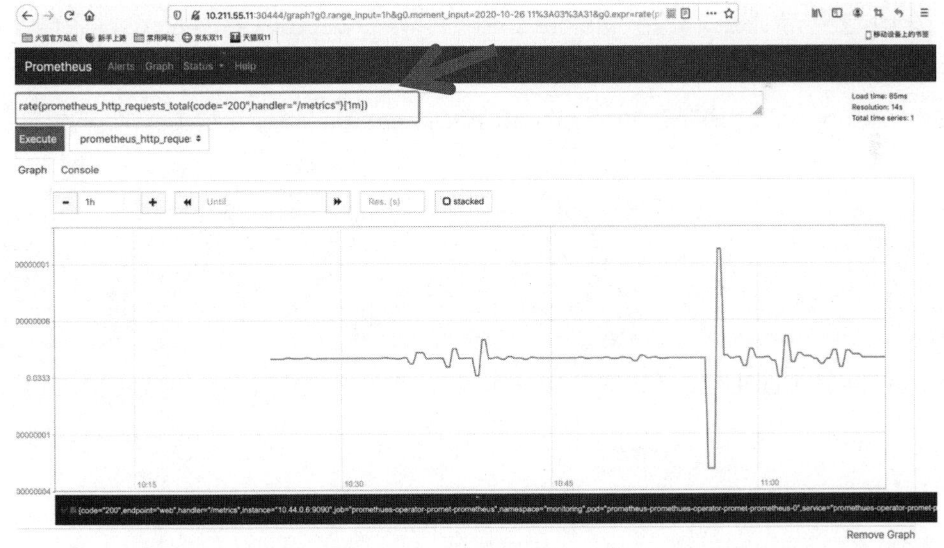

图 10-8

由此说明，Promethues 监控系统已经开始运行，并采集了相关的 Metrics（度量指标）数据。

（5）单击界面中的 Alerts 菜单，可以看到 Promethues 中的相关告警信息，如图 10-9 所示。

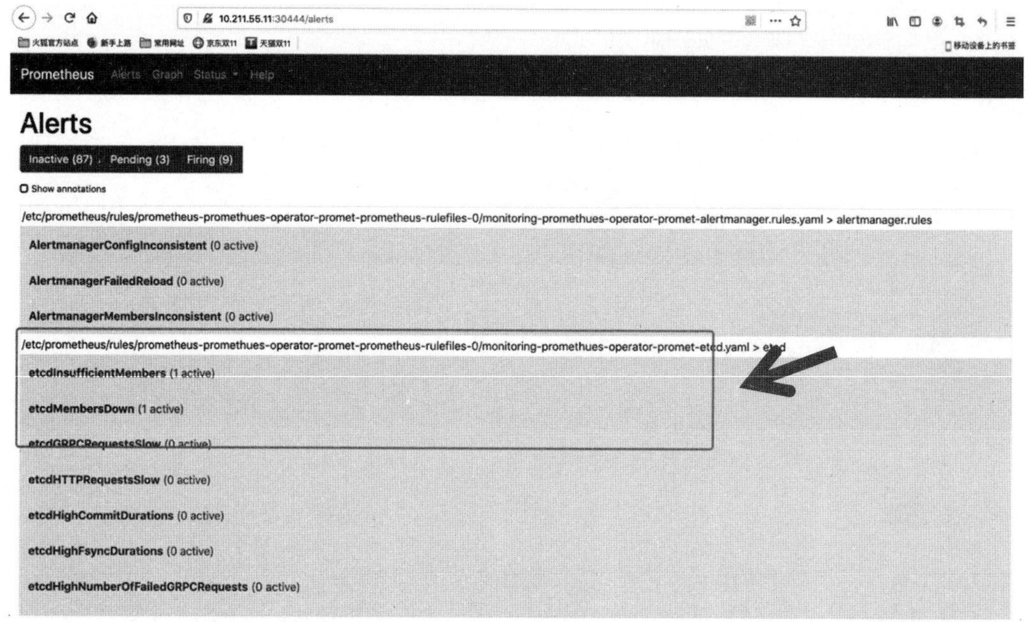

图 10-9

从图 10-9 中可以看到，在 Promethues 中存在两项关于 Kubernetes etcd 组件的告警信息。对于告警信息，也可以通过 Promethues 内置的 Alertmanager UI 进行查看。

使用 nodeport 的方式设置 Promethues Alertmanager UI 服务的集群外访问端口（例如 30445）。命令如下：

```
$ kubectl  patch svc promethues-operator-promet-alertmanager -n monitoring -p
'{"spec":{"type":"NodePort","ports":[{"port":9093,"targetPort":9093,"nodePort":30445}]}}'
service/promethues-operator-promet-alertmanager patched
```

在浏览器中输入 URL 后，可以看到 Alertmanager UI 中的告警信息，如图 10-10 所示。

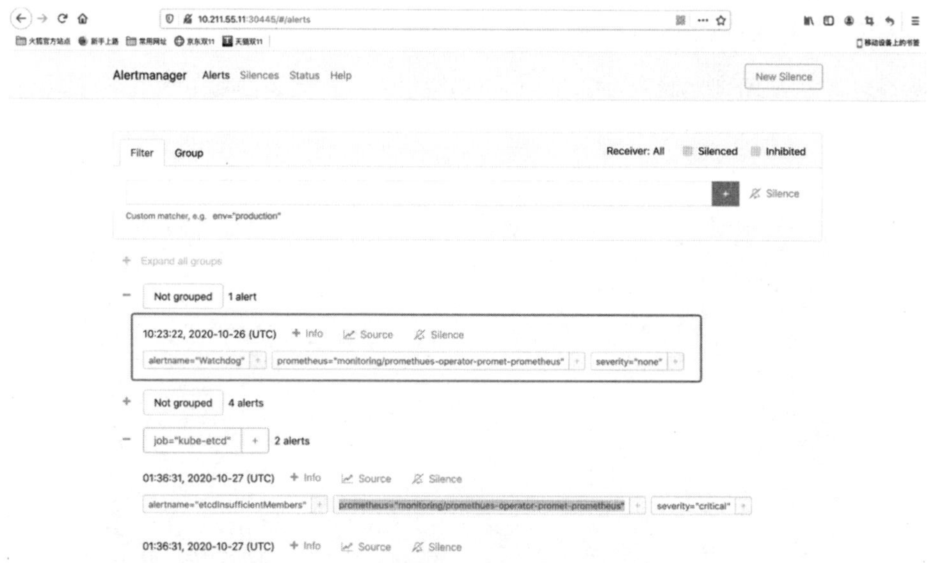

图 10-10

10.2.4　步骤 3：部署 Grafana 可视化监控系统

Grafana 是一个强大的跨平台的开源度量分析和可视化工具，可以将采集的 Metrics（度量指标）数据方便地定制成图形展示界面。Grafana 被广泛用于实现时间序列数据和应用分析的可视化。Grafana 支持多种数据源，如 InfluxDB、OpenTSDB、ElasticSearch 及 Prometheus。

在 Kubernetes 安装 Prometheus-Operator 时，Grafana 实际上就已经被部署并运行了。接下来通过 Kubernetes 的命令查询 Grafana 所运行的 Pod，并设置 Grafana 服务的集群外访问端口。

（1）查看 Grafana 运行的 Pod 信息。命令如下：

```
$ kubectl -n monitoring get svc
```

（2）设置 Grafana 服务的访问端口。

使用“nodeport”的方式设置 Grafana 服务的集群外访问端口（例如 30441）。命令如下：

```
#使用“nodeport”的方式将promethues-operator-grafana暴露在集群外,并指定使用30441端口
$ kubectl  patch svc promethues-operator-grafana -n monitoring -p '{"spec":{"type":"NodePort","ports":[{"port":80,"targetPort":3000,"nodePort":30441}]}}'
```

需要注意，由于 Grafana 的应用运行的端口默认为 80，为避免环境冲突，在这里映射时将容器目标端口指定为 3000，并最终将节点端口映射为 30441。

（3）在浏览器输入 Grafana 的访问 URL。

如果映射正常，则在浏览器输入 Grafana 的访问 URL 后，会返回 Grafana 的登录界面，如图 10-11 所示。

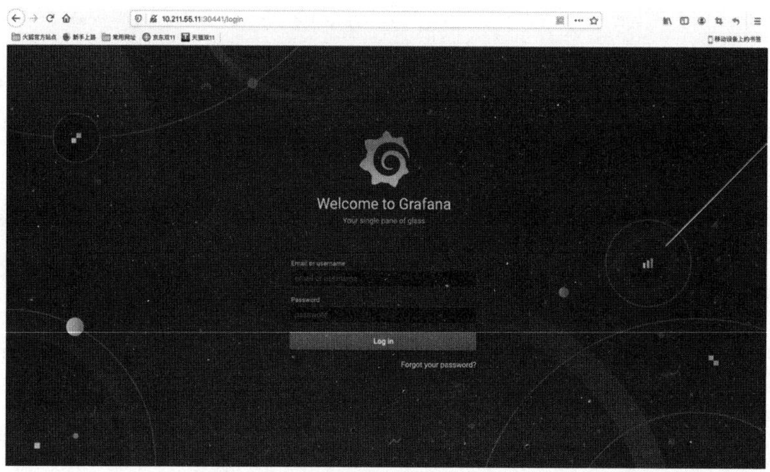

图 10-11

输入默认的登录账号/密码：admin/prom-operator。进入后 Grafana 的界面如图 10-12 所示。

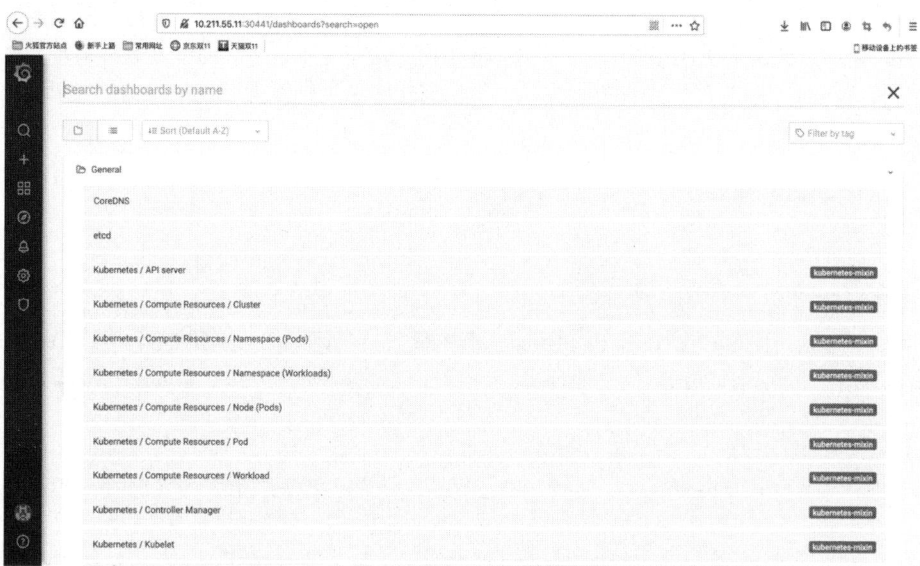

图 10-12

可以看到，部署完成的 Grafana 已经默认内置了许多针对 Kubernetes 平台的企业级监控 Dashboard。例如，针对 Kubernetes 集群组件的“Kubernetes/API server”“Kubernetes/Kubelet”，以及针对 Kubernetees 计算资源的“Kubernetes/Compute Resources/Pod”“Kubernetes/Compute Resources/Workload”等。

这里找一个针对 Kubernetes 物理节点（Nodes）的监控 Dashboard，打开后看到的监控效果如图 10-13 所示。

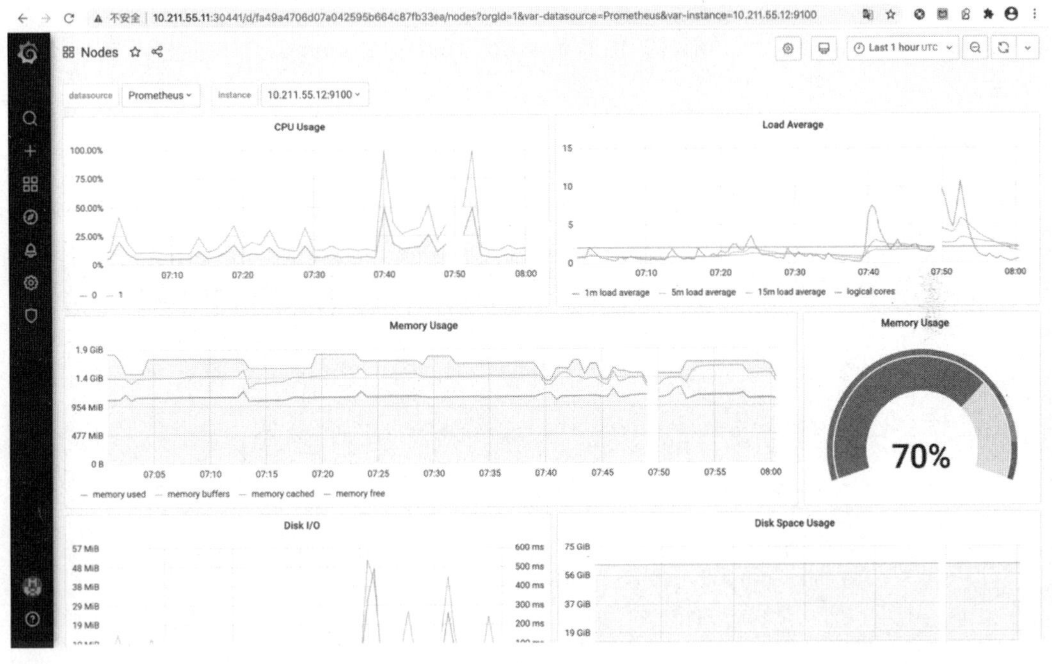

图 10-13

在如图 10-13 所示的 Dashboard 中展示了 Kubernetes 集群所在的各物理节点的 CPU、负载、内存、磁盘 I/O、磁盘空间、网络传输等硬件资源的使用情况。从这些丰富的视图可以看出 Grafana 强大的监控指标分析及可视化能力。

10.2.5　步骤 4：将 Spring Cloud 微服务接入 Prometheus

一般来说，使用 Prometheus 对 Kubernetes 集群进行监控，Metrics（度量指标）数据的来源主要有以下 3 种：

- 宿主机的监控数据。这部分数据是借助 Node Exporter 工具，以 DaemonSet 的方式运行在宿主机上采集得到的。
- Kubernetes 的 API Server、Kubelet 等组件的“/metrics”API。除常规的 CPU、内存

等资源信息外，这部分数据还包括各个组件的核心监控指标。例如对 API Server 来说，它会在“/metrics” API 中暴露各个 Controller 的工作队列长度、QPS 和延迟数据等，而这些信息是检查 Kubernetes 集群工作情况的主要依据。

- Kubernetes 资源相关的监控数据。这部分数据主要包括了 Pod、Node、容器、Service 等 Kubernetes 核心资源的 Metrics（度量指标）数据。

而对于部署在 Kubernetes 中的 Spring Cloud 微服务，主要的监控数据是与容器相关的 Metrics（度量指标）数据。

接下来介绍 Promethues 对部署在 Kubernetes 集群中的 Spring Cloud 微服务实施监控的具体方法。

1. 构建 Spring Cloud 微服务示范工程

为了演示 Spring Cloud 微服务接入 Prometheus 监控系统的场景，这里新创建一个 Spring Cloud 微服务示例工程。

（1）利用第 2.3.1 节介绍的方法创建一个 Maven 工程，完成后的工程代码结构如图 10-14 所示。

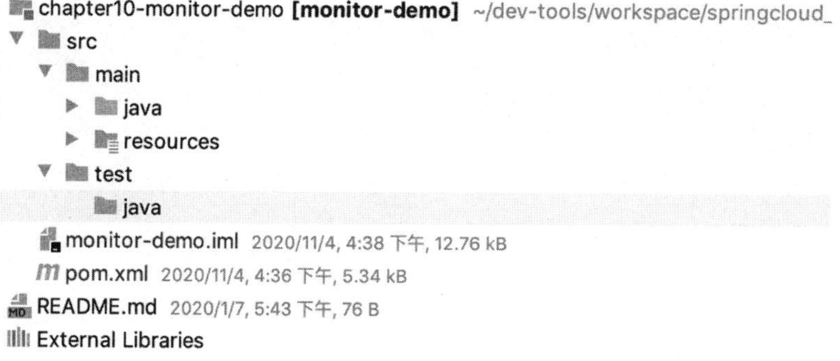

图 10-14

（2）引入 Spring Cloud 微服务的核心依赖。

这里可以参考 2.5.2 节中的具体步骤。

（3）考虑到后续内容还将基于本示例服务进行分布式链路追踪之类的测试，所以继续在 pom.xml 文件中集成访问 MySQL 数据库及 Redis 缓存等所需要的依赖。代码如下：

```
<!--引入 MyBatis 依赖-->
<dependency>
    <groupId>org.mybatis.spring.boot</groupId>
    <artifactId>mybatis-spring-boot-starter</artifactId>
    <version>2.0.1</version>
```

```
</dependency>
<!--引入 Druid 连接池依赖-->
<dependency>
    <groupId>com.alibaba</groupId>
    <artifactId>druid</artifactId>
    <version>1.0.28</version>
</dependency>
<!--引入 MySQL 数据库驱动程序连接包-->
<dependency>
    <groupId>mysql</groupId>
    <artifactId>mysql-connector-java</artifactId>
    <scope>runtime</scope>
</dependency>

<!--引入 Redis 依赖-->
<dependency>
    <groupId>org.springframework.boot</groupId>
    <artifactId>spring-boot-starter-data-redis</artifactId>
</dependency>
<dependency>
    <groupId>org.springframework.data</groupId>
    <artifactId>spring-data-redis</artifactId>
    <version>2.1.8.RELEASE</version>
    <scope>compile</scope>
</dependency>

<!--引入 lombok 开发工具包-->
<dependency>
    <groupId>org.projectlombok</groupId>
    <artifactId>lombok</artifactId>
</dependency>
```

（4）在工程代码的 resources 目录新建一个基础性配置文件——bootstrap.yml。配置文件中的代码如下：

```
spring:
  application:
    name: chapter10-monitor-demo
  profiles:
    active: debug
  cloud:
    consul:
      discovery:
        preferIpAddress: true
```

```
        instance-id:
${spring.application.name}:${spring.cloud.client.ipAddress}:${spring.applica
tion.instance_id:${server.port}}:@project.version@
        healthCheckPath: /actuator/health
  server:
    port: 9092
```

（5）Spring Boot 不会默认加载 bootstrap.yml 这个文件，所以需要在 pom.xml 中添加 Maven 资源相关的配置，具体参考 2.5.2 节内容。

（6）为微服务工程继续创建一个 test 环境配置文件——application-test.yml，用来配置测试环境 MySQL 及 Redis 的连接信息，代码如下：

```
spring:
  cloud:
    #独立部署的 Consul 注册中心地址（根据自己本地实际环境填写）
    consul:
      host: 10.211.55.2
      port: 8500

  datasource:
    #独立部署的 MySQL 数据库地址（根据自己本地实际环境填写）
    url: jdbc:mysql://10.211.55.2:3306/monitor_test
    username: root
    password: 123456
    type: com.alibaba.druid.pool.DruidDataSource
    driver-class-name: com.mysql.jdbc.Driver
    separator: //

  #独立部署的 Redis 服务地址（根据自己本地实际环境填写）
  redis:
    host: 10.211.55.2
    port: 6379
    password: 123456

# Logging
logging.level.org.springframework: INFO
```

（7）创建本 Spring Cloud 示范微服务的入口程序类。代码如下：

```
package com.wudimanong.monitor;

import org.springframework.boot.SpringApplication;
import org.springframework.boot.autoconfigure.SpringBootApplication;
import org.springframework.cloud.client.discovery.EnableDiscoveryClient;

@EnableDiscoveryClient
```

```
@SpringBootApplication
public class MonitorDemoApplication {

    public static void main(String[] args) {
        SpringApplication.run(MonitorDemoApplication.class, args);
    }
}
```

2. 编写微服务测试接口

为了方便后续验证测试，这里继续编写一个简单的 HTTP 测试接口。

（1）定义测试接口的 Controller 层。代码如下：

```
package com.wudimanong.monitor.controller;

import com.wudimanong.monitor.service.MonitorService;
...
import org.springframework.web.bind.annotation.RestController;

@RestController
@RequestMapping("/monitor")
public class MonitorController {

    @Autowired
    private MonitorService monitorServiceImpl;

    @GetMapping("/test")
    public String monitorTest(@RequestParam("name") String name) {
        monitorServiceImpl.monitorTest(name);
        return "监控示范工程测试接口返回->OK!";
    }
}
```

（2）定义业务层（Service 层）接口类 MonitorService。代码如下：

```
package com.wudimanong.monitor.service;
public interface MonitorService {
    /**
     * 监控测试代码
     */
    String monitorTest(String name);
}
```

（3）实现业务层（Service 层）接口类的方法。代码如下：

```
package com.wudimanong.monitor.service.impl;
import com.wudimanong.monitor.dao.mapper.TestInfoDao;
```

```
import com.wudimanong.monitor.dao.model.TestInfoPO;
...
import org.springframework.stereotype.Service;

@Service
public class MonitorServiceImpl implements MonitorService {
    /**
     * 持久层（Dao 层）组件
     */
    @Autowired
    private TestInfoDao testInfoDao;
    /**
     * Redis 访问组件
     */
    @Autowired
    private RedisTemplate redisTemplate;
    /**
     * 保存方法
     */
    @Override
    public String monitorTest(String name) {
        TestInfoPO testInfoPO = new TestInfoPO();
        testInfoPO.setName(name);
        testInfoPO.setCreateTime(new Timestamp(System.currentTimeMillis()));
        testInfoPO.setUpdateTime(new Timestamp(System.currentTimeMillis()));
        //插入数据库
        testInfoDao.saveTestInfo(testInfoPO);
        //插入缓存
        redisTemplate.opsForValue().set(name, testInfoPO);
        return name;
    }
}
```

上述代码的主要逻辑是：将数据插入一张测试表，并将测试数据缓存到 Redis 中。

（4）实现测试接口的持久层（Dao 层）。

定义测试表的数据库实体类。代码如下：

```
package com.wudimanong.monitor.dao.model;
import java.io.Serializable;
import java.sql.Timestamp;
import lombok.Data;
@Data
public class TestInfoPO implements Serializable {
    private Integer id;
    private String name;
```

```
    private Timestamp createTime;
    private Timestamp updateTime;
}
```

定义测试表的持久层（Dao 层）接口。代码如下：

```
package com.wudimanong.monitor.dao.mapper;
import com.wudimanong.monitor.dao.model.TestInfoPO;
...
import org.springframework.stereotype.Repository;
@Repository
@Mapper
public interface TestInfoDao {
    /**
     *保存数据的方法
     */
    @Insert("insert into test_info(name,create_time,update_time)
values(#{name},#{createTime,jdbcType=TIMESTAMP},#{updateTime,jdbcType=TIMEST
AMP})")
    int saveTestInfo(TestInfoPO testInfoPO);
}
```

如上所示，该持久层（Dao 层）接口中只有一个保存数据的方法——向数据库中插入数据。

创建测试信息表的具体 SQL 语句如下：

```
create table test_info
(
   id                  bigint not null auto_increment,
   name             varchar(11) comment '测试名称',
   create_time       timestamp default current_timestamp comment '创建时间',
   update_time       timestamp default current_timestamp comment '更新时间',
   primary key (id)
);
alter table test_info comment '测试信息表';
```

至此，完成了微服务示例代码的编写。运行微服务程序，如果该接口能够被正常访问，则说明示范工程搭建成功了。

3. 配置微服务的 Prometheus 指标采集客户端

为了 Prometheus 能够正常采集到 Spring Cloud 微服务的 Metrics（度量指标）数据，需要引入 Prometheus 指标采集客户端的依赖。在微服务工程的 pom.xml 文件中引入如下依赖：

```
<dependency>
    <groupId>io.micrometer</groupId>
    <artifactId>micrometer-registry-prometheus</artifactId>
</dependency>
```

在 bootstrap.yml 文件中，增加开启微服务 Metrics（度量指标）采集端点的配置。代码如下：

```
# 开启Metrics指标采集端点
management:
  endpoint:
    metrics:
      enabled: true
    prometheus:
      enabled: true
  endpoints:
    web:
      exposure:
        include: '*'
  metrics:
    export:
      prometheus:
        enabled: true
```

至此，涉及示例工程 Prometheus 采集端点相关的逻辑就完成了。Prometheus 可以通过采集端点获取 Spring Boot 默认暴露的 Metrics（度量指标）数据。

在 Spring Boot 中，Spring Boot 默认暴露的 Metrics（度量指标）数据是通过“spring-boot-starter-actuator”组件来生成的。

4. 自定义微服务的 Prometheus 监控指标

前面通过集成 Prometheus 指标采集客户端，已经可以获取 Spring Boot 默认暴露的 Metrics 了。但是，Spring Boot 默认暴露的 Metrics 数量及类型是有限的，如果要针对 Spring Cloud 微服务应用建立更丰富的指标维度（例如在 10.1.2 节提及的指标类型），则还需要配置 Prometheus 指标采集客户端（micrometer-registry-prometheus）提供的相关指标类型（如统计 TP 值）。

接下来将演示在 Spring Cloud 微服务中，以更加优雅的方式来自定义 Prometheus 监控指标。

（1）自定义监控指标的配置注解。

在 Spring Cloud 微服务中，对于程序运行信息的收集（如指标、日志），比较常用的方法是通过 Spring 的 AOP 代理拦截来实现的。但这种方式会损耗一定的系统性能。所以，在设计自定义 Prometheus 监控指标的方式时，可以将是否上报指标的选择权交给开发人员。而从易用性角度来说，一般可以通过注解来实现。例如，定义 TP 值采集的注解代码如下：

```
package com.wudimanong.monitor.metrics.annotation;

import java.lang.annotation.ElementType;
```

```
...
import java.lang.annotation.Target;

@Target({ElementType.METHOD})
@Retention(RetentionPolicy.RUNTIME)
@Inherited
public @interface Tp {

    String description() default "";
}
```

上述代码定义了一个用于标注上报计时器指标类型的注解。如果想统计接口的 TP90、TP99 这样的分位值指标，则可以使用该注解来暴露相应 Metrics（度量指标）数据。

除此之外，还可以定义暴露其他指标类型的注解，例如：

```
package com.wudimanong.monitor.metrics.annotation;

import java.lang.annotation.ElementType;
import java.lang.annotation.Inherited;
import java.lang.annotation.Retention;
import java.lang.annotation.RetentionPolicy;
import java.lang.annotation.Target;

@Target({ElementType.METHOD})
@Retention(RetentionPolicy.RUNTIME)
@Inherited
public @interface Count {

    String description() default "";
}
```

上述代码定义了一个用于上报计数器类型指标的注解。如果要统计接口的平均响应时间、接口的请求量之类的指标，则可以使用该注解。

如果觉得分别定义不同指标类型的注解比较麻烦，希望某些接口的各种指标类型都上报到 Prometheus，则可以定义一个通用注解，以实现同时上报多个指标类型，例如：

```
package com.wudimanong.monitor.metrics.annotation;

import java.lang.annotation.ElementType;
...
import java.lang.annotation.Target;

@Target({ElementType.METHOD})
@Retention(RetentionPolicy.RUNTIME)
```

```
@Inherited
public @interface Monitor {

    String description() default "";
}
```

总之，无论是分别定义某种特定指标的注解，还是定义一个通用的暴露多种指标注解，其目标都是希望能以更灵活的方式来扩展 Spring Cloud 微服务应用的监控指标类型。

（2）实现自定义监控指标注解 AOP 代理的逻辑。

在步骤（1）中定义了上报不同指标类型的注解。而注解的具体实现逻辑，可以通过定义一个通用的 AOP 代理类来实现，代码如下：

```
package com.wudimanong.monitor.metrics.aop;

import com.wudimanong.monitor.metrics.Metrics;
...
import org.springframework.stereotype.Component;

@Aspect
@Component
public class MetricsAspect {
    /**
     * Prometheus 指标管理
     */
    private MeterRegistry registry;
    private Function<ProceedingJoinPoint, Iterable<Tag>>
tagsBasedOnJoinPoint;
    public MetricsAspect(MeterRegistry registry) {
        this.init(registry, pjp -> Tags
                .of(new String[]{"class",
pjp.getStaticPart().getSignature().getDeclaringTypeName(), "method",
                        pjp.getStaticPart().getSignature().getName()}));
    }

    public void init(MeterRegistry registry, Function<ProceedingJoinPoint,
Iterable<Tag>> tagsBasedOnJoinPoint) {
        this.registry = registry;
        this.tagsBasedOnJoinPoint = tagsBasedOnJoinPoint;
    }

    /**
     * @Tp 指标配置注解
     */
    @Around("@annotation(com.wudimanong.monitor.metrics.annotation.Tp)")
```

```
    public Object timedMethod(ProceedingJoinPoint pjp) throws Throwable {
        Method method = ((MethodSignature) pjp.getSignature()).getMethod();
        method = pjp.getTarget().getClass().getMethod(method.getName(),
method.getParameterTypes());
        Tp tp = method.getAnnotation(Tp.class);
        Timer.Sample sample = Timer.start(this.registry);
        String exceptionClass = "none";
        try {
            return pjp.proceed();
        } catch (Exception ex) {
            exceptionClass = ex.getClass().getSimpleName();
            throw ex;
        } finally {
            try {
                String finalExceptionClass = exceptionClass;
                //创建定义计数器，并设置指标的 tag 信息（名称可以自定义）
                Timer timer = Metrics.newTimer("tp.method.timed",
                        builder -> builder.tags(new String[]{"exception",
finalExceptionClass})
                                .tags(this.tagsBasedOnJoinPoint.apply(pjp)).ta
g("description", tp.description())
                                .publishPercentileHistogram().register(this.re
gistry));
                sample.stop(timer);
            } catch (Exception exception) {
            }
        }
    }
    /**
     * @Count 指标配置注解
     */
    @Around("@annotation(com.wudimanong.monitor.metrics.annotation.Count)")
    public Object countMethod(ProceedingJoinPoint pjp) throws Throwable {
        Method method = ((MethodSignature) pjp.getSignature()).getMethod();
        method = pjp.getTarget().getClass().getMethod(method.getName(),
method.getParameterTypes());
        Count count = method.getAnnotation(Count.class);
        String exceptionClass = "none";
        try {
            return pjp.proceed();
        } catch (Exception ex) {
            exceptionClass = ex.getClass().getSimpleName();
            throw ex;
        } finally {
            try {
                String finalExceptionClass = exceptionClass;
```

```
                //创建定义计数器，并设置指标的 Tags 信息（名称可以自定义）
                Counter counter = Metrics.newCounter("count.method.counted",
                        builder -> builder.tags(new String[]{"exception",
finalExceptionClass})
                                .tags(this.tagsBasedOnJoinPoint.apply(pjp)).ta
g("description", count.description())
                                .register(this.registry));
                counter.increment();
            } catch (Exception exception) {
            }
        }
    }
    /**
     * @Monitor 指标配置注解
     */
    @Around("@annotation(com.wudimanong.monitor.metrics.annotation.Monitor)")
    public Object monitorMethod(ProceedingJoinPoint pjp) throws Throwable {
        Method method = ((MethodSignature) pjp.getSignature()).getMethod();
        method = pjp.getTarget().getClass().getMethod(method.getName(),
method.getParameterTypes());
        Monitor monitor = method.getAnnotation(Monitor.class);
        String exceptionClass = "none";
        try {
            return pjp.proceed();
        } catch (Exception ex) {
            exceptionClass = ex.getClass().getSimpleName();
            throw ex;
        } finally {
            try {
                String finalExceptionClass = exceptionClass;
                //计时器 Metric
                Timer timer = Metrics.newTimer("tp.method.timed",
                        builder -> builder.tags(new String[]{"exception",
finalExceptionClass})
                                .tags(this.tagsBasedOnJoinPoint.apply(pjp)).ta
g("description", monitor.description())
                                .publishPercentileHistogram().register(this.re
gistry));
                Timer.Sample sample = Timer.start(this.registry);
                sample.stop(timer);

                //计数器 Metric
                Counter counter = Metrics.newCounter("count.method.counted",
                        builder -> builder.tags(new String[]{"exception",
finalExceptionClass})
                                .tags(this.tagsBasedOnJoinPoint.apply(pjp)).ta
g("description", monitor.description())
```

```
                        .register(this.registry));
                counter.increment();
            } catch (Exception exception) {
            }
        }
    }
}
```

上述代码完整地实现了前面所定义的指标配置注解的逻辑。其中，@Monitor 注解的逻辑是@Tp 和@Count 注解逻辑的整合。如果还需要定义其他指标类型，则可以在此基础上继续扩展。

需要注意，在上述逻辑实现中，对于“Timer”及“Counter”等类型指标的构建，这里并没有直接使用“micrometer-registry-prometheus”依赖包中的构建对象，而是通过自定义的 Metrics.newTimer()方法来实现的，这主要是希望以更简洁、灵活的方式去实现指标的暴露。具体代码如下：

```
package com.wudimanong.monitor.metrics;

import io.micrometer.core.instrument.Counter;
...
import org.springframework.context.ApplicationContextAware;

public class Metrics implements ApplicationContextAware {

    private static ApplicationContext context;

    @Override
    public void setApplicationContext(@NonNull ApplicationContext
applicationContext) throws BeansException {
        context = applicationContext;
    }

    public static ApplicationContext getContext() {
        return context;
    }

    public static Counter newCounter(String name, Consumer<Builder> consumer) {
        MeterRegistry meterRegistry = context.getBean(MeterRegistry.class);
        return new CounterBuilder(meterRegistry, name, consumer).build();
    }

    public static Timer newTimer(String name, Consumer<Timer.Builder>
consumer) {
        return new TimerBuilder(context.getBean(MeterRegistry.class), name,
consumer).build();
```

```
        }
    }
```

上述代码通过接入 Spring 容器上下文获取了 MeterRegistry 实例，并以此来构建像 Counter、Timer 这样的指标类型对象。这里之所以将获取方法定义为静态的，主要是便于未来在业务代码中进行引用。

此外，Metrics 类中涉及的构造器 CounterBuilder 及 TimerBuilder 的定义如下。

①构造器 CounterBuilder 的定义如下：

```java
package com.wudimanong.monitor.metrics;

import io.micrometer.core.instrument.Counter;
import io.micrometer.core.instrument.Counter.Builder;
import io.micrometer.core.instrument.MeterRegistry;
import java.util.function.Consumer;

public class CounterBuilder {

    private final MeterRegistry meterRegistry;

    private Counter.Builder builder;

    private Consumer<Builder> consumer;

    public CounterBuilder(MeterRegistry meterRegistry, String name,
Consumer<Counter.Builder> consumer) {
        this.builder = Counter.builder(name);
        this.meterRegistry = meterRegistry;
        this.consumer = consumer;
    }

    public Counter build() {
        consumer.accept(builder);
        return builder.register(meterRegistry);
    }
}
```

②构造器 TimerBuilder 的定义如下：

```java
package com.wudimanong.monitor.metrics;

import io.micrometer.core.instrument.MeterRegistry;
import io.micrometer.core.instrument.Timer;
import io.micrometer.core.instrument.Timer.Builder;
import java.util.function.Consumer;
```

```
public class TimerBuilder {

    private final MeterRegistry meterRegistry;

    private Timer.Builder builder;

    private Consumer<Builder> consumer;

    public TimerBuilder(MeterRegistry meterRegistry, String name, 
Consumer<Timer.Builder> consumer) {
        this.builder = Timer.builder(name);
        this.meterRegistry = meterRegistry;
        this.consumer = consumer;
    }

    public Timer build() {
        this.consumer.accept(builder);
        return builder.register(meterRegistry);
    }
}
```

之所以将构造器代码单独定义，主要是从代码的优雅性来考虑。如果涉及其他指标类型的构造，也可以通过类似的方法进行扩展。

（3）编写自定义指标注解的配置类。

在前面的步骤中，已经自定义了几个指标注解，并实现了具体暴露逻辑。为了使其在Spring Boot环境中运行，还需要编写如下配置类：

```
package com.wudimanong.monitor.metrics.config;

import com.wudimanong.monitor.metrics.Metrics;
...
import org.springframework.core.env.Environment;

@Configuration
public class CustomMetricsAutoConfiguration {

    @Bean
    @ConditionalOnMissingBean
    public MeterRegistryCustomizer<MeterRegistry> 
meterRegistryCustomizer(Environment environment) {
        return registry -> {
```

```
            registry.config()
                    .commonTags("application",
environment.getProperty("spring.application.name"));
        };
    }

    @Bean
    @ConditionalOnMissingBean
    public Metrics metrics() {
        return new Metrics();
    }
}
```

上述代码主要约定了暴露 Prometheus 指标信息中所携带的应用名称，并配置了自定义 Metrics 类的实例。

（4）业务代码的使用方式及效果。

如果要在 Spring Cloud 微服务中暴露自定义的 Prometheus 监控指标，则可以通过在接口的 Controller 层添加相关自定义注解来实现，代码如下：

```
package com.wudimanong.monitor.controller;

import com.wudimanong.monitor.metrics.annotation.Count;
...
import org.springframework.web.bind.annotation.RestController;

@RestController
@RequestMapping("/monitor")
public class MonitorController {

    @Autowired
    private MonitorService monitorServiceImpl;

    //监控指标注解使用
    //@Tp(description = "/monitor/test")
    //@Count(description = "/monitor/test")
    @Monitor(description = "/monitor/test")
    @GetMapping("/test")
    public String monitorTest(@RequestParam("name") String name) {
        monitorServiceImpl.monitorTest(name);
        return "监控示范工程测试接口返回->OK!";
    }
}
```

如上述代码所示，在实际的业务编程中，可以通过注解的方式来配置接口所暴露的 Prometheus

监控指标。

启动微服务程序，通过访问服务的“/actuator/prometheus”指标采集端点，来查看暴露的指标数据，如图 10-15 所示。

```
10.211.55.12:30002/actuator/prometheus
jvm_gc_pause_seconds_sum{action="end of minor GC",application="chapter10-monitor-demo",cause="Allocation Failure",} 0.542
jvm_gc_pause_seconds_count{action="end of major GC",application="chapter10-monitor-demo",cause="Metadata GC Threshold",} 1.0
jvm_gc_pause_seconds_sum{action="end of major GC",application="chapter10-monitor-demo",cause="Metadata GC Threshold",} 0.128
# HELP jvm_gc_pause_seconds_max Time spent in GC pause
# TYPE jvm_gc_pause_seconds_max gauge
jvm_gc_pause_seconds_max{action="end of minor GC",application="chapter10-monitor-demo",cause="Allocation Failure",} 0.0
jvm_gc_pause_seconds_max{action="end of major GC",application="chapter10-monitor-demo",cause="Metadata GC Threshold",} 0.0
# HELP jvm_threads_daemon_threads The current number of live daemon threads
# TYPE jvm_threads_daemon_threads gauge
jvm_threads_daemon_threads{application="chapter10-monitor-demo",} 30.0
# HELP tomcat_sessions_active_max_sessions
# TYPE tomcat_sessions_active_max_sessions gauge
tomcat_sessions_active_max_sessions{application="chapter10-monitor-demo",} 0.0
# HELP jvm_buffer_memory_used_bytes An estimate of the memory that the Java virtual machine is using for this buffer pool
# TYPE jvm_buffer_memory_used_bytes gauge
jvm_buffer_memory_used_bytes{application="chapter10-monitor-demo",id="mapped",} 0.0
jvm_buffer_memory_used_bytes{application="chapter10-monitor-demo",id="direct",} 81921.0
# HELP tomcat_sessions_active_current_sessions
# TYPE tomcat_sessions_active_current_sessions gauge
tomcat_sessions_active_current_sessions{application="chapter10-monitor-demo",} 0.0
# HELP tp_method_timed_seconds
# TYPE tp_method_timed_seconds histogram
tp_method_timed_seconds_bucket{application="chapter10-monitor-
demo",class="com.wudimanong.monitor.controller.MonitorController",description="/monitor/test",exception="none",method="monitorTest",le="0.001",}
175.0
tp_method_timed_seconds_bucket{application="chapter10-monitor-
demo",class="com.wudimanong.monitor.controller.MonitorController",description="/monitor
/test",exception="none",method="monitorTest",le="0.001048576",} 175.0
tp_method_timed_seconds_bucket{application="chapter10-monitor-
demo",class="com.wudimanong.monitor.controller.MonitorController",description="/monitor
/test",exception="none",method="monitorTest",le="0.001398101",} 175.0
tp_method_timed_seconds_bucket{application="chapter10-monitor-
demo",class="com.wudimanong.monitor.controller.MonitorController",description="/monitor
/test",exception="none",method="monitorTest",le="0.001747626",} 175.0
tp_method_timed_seconds_bucket{application="chapter10-monitor-
demo",class="com.wudimanong.monitor.controller.MonitorController",description="/monitor
/test",exception="none",method="monitorTest",le="0.002097151",} 175.0
tp_method_timed_seconds_bucket{application="chapter10-monitor-
demo",class="com.wudimanong.monitor.controller.MonitorController",description="/monitor
/test",exception="none",method="monitorTest",le="0.002446676",} 175.0
tp_method_timed_seconds_bucket{application="chapter10-monitor-
demo",class="com.wudimanong.monitor.controller.MonitorController",description="/monitor
/test",exception="none",method="monitorTest",le="0.002796201",} 175.0
tp_method_timed_seconds_bucket{application="chapter10-monitor-
demo",class="com.wudimanong.monitor.controller.MonitorController",description="/monitor
```

图 10-15

从图 10-15 可以看到，在自定义监控指标（TP 值）数据后，就可以通过指标采集端点来将其暴露给 Prometheus。在 10.2.7 节中将利用这些指标数据来构建微服务的可视化监控视图。

5. 将微服务部署至 Kubernetes 集群中

接下来将 Spring Cloud 微服务部署到 Kubernetes 集群中。

（1）将用 Spring Boot 构建的微服务打包为 Docker 镜像。

在基于 Maven 构建的微服务工程中，可以通过引入 Docker 打包插件，来自动将应用打包成 Docker 镜像。

①在工程 pom.xml 文件中配置打包插件，代码如下：

```
<properties>
    <!--定义 Docker 镜像仓库地址-->
    <docker.repository>10.211.55.2:8080</docker.repository>
    <!--定义项目名称作为镜像名称生成的组成部分-->
    <app.name>chapter10-monitor-demo</app.name>
```

```
    </properties>
    <build>
        <plugins>
            ...
            <!--发布 Maven 插件-->
            <plugin>
                <groupId>org.apache.maven.plugins</groupId>
                <artifactId>maven-deploy-plugin</artifactId>
                <configuration>
                    <skip>true</skip>
                </configuration>
            </plugin>
            <!--添加将 Java 应用打包为 Docker 镜像的 Maven 插件-->
            <plugin>
                <groupId>com.spotify</groupId>
                <artifactId>dockerfile-maven-plugin</artifactId>
                <version>1.4.13</version>
                <executions>
                    <execution>
                        <id>build-image</id>
                        <phase>package</phase>
                        <goals>
                            <goal>build</goal>
                            <goal>push</goal>
                        </goals>
                    </execution>
                </executions>
                <configuration>
                    <!--指定 Dockerfile 文件位置-->
                    <dockerfile>docker/Dockerfile</dockerfile>
                    <repository>${docker.repository}/springcloud-action/
${app.name}</repository>
                    <!--<tag>${project.version}</tag>-->
                    <buildArgs>
                        <!--向 Dockerfile 文件传递构建参数-->
                        <JAR_FILE>target/${project.build.finalName}.jar</JAR_FILE>
                    </buildArgs>
                </configuration>
            </plugin>
        </plugins>
        ...
    </build>
```

上述代码，在微服务的 pom.xml 文件中添加了“dockerfile-maven-plugin”插件配置，并通过该插件的“configuration”属性配置了镜像打包命令文件 Dockerfile。该文件被存储在项目工程的 Docker 文件目录下。

②打包命令文件 Dockerfile 的内容如下：

```
FROM openjdk:8u191-jre-alpine3.9
ENTRYPOINT ["/usr/bin/java", "-jar", "/app.jar"]
ARG JAR_FILE
ADD ${JAR_FILE} /app.jar
EXPOSE 8080
```

这是一个简洁的 Docker 镜像构建命令文件。根据前面配置，在将项目打包为 Docker 镜像后，会自动将镜像 Push 到私有镜像仓库。

可以在项目工程中执行 Maven 的构建命令“mvn clean package -X”，这样在项目构建完成后会自动将构建的 Docker 镜像上传至在打包插件中所配置的镜像仓库中。而 Docker 镜像仓库的搭建及登录方式可参考第 9.4 节的内容。

在 Docker 镜像上传成功后，登录镜像仓库管理界面，可以看到上传的镜像信息如图 10-16 所示。

图 10-16

（2）编写微服务的 Kubernetes 部署文件。

在将应用上传到镜像仓库后，就可以通过编写微服务的 Kubernetes 部署文件，来将 Spring Cloud 微服务部署到 Kubernetes 集群中。部署文件的内容如下：

```
apiVersion: apps/v1
kind: Deployment
metadata:
```

```
    name: chapter10-monitor-demo
  spec:
    selector:
      matchLabels:
        app: chapter10-monitor-demo
    replicas: 1
    #设置滚动升级策略
    #Kubernetes 在设置的时间后才开始进行升级，例如 5s
    minReadySeconds: 5
    strategy:
      type: RollingUpdate
      rollingUpdate:
        #在升级过程中，最多可以比原先设置多出的 Pod 数量
        maxSurge: 1
        #在升级过程中，Deployment 控制器最多可以删除多少个旧 Pod，主要用于提供缓冲时间
        maxUnavailable: 1
    template:
      metadata:
        labels:
          app: chapter10-monitor-demo
      spec:
        containers:
          - name: chapter10-devops-demo
            image:
10.211.55.2:8080/springcloud-action/chapter10-monitor-demo:latest
            env:
              #设置 Spring Boot 配置环境
              - name: SPRING_PROFILES_ACTIVE
                value: test
              - name: SERVER_PORT
                value: "8080"
---
apiVersion: v1
kind: Service
metadata:
  name: chapter10-monitor-demo
  #设置服务 label 标签，后面 Prometheus 会依据 label 机制来采集微服务监控数据
  labels:
    svc: chapter10-monitor-demo
spec:
  selector:
    app: chapter10-monitor-demo
  ports:
    - name: http
      #Service 在集群中暴露的端口（用于 Kubernetes 服务间的访问）
```

```
    port: 8080
    #Pod 上的端口（与制作容器时暴露的端口一致，在微服务工程代码中指定的端口）
    targetPort: 8080
    #K8s 集群外部访问的端口（外部机器访问）
    nodePort: 30002
  type: NodePort
```

在上述部署文件中，定义了 Kubernetes 的 Deployment 编排资源，以及 Service 服务资源。其中，配置的“image”镜像地址就是步骤（1）中 Docker 镜像在镜像仓库中的地址。

部署文件的路径在代码工程的“/kubernetes/deploy-prometheus.yml”目录下。

（3）发布微服务应用。

①将微服务部署到 Kubernetes 集群，命令如下：

```
$ kubectl apply -f kubernetes/deploy-prometheus.yml
```

如果发布成功，则可以通过命令在 Kubernetes 集群中查看发布的 Pod 资源。命令如下：

```
# kubectl get pods
NAME                                  READY   STATUS    RESTARTS   AGE
chapter10-monitor-demo-755f6b8bb5-k67fp   1/1     Running   2          17h
```

如上所示，可以看到微服务应用的 Pod 资源已经成功启动。

②查看 Service 资源，命令如下。

```
# kubectl get svc
NAME                    TYPE        CLUSTER-IP      EXTERNAL-IP   PORT(S)       AGE
chapter10-monitor-demo   NodePort    10.110.237.38    <none>
8080:30002/TCP   24d
kubernetes              ClusterIP   10.96.0.1       <none>        443/TCP       25d
```

如上所示，可以看到微服务的 Service 资源也成功定义了。

③访问微服务的测试接口。

因为在编写微服务的部署文件时，已经配置了 Service 的 NodePort 端口访问方式，所以可以在 Kubernetes 集群之外访问微服务的测试接口。测试效果如下：

```
$ curl http://10.211.55.12:30002/monitor/test?name=wudimanong
监控示范工程测试接口返回->OK!
```

可以看到，通过 Kubernetes 集群的“IP 地址 + NodePort 端口”，成功地调用了小标题“1.”中在构建微服务示范工程时所编写的测试接口。这说明微服务已经成功运行在 Kubernetes 集群中。

10.2.6 步骤 6：使用 ServiceMonitor 管理监控目标

前面已经成功将 Spring Cloud 微服务部署在 Kubernetes 集群中了。那么部署在 Kubernetes 中的 Prometheus，如何才能实现对 Spring Cloud 微服务的监控呢?

可以通过 Prometheus Operator 的自定义资源 ServiceMonitor 来配置管理 Prometheus 的监控目标列表，并以此实现 Prometheus 对 Spring Cloud 微服务的监控。

（1）查看 Kubernetes 已经存在的 ServiceMonitor 资源定义。

在部署 Prometheus Operator 时，在系统中就已经存在了一些与 Kubernetes 集群相关的 ServiceMonitor 的资源定义，查询命令如下：

```
# kubectl get servicemonitors -n monitoring
NAME                                                  AGE
promethues-operator-promet-alertmanager               108m
promethues-operator-promet-apiserver                  108m
promethues-operator-promet-coredns                    108m
promethues-operator-promet-grafana                    108m
promethues-operator-promet-kube-controller-manager 108m
promethues-operator-promet-kube-etcd                  108m
promethues-operator-promet-kube-proxy                 108m
promethues-operator-promet-kube-scheduler             108m
promethues-operator-promet-kube-state-metrics         108m
promethues-operator-promet-kubelet                    108m
promethues-operator-promet-node-exporter              108m
promethues-operator-promet-operator                   108m
promethues-operator-promet-prometheus                 108m
```

可以看到，在部署 Prometheus Operator 后，系统已经自动创建了针对 Kubernetes 及 Prometheus 自身组件的诸多 ServiceMonitor，这些组件也已经处在 Prometheus 的监控范围内。

这也是前面可以通过 Prometheus 监控界面能够看到许多 Kubernetes 组件的监控指标数据的原因。

（2）编写 ServiceMonitor 文件来配置 Prometheus 对 Spring Cloud 微服务的监控。

在 Kubernetes 的主机目录中创建 Prometheus 监控目标的管理文件（如 serviceMonitor.yaml）。代码如下：

```
prometheus:
  additionalServiceMonitors:
    - name: chapter10-monitor-demo
      selector:
        matchLabels:
```

```
        #对应于发布文件中设置的 label 标签
        svc: chapter10-monitor-demo
    namespaceSelector:
      #监控的 K8s 命名空间
      matchNames:
        - default
    endpoints:
      #对应于发布文件中 Service 资源所设置的 HTTP 端口设置
      - port: http
        #Prometheus 指标采集的 endpoint 端口
        path: /actuator/prometheus
        scheme: http
  #多个服务可以重复上述配置
```

如上所示，通过“additionalServiceMonitors”属性来追加 ServiceMonitor 的配置，而“-name”属性的值则是微服务部署在 Kubernetes 集群中的服务名。

ServiceMonitor 通过 Labels 标签来选取相应的 Service Endpoint，从而让 Prometheus Server 通过选取的 Service Endpoint 来采集被监控微服务的 Metrics（度量指标）信息。所以，在上述配置“selector.mathchLabels.svc”中所指定的 Service 名称，应该与 10.2.5 节中小标题“4.”中编写的微服务 Kubernetes 部署文件中的 Labels 标签定义相对应。

这里回顾一下微服务 Kubernetes 部署文件中的相关定义，代码如下：

```
...
---
apiVersion: v1
kind: Service
metadata:
  name: chapter10-monitor-demo
  #设置服务 label 标签，后面 Prometheus 会依据 label 机制来采集微服务的监控数据
  labels:
    svc: chapter10-monitor-demo
spec:
  selector:
    app: chapter10-monitor-demo
  ports:
    - name: http
      #Service 在集群中暴露的端口（用于 Kubernetes 服务间的访问）
      port: 8080
      #Pod 上的端口（与制作容器时暴露的端口一致，在微服务工程代码中指定的端口）
      targetPort: 8080
      #K8s 集群外部访问的端口（外部机器访问）
```

```
    nodePort: 30002
  type: NodePort
```

如上所示，在 Service 资源的属性 metadata.labels.svc 中定义的标签信息，就是 ServiceMonitor 所匹配的名称。

而在前面 ServiceMonitor 文件中配置的数据采集端点的地址“/actuator/prometheus”就是 10.2.5 节“3.”小标题中引入的 Prometheus 指标采集客户端依赖所定义的接口路径。

（3）通过 Helm 指令更新 Prometheus 的 ServiceMonitor 配置。

```
  # helm upgrade promethues-operator stable/prometheus-operator
--values=serviceMonitor.yaml -n monitoring

  WARNING: This chart is deprecated
  manifest_sorter.go:192: info: skipping unknown hook: "crd-install"
  manifest_sorter.go:192: info: skipping unknown hook: "crd-install"
  manifest_sorter.go:192: info: skipping unknown hook: "crd-install"
  manifest_sorter.go:192: info: skipping unknown hook: "crd-install"
  manifest_sorter.go:192: info: skipping unknown hook: "crd-install"
  manifest_sorter.go:192: info: skipping unknown hook: "crd-install"
  Release "promethues-operator" has been upgraded. Happy Helming!
  NAME: promethues-operator
  LAST DEPLOYED: Mon Mar 22 11:50:39 2021
  NAMESPACE: monitoring
  STATUS: deployed
  REVISION: 2
  NOTES:
  *******************
  *** DEPRECATED ****
  *******************
  * stable/prometheus-operator chart is deprecated.
  * Further development has moved to
https://github.com/prometheus-community/helm-charts
  * The chart has been renamed kube-prometheus-stack to more clearly reflect
  * that it installs the `kube-prometheus` project stack, within which
Prometheus
  * Operator is only one component.

  The Prometheus Operator has been installed. Check its status by running:
    kubectl --namespace monitoring get pods -l "release=promethues-operator"

  Visit https://github.com/coreos/prometheus-operator for instructions on how
  to create & configure Alertmanager and Prometheus instances using the
Operator.
```

可以看到，更新指令执行成功（如将部分组件的 type 修改成了 NodePort，则可能会提示更新失败，但不影响，具体以实际效果为准）。此时，Spring Cloud 微服务的指标采集端点就被添加到 Prometheus 的 Targets 列表了。

（4）查看 Prometheus 控制台的 Targets 列表，如图 10-17 所示。

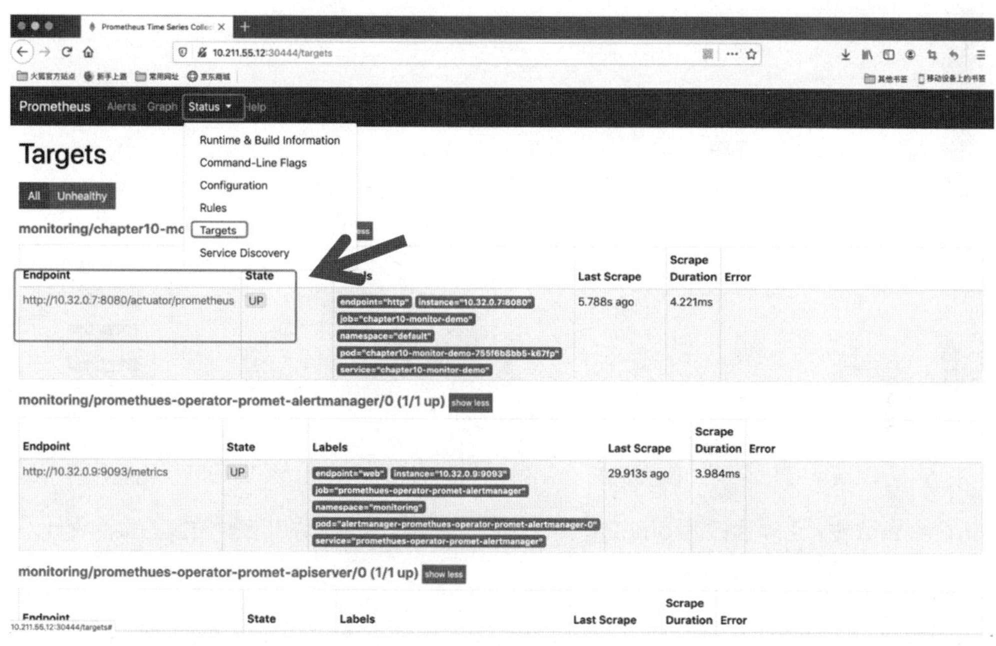

图 10-17

从图 10-17 中可以看出，此时 Spring Cloud 微服务已经被成功添加进 Prometheus 的 Targets 列表了，微服务指标采集接口状态也运行正常。

10.2.7　步骤 7：构建基于 Grafana 的可视化监控界面

通过 10.2.6 节的操作，此时 Prometheus 已经将部署在 Kubernetes 集群中的 Spring Cloud 微服务监控起来。

接下来，通过 Grafana 来定制微服务的监控视图。这里以最常见的“QPS”指标、“TP90/TP95”分位线，以及接口的平均响应时间为例。

1. 微服务总体 QPS 指标监控视图

（1）打开在 10.2.4 节中部署的 Grafana 监控界面，然后单击“+”来创建一个 Dashboard，如图 10-18 所示。

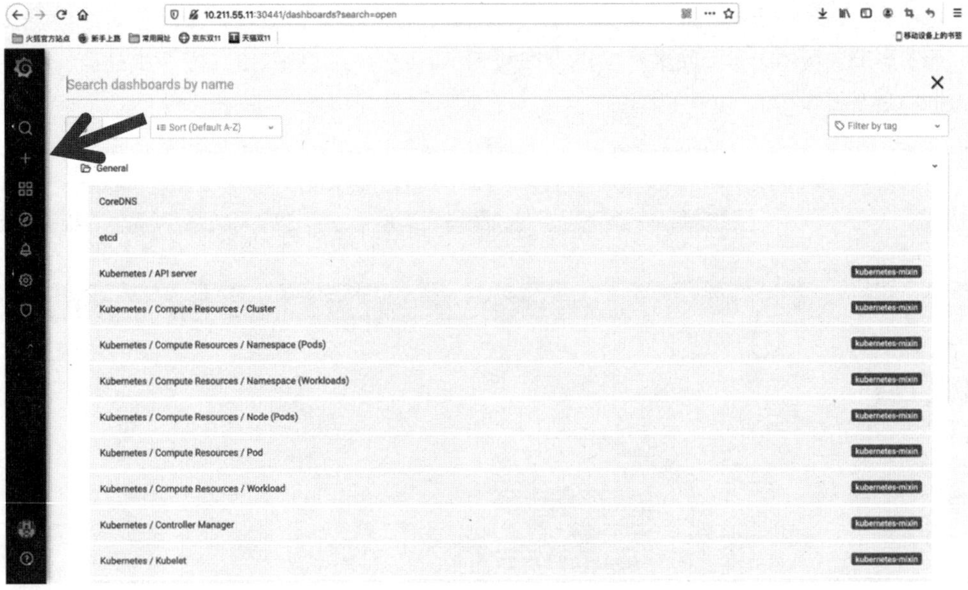

图 10-18

（2）定义 PromeQL 来统计微服务的 QPS 指标，并通过 Grafana 实现可视化监控界面。

在 Grafana 中，可以通过 PromeQL 实现对指标查询、检索，以及多种维度的灵活统计，并实现可视化展示，如图 10-19 所示。

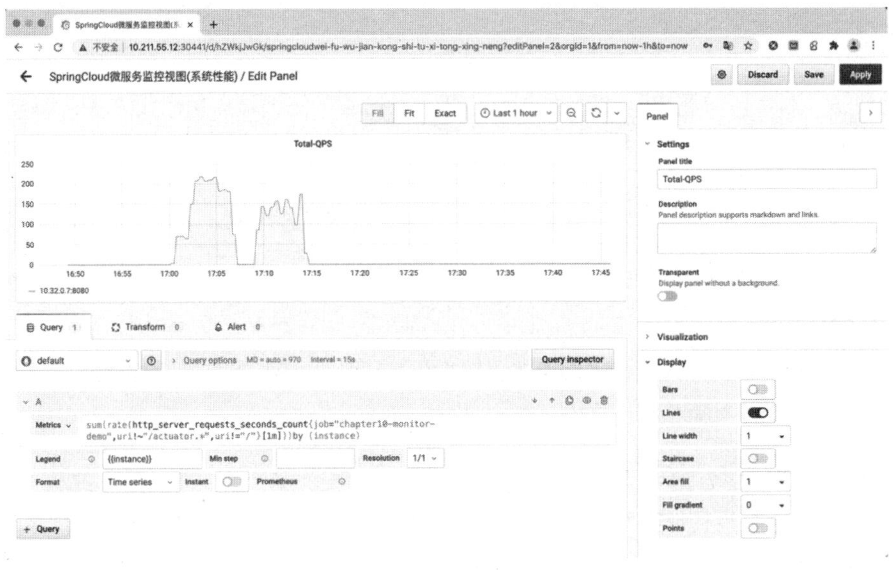

图 10-19

图 10-19 中的语句，就是以每 1 分钟的频率统计微服务“chapter10-monitor-demo”中除“/actuator.*”接口外的总体请求量。PromeQL 代码如下：

```
    sum(rate(http_server_requests_seconds_count{job="chapter10-monitor-demo"
,uri!~"/actuator.*",uri!="/"}[1m]))by (instance)
```

在图 10-19 统计中所使用的指标“http_server_requests_seconds_count”是由 Spring Boot 默认监控端点暴露的。

2. 微服务接口分位值（TP90/TP95）监控视图

接下来，通过在 10.2.5 节“3.”小标题中自定义的指标来统计示范微服务接口“/monitor/test”的 TP90/TP95 分位值，如图 10-20 所示。

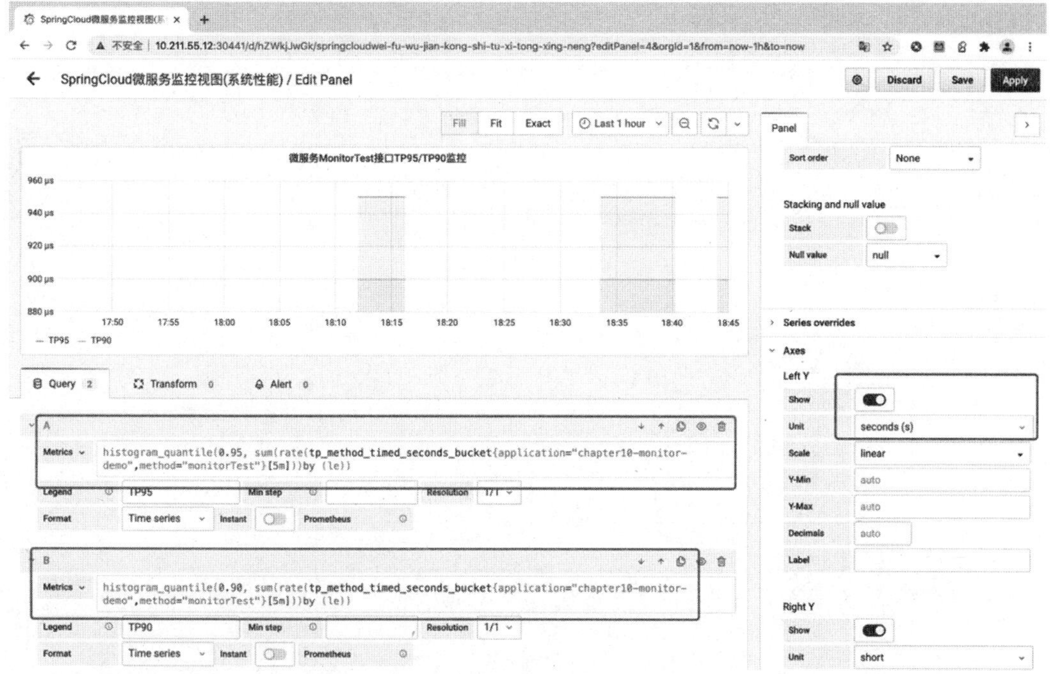

图 10-20

涉及的 PromeQL 代码如下：

```
    PromeQL-A:
    histogram_quantile(0.95,
sum(rate(tp_method_timed_seconds_bucket{application="chapter10-monitor-demo"
,method="monitorTest"}[5m]))by (le))
```

```
PromeQL-B:
histogram_quantile(0.90,
sum(rate(tp_method_timed_seconds_bucket{application="chapter10-monitor-demo"
,method="monitorTest"}[5m]))by (le))
```

在 Grafana 中，可以同时定义多个 PromeQL 来统计不同的监控指标。在上述代码中，分别通过 Prometheus 所提供的 histogram_quantile()函数统计了接口方法 monitorTest()的 TP90 及 TP95 分位值。所使用的指标，正是在 10.2.5 节中自定义暴露的“tp_method_timed_xx”指标类型。

3. 微服务接口平均响应时间监控视图

接下来通过自定义的计数器指标类型，来构建针对微服务测试接口方法“monitorTest”平均响应时间的可视化监控视图，如图 10-21 所示。

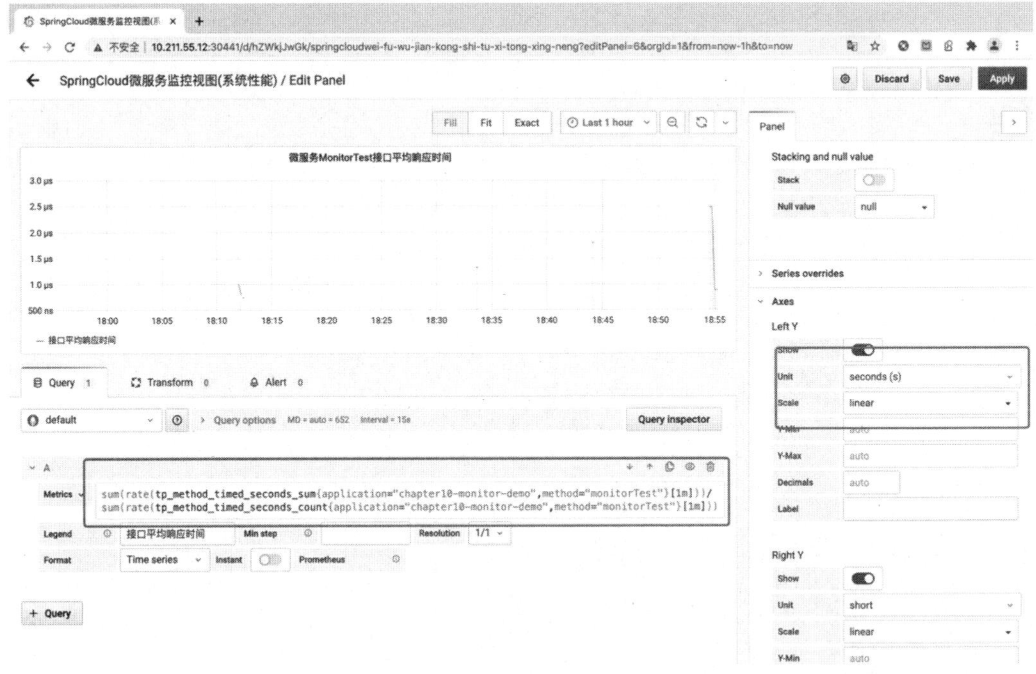

图 10-21

图 10-21 中涉及的 PromeQL 代码如下：

```
sum(rate(tp_method_timed_seconds_sum{application="chapter10-monitor-demo
",method="monitorTest"}[1m]))/
```

```
sum(rate(tp_method_timed_seconds_count{application="chapter10-monitor-demo",
method="monitorTest"}[1m]))
```

通过上述 PromeQL 代码可以看出，接口平均响应时间的计算是通过一段时间内“总的接口耗时/总的接口请求数”来计算的。

除上述所演示的几种常见的监控指标外，在实际的业务场景中，还可以通过扩展更多的自定义指标，以及借助更多维度的 PromeQL 查询语句，来构建更加丰富的可视化监控视图。

本实例最终构建的可视化监控视图如图 10-22 所示。

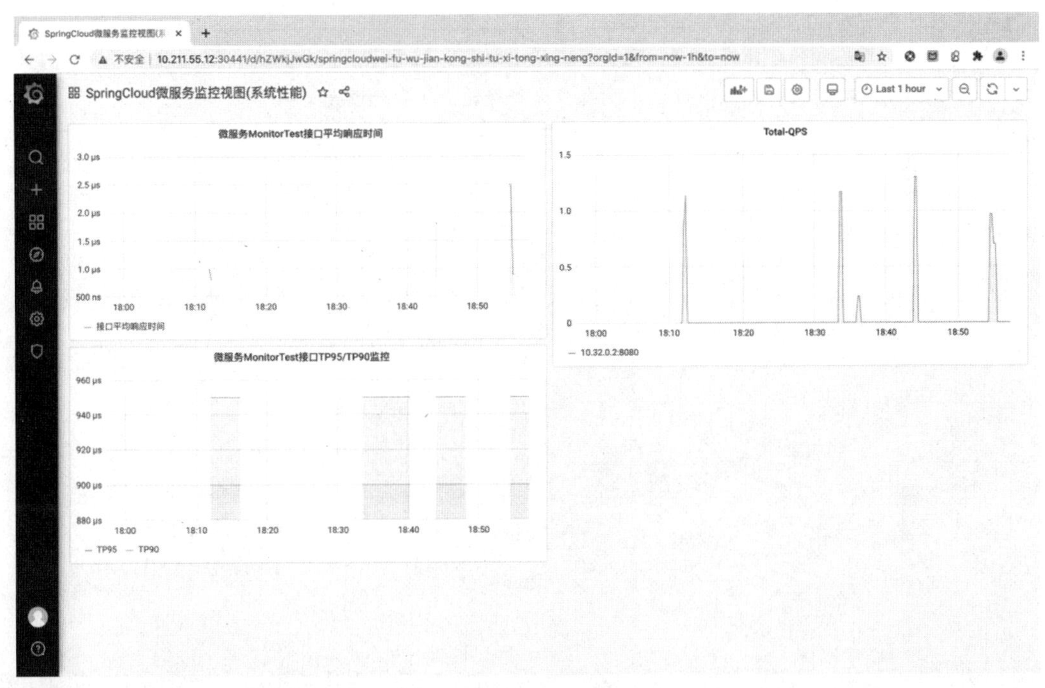

图 10-22

10.3 【实战】构建微服务分布式链路追踪系统

在 10.2 节中构建了基于 Prometheus 的微服务度量指标监控系统。对于像微服务这样链路复杂的分布式系统来说，构建基于调用链的分布式链路追踪系统，是快速定位问题、确保微服务系统稳定运行的重要手段。

接下来以 SkyWalking 为例，介绍分布式链路追踪系统在微服务体系中的实际应用。

10.3.1 认识分布式链路追踪

在介绍分布式链路追踪系统之前，需要先理解什么是链路追踪。

从 10.1 节的介绍中可以知道，监控系统的观测数据主要来源于：统计指标、日志及链路追踪。

而这些观测数据从类型上又可以分为以下两种。

- 请求级别的数据：来源于真实的请求，例如 HTTP 调用、RPC 调用等。本节要介绍的链路追踪就是这种类型。
- 聚合级别的数据：对接口请求的度量指标数据的聚合，例如 QPS、CPU 使用率等数值。

日志和统计指标数据，既可以是请求级别，也可以是聚合级别，因为它们可能来自源于真实的请求，也可能是在系统自身诊断时记录下来的信息。

链路追踪就是将请求链路的完整行为记录下来，以便通过可视化的形式实现调用链路的查询、性能分析、依赖关系，以及拓扑图等分布式链路追踪的功能，如图 10-23 所示。

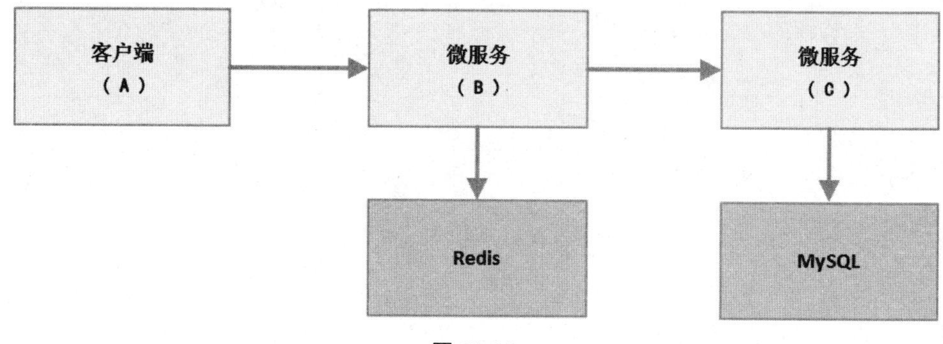

图 10-23

在图 10-23 中，假设一次接口调用共有两个微服务参与，调用关系是：A→B→C。其中，B 服务调用了 Redis 服务，C 服务调用了 MySQL 数据库。所以，链路追踪就是详细记录：A→B（B→Redis）→C（C→MySQL）这条链路上的调用信息，例如接口响应结果、耗时等。

那么链路追踪的数据到底是怎么记录的呢？接下来以图 10-23 中的调用链为例，来分析下链路追踪数据的具体组成和传递形式，如图 10-24 所示。其中分布式链路追踪的对象就是每次调用所产生的链路（Trace），①～⑧所表示的就是一条完整的链路（Trace），系统会通过唯一的标识（TraceId）来进行记录。

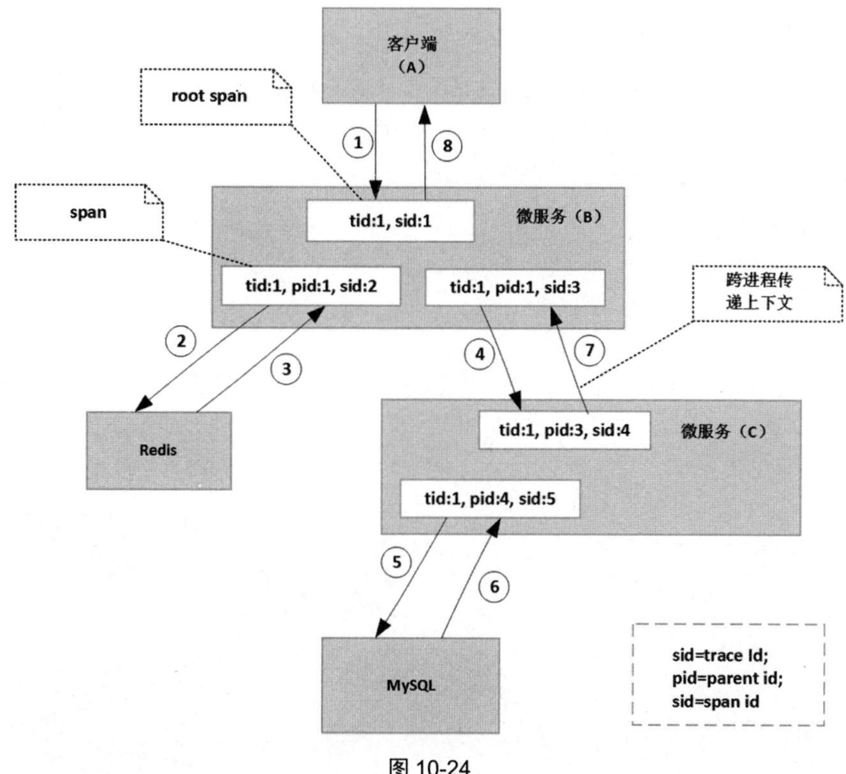

图 10-24

链路中的每一个依赖调用，都会生成一个调用踪迹信息（Span）。最开始生成的 Span 被叫作根 Span（Root Span），后续生成的 Span 都会将前一个 Span 的标识（Sid）作为本 Span 信息的父 ID（Pid）。

以此类推，Span 信息会随着调用链路的执行在进程内或跨进程进行传递。通过 Span 数据链，能将每次调用链路所产生的踪迹信息串联起来。而在每一个 Span 上附着的日志信息（Annotation）就是调用链路监控及分析的数据来源。

你可能会有疑问：监控这么大的数据量，是不是会很消耗系统资源？的确如此。所以，大部分链路追踪系统都会存在一个“采样率”（Sampling）的概念，用来控制系统采集链路信息的比例，从而提升系统性能。

因为很多时候大量链路信息都是相同的，需要关注的可能也只是相对耗时较高、出错次数较多的链路，所以并没有必要进行 100%的采集。

10.3.2 认识 SkyWalking

SkyWalking 是一款优秀的开源 APM（Application Performance Management）系统。它不仅支持分布式链路追踪、链路分析等功能，还支持性能指标分析、应用和服务的依赖性分析、服务的拓扑图分析，以及报警等与应用性能监控相关的功能。

1. 从数据收集来看

从数据收集来看，SkyWalking 支持多种不同的数据来源及格式，例如 Java、PHP 和 Python 等语言的应用都可以通过相应的无侵入式探针（Agent）接入 SkyWalking。

除此之外，SkyWalking 的新版本还支持对以 Istio 为代表的 Service Mesh（服务网格）控制面和数据面进行监控。SkyWalking 的架构如图 10-25 所示。

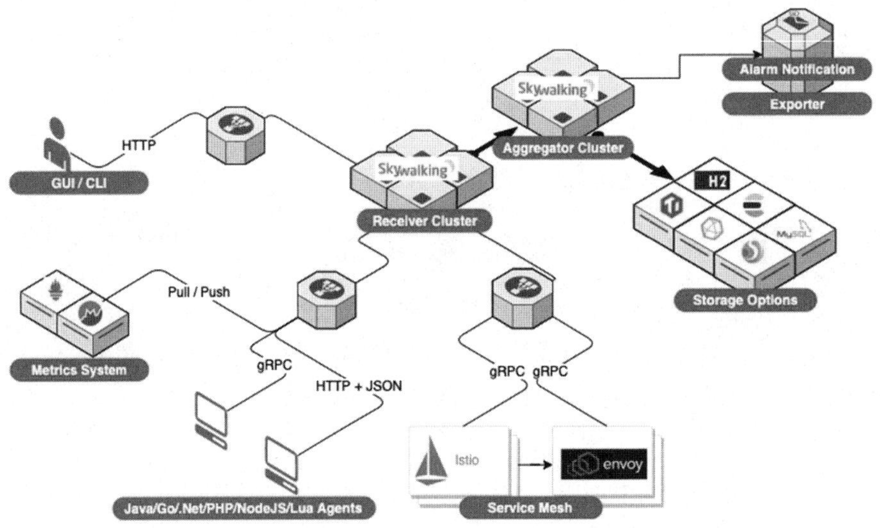

图 10-25

SkyWalking 由链路收集服务器（Receiver Cluster）、聚合服务器（Aggregator Cluster）组成。其中，链路收集服务器（Receiver Cluster）是整个后端服务接入的入口，主要用于收集接入服务的各种指标及链路信息。聚合服务器（Aggregator Cluster）则用于汇总、聚合链路收集服务器（Receiver Cluster）收集到的数据，并最终将聚合数据存储到数据库中。这些聚合数据将用于设置告警，或被 GUI/CLI 等可视化系统以 HTTP 的形式访问并进行可视化展示。

SkyWalking 具体的存储方式可以有多种，例如 ElasticSearch、MySQL、TIDB 等，可以根据实际需要进行选择。

2. 从数据采集逻辑来看

SkyWalking 支持多种语言探针及项目协议（见下方所列），能够覆盖目前主流的分布式技术栈。

- Metrics System：统计系统。支持直接从 Prometheus 中拉取度量指标数据到 SkyWalking，也支持程序自身通过 micrometer 推送数据。
- Agents：业务探针。在各个业务系统中集成接入探针（Agent），以实现链路数据采集。SkyWalking 支持 Java、Go、.NET、PHP、Node.js、Python、Nginx LUA 等的探针。除此之外，它还支持通过 gRPC 的方式来传递数据。
- Service Mesh：通过特定的 Service Mesh 协议来采集数据面、控制面的数据，以此实现对 Service Mesh（服务网格）系统链路数据的观测。

最近这几年，SkyWalking 发展得非常快，社区也非常活跃。在微服务链路追踪、应用性能监控领域，它被使用得也越来越广泛，感兴趣的读者可以深入了解一下。

10.3.3　步骤 1：部署 SkyWalking

SkyWalking 的部署主要涉及 SkyWalking OAP Server 和 SkyWalking UI，根据实际需要可以将它们部署在物理机、虚拟机或者 Kubernetes 集群中。本节将 SkyWalking 的后端 OAP Server 及 SkyWalking-UI 分别部署到 Kubernetes 集群中。

在 Kubernetes 集群中部署 SkyWalking，可以通过官方提供的“charts”采用 Helm 方式安装，也可以手动编写 Kubernetes 部署文件。

关于 Helm 及 charts 的概念，可以参考 10.2.2 节的内容。

为了便于学习，这里采用手动编写 Kubernetes 部署文件的方式来部署 SkyWalking OAP Server 和 SkyWalking UI。

1. 创建 SkyWalking 的 Kubernetes 命名空间

（1）在 Kubernetes 集群中，创建一个单独运行 SkyWalking 的 Namespace。命令如下：

```
#在通过 kubectl 连接 Kubernetes 集群后，执行创建 Namespace 的命令
$ kubectl create ns skywalking
```

（2）查看 Namespace 是否创建成功。命令如下：

```
#查看 Namespace 的创建情况
$ kubectl get ns
NAME                      STATUS   AGE
default                   Active   10d
kube-node-lease           Active   10d
kube-public               Active   10d
kube-system               Active   10d
kubernetes-dashboard      Active   10d
skywalking                Active   46s
```

可以看到 Kubernetes 的“skywalking”命名空间已经成功创建。

2. 编写 SkyWalking OAP Server 及 SkyWalking UI 的 Kubernetes 部署文件

在编写具体的 Kubernetes 部署文件时，需要指定 SkyWalking OAP Server 及 SkyWalking UI 的容器镜像。一般来说，SkyWalking 的安装包可以通过源码编译，也可以直接使用官方已经打包好的镜像。

接下来采用官方 Docker 镜像仓库中已经打包好的镜像，来指定 Kubernetes 部署文件中镜像的名称。

（1）在 Docker Hub 官方镜像仓库中，分别找到了 SkyWalking OAP Server 及 SkyWalking UI 的容器镜像版本。

在 Docker Hub 官方镜像仓库中查找 SkyWalking UI 的镜像，如图 10-26 所示。

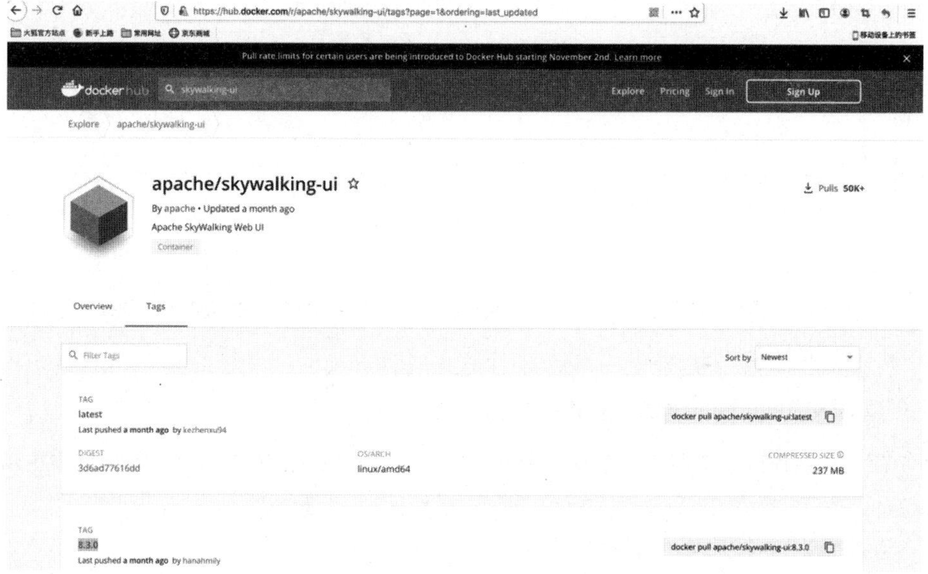

图 10-26

在 Docker Hub 官方镜像仓库中查找 SkyWalking OAP Server 的镜像，如图 10-27 所示。

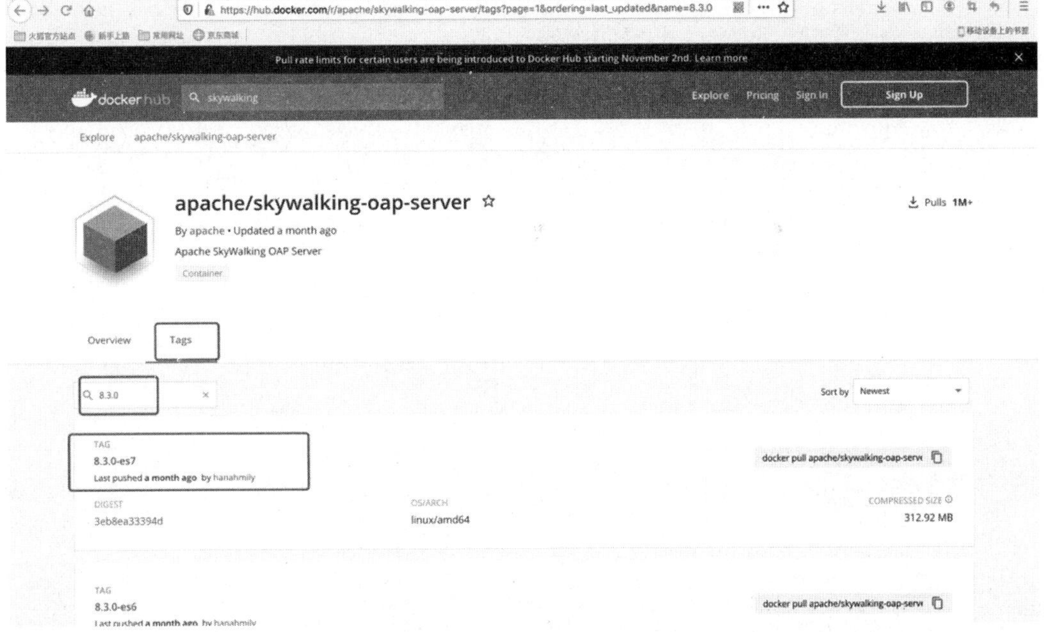

图 10-27

（2）编写 SkyWalking OAP Server 的 Kubernetes 部署文件。

编写用于部署 SkyWalking OAP Server 的 Kubernetes 部署文件“skywalking-aop.yml”。具体内容如下：

```
apiVersion: apps/v1
kind: Deployment
metadata:
  name: oap
  namespace: skywalking
spec:
  replicas: 1
  selector:
    matchLabels:
      app: oap
      release: skywalking
  template:
    metadata:
      labels:
        app: oap
        release: skywalking
```

```
    spec:
      containers:
        - name: oap
          #指定 OAP Server 容器镜像及版本信息
          image: apache/skywalking-oap-server:8.3.0-es7
          imagePullPolicy: IfNotPresent
          ports:
            - containerPort: 11800
              name: grpc
            - containerPort: 12800
              name: rest
---
apiVersion: v1
kind: Service
metadata:
  name: oap
  namespace: skywalking
  labels:
    service: oap
spec:
  ports:
    #restful 端口
    - port: 12800
name: rest
     #rpc 端口
    - port: 11800
      name: grpc
    - port: 1234
      name: page
  selector:
    app: oap
```

以上是一个标准的 Kubernetes 部署文件，具体含义可查阅 Kubernetes 的相关资料。

（3）编写 SkyWalking UI 的 Kubernetes 部署文件“skywalking-ui.yml”。具体内容如下：

```
apiVersion: apps/v1
kind: Deployment
metadata:
  name: ui-deployment
  namespace: skywalking
  labels:
    app: ui
```

```
spec:
  replicas: 1
  selector:
    matchLabels:
      app: ui
  template:
    metadata:
      labels:
        app: ui
    spec:
      containers:
        - name: ui
          image: apache/skywalking-ui:8.3.0
          ports:
            - containerPort: 8080
              name: page
          env:
            - name: SW_OAP_ADDRESS
              value: oap:12800
---
apiVersion: v1
kind: Service
metadata:
  name: ui
  namespace: skywalking
  labels:
    service: ui
spec:
  ports:
    - port: 8080
      name: page
      nodePort: 31234
  type: NodePort
  selector:
    app: ui
```

（4）执行 Kubernetes 部署命令，将 SkyWalking OAP Server 及 SkyWalking-UI 部署到 Kuberntes 集群中。

通过步骤（2）、步骤（3）编写的 Kubernetes 部署文件执行 Kubernetes 部署命令：

```
#进入 Kubernetes 部署文件的存储目录下，执行一次部署全部文件的命令
$ kubectl apply -f .
deployment.apps/oap created
service/oap created
deployment.apps/ui-deployment created
service/ui created
```

在部署命令执行完成后，查看具体部署的情况。命令如下：

```
#查看 Kubernetes 集群中 Skywalking 命名空间中的 Pod、Service 资源对象的运行情况
$ kubectl get all -n skywalking
NAME                                     READY   STATUS    RESTARTS   AGE
pod/oap-5f6d6bc4f6-k4mvv                 1/1     Running   0          36h
pod/ui-deployment-868c66449d-fffrt       1/1     Running   0          36h

NAME          TYPE        CLUSTER-IP       EXTERNAL-IP   PORT(S)          AGE
service/oap   ClusterIP   10.110.112.244   <none>
12800/TCP,11800/TCP,1234/TCP   36h
service/ui    NodePort    10.100.154.93    <none>        8080:31234/TCP   36h

NAME                               READY   UP-TO-DATE   AVAILABLE   AGE
deployment.apps/oap                1/1     1            1           36h
deployment.apps/ui-deployment      1/1     1            1           36h

NAME                                       DESIRED   CURRENT   READY   AGE
replicaset.apps/oap-5f6d6bc4f6             1         1         1       36h
replicaset.apps/ui-deployment-868c66449d   1         1         1       36h
```

可以看到，SkyWalking OAP Server 及 SkyWalking UI 服务都已经正常运行。

如果是第一次部署，则拉取镜像的过程可能会比较慢一点。此外，如果在部署过程中存在问题，则可以查看 Pod 对象的运行日志来排查问题，例如：

```
#查看 SkyWalking OAP Server 的对应 Pod 对象的运行日志
$ kubectl logs pod/oap-5f6d6bc4f6-k4mvv -n skywalking
```

（5）访问 SkyWalking UI 的 Web 界面。

经过前面的步骤，已经成功将 SkyWalking OAP Server 及 SkyWalking UI 服务运行在 Kubernetes 集群中了。接下来，通过 SkyWalking UI 服务的映射端口（在 Kubernetes 部署文件中定义是 31234 端口）来访问 SkyWalking 的管理界面，如图 10-28 所示。

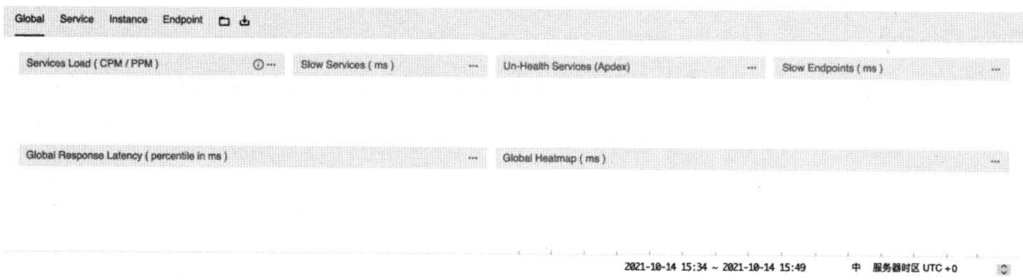

图 10-28

具体通过“http://NodeIP: 31234”进行访问，例如“http://10.211.55.12:31234”。

这里的 IP 地址为 Kubernetes 集群向外暴露的节点入口 IP 地址。如果不知道 Kubernetes 集群节点入口 IP 地址，则可以通过以下命令进行查看：

```
#查询 SkyWalking-UI 所部署的 Kubernetes 集群 Node 节点的 IP 地址
$ kubectl describe node kubernetes
Name:               kubernetes
Roles:              master
...
Addresses:
  InternalIP:  10.211.55.12
  Hostname:    kubernetes
...
```

可以看到 SkyWalking 服务运行成功。由于还没有接入服务，所以还看不到有任何监控数据。

10.3.4　步骤 2：将 Spring Cloud 微服务接入 SkyWalking

Spring Cloud 微服务可以通过 Java Agent（探针）的方式接入 SkyWalking。在 Spring Cloud 微服务中集成 Java Agent（探针）主要有以下 3 种方式：

- 使用官方提供的基础镜像。
- 将 Java Agent（探针）的依赖包构建到已存在的基础镜像中。
- 通过 SideCar 模式来挂载 Java Agent（探针）的依赖包。

如果微服务部署在 Kubernetes 集群中，则采用 SideCar 模式来挂载 Java Agent（探针）的依赖包会更加方便。因为，这种方式不需要修改原来的基础镜像，也不需要重新构建新的服务镜像，而是通过共享“volume”将 Java Agent（探针）依赖包的相关文件直接挂载到已经存在的服务镜像中。

接下来，通过 SideCar 模式将 Spring Cloud 微服务接入 SkyWalking。

1. 构建 SkyWalking Java Agent（探针）镜像

（1）下载 SkyWalking 的官方发行包，并将其解压缩到服务器指定目录下。命令如下：

```
#下载 Skywalking-8.3.0 for es7 版本的发布包，与部署的 SkyWalking OAP Server 的版本一致
$ wget
https://mirror.bit.edu.cn/apache/skywalking/8.3.0/apache-skywalking-apm-es7-8.3.0.tar.gz
```

```
#将下载的发布包解压缩到当前目录下
$ tar -zxvf apache-skywalking-apm-es7-8.3.0.tar.gz
```

（2）编写构建 SkyWalking Java Agent（探针）依赖包 Docker 镜像的 Dockerfile 文件。

在步骤（1）解压缩的 SkyWalking 发行包的同级目录下编写 Dockerfile 文件，具体内容如下：

```
FROM busybox:latest
ENV LANG=C.UTF-8
RUN set -eux && mkdir -p /usr/skywalking/agent
add apache-skywalking-apm-bin-es7/agent /usr/skywalking/agent
WORKDIR /
```

在上述 Dockefile 文件中，使用是的“bosybox”镜像，而不是 SkyWalking 的发行镜像。这样可以确保构建出来的 Docker 镜像最小。

（3）执行 Docker 镜像构建命令：

```
#执行镜像构建命令
$ docker build . -t springcloud-action/skywalking-agent-sidecar:8.3.0

Sending build context to Docker daemon  556.5MB
Step 1/5 : FROM busybox:latest
latest: Pulling from library/busybox
d60bca25ef07: Pull complete
Digest:
sha256:49dae530fd5fee674a6b0d3da89a380fc93746095e7eca0f1b70188a95fd5d71
Status: Downloaded newer image for busybox:latest
 ---> a77dce18d0ec
Step 2/5 : ENV LANG=C.UTF-8
 ---> Running in e95b4c25ebf3
Removing intermediate container e95b4c25ebf3
 ---> 83f22bccb6f3
Step 3/5 : RUN set -eux && mkdir -p /usr/skywalking/agent
 ---> Running in 49c2eac2b6ab
+ mkdir -p /usr/skywalking/agent
Removing intermediate container 49c2eac2b6ab
 ---> 89cf3ce8238e
Step 4/5 : add apache-skywalking-apm-bin/agent /usr/skywalking/agent
 ---> 91fe5f06948f
Step 5/5 : WORKDIR /
 ---> Running in 6a64553f1870
Removing intermediate container 6a64553f1870
 ---> 7e73ddba48bb
Successfully built 7e73ddba48bb
```

```
Successfully tagged springcloud-action/skywalking-agent-sidecar:8.3.0
```

（4）通过命令查看本地构建的镜像，确保镜像构建成功。命令如下：

```
#查看本地镜像信息
$ docker images
REPOSITORY      TAG                   IMAGE ID          CREATED           SIZE
springcloud-action/skywalking-agent-sidecar   8.3.0   7e73ddba48bb   2
minutes ago   32.2MB
...
```

2. 将打包的 Java Agent（探针）镜像推送到 Harbor 镜像仓库中

为了便于后续微服务直接集成已经构建好的 Java Agent(探针)镜像，需要将其 Push 至 Harbor 私有镜像仓库中。

（1）登录 Harbor 私有镜像仓库。命令如下：

```
#登录镜像仓库，输入用户账号及密码
$ docker login http://10.211.55.2:8080
Username: admin
Password:
Login Succeeded
```

有关 Harbor 私有镜像仓库的安装及登陆操作可参考本书 9.4 节的内容。

（2）将“1.”小标题中构建 Docker 镜像打上 tag。命令如下：

```
$ docker tag springcloud-action/skywalking-agent-sidecar:8.3.0
10.211.55.2:8080/springcloud-action/skywalking-agent-sidecar
```

（3）查看已经打上 tag 的镜像信息。命令如下：

```
$ docker images
REPOSITORY              TAG              IMAGE ID        CREATED      SIZE
springcloud-action/skywalking-agent-sidecar          8.3.0
e21040c57e42        2 weeks ago         32.2MB
10.211.55.2:8080/springcloud-action/skywalking-agent-sidecar     latest
e21040c57e42        2 weeks ago         32.2MB
...
```

（4）将打上 tag 的 Java Agent（探针）的镜像推送至 Harbor 私有镜像仓库中。命令如下：

```
#将镜像推送至 Harbor 私有镜像仓库中
$ docker push 10.211.55.2:8080/springcloud-action/skywalking-agent-sidecar
```

```
    The push refers to repository
[10.211.55.2:8080/springcloud-action/skywalking-agent-sidecar]
    e80d641c3ed9: Layer already exists
    11fe582bd430: Layer already exists
    1dad141bdb55: Layer already exists
    latest: digest:
sha256:b495c18c3ae35f563ad4db91c3db66f245e6038be0ced635d16d0e3d3f3bcb80 size:
946
```

（5）进入 Harbor 仓库管理界面查看刚推送的镜像，如图 10-29 所示：

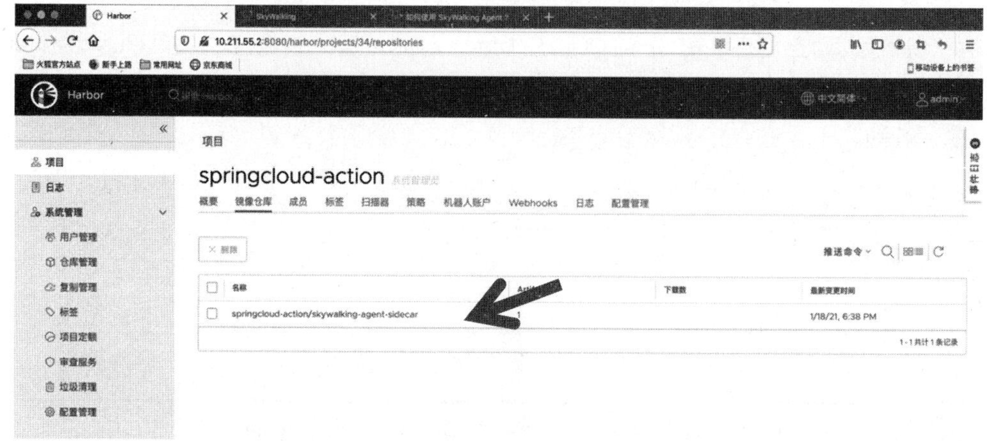

图 10-29

3. 构建 Spring Cloud 微服务镜像，并推送至 Harbor 镜像仓库中

将要接入 SkyWalking 的微服务打包成 Docker 镜像，并上传至 Harbor 私有镜像仓库中。

（1）通过 Maven 构建方式打包 Spring Cloud 微服务镜像。命令如下：

```
    # Maven 项目构建，会自动根据 pom.xml 中的相关插件配置进行 Docker 镜像构建
    $ mvn clean install -X
```

（2）在本地查看构建的微服务镜像信息。命令如下：

```
    $ docker images
    REPOSITORY                                                          TAG
IMAGE ID            CREATED             SIZE
    10.211.55.2:8080/springcloud-action/chapter10-monitor-demo       latest
3ae132cdfeb7        12 seconds ago      121MB
    10.211.55.2:8080/springcloud-action/skywalking-agent-sidecar     latest
e21040c57e42        2 weeks ago         32.2MB
    springcloud-action/skywalking-agent-sidecar                        8.3.0
e21040c57e42        2 weeks ago         32.2MB
    ...
```

（3）将步骤（2）中构建的微服务镜像也推送至 Harbor 私有镜像镜像仓库中。命令如下：

```
$ docker push 10.211.55.2:8080/springcloud-action/chapter10-monitor-demo
The push refers to repository
[10.211.55.2:8080/springcloud-action/chapter10-monitor-demo]
5f3427edfc10: Pushed
925523484e00: Layer already exists
344fb4b275b7: Layer already exists
bcf2f368fe23: Layer already exists
latest: digest:
sha256:b424180c56b28a9a7704a1f6476f4247fad12cc27721c21fce32149a8f344dee size:
1159
```

（4）在 Harbor 私有仓库中查看微服务镜像是否成功上传，如图 10-30 所示。

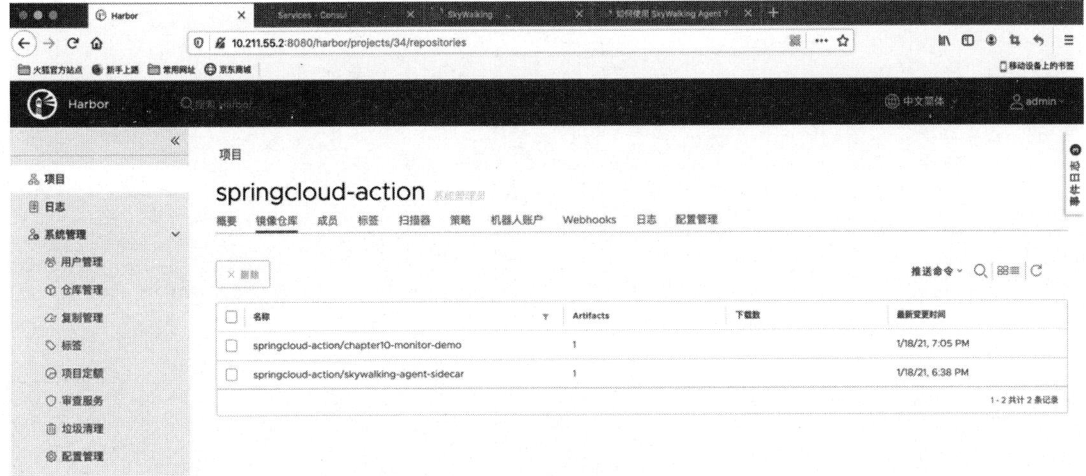

图 10-30

4. 微服务 Kubernetes 部署文件集成 SkyWalking Java Agent

用 SideCar 模式挂载 SkyWalking Java Agent（探针）依赖包，主要是通过 Kubernetes 的初始化容器 initContainers 来实现的。initContainers 是一种专用容器，在服务容器启动之前运行，主要用于在完成服务启动前的必要初始化工作。

接下来，改造 Spring Cloud 微服务的 Kubernetes 部署文件，并将其接入 SkyWalking。改造后的部署文件的内容如下：

```
apiVersion: apps/v1
kind: Deployment
metadata:
  name: chapter10-monitor-demo
```

```
  spec:
    selector:
      matchLabels:
        app: chapter10-monitor-demo
    replicas: 1
    #设置滚动升级策略
    #Kubernetes 在等待设置的时间后才开始进行升级，例如 5s
    minReadySeconds: 5
    strategy:
      type: RollingUpdate
      rollingUpdate:
        #在升级过程中最多可以比原先设置多出的 Pod 数量
        maxSurge: 1
        #在升级过程中 Deployment 控制器最多可以删除多少个旧 Pod，主要用于提供缓冲时间
        maxUnavailable: 1
    template:
      metadata:
        labels:
          app: chapter10-monitor-demo
      spec:
        #构建初始化镜像（通过初始化镜像的方式集成 SkyWalking Agent）
        initContainers:
          - image:
10.211.55.2:8080/springcloud-action/skywalking-agent-sidecar:latest
            name: sw-agent-sidecar
            imagePullPolicy: IfNotPresent
            command: ["sh"]
            args:
              [
                "-c",
                "mkdir -p /skywalking/agent && cp -r /usr/skywalking/agent/*
/skywalking/agent",
              ]
            volumeMounts:
              - mountPath: /skywalking/agent
                name: sw-agent
        containers:
          - name: chapter10-devops-demo
            image:
10.211.55.2:8080/springcloud-action/chapter10-monitor-demo:latest
            env:
              #这里通过 JAVA_TOOL_OPTIONS，而不是 JAVA_OPTS 来配置启动参数，是因为使用
JAVA_TOOL_OPTIONS 不需要给“agent”命令加上 JVM 启动参数就能实现 Java Agent（探针）的集成
              - name: JAVA_TOOL_OPTIONS 给
                value: -javaagent:/usr/skywalking/agent/skywalking-agent.jar
```

```
            - name: SW_AGENT_NAME
              value: chapter10-devops-demo
            - name: SW_AGENT_COLLECTOR_BACKEND_SERVICES
              # FQDN: servicename.namespacename.svc.cluster.local
              value: oap.skywalking:11800
            - name: SERVER_PORT
              value: "8080"
            - name: SPRING_PROFILES_ACTIVE
              value: test
          volumeMounts:
            - mountPath: /usr/skywalking/agent
              name: sw-agent
      volumes:
        - name: sw-agent
          emptyDir: {}
---
apiVersion: v1
kind: Service
metadata:
  name: chapter10-monitor-demo
  labels:
    svc: chapter10-monitor-demo
spec:
  selector:
    app: chapter10-monitor-demo
  ports:
    - name: http
      #Service 在集群中暴露的端口（用于 Kubernetes 服务间的访问）
      port: 8080
      #Pod 上的端口（与制作容器时暴露的端口一致，在微服务工程代码中指定的端口）
      targetPort: 8080
      #K8s 集群外部访问的端口（外部机器访问）
      nodePort: 30002
  type: NodePort
```

在微服务“chapter10-devops-demo”的 Kubernetes 部署文件中，主要是通过共享“volume”的方式来挂载 Java Agent（探针）的依赖包。其中，“initContainers”通过“skywalking-agent”卷将 “skywalking-agent-sidecar”镜像挂载到“/skywalking/agent”目录下，并将“agent”目录中的文件复制到“/skywalking/agent”目录下。这样，微服务容器启动时通过挂载“skywalking-agent”卷，并将该卷挂载到容器的“/usr/skywalking/agent”目录下，就可以实现微服务集成 SkyWalking Java Agent（探针）、接入 SkyWalking 的逻辑。

将微服务通过 Java Agent（探针）接入 SkyWalking，还可以通过在启动命令中加入 JVM 参数（例如“-javaagent:/usr/skywalking/agent/skywalking-agent.jar”）来实现。

可以在定义微服务镜像打包的 Dockerfile 文件中，通过“ENTRYPOINT”来配置，例如：

```
ENTRYPOINT [ "sh", "-c", "java ${JAVA_OPTS} -javaagent:/app/agent/skywalking-agent.jar -Dskywalking.collector.backend_service=${SW_AGENT_COLLECTOR_BACKEND_SERVICES} -Dskywalking.agent.service_name=${SW_AGENT_NAME} -Dskywalking.agent.instance_name=${HOSTNAME} -Djava.security.egd=file:/dev/./urandom -jar /app/app.jar $PROFILE" ]
```

但这种方式需要在 Dockerfile 文件中额外设置相关的 JVM 参数。这里为了方便，通过“JAVA_TOOL_OPTIONS”，以 JVM 环境变量的方式代替用“JAVA_OPTS”设置 JVM 启动参数的方式。

5. 将微服务部署至 Kubernetes 集群中，并验证其是正常接入 SkyWalking

（1）进入微服务 Kubenetes 部署文件所在目录下，执行如下发布命令：

```
$ kubectl apply -f deploy-skywalking.yml
deployment.apps/chapter10-monitor-demo created
service/chapter10-monitor-demo created
```

（2）由于在部署微服务时并未特别指定命名空间，所以，可以直接在默认的命名空间中查看运行的 Pod 信息。命令如下：

```
$ kubectl get pods
NAME                                     READY   STATUS    RESTARTS   AGE
chapter10-monitor-demo-5767d54f5-vfqqf   1/1     Running   0          96m
```

（3）为了验证微服务在 Kubernetes 中运行时是否集成了 SkyWalking Java Agent（探针），可以通过以下命令查看容器的具体日志。

```
$ kubectl logs chapter10-monitor-demo-5767d54f5-vfqqf
```

查看到的应用的启动日志如下：

```
Picked up JAVA_TOOL_OPTIONS:
-javaagent:/usr/skywalking/agent/skywalking-agent.jar
DEBUG 2021-02-20 08:03:22:220 main AgentPackagePath : The beacon class location is jar:file:/usr/skywalking/agent/skywalking-agent.jar!/org/apache/skywalking/apm/agent/core/boot/AgentPackagePath.class.
INFO 2021-02-20 08:03:22:222 main SnifferConfigInitializer : Config file found in /usr/skywalking/agent/config/agent.config.
2021-02-20 08:03:39.157  INFO 1 --- [           main]
trationDelegate$BeanPostProcessorChecker : Bean
'org.springframework.cloud.autoconfigure.ConfigurationPropertiesRebinderAuto
```

```
Configuration' of type
[org.springframework.cloud.autoconfigure.ConfigurationPropertiesRebinderAuto
Configuration$$EnhancerBySpringCGLIB$$282a8554] is not eligible for getting
processed by all BeanPostProcessors (for example: not eligible for auto-proxying)
   ...
   2021-02-20 08:03:54.431  INFO 1 --- [           main]
o.s.b.a.e.web.EndpointLinksResolver      : Exposing 2 endpoint(s) beneath base
path '/actuator'
   2021-02-20 08:03:54.851  INFO 1 --- [           main]
o.s.b.w.embedded.tomcat.TomcatWebServer  : Tomcat started on port(s): 8080
(http) with context path ''
   2021-02-20 08:03:54.892  INFO 1 --- [           main]
o.s.c.c.s.ConsulServiceRegistry          : Registering service with consul:
NewService{id='monitor-demo-spring-cloud-client-ipAddress-8080-1-0-SNAPSHOT',
name='monitor-demo', tags=[secure=false], address='10.32.0.16', meta=null,
port=8080, enableTagOverride=null, check=Check{script='null', interval='10s',
ttl='null', http='http://10.32.0.16:8080/actuator/health', method='null',
header={}, tcp='null', timeout='null', deregisterCriticalServiceAfter='null',
tlsSkipVerify=null, status='null'}, checks=null}
   2021-02-20 08:03:55.070  INFO 1 --- [           main]
c.w.monitor.MonitorDemoApplication       : Started MonitorDemoApplication in
24.584 seconds (JVM running for 32.992)
```

从中可以看出：服务在启动时已经识别到了在 Kubernetes 部署文件中通过“JAVA_TOOL_OPTIONS”设置的参数“-javaagent:/usr/skywalking/agent/skywalking-agent.jar”，并找到了 SkyWalking Java Agent（探针）相关的配置。这说明微服务已经通过 Java Agent（探针）接入了 SkyWalking 服务，并成功运行。

10.3.5 步骤 3：通过 SkyWalking UI 追踪分布式链路

接下来通过调用 10.2.5 节中编写的微服务测试接口来产生 SkyWalking 监控数据，并通过 SkyWalking UI 对上报的链路数据进行追踪。

（1）以一定频率访问微服务的测试接口。例如：

```
http://10.211.55.12:30002/monitor/test?name=wudimanong
```

以上是在将微服务部署到 Kubernetes 集群后通过 Service 映射的 nodePort 端口。

（2）在访问微服务测试接口后，刷新 SkyWalking UI 界面，可以看到 Spring Cloud 微服务已经通过 Java Agent（探针）向 SkyWalking 上报了监控数据，如图 10-31 所示。

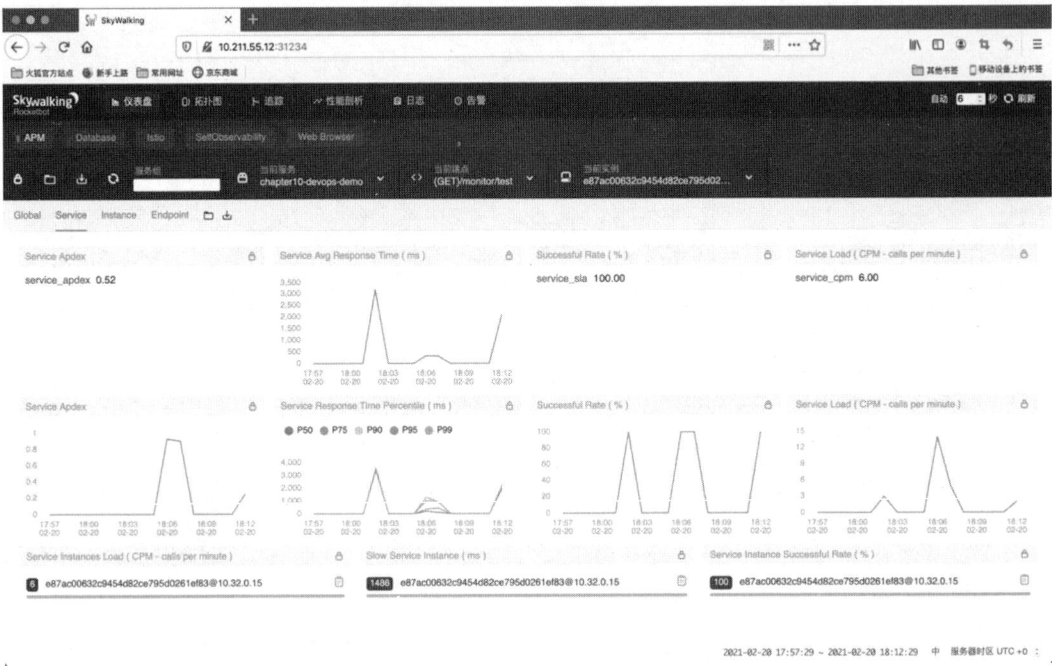

图 10-31

（3）单击上方的“追踪”菜单后，可以看到每一次请求调用所经历的链路过程，如图 10-32 所示。

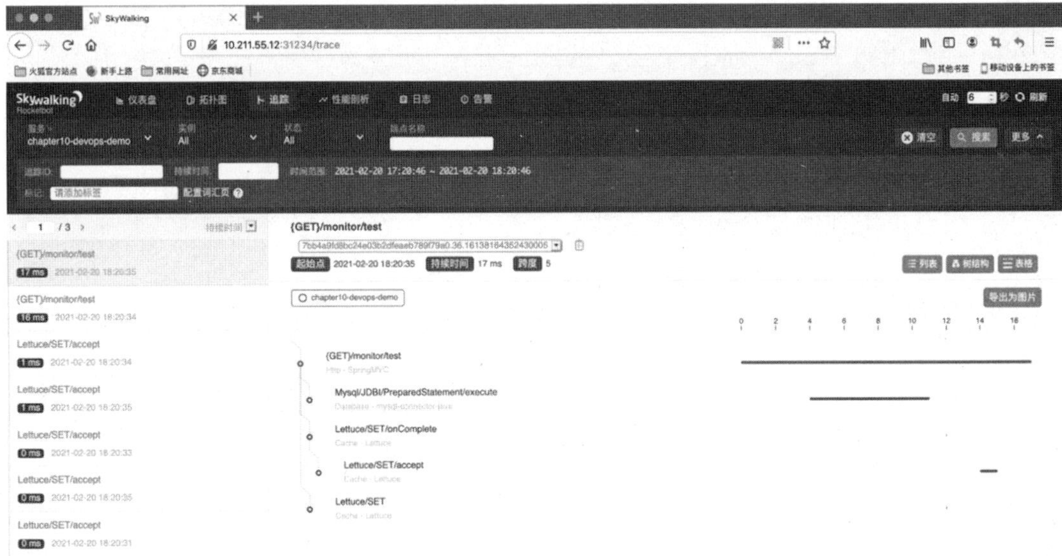

图 10-32

在图 10-32 中可以看到，在微服务测试接口请求链路的详情中清晰地展示了每个链路调用的耗时、应答等详细信息。

10.4　本章小结

本章系统地介绍了在 Spring Cloud 微服务中构建多维度监控体系的理论、方法及实践，具有很强的实战意义。其中涉及的 Prometheus、Grafana、SkyWalking 等开源监控产品，也都是目前业界在 Metrics（度量指标）、Tracing（调用链）监控领域主流的构建方案。

在实际的应用场景中，日志监控也是很重要的一部分。由于篇幅的原因，本章并没有具体演示，感兴趣的读者可以研究一下主流的 ELK（Elasticsearch + Logstash + Kabana）方案。